# microRNA Modulation: A Promising Combination Therapy for Cancer

Shikha Tiwari • Dhiraj Bhatia
Amit K. Yadav
Editors

# microRNA Modulation: A Promising Combination Therapy for Cancer

Springer

*Editors*
Shikha Tiwari
School of Sciences
Sanjeev Agrawal Global Educational
(SAGE) University
Bhopal, Madhya Pradesh, India

Dhiraj Bhatia
Department of Biological Sciences
and Engineering
Indian Institute of Technology Gandhinagar
Gandhinagar, Gujarat, India

Amit K. Yadav
Department of Biological Sciences
and Engineering
Indian Institute of Technology
Gandhinagar
Gandhinagar, Gujarat, India

ISBN 978-981-95-5988-6 ISBN 978-981-95-5989-3 (eBook)
https://doi.org/10.1007/978-981-95-5989-3



This Springer imprint is published by the registered company Springer Nature Singapore Pte Ltd.
The registered company address is: 152 Beach Road, #21-01/04 Gateway East, Singapore 189721, Singapore

*This book is lovingly dedicated to all the patients and families who have faced cancer with extraordinary courage, resilience, and hope. Your strength in the face of adversity has been a constant reminder of why we pursue research, innovation, and better therapies. You inspire every step of our work.*
*We also dedicate this book to our families—the silent pillars of support—who have stood by us through long nights of research, writing, and revision. Your patience, encouragement, and unwavering belief in us have made this journey not only possible but meaningful.*
*To the teachers, mentors, and colleagues whose guidance has shaped our understanding, your wisdom and generosity have been invaluable.*
*Finally, we dedicate this work to the future generation of scientists, clinicians, and caregivers. May this book spark your curiosity, fuel your passion, and inspire your dedication to transforming cancer care for the better.*

# Foreword

In the evolving landscape of oncology, microRNAs have emerged as powerful regulators of cancer biology, offering remarkable potential for both understanding disease mechanisms and designing innovative therapies. *microRNA Modulation: A Promising Combination Therapy for Cancer* provides a lucid, comprehensive, and timely exploration of this field, bridging the gap between molecular research and clinical application.

The authors have successfully captured the dual nature of miRNAs—both as tumor suppressors and oncogenic drivers—and the immense therapeutic possibilities that arise from their modulation. What makes this book particularly valuable is its holistic approach: it celebrates scientific promise while thoughtfully addressing the practical challenges, including delivery methods, off-target effects, and safety considerations.

Beyond the science, this book reflects the authors' passion and commitment to improving patient outcomes. The chapters are written with clarity, insight, and empathy, making them accessible to researchers, clinicians, and students alike.

It is my honor to endorse this book. I believe it will not only serve as an essential reference for current practitioners and researchers but will also inspire the next generation of scientists dedicated to advancing cancer therapy. The meticulous scholarship, combined with a humanistic perspective, ensures that this work will make a lasting contribution to the field.

Professor, Department of Biochemistry,
All India Institute of Medical Sciences
Bhopal, Madhya Pradesh, India

Ashok Kumar

# Preface

*microRNA Modulation: A Promising Combination Therapy for Cancer* offers all the leaflets of research in the area of combinatorial cancer therapy approaches, and significance of miRNA modulation in improving anticancer therapy outcomes. The book enlightens the future "next generation cancer therapy," with the potential of translation into clinics. Cancer is a lethal disease and a global health concern that imprints a significant proportion of disease and financial burden. In spite of progressive advancement in cancer therapeutics, continuous efforts are needed for the development of new approaches to avoid the side effects, non-targeted action, and better prognosis. After human genome sequencing and in the era of "transcriptomics," our focus has been increased in small noncoding RNAs (ncRNAs). Among these microRNA (miRNA) are drawing more attention because of its ability to effect more than 200 mRNA post-transcriptionally, and are key modulator of cancer initiation, advancement, and progression.

The biogenesis and expression of miRNA is under control of multiple molecular mechanisms, and also represents site-specific nature. The therapeutic efficacy of miRNA modulation has been proven in vitro and preclinical studies for multiple diseases including cancer. Currently, researchers are focusing on improving cancer treatment by optimizing miRNA delivery to tumor sites, reducing "off-target" action, minimizing toxicities to normal cells, and clinical translations. miRNA-based combination therapy aimed to combine traditional chemotherapy, radiotherapy, and immunotherapy with targeted modulation of tumor micro environment or cell senescence, apoptosis, cellular metabolism, and angiogenesis to decrease the drug load and toxicity.

The book also covers current status and latest targeted miRNA/siRNA delivery methods, clinical trials, challenges, and advances in miRNA based therapeutics, in cancer. The book is a great addition to efforts of oncologist's common goal of completely eradicating cancer-associated mortality and disability. The potential audience for this book may be clinicians, biochemists, molecular biologists, pathologists, medical oncologists, radiation oncologists, pharmacologists, medical practitioners, bioengineers, academic institutes, etc.

Bhopal, Madhya Pradesh, India — Shikha Tiwari
Gandhinagar, Gujarat, India — Dhiraj Bhatia
Gandhinagar, Gujarat, India — Amit K. Yadav

# Acknowledgments

The journey of creating this book has been one of discovery, reflection, and immense gratitude. First and foremost, we offer our heartfelt thanks to God, whose grace and guidance have been our constant source of strength, inspiration, and perseverance throughout this endeavor. Without this divine blessing, neither the ideas nor the energy to pursue this work would have been possible.

We are deeply indebted to the researchers, clinicians, and mentors whose groundbreaking work on microRNAs in cancer has laid the foundation for this book. Their dedication, curiosity, and tireless pursuit of knowledge continue to inspire us and countless others in the scientific community.

We extend our sincere gratitude to the editorial and production teams at Springer Nature. Your patience, guidance, and meticulous attention to detail have been invaluable in transforming our manuscript into a coherent and polished work, ensuring that the knowledge within reaches its audience effectively.

On a personal level, we owe profound thanks to our families and friends, whose unwavering support, encouragement, and understanding have carried us through long hours of research, writing, and revision. Your faith in us and constant reassurance have been an indispensable part of this journey.

Lastly, we are grateful to all those whose names may not appear here but whose contributions—through discussions, collaborations, and encouragement—have enriched this work. To each one of you, we extend our deepest appreciation.

# Contents

# Editors and Contributors

## About the Editors

**Shikha Tiwari** is an Associate Professor in the School of Sciences at the Sanjeev Agrawal Global Educational University, Bhopal, Madhya Pradesh, India. Her research centers on the role of non-coding RNAs in the initiation and progression of oral cancer, with emphasis on identifying miRNA-based biomarkers and functional SNPs linked to oral squamous cell carcinoma. She has extensive expertise in molecular oncology, animal cell culture, genomics, and bioinformatics. She has published widely in high-impact journals, contributed to several international books, and presented her work at national and international conferences. Her current work advances early diagnosis and therapeutic strategies in cancer biology.

**Dhiraj Bhatia** is an Associate Professor in the Department of Biological Sciences and Engineering at the Indian Institute of Technology Gandhinagar and a member of INYAS–INSA. He obtained his PhD from NCBS-TIFR, Bangalore, followed by postdoctoral training at Institut Curie, Paris, as an HFSP and Curie Fellow. His research focuses on structural DNA nanotechnology, advanced chemical biology tools, and stem cell engineering, with applications in regenerative medicine and nanomedicine. He has authored over 150 publications, delivered more than 100 invited talks worldwide, and is co-founder of Qnanosol Biotech Pvt. Ltd., translating nanoscience to biomedical applications.

**Amit Kumar Yadav** is an Early Career Fellow at the Department of Biological Sciences and Engineering, Indian Institute of Technology Gandhinagar. He earned his PhD in Nanoscience from Jawaharlal Nehru University, New Delhi, as a Prime Minister Research Fellow. His research focuses on nanobiotechnology, nanobiosensors, aptamers, and nanodiagnostics for cancer and infectious disease biomarkers, with strong expertise in electrochemistry and 2D materials. He has authored over 45 publications, contributed multiple book chapters, and holds a granted Indian patent in biosensor technology. His work integrates materials engineering and biotechnology toward innovative, low-cost diagnostic and healthcare solutions.

## Contributors

**Pranjal Agarwal** Department of Biological Sciences and Engineering, Indian Institute of Technology Gandhinagar, Gandhinagar, Gujarat, India

**Parwez Ahmad** RECETOX, Faculty of Science, Masaryk University, Brno, Czech Republic

**Tanvir Ahmed** Department of Pharmaceutical Sciences, School of Health and Life Sciences, North South University, Dhaka, Bangladesh

**Kazi Tasnuva Alam** Department of Pharmaceutical Sciences, School of Health and Life Sciences, North South University, Dhaka, Bangladesh

**Aayushi Arora** Department of Microbiology, Anatomy, Physiology, and Pharmacology, La Trobe Institute for Molecular Science, School of Agriculture, Biomedicine and Environment, La Trobe University, Melbourne, VIC, Australia

**Tinku Basu** Amity Centre for Nanomedicine, Amity University, Noida, UP, India

**Dhiraj Bhatia** Department of Biological Sciences and Engineering, Indian Institute of Technology Gandhinagar, Gandhinagar, Gujarat, India

**Navneet Chaudhary** Nano for Energy, Department of Physics, Umeå University, Umeå, Sweden

**Deepika Chauhan** Amity Centre for Nanomedicine, Amity University, Noida, UP, India

**Sandesh Chibber** Department of Biotechnology, Mehsana Urban Institute of Sciences, Faculty of Science, Ganpat University, Mehsana, Gujarat, India

**Foram Lomesh Contractor** Department of Biotechnology, Parul Institute of Technology, Parul University, Vadodara, Gujarat, India

**Satheesh Ellipilli** Chemical Biology Lab, Department of Chemistry, School of Engineering and Sciences, SRM University-AP, Amaravati, Andhra Pradesh, India

**Rekha Jalandra** Department of Pharmacology & Immunology, Medical University of South Carolina, Charleston, SC, USA

**Archana Joon** Department of Pediatrics and Department of Microbiology & Immunology, Albert Einstein College of Medicine, New York, NY, USA

**Sahil Kohli** Department of Chemistry, School of Engineering and Technology, Manav Rachna International Institute of Research and Studies, Faridabad, Haryana, India

**Abhishek Kumar** Albert Einstein College of Medicine, Bronx, NY, USA

**Ashutosh Kumar** Department of Electrical Engineering, University of Notre Dame, Notre Dame, IN, USA

**Smita Kumari** College of Pharmacy, The Ohio State University, Columbus, OH, USA

**Sumana Kumar** Department of Microbiology, JSS Academy of Higher Education and Research, Mysuru, Karnataka, India

**Yogesh Kumar** Department of Biotechnology, Mehsana Urban Institute of Sciences, Faculty of Science, Ganpat University, Mehsana, Gujarat, India

**Alisha Lalhall** Centre for Nanoscience and Nanotechnology (CNSNT), Panjab University, Chandigarh, India

**Unnati Modi** School of Biotechnology & Bioengineering, Institute of Advanced Research, Gandhinagar, Gujarat, India

**Rudraksh Mukherjee** Department of Mathematics, Bioinformatics and Computer Applications, Maulana Azad National Institute of Technology (MANIT), Bhopal, Madhya Pradesh, India

**Bharti Nayak** School of Sciences, Sanjeev Agrawal Global Educational University, Bhopal, India

**Karan Pankajbhai Panchal** Sarvajanik College of Engineering and Technology, Athwa, Surat, Gujarat, India

**Puja Patel** Department of Animal Health and Management, Alagappa University, Karaikudi, TN, India

**Jugal Patil** Department of Biological Sciences and Engineering, Indian Institute of Technology Gandhinagar, Gandhinagar, Gujarat, India

**Jugal R. Patil** Department of Biological Sciences and Engineering, Indian Institute of Technology Gandhinagar, Gandhinagar, Gujarat, India

**Jogendra Singh Pawar** College of Pharmacy, The Ohio State University, Columbus, OH, USA
The James Cancer Hospital and Solove Research Institute, Comprehensive Cancer Center, The Ohio State University, Columbus, OH, USA

**Kotthapalli Prashanth** Department of Microbiology, JSS Academy of Higher Education and Research, Mysuru, Karnataka, India

**Antonio Puertas-Segura** Department of Chemical Engineering, Universitat Politècnica de Catalunya, Barcelona, Spain

**Sagar Rana** Academy of Scientific and Innovative Research (AcSIR), Ghaziabad, India
CSIR-Central Scientific Instruments Organization (CSIO), Sector-30C, Chandigarh, India

**Garima Rathee** Department of Chemical Engineering, Universitat Politècnica de Catalunya, Barcelona, Spain

**Jahnavi Sharma** Department of Biochemistry, All India Institute of Medical Sciences, Bhopal, India

**Payal Sharma** Centre of Biotechnology, Nana JI Deshmukh Veterinary Science University, Jabalpur, India

**Anusha Singhal** Albert Einstein College of Medicine, Bronx, NY, USA

**Raghu Solanki** Department of Biological Sciences and Engineering, Indian Institute of Technology Gandhinagar, Gandhinagar, Gujarat, India

**Joanna Szczepanek** Centre for Modern Interdisciplinary Technologies, Nicolaus Copernicus University, Torun, Poland

**Zarin Tasnim Tisha** School of Chemistry and Chemical Sciences, University of Surrey, Guildford, UK

**Shikha Tiwari** School of Sciences, Sanjeev Agrawal Global Educational University, Bhopal, India
Department of Biochemistry, All India Institute of Medical Sciences, Bhopal, India

**Deepika Tuteja** Institute of Nuclear Medicine and Allied Sciences, DRDO, New Delhi, India

**Abhi Tyagi** Department of Mathematics, Bioinformatics and Computer Applications, Maulana Azad National Institute of Technology (MANIT), Bhopal, Madhya Pradesh, India

**Swetha Vaishnavi** Chemical Biology Lab, Department of Chemistry, School of Engineering and Sciences, SRM University-AP, Amaravati, Andhra Pradesh, India

**Bazil Gulzar Wani** Amity Centre for Nanomedicine, Amity University, Noida, UP, India

**Amit K. Yadav** Department of Biological Sciences and Engineering, Indian Institute of Technology Gandhinagar, Gandhinagar, Gujarat, India

# The Biogenesis, Regulation, and Canonical and Noncanonical Functional Machinery of miRNAs

1

Yogesh Kumar, Parwez Ahmad, and Sandesh Chibber

**Abstract**

MicroRNAs (miRNAs) are small noncoding RNAs that play a central role in posttranscriptional gene regulation across eukaryotes. This book chapter explores the biogenesis, regulatory mechanisms, and both canonical and noncanonical functions of miRNAs. The biogenesis of miRNAs involves sequential processing by the Drosha-DGCR8 complex and Dicer, followed by incorporation into Argonaute (AGO)-containing RNA-induced silencing complexes (RISCs). Precise strand selection and guide strand retention are mediated by thermodynamic asymmetry, RNA-binding proteins, and ATP-dependent helicases. AGO proteins facilitate target mRNA silencing via cleavage or translational repression, driven by the seed sequence of the miRNA. Beyond their classical role in gene silencing, miRNAs are also implicated in chromatin modification, immune responses, and alternative splicing. Regulation of miRNA function occurs at multiple levels, including RNA editing, 3′ ends tailing, and turnover by exonucleases such as Eri1 and SDNs. Posttranslational modifications of AGO proteins further influence RISC activity, localization, and stability. Feedback loops between miRNA availability and AGO abundance maintain homeostasis within the silencing machinery. This chapter underscores the complexity and versatility of miRNA-mediated pathways and highlights their relevance in development, epigenetic inheritance, stress responses, and disease. Understanding the integrated regulation and multifaceted roles of miRNAs provides new insights into gene regulatory networks and therapeutic opportunities.

Y. Kumar (✉) · S. Chibber
Department of Biotechnology, Mehsana Urban Institute of Sciences, Faculty of Science, Ganpat University, Mehsana, Gujarat, India

P. Ahmad
RECETOX, Faculty of Science, Masaryk University, Brno, Czech Republic





S. Tiwari et al. (eds.), *microRNA Modulation: A Promising Combination Therapy for Cancer*, https://doi.org/10.1007/978-981-95-5989-3_1

**Keywords**

MicroRNAs · Argonaute proteins · RNA-induced silencing complex · miRNA biogenesis · Canonical pathway · Noncanonical miRNA functions · RNA editing

## Abbreviations

| | |
|---|---|
| AGO | Argonaute |
| AS | Antisense |
| ATP | Adenosine Triphosphate |
| C-P4H(I) | Type I Collagen Prolyl-4-Hydroxylase |
| DGCR8 | DiGeorge Syndrome Critical Region Gene 8 |
| HSP90 | Heat Shock Protein 90 |
| MAPK | Mitogen-Activated Protein Kinase |
| miRLC | microRNA Loading Complex |
| miRNA | microRNA |
| miRNA* | Passenger strand of microRNA duplex |
| miRNP | microRNA Ribonucleoprotein |
| PACT | Protein Activator of PKR |
| PAZ | Piwi/Argonaute/Zwille |
| P-bodies | Processing Bodies |
| piRNA | Piwi-Interacting RNA |
| PIWI | P-element Induced Wimpy Testis |
| PRC2 | Polycomb Repressive Complex 2 |
| pre-miRNA | Precursor microRNA |
| pri-miRNA | Primary microRNA |
| PTGS | Post-Transcriptional Gene Silencing |
| RDE-1 | RNAi Deficient-1 |
| RISC | RNA-Induced Silencing Complex |
| RNAi | RNA Interference |
| SDN | Small RNA Degrading Nuclease |
| siRNA | Small Interfering RNA |
| SNP | Single Nucleotide Polymorphism |
| TGS | Transcriptional Gene Silencing |
| TRBP | TAR RNA Binding Protein |
| UTR | Untranslated Region |

**Learning Objectives**

- Understand the biogenesis of canonical and noncanonical miRNAs.
- Identify key proteins involved in miRNA processing (Drosha, DGCR8, Dicer, Argonaute).
- Explore how miRNAs regulate gene expression and cellular functions.

- Analyze the regulation of miRNA pathways at multiple levels.
- Recognize the broader roles of the microprocessor complex in gene regulation.

## 1.1 Introduction

MicroRNAs (miRNAs) are tiny RNA molecules, usually 18–24 nucleotides long, that help control gene expression by regulating mRNA after it's been transcribed. They are crucial gene regulators in various biological processes and host–pathogen interactions. It is also present in biological fluids such as serum and blood, which makes it a prognostic tool. These small noncoding RNAs bind to target mRNA sequences, often leading to repression of gene expression. Several methods exist for identifying miRNA target sites, but simply detecting these sites is not enough to predict target regulation. The regulation of miRNA targets is complex, and recent studies have revealed a new layer of control, where targets can also influence the function and levels of miRNAs. This reciprocal regulation challenges our understanding of miRNA's role in gene expression and has significant implications for therapeutic applications (Nunes et al. 2025; Pasquinelli 2012). Over the past decade, microRNAs have become essential regulators of gene expression in animals. Biochemical networks, which often appear robust, are influenced by miRNAs. These networks are typically resistant to variations in molecule concentrations or reaction constants. However, under stress or genetic changes, the role of miRNAs in network robustness becomes more evident. Advances in research have uncovered features of miRNA organization that promote this robustness, showing their critical role in maintaining stability under changing conditions (Pasquinelli 2012; Peláez and Carthew 2012). Small RNAs, ranging from 20 to 30 nucleotides, target both chromatin and mRNA, offering extensive surveillance of the genome and transcriptome. High-throughput sequencing has unveiled a broad spectrum of small RNAs in eukaryotic cells. These RNAs can be categorized into microRNAs, endogenous small interfering RNAs (esiRNAs), and Piwi-interacting RNAs (piRNAs), based on their biogenesis and associated Argonaute proteins. Understanding the generation of these small RNAs is essential for exploring their role in gene regulation (Kim et al. 2009). In *Caenorhabditis elegans*, lin-4 plays a vital role in the temporal control of development. It negatively regulates the lin-14 protein to ensure proper timing during larval stages. Studies have shown that lin-4 does not encode a protein but functions through antisense RNA–RNA interactions to regulate lin-14 translation. Additionally, new tools have been developed to analyze conserved miRNA target sites in human 3′ UTRs, revealing a vast number of conserved targets and advancing our understanding of miRNA-mediated regulation (Friedman et al. 2009; Lee et al. 1993).

## 1.2 Biogenesis of miRNAs

Biogenesis of miRNA begins with the processing of RNA polymerase II/III transcripts, occurring during or after transcription. Half of miRNAs are intragenic, originating from introns or exons of protein-coding genes, while the other half are intergenic and transcribed independently. Some miRNAs are transcribed as clusters and classified as a family. Biogenesis of miRNA occurs via canonical or noncanonical pathways. Canonical miRNAs follow the classical pathway of biogenesis. Both canonical and noncanonical miRNAs play essential roles in gene regulation, with noncanonical miRNAs contributing to more diverse and context-dependent regulatory networks. Understanding both types is crucial for fully appreciating the complexity of miRNA-mediated gene regulation (de Rie et al. 2017; Ha and Kim 2014). A typical illustration and mode of action of miRNA biogenesis is given in Fig. 1.1.

**Canonical Pathway**

MicroRNAs (miRNAs) are initially transcribed as long primary miRNA (pri-miRNA) transcripts, which are processed in the nucleus into precursor miRNA (pre-miRNA) hairpins. The Microprocessor complex, consisting of Drosha and DGCR8, is essential for this processing. The efficiency of miRNA processing varies across different pri-miRNAs, and processing is often a better predictor of miRNA abundance than the primary transcription itself. In addition to its variability, miRNA processing is largely stable across different cell lines, suggesting sequence-specific factors play a role in processing. Short sequence features near the

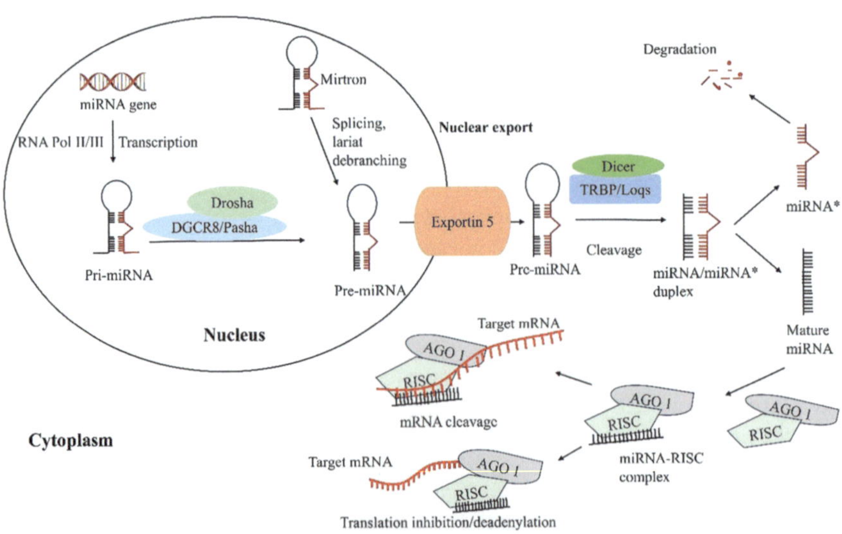


**Fig. 1.1** A typical illustration of canonical and noncanonical miRNA biogenesis and mode of action

pre-miRNA hairpin also contribute to processing efficiency (Conrad et al. 2014; Gregory et al. 2004). Understanding miRNA biogenesis also requires studying their transcription. A large number of miRNAs are derived from intergenic noncoding loci, but identifying their promoters has been a challenging task. To address this, the microTSS algorithm was developed, which predicts miRNA transcription start sites (TSSs) with high accuracy. This algorithm integrates RNA sequencing data and transcription marks to identify tissue-specific promoters. The utility of microTSS was validated using a Drosha-null mouse model, highlighting the importance of divergent transcription and the overlap of pri-miRNAs with long noncoding RNAs (Georgakilas et al. 2014).

DGCR8 identifies an N6-methyladenylated GGAC motif, along with other sequences, within the pri-miRNA. Meanwhile, Drosha cuts the pri-miRNA duplex at the base of its characteristic hairpin structure, generating a 2-nucleotide 3′ overhang on the pre-miRNA (Alarcón et al. 2015). miRNAs are regulated by various factors, including Exportin-5, which is essential for the nuclear export of pre-miRNAs. This protein binds pre-miRNAs in the presence of Ran-GTP, enabling their transport to the cytoplasm for further processing by RNase III endonuclease Dicer (Denli et al. 2004). Furthermore, miRNA-guided repression of target mRNAs involves the Argonaute protein complex, with specific binding affinities and noncanonical target sites influencing miRNA activity. Recent advancements in understanding these processes have provided a more detailed framework for integrating miRNAs into gene-regulatory networks, which will help elucidate their roles in biological functions across different organisms (McGeary et al. 2019; Yi et al. 2003). MiRNA processing involves the removal of the terminal loop, resulting in a mature miRNA duplex. The direction of the miRNA strand determines the mature miRNA form. The 5p strand originates from the 5′ end, while the 3p strand comes from the 3′ end of the pre-miRNA hairpin. Both strands can be loaded into Argonaute (AGO) proteins in an ATP-dependent manner (Yoda et al. 2010; Zhang et al. 2004). The proportion of AGO-loaded 5p or 3p strands varies significantly depending on the cell type or environment. The selection of the 5p or 3p strand is influenced by the thermodynamic stability at the 5′ ends of the duplex or the presence of a 5′ uracil at nucleotide position 1 (Khvorova et al. 2003; Meijer et al. 2014). Generally, the strand with lower 5′ stability or 5′ uracil is loaded into AGO and becomes the guide strand. The unloaded strand is called the passenger strand, which is unwound or degraded depending on its complementarity and AGO2 interaction (Ha and Kim 2014).

**Noncanonical Pathway**

Multiple noncanonical miRNA biogenesis pathways have been identified, expanding the understanding of miRNA processing beyond the canonical Drosha/Dicer pathway. These alternative pathways typically involve different protein combinations and can be grouped into Drosha-/DGCR8-independent and Dicer-independent pathways. MicroRNAs (miRNAs) are short, endogenous RNAs that regulate gene expression by targeting complementary messenger RNAs. In animals, canonical miRNA biogenesis involves two major steps. First, Drosha cleaves the primary transcript to release a pre-miRNA hairpin, and then Dicer cleaves the hairpin to generate a miRNA:miRNA* duplex. This duplex is processed further to produce a mature miRNA that participates in the RNA-induced silencing complex. However, alternative pathways also contribute to miRNA biogenesis. One such pathway involves mirtrons, which are pre-miRNAs formed from debranched introns. These mirtrons mimic the structure of pre-miRNAs and enter the miRNA processing pathway without Drosha cleavage. Studies have identified several mirtrons in species like *Drosophila melanogaster* and *C. elegans*, suggesting that mirtrons may have evolved early in the development of miRNAs, particularly in species with abundant introns similar in size to pre-miRNAs (Babiarz et al. 2008; Ruby et al. 2007). Another alternative biogenesis pathway bypasses the traditional microprocessor complex. Some miRNAs, such as 7-methylguanosine (m7G)-capped pre-miRNAs, are processed via the PHAX-exportin 1 pathway. These pre-miRNAs are exported from the nucleus without Drosha cleavage and processed by Dicer in the cytoplasm, with only the 3p strand efficiently loaded onto Argonaute to form a functional microRNP. This pathway demonstrates a unique mode of miRNA biogenesis that involves different steps for synthesis, transport, and strand selection compared to the canonical model (Xie et al. 2013). Additionally, some miRNAs, like miR-451, are processed independently of Dicer. Instead, Drosha generates a short pre-miRNA hairpin, which is directly cleaved by AGO2, and the 3′ end is resected to complete maturation. This processing pathway has been observed in cells lacking Dicer, where miR-451 can still mature, but other miRNAs cannot. The ability to express miR-451 in Dicer-deficient cells highlights an alternative mechanism for miRNA maturation that could be used for gene silencing and potential therapeutic applications in Dicer knockout models. Finally, the catalytic activity of Argonaute proteins, particularly AGO2, plays a significant role in miRNA biogenesis. AGO2, which retains catalytic competence, is essential for the processing of certain miRNAs like miR-451. Mutations in AGO2 can disrupt this processing, demonstrating the conserved nature of Argonaute's enzymatic function in vertebrate development and miRNA maturation (Cheloufi et al. 2010; Yang et al. 2010). The noncanonical NF–κB signaling pathway plays a crucial role in regulating cell survival, growth, maturation, and differentiation in response to various biological signals. These functions are vital for the proper formation and upkeep of hematopoietic cells and immune system tissues. When this pathway is disrupted, it can contribute to the development of immune-related disorders and certain cancers (Rodriguez et al. 2024).

## 1.3 miRNA Gene and Families

miRBase catalogs distribute microRNA sequences from 271 species, including 38,589 hairpin precursors and 48,860 mature miRNAs in its latest release (v22). Recent improvements provide enhanced annotations and better functional data for miRNA products. Through 1,493 small RNA sequencing datasets and 5.5 billion read mappings, miRBase validates 20–65% of annotations in various animal genomes. miRBase offers a variety of resources to advance miRNA genomics, including genomic coordinates for miRNAs, miRNA clusters, and their overlap with annotated transcripts (Chiang et al. 2010; Kozomara et al. 2019). Recent sequencing of small RNAs from mouse tissues has confirmed 398 annotated miRNA genes and identified 108 novel miRNAs. Many previously proposed miRNAs failed to yield small RNAs with miRNA-like features, suggesting that earlier estimates of conserved mammalian miRNAs were inflated. This highlights the need for careful miRNA validation and provides an updated list of confirmed murine miRNAs (Chiang et al. 2010). Furthermore, the study of miRNA evolution across metazoans reveals that miRNAs are continuously added to genomes with rare mutations and losses. However, shifts in the seed sequence and posttranscriptional editing at the 5′ end of miRNAs add complexity to their regulatory roles. The miR-17 approximately 92 cluster, a highly conserved family, plays critical roles in lung development and B cell differentiation. Knockout studies using TALEN technology have further revealed the distinct functions of miR-141 and miR-200c, underscoring the importance of strict seed-pairing for miRNA–target interactions (Kim et al. 2013; Ventura et al. 2008).

## 1.4 Regulation of miRNA Biogenesis

miRNA biogenesis is crucial for proper cellular function and is regulated at multiple stages, including transcription, nuclear processing, export, cytoplasmic pre-miRNA processing, RNA-induced silencing complex formation, and intrinsic regulatory mechanisms. Posttranscriptional regulators of miRNA biogenesis in animals are given in Table 1.1.

### 1.4.1 Transcriptional Regulation and Diversity of miRNA Biogenesis

MicroRNAs play crucial roles in both normal human physiology and disease, influencing various cellular processes. However, the transcriptional regulation of miRNAs remains poorly understood, primarily because the promoters of most miRNAs are not well characterized. Recent research identified the proximal promoters of 175 human miRNAs by combining nucleosome mapping with chromatin signatures, providing new insights into miRNA gene regulation. This study found that one-third of intronic miRNAs have independent transcription initiation regions, distinct from their host gene promoters, challenging the previous assumption that intronic miRNAs are exclusively regulated by their host genes (Monteys et al. 2010; Ozsolak et al. 2008). Further analysis revealed that certain intronic miRNAs are regulated by RNA polymerase II (RNA Pol II) and RNA polymerase III (RNA Pol III). Intronic regions containing miRNAs and their upstream RNA Pol II (miR-107, miR-126, miR-208b, miR-548f-2, miR-569, and miR-590) or RNA Pol III (miR-566 and miR-128-2) sequences were cloned into a promoterless plasmid,

**Table 1.1** Posttranscriptional regulators of miRNA biogenesis in animals

| Regulator | Function | Target step | Effect | Example miRNAs affected | References |
|---|---|---|---|---|---|
| Lin28 (Lin28A/B) | RNA-binding protein that blocks processing | Blocks pri- and pre-miRNA processing by Microprocessor | Inhibits maturation | *let-7* family | Piskounova et al. (2011) |
| hnRNP A1 | RNA-binding protein | Promotes Drosha processing | Promotes cleavage into pre-miRNA | *Pri-miR-18a, pri-Let-7a* | Guil and Cáceres (2007), Michlewski et al. (2008) |
| KSRP | KH-type splicing regulatory protein | Binds to pri- and pre-miRNAs to aid Drosha/Dicer processing | Promotes maturation | *let-7*, *miR-1*, *miR-206* | Amirouche et al. (2014), Trabucchi et al. (2009) |
| p68/p72 (DDX5/ DDX17) | DEAD-box RNA helicases | Co-factors of Drosha/ DGCR8 complex | Promote pri-miRNA processing | General, non-specific | Blahna and Hata (2012), Finnegan and Pasquinelli (2013) |
| SMAD proteins | Transcription factors activated by TGF-β signaling | Associate with Drosha complex | Promote pri-miRNA processing | *miR-21*, *miR-199a* | Blahna and Hata (2012), Lin et al. (2009) |
| p53 | Tumor suppressor transcription factor | Interacts with Drosha complex via p68 | Enhances pri-miRNA processing | miR-143, miR-145, *miR-16-1* | Suzuki et al. (2009) |
| TRBP (TAR RNA-binding protein) | Dicer co-factor | Facilitates Dicer processing | Promotes pre-miRNA to miRNA | *let-7* | Benoit et al. (2013), Fareh et al. (2016) |
| ADAR (Adenosine Deaminases Acting on RNA) | RNA-editing enzyme | Edits pri–/pre-miRNA, affecting structure | Can inhibit or alter miRNA maturation | *miR-376*, *miR-142* | Das and Carmichael (2007), Widmark et al. (2022) |
| MCPIP1 (Zc3h12a) | RNase that cleaves pre-miRNAs | Degrades pre-miRNA | Inhibits maturation | *miR-146a,* *miR*-135b, *miR*-146a, *miR*-21, *miR*-155, *miR*-143, and *miR*--145 | Suzuki et al. (2011) |

| | | | | | |
|---|---|---|---|---|---|
| TUTases (e.g., TUT4, TUT7) | Terminal uridylyl transferases | Add uridines to pre-miRNA | Promote degradation (esp. with Lin28) | *let-7* family | Zhang et al. (2022) |
| XPO5 (Exportin-5) | Nuclear export protein | Transports pre-miRNA to the cytoplasm | Promotes export | All pre-miRNAs | Yi et al. (2003) |

confirming that miRNA expression occurs independently of host gene transcription. Additionally, high-throughput sequencing techniques helped identify specific transcription factors, such as MITF, that regulate miRNA expression, particularly in melanoma cells. These findings highlight the complexity of miRNA regulation and its potential relevance in disease settings (Monteys et al. 2010). The structural features of miRNA genes also play an important role in their transcription. Pri-miRNA precursors, which are the initial transcripts of miRNAs, are capped and polyadenylated, resembling messenger RNAs (mRNAs). This finding was confirmed by sequencing the full-length pri-miR-21 transcript, which is about 3433 nucleotides long. The transcription of miRNAs is primarily carried out by RNA polymerase II, as demonstrated by the inhibition of pri-miRNA expression upon treatment with α-amanitin, a selective inhibitor of RNA Pol II. Chromatin immunoprecipitation (ChIP) analyses further confirmed the association of RNA Pol II with miRNA promoters, providing direct evidence of RNA Pol II-mediated miRNA transcription (Cai et al. 2004; Lee et al. 2004; Monteys et al. 2010). Additionally, recent studies have uncovered a diverse range of miRNA biogenesis pathways, extending beyond the canonical models. For example, some miRNAs are processed via noncanonical pathways, such as mirtrons, which are processed directly from the introns of host genes without Drosha cleavage. These noncanonical miRNAs were identified in mouse embryonic stem cells (mESCs) lacking DGCR8, a key component of the Microprocessor complex. Furthermore, small interfering RNAs (siRNAs), which share some similarities with miRNAs, can also be generated from long hairpin RNAs, adding another layer of complexity to the regulation of small RNA molecules in mammalian cells (Babiarz et al. 2008). The regulation of miRNA biogenesis is not limited to transcriptional mechanisms. Several studies have also revealed that miRNA expression is modulated by various signaling pathways and transcription factors. Growth factors, in particular, play an essential role in regulating the expression of miRNAs during development and disease progression. Aberrant miRNA expression has been linked to a range of diseases, including cancer and cardiovascular disorders, underscoring the importance of understanding the regulatory networks governing miRNA function (Davis and Hata 2010).

### 1.4.2 Nuclear Processing of miRNA

After transcription, nuclear processing events for microRNAs (miRNAs) begin with the cleavage of primary miRNA (pri-miRNA) transcripts by the Microprocessor complex, which includes Drosha and the RNA-binding protein Pasha. This initial step generates approximately 70-nucleotide precursor miRNAs (pre-miRNAs) from the longer pri-miRNA transcripts. The processing of pri-miRNAs into pre-miRNAs occurs in the nucleus, setting the stage for the next phase. Once pre-miRNAs are formed, they are exported to the cytoplasm for further processing. The second processing step occurs in the cytoplasm, where pre-miRNAs are further cleaved into mature miRNAs by Dicer (Denli et al. 2004; Lee et al. 2002). These miRNAs regulate gene expression at the posttranscriptional level. Additionally, miRNA expression regulation involves nuclear export, which is mediated by the pre-miRNA. The clustering of miRNA genes suggests polycistronic expression. A novel gene, DGCR8, identified in the DiGeorge syndrome region, plays a role in this biogenesis process. It exhibits ubiquitous expression across tissues and may contribute to congenital defects in DiGeorge syndrome (Shiohama et al. 2003).

### 1.4.3 Microprocessor Machinery: From Pri-miRNA to Gene Control

The microprocessor complex, comprising Drosha and DGCR8, is crucial for the biogenesis of microRNAs in eukaryotes. The complex initiates the processing of primary miRNA transcripts into precursor miRNA, which is further processed into mature miRNAs, essential regulators of gene expression. Drosha, a nuclear RNase III enzyme, is the catalytic component of this complex, while DGCR8 serves as a cofactor that facilitates substrate recognition and cleavage. Together, they cleave the stem-loop structures of pri-miRNAs in a highly coordinated manner, ensuring proper miRNA maturation. Drosha is known for its role in cleaving pri-miRNA substrates. Its structure features two RNase III domains, A and B, responsible for cleaving the 3′ and 5′ strands of the miRNA stem, respectively. The enzyme forms a large complex, with a molecular mass of approximately 650 kDa, indicative of its association with other factors such as DGCR8. The interaction between Drosha and DGCR8 is essential for miRNA processing. Drosha's RNase III domains act as the catalytic core, while DGCR8 binds to the pri-miRNA, stabilizing the complex and ensuring precise cleavage at the stem–loop junction (Han et al. 2004).

DGCR8 plays a pivotal role in the recognition of pri-miRNA substrates, specifically through its two double-stranded RNA-binding domains (dsRBDs). These domains recognize and bind to the double-stranded regions of the pri-miRNA, providing the molecular scaffold that anchors the substrate in the Drosha–DGCR8 complex. Structural studies have demonstrated that DGCR8, but not Drosha, directly binds to pri-miRNAs, with its dsRBDs crucial for this recognition. In this complex, DGCR8 appears to function as the molecular ruler, determining the exact location of cleavage based on the distance from the stem–ssRNA junction (Han et al. 2006). Additionally, DGCR8's N-terminal region, although necessary for its nuclear localization, does not contribute directly to pri-miRNA processing (Yeom et al. 2006). Further investigations into the biochemical properties of DGCR8 revealed its ability to bind heme, which significantly affects its activity in pri-miRNA processing. The binding of ferric heme (Fe(III)) enhances the dimerization and activity of DGCR8, promoting its interaction with Drosha and subsequent cleavage of pri-miRNA. On the other hand, reduction of the heme to the ferrous (Fe(II)) state impairs this processing activity, suggesting that heme plays an essential role in modulating the function of DGCR8. This discovery underscores a potential link between intracellular heme availability and miRNA processing efficiency, which could have broader implications for cellular function and disease (Barr et al. 2012; Faller et al. 2007).

The functional interplay between Drosha and DGCR8 is further elucidated through studies on their subcellular localization. Drosha's nuclear localization is facilitated by phosphorylation events at Serine300 and Serine302, which are necessary for its proper translocation into the nucleus (Tang et al. 2010). In contrast, DGCR8's nuclear localization is influenced by RNA transcription, and its interaction with proteins such as Nucleolin and ILF3 suggests its involvement in RNA processing and transport (Shiohama et al. 2007). These interactions point to the complexity of the microprocessor complex, as DGCR8 not only aids in pri-miRNA recognition but also participates in larger RNA regulatory networks.

The mechanism of action of Drosha and DGCR8 is further characterized by their cooperative binding to pri-miRNAs. Studies have shown that DGCR8 forms higher-order structures upon binding to pri-miRNAs, including a trimerization process that enhances its recognition and processing activity. This cooperative binding, particularly in the case of pri-mir-30a, demonstrates the high specificity of DGCR8 for its miRNA substrates (Faller et al. 2010). Moreover, computational models and biochemical analyses have contributed to our understanding of the structural basis for pri-miRNA recognition by DGCR8, offering insights into how these molecular interactions drive the microRNA maturation process (Sohn et al. 2007). The Drosha–DGCR8 complex is a central player in miRNA biogenesis, responsible for the initial and crucial cleavage of pri-miRNAs. The catalytic activity of Drosha, in concert with the substrate recognition by DGCR8, ensures the accurate maturation of miRNAs. Recent studies have also highlighted the importance of heme in regulating DGCR8 activity and its potential connection to intracellular signaling pathways. The intricate interplay between Drosha, DGCR8, and other RNA-binding proteins underscores the complexity of miRNA maturation, which is essential for the regulation of gene expression and cellular function (Weitz et al. 2014).

### 1.4.4 Expanding Roles of Drosha: DGCR8 and Gene Expression via the Microprocessor Complex

The cleavage of pri-miRNAs by Drosha is a precise process, facilitated by its interaction with the DGCR8 protein, which plays a key role in RNA substrate recognition. Drosha, the catalytic enzyme, cleaves the double-stranded RNA (dsRNA) regions, while DGCR8 binds to the flanking single-stranded RNA (ssRNA) segments to stabilize the complex and help define the cleavage site. Notably, the terminal loop of pri-miRNAs is not essential for cleavage, although the adjacent ssRNA regions are critical for the binding and processing steps (Han et al. 2006). A typical metazoan pri-miRNA contains a 33-nucleotide stem, which forms a hairpin structure, and a terminal loop. The cleavage site is determined primarily by the distance from the stem–ssRNA junction. Specifically, the distance from the lower stem–ssRNA junction to the cleavage site is approximately 11 base pairs, a measurement that is essential for the accuracy of Drosha cleavage (Zeng et al. 2005). In some cases, variations in the sequence or structure of the stem may affect the precise location of the cleavage site, suggesting that the Microprocessor complex is capable of fine-tuning its cleavage activity in response to subtle differences in RNA structure. Additionally, research suggests that Drosha cleavage is influenced by the structure of both the upper and lower stem–ssRNA junctions, with precise measurements of these distances crucial for effective cleavage (Ma et al. 2013).

One of the challenges in understanding pri-miRNA processing lies in distinguishing these primary transcripts from other hairpin-containing RNA molecules within the cell. Various determinants, including specific RNA motifs, must be present for efficient processing. For instance, a specific SRp20-binding motif found downstream of pri-miRNAs in bilaterian animals, but absent in nematodes, plays a significant role in distinguishing pri-miRNAs from non-miRNA hairpins. Moreover, experimental studies indicate that primary-sequence determinants, including pairing in the stem region, are vital for the recognition of pri-miRNAs by the processing machinery (Auyeung et al. 2013).

The microprocessor's activity extends beyond miRNA biogenesis, as it has been implicated in the regulation of mRNA stability. Drosha and DGCR8 can regulate each other posttranscriptionally, with Drosha cleaving the hairpin structures within the DGCR8 mRNA, leading to its destabilization. DGCR8, in turn, stabilizes Drosha, suggesting a feedback loop that maintains homeostasis in miRNA biogenesis (Han et al. 2009). Additionally, this regulatory relationship has broader implications, as microarray analyses suggest that mRNA targets might be downregulated in a miRNA-independent manner through the action of the Microprocessor complex. Beyond its classical role in miRNA maturation, Drosha–DGCR8 also participates in mRNA stability control, which could influence gene expression patterns. For example, the cleavage of hairpins in Neurog2 mRNA by Drosha contributes to the destabilization of these transcripts, which is crucial for maintaining stem cell identity during neurogenesis. This process highlights the importance of miRNA-independent functions of the microprocessor, which facilitates the regulation of proneural mRNAs, ensuring proper neural stem cell (NSC) maintenance (Knuckles et al. 2012).

Moreover, the Drosha–DGCR8 complex also has important roles in controlling gene expression in response to specific physiological conditions. For example, a posttranscriptional switch in primate cells regulates the mutually exclusive expression of miR-198 and FSTL1. This switch is controlled by the KH-type splicing regulatory protein (KSRP), which facilitates the processing of miR-198. Upon wounding, transforming growth factor-β (TGF-β) signaling downregulates KSRP, allowing the switch to FSTL1 expression, which promotes keratinocyte migration essential for wound healing. A failure in this switch, as seen in chronic diabetic ulcers, leads to impaired wound healing and defective migration due to persistent miR-198 expression (Sundaram et al. 2013). Furthermore, studies on the cellular regulation of miRNA biogenesis reveal that Drosha cleavage can occur co-transcriptionally. This means that pri-miRNAs are processed while still associated with the chromatin, enhancing the efficiency of miRNA production. Pri-miRNAs retained at transcription sites due to disrupted polyadenylation signals are more efficiently processed into pre-miRNAs, suggesting a coupling between transcription and miRNA maturation (Pawlicki and Steitz 2008). This process is further influenced by the interaction of the microprocessor with transcription factors and splicing factors, demonstrating that miRNA processing is intricately linked with other RNA processing events within the cell. The role of the Drosha–DGCR8 complex in miRNA biogenesis and its broader influence on gene expression are fundamental to cellular function, contributing not only to the regulation of miRNAs but also to the modulation of mRNA stability and transcriptional outcomes (Mori et al. 2014; Morlando et al. 2008).

### 1.4.5 Regulation of Microprocessor Activity in miRNA Biogenesis

It is evident that the regulation of microprocessor activity during miRNA biogenesis is a complex process involving multiple proteins and pathways. Drosha, a type III RNase, initiates this process by cleaving pri-miRNAs into pre-miRNAs within the nucleus. It is crucial to note that Drosha works in conjunction with its double-stranded RNA-binding partner, Pasha/DGCR8, to recognize and cleave these pri-miRNA hairpins. There are instances where Drosha regulates RNAs not annotated as miRNAs, indicating additional roles beyond canonical miRNA processing (Kadener et al. 2009). It is observed that the microprocessor complex (MC) comprising Drosha and DGCR8 is essential for this initial cleavage step. Several regulatory mechanisms affect the microprocessor complex function. It is shown that DGCR8 is subject to phosphorylation at multiple sites, including those targeted by the mitogenic ERK/MAPK pathway. It is demonstrated that phosphorylation of DGCR8 influences protein stability and subsequently affects cellular proliferation (Herbert et al. 2013). There is evidence that Drosha itself is also phosphorylated, with GSK3β identified as the kinase responsible for phosphorylating Drosha at Ser 300 and 302, which is critical for its nuclear localization (Tang et al. 2011). It is also found that histone deacetylase 1 (HDAC1) enhances miRNA processing by increasing the affinity of DGCR8 to pri-miRNAs through deacetylation (Wada et al. 2012). It is observed that Drosha protein levels are regulated by both ubiquitination and acetylation. It is shown that acetylation of Drosha competes with its ubiquitination, thereby inhibiting proteasome-mediated degradation and increasing Drosha protein levels (Tang et al. 2013).

There are also instances where other proteins influence microprocessor function. It is shown that MeCP2, a protein associated with neurodevelopmental disorders, directly binds to DGCR8 and interferes with the assembly of the Drosha–DGCR8 complex, thereby suppressing nuclear miRNA processing (Cheng et al. 2014). It is observed that TGF-β and BMP signaling promote the processing of pri-miR-21 through the DROSHA complex, with SMAD signal transducers and the RNA helicase p68 playing crucial roles (Davis et al. 2008). It is demonstrated that Smads directly bind to a consensus sequence (R-SBE) within the stem region of pri-T/B-miRs, facilitating Drosha-mediated processing (Davis et al. 2010).

In light of exploring the nuclear biogenesis of miRNA by BioID analysis, Drosha, the catalytic core of the microprocessor complex, was found to interact closely with

SFPQ, a multifunctional RNA-binding protein involved in RNA processing. Unlike other paraspeckle-associated proteins or NEAT1 RNA, SFPQ uniquely influences both primary and mature miRNA levels, suggesting a paraspeckle-independent function. Loss of SFPQ alters transcription and stability of miRNA host genes. Importantly, SFPQ shields the oncogenic miR-17 ~ 92 cluster from degradation via the NEXT-exosome complex, linking it to cancer-associated miRNA overexpression (Thivierge et al. 2024). It is shown that p53, a tumor suppressor, enhances the posttranscriptional maturation of several miRNAs by interacting with the Drosha processing complex through p68 (Suzuki et al. 2009). It is revealed that TDP-43, an RNA-binding protein, controls the stability of Drosha protein during neuronal differentiation and is required for Drosha-mediated regulation of specific genes (Di Carlo et al. 2013). It is also found that TDP-43 facilitates the production of pre-miRNAs by interacting with the Drosha complex and binding directly to pri-miRNAs, as well as promoting pre-miRNA processing in the cytoplasm (Kawahara and Mieda-Sato 2012). It is shown that hnRNP A1, another RNA-binding protein, binds to pri-miR-18a and facilitates its processing by Drosha (Guil and Cáceres 2007). It is further demonstrated that hnRNP A1 induces a relaxation at the stem of pri-miR-18a, creating a more favorable cleavage site for Drosha, and that conserved loops in pri-miRNAs are essential for processing (Michlewski et al. 2008). It is observed that KSRP, an RNA-binding protein, serves as a component of both Drosha and Dicer complexes and regulates the biogenesis of a subset of miRNAs by binding to the terminal loop of miRNA precursors (Trabucchi et al. 2009).

### 1.4.6 Nuclear Export

The nuclear export of pre-miRNAs plays a crucial role in the biogenesis and function of miRNAs, which regulate gene expression in eukaryotic cells. Initially, miRNAs are transcribed as long primary transcripts that are processed by the microprocessor complex in the nucleus. This processing produces a stem–loop structure, termed pre-miRNA, which is approximately 70 nucleotides in length. Once processed, these pre-miRNAs must be exported to the cytoplasm for further maturation into functional miRNAs. The export of pre-miRNAs is a carrier-mediated, saturable process that is sensitive to nuclear RanGTP depletion, suggesting the involvement of a RanGTP-dependent transport pathway (Bohnsack et al. 2004). Exportin-5 (Exp5), a member of the karyopherin family of proteins, is the key nuclear export receptor for pre-miRNAs. This protein interacts specifically with pre-miRNAs in a sequence-independent manner, recognizing the structural features of the stem–loop structure. The export process is facilitated by the binding of RanGTP, a small GTPase, to Exportin-5. This interaction is critical for the efficient export of pre-miRNAs, as depletion of Exp5 results in reduced levels of mature miRNAs. Importantly, Exp5 is not only essential for miRNA biogenesis but also helps coordinate the processing steps that occur in the nucleus and cytoplasm (Lund et al. 2004; Yi et al. 2003).

The molecular details of how Exp5 binds pre-miRNAs have been elucidated through structural studies. The structure of the pre-miRNA export complex reveals that Exp5, in association with RanGTP, recognizes the 2-nucleotide 3′ overhang and the double-stranded stem of the pre-miRNA. This interaction is facilitated by weak, broad interactions with the stem and stronger, specific interactions with the 3′ overhang through hydrogen bonds and ionic contacts. These structural insights confirm that Exp5's recognition of pre-miRNAs does not depend on the specific RNA sequence but instead relies on the overall structure of the RNA (Okada et al. 2009). Furthermore, Exp5-mediated nuclear export is not limited to miRNAs. Other RNAs with similar structural motifs, such as the adenovirus VA1 RNA, are also transported by Exp5 in a RanGTP-dependent manner. This indicates that Exp5 has a broader role in RNA export beyond miRNAs. The dependence of pre-miRNA export on Exp5 is consistent across various systems, including *Xenopus laevis* oocytes and human cells (Gwizdek et al. 2003; Zeng and Cullen 2004).

In addition to its role in miRNA biogenesis, Exp5 is involved in the regulation of gene expression during the cell cycle. The expression of Exp5 is elevated during cell cycle entry, which is critical for global miRNA elevation and proper cell cycle progression. Inhibition of Exp5 induction disrupts miRNA biogenesis and causes defects in the G1/S transition, underscoring the importance of Exp5 in coordinating gene expression during cell proliferation (Iwasaki et al. 2013). Recent studies have also highlighted the role of Exp5 in the DNA damage response. Following DNA damage, the nuclear export of pre-miRNAs is accelerated through an ATM-dependent mechanism. This involves the phosphorylation of Nup153, a component of the nuclear pore, which enhances its interaction with Exp5 and promotes the nuclear export of pre-miRNAs. This finding underscores the dynamic regulation of pre-miRNA export, linking it to critical cellular processes such as the DNA damage response (Wan et al. 2013).

In certain human cancers, defects in Exp5 function can lead to impaired miRNA export, resulting in the accumulation of pre-miRNAs in the nucleus. Such defects are often associated with mutations in the C-terminal region of Exp5, which is crucial for forming the pre-miRNA/Exp5/RanGTP complex. The inactivation of Exp5 in these cancers leads to reduced miRNA processing, diminished miRNA-target inhibition, and contributes to tumorigenesis. Restoration of Exp5 function in these cells can reverse the defect and restore proper miRNA biogenesis, suggesting that Exp5 could have tumor-suppressor properties (Melo et al. 2010). Exportin-5 interaction with pre-miRNAs is dependent on the structural features of the RNA rather than specific sequences, highlighting its flexibility in recognizing various pre-miRNAs. By mediating pre-miRNA export and coordinating nuclear and cytoplasmic processing, Exp5 is essential not only for miRNA maturation but also for regulating cellular processes such as the cell cycle and DNA damage response. Its dysfunction, whether through genetic mutations or cellular stress, can have significant consequences for gene expression and is linked to diseases, including cancer (Xie et al. 2013).

### 1.4.7 Cytoplasmic Pre-miRNA Processing: Dicer and Its Cofactors

Cytoplasmic pre-miRNA processing is a key event in the regulation of gene expression through RNA interference (RNAi). RNAi is a conserved mechanism where double-stranded RNAs (dsRNA) are processed into small RNA fragments, typically 21–23 nucleotides, which guide the destruction of specific messenger RNAs (mRNAs). The enzyme Dicer, a member of the RNase III family, plays a central role in cleaving dsRNA into these small fragments and is critical for both RNAi and the biogenesis of small temporal RNAs (stRNAs). Dicer is conserved across various species, including worms, flies, plants, fungi, and mammals (Bernstein et al. 2001).

In *C. elegans*, Dicer's function is crucial for the maturation of stRNAs such as lin-4 and let-7, which regulate developmental timing. Disruption of genes encoding Dicer and its homologs, such as alg-1 and alg-2, leads to phenotypic defects similar to those seen in mutants of lin-4 and let-7, demonstrating that common processing machinery is involved in both RNAi and endogenous gene regulation (Grishok et al. 2001). Moreover, in *D. melanogaster* and *C. elegans*, Dicer-mediated processing of precursor RNAs produces mature stRNAs, indicating that the RNAi and stRNA pathways intersect (Hutvágner et al. 2001). Mutations in Dicer genes cause defects in RNAi and other developmental processes. For example, in *C. elegans*, dcr-1 mutants exhibit defects in RNAi and abnormal germline development, including sterility, underscoring the importance of Dicer in these processes (Ketting et al. 2001; Knight and Bass 2001). Additionally, noncanonical mechanisms of microRNA biogenesis that do not rely on Dicer have been discovered, indicating the complexity of small RNA processing pathways (Yang and Lai 2011). In cancer, mutations in Dicer suggest it functions as a haploinsufficient tumor suppressor gene. Loss of one Dicer allele impairs miRNA processing and leads to reduced survival in tumor models, suggesting that Dicer may play a role in tumor suppression (Kumar et al. 2009).

#### 1.4.7.1 Dicer in RNA Processing and Gene Silencing

Dicer is a specialized ribonuclease that plays a pivotal role in RNA interference by processing long double-stranded RNA (dsRNA) and pre-miRNAs into biologically active small RNA fragments, such as siRNAs and miRNAs (Zhang et al. 2002). Dicer's enzymatic activity is fundamental to the initiation of RNAi, where it cleaves these RNA substrates into small fragments typically around 22 nucleotides in length. The structure of Dicer is multi-domain, with key components including the PAZ domain, which binds to the end of dsRNA, and two catalytic RNase III domains that facilitate cleavage of the substrate (Macrae et al. 2006; Zhang et al. 2002). These structural features allow Dicer to measure the length of RNA substrates and cleave them at specific positions, a process described as functioning like a molecular ruler (MacRae et al. 2007).

There are distinct isoforms of Dicer that perform specialized functions in various organisms. In *D. melanogaster*, for instance, Dicer-1 and Dicer-2 process different RNA substrates. Dicer-1 is responsible for processing precursor miRNAs into miRNA-miRNA* duplexes, whereas Dicer-2 processes long dsRNA into siRNA duplexes. The specificity of Dicer-2 for long dsRNA is enhanced by inorganic phosphate and a partner protein, R2D2, while Dicer-1 recognizes pre-miRNAs through its N-terminal helicase domain, which interacts with the single-stranded terminal loop of pre-miRNAs. This interaction enables Dicer-1 to carefully assess the authenticity of the pre-miRNA structure by measuring the size of the loop and the distance between the 3′ overhang and the loop. This unique recognition mechanism underscores the specificity with which Dicer-1 processes miRNAs (Tsutsumi et al. 2011).

In addition to these functions, Dicer's activity can be influenced by structural features of the RNA substrates, including the 3′ overhang and sequence composition. Dicer's cleavage position within dsRNA is strongly affected by the 3′ end structure, and the enzyme preferentially cleaves dsRNAs near their termini, ensuring precise processing of RNA molecules (Vermeulen et al. 2005). Dicer also recognizes bulge or loop structures within RNA, which contribute to its ability to process pre-miRNAs more accurately, as observed with short hairpin RNAs (shRNAs). This recognition mechanism is important not only for endogenous miRNA processing but also for optimizing the design of therapeutic shRNAs, where reducing off-target effects is critical (Gu et al. 2012).

Moreover, Dicer interacts with other proteins such as TRBP and AGO2 to facilitate the loading of processed siRNAs and miRNAs onto the RNA-induced silencing complex (RISC). This interaction is crucial for the transfer of guide RNAs from Dicer to AGO2, completing the RNAi pathway. Studies using electron microscopy (EM) have revealed that human Dicer, in complex with TRBP, adopts an L-shaped structure, with the N-terminal DExH/D domain interacting with TRBP and the catalytic domains interacting with AGO2. This structural arrangement is essential for efficient siRNA processing and RISC assembly (Lau et al. 2009; Wang et al. 2009). The structural data from human Dicer complexes, including crystal structures of its "platform-PAZ-connector helix" cassette, provide insights into the binding mechanism, with pockets for 3′ overhangs and phosphates crucial for processing siRNAs (Tian et al. 2014).

#### 1.4.7.2 Dicer Regulation by TRBP, Loquacious, and PACT

The processing of miRNAs, in particular, involves a stepwise cascade of actions that convert pri-miRNA transcripts into mature miRNAs, which then regulate gene expression by silencing target mRNAs. Initially, the pri-miRNAs are processed by the Drosha–Pasha/DGCR8 complex to form pre-miRNAs. These pre-miRNAs are further processed by Dicer to produce mature miRNAs. Dicer itself, however, does not act in isolation; it requires many cofactors to facilitate its function and ensure precise processing of its RNA substrates (Chendrimada et al. 2005; Förstemann et al. 2005). In *D. melanogaster*, the protein Loquacious (Loqs) is an essential cofactor for Dicer-1, the enzyme responsible for processing pre-miRNAs. Loqs, which is homologous to the human TRBP (trans-activation response RNA-binding protein), interacts directly with Dicer-1, enhancing its ability to process pre-miRNAs. Depletion of Loqs results in the accumulation of precursor miRNAs and a failure to process them into mature miRNAs. Moreover, Loqs is critical for miRNA-directed gene silencing, and its loss leads to defective RNA interference in various biological contexts, including transgene silencing and stem cell maintenance. The association between Loqs and Dicer-1 ensures efficient miRNA production and facilitates the biogenesis of miRNAs that regulate gene expression (Förstemann et al. 2005; Saito et al. 2005).

Another important cofactor in Dicer regulation is R3D1-L, a novel double-stranded RNA-binding protein that forms a stable complex with Dicer-1 in *Drosophila*. The depletion of R3D1-L results in a defect in miRNA production and causes severe sterility in both male and female flies. R3D1-L appears to enhance the activity of Dicer-1 in processing pre-miRNAs and is thus critical for miRNA biogenesis and reproductive development (Jiang et al. 2005). In mammals, the interaction between Dicer and TRBP also plays a crucial role in miRNA and siRNA processing. TRBP, which contains three double-stranded RNA-binding domains, enhances Dicer's ability to process both miRNAs and small siRNAs. TRBP is required for the proper loading of siRNAs onto the RNA-induced silencing complex (RISC), a key step in RNA interference. The complex between Dicer and TRBP also interacts with AGO2, a core component of RISC, facilitating the transfer of the processed small RNAs to AGO2 for subsequent gene silencing. Disruption of the Dicer–TRBP interaction results in the destabilization of Dicer and a loss of miRNA biogenesis, underscoring the importance of TRBP in regulating Dicer's activity and efficiency (Chendrimada et al. 2005; Haase et al. 2005).

Furthermore, the activity of Dicer is influenced by the presence of additional cofactors such as PACT (protein activator of the interferon-induced protein kinase PKR), which also interacts with Dicer and AGO2 in the RNA silencing pathway. While PACT does not affect the cleavage step of pre-miRNAs, it modulates the accumulation of mature miRNAs and enhances siRNA-mediated RNA interference. Depletion of PACT reduces the efficiency of RNAi, suggesting that it plays an important role in the assembly of RISC and the regulation of miRNA processing (Lee et al. 2006). Additionally, the regulation of Dicer's activity is fine-tuned by the choice of cleavage sites, which are influenced by its binding partners. For example, alternative splicing of Loquacious produces different isoforms (Loqs-PA, Loqs-PB, and Loqs-PD), with each isoform modulating Dicer's cleavage activity. Specifically, Loqs-PB influences the cleavage of pre-miRNA-307a, generating longer miRNA isoforms with distinct seed sequences and different target specificities. Such isoforms may repress specific mRNAs, demonstrating how cofactors like Loqs can modulate Dicer's cleavage sites, thus influencing the functional output of miRNA biogenesis. These findings underscore how the interaction between Dicer and its cofactors can modify miRNA functionality and target selection (Fukunaga et al. 2012). Finally, Dicer's product processing is further regulated by the structural properties of its RNA substrates. For instance, the presence of double-stranded RNA-binding proteins like TRBP can accelerate the processing of pre-miRNAs over pre-siRNAs, highlighting the importance of RNA structure in determining processing efficiency. The interaction of TRBP with Dicer influences dicing kinetics and facilitates the generation of miRNA isoforms (isomiRs), which can alter target specificity and gene silencing patterns. This underscores the pivotal role of cofactors like TRBP in regulating Dicer's activity and shaping the functional diversity of the RNAi pathway (Chakravarthy et al. 2010; Lee and Doudna 2012).

### 1.4.8 RNA-Induced Silencing Complex (RISC) Formation

The Argonaute protein family plays a central role in the RNA interference pathway, facilitating gene silencing by binding small RNA molecules that guide the complex to complementary target RNA sequences. In *C. elegans*, 27 AGO genes contribute to various biological processes, including RNAi, fertility, and chromosome segregation, with each AGO protein performing distinct roles within the gene-silencing machinery. AGO proteins interact with small RNAs such as small interfering RNAs and Piwi-interacting RNAs, facilitating sequence-specific RNA cleavage. Recent findings demonstrate that RDE-1, an AGO protein, specifically recognizes trigger-derived sense and antisense RNAs to initiate RNAi, while other AGOs engage with amplified siRNAs to mediate downstream silencing processes, suggesting a highly coordinated mechanism within the RNAi pathway (Yigit et al. 2006). The structure of AGO proteins is composed of several highly conserved domains, including the PIWI and PAZ domains, which are critical for their RNA-binding and cleavage activities. The PIWI domain, named for the "Piwi" protein family, adopts a ribonuclease H-like fold, which is essential for catalyzing RNA cleavage in the RNA-induced silencing complex (Parker et al. 2005; Yuan et al. 2005). The PAZ domain, located adjacent to the PIWI domain, binds the 3′ end of the small RNA guide strand, positioning it for precise recognition and cleavage of target RNA. The interplay between these two domains enables AGO proteins to execute their catalytic function, with the PIWI domain containing key aspartate residues that coordinate divalent cations for RNA cleavage (Song et al. 2004; Wang et al. 2008a, b). In structural terms, AGO proteins have been studied extensively through X-ray crystallography, providing insights into their molecular architecture. The crystal structure of *Thermus thermophilus* AGO reveals a crescent-shaped structure formed by the PAZ and PIWI domains, with a nucleic-acid-binding channel between them that accommodates the small RNA guide strand (Wang et al. 2008a, b). This channel widens upon the formation of the ternary complex, which consists of AGO, the guide RNA, and the target RNA. The guide RNA, specifically the 5′-phosphorylated strand, binds within a basic pocket in the PAZ domain, anchoring the guide to the AGO protein. The target RNA, in turn, aligns with the guide RNA in a sequence-specific manner, with mismatches in the seed region (positions 2–8 of the guide) affecting the stability and catalytic activity of the complex (Wang et al. 2008a, b).

Structural studies of AGO proteins in complex with guide RNAs have further elucidated their catalytic mechanisms. The 5′-phosphorylated end of the guide RNA is anchored in a conserved basic pocket, while the target RNA pairs with the guide through Watson-Crick base pairing. Notably, the first few nucleotides of the guide RNA form an A-helix, stabilizing the interaction between the guide and the target (Ma et al. 2005; Wang et al. 2008a, b). The catalytic activity of AGO proteins is dependent on the precise alignment of the guide RNA and target RNA, with specific arginine residues playing a crucial role in stabilizing the cleavage site. The mechanism of RNA cleavage by AGO proteins involves a four-step catalytic cycle, beginning with the recognition of the guide RNA and target RNA, followed by cleavage of the target at the scissile phosphate. This process is facilitated by the PIWI domain, which contains an RNase H-like active site. Upon binding of the guide RNA to AGO, conformational changes in the protein allow for the alignment of the target RNA, positioning it for cleavage. This catalytic process is highly specific, with modifications in the guide or target RNA, such as bulges or mismatches, influencing the efficiency and accuracy of the cleavage (Wang et al. 2008a, b; Yuan et al. 2005). The structural and functional diversity of AGO proteins extends beyond their role in RNAi. In addition to their involvement in RNA-mediated gene silencing, AGOs also participate in processes such as epigenetic inheritance, antiviral defense, and regulation of chromosomal domains. In plants and animals, small RNAs, guided by AGOs, regulate gene expression and influence the inheritance of epigenetic marks, including DNA methylation. This highlights the versatility of AGO proteins in mediating diverse regulatory processes across eukaryotes (Aravin et al. 2007; Brennecke et al. 2008).

#### 1.4.8.1 Strand Selection and Sorting in RISC Assembly

Both miRNAs and siRNAs are processed and loaded into RISCs through similar pathways, although with some distinct steps and protein requirements. In *D. melanogaster*, AGO1 and AGO2 are key Argonaute proteins that participate in RISC loading and miRNA function. AGO1 is mainly involved in the regulation of miRNA target transcripts, whereas AGO2 mediates siRNA-directed RNA interference (Förstemann et al. 2007; Okamura et al. 2004). The sorting of miRNAs and siRNAs into these distinct complexes occurs at the stage of RISC assembly, where specific molecular characteristics of the small RNA duplex influence the selection of Argonaute proteins. In *D. melanogaster*, the loading of miRNAs into AGO1 is highly selective, with the miRNA* (passenger strand) typically being degraded, while the guide strand is incorporated into the complex. AGO2, on the other hand, can load both the guide and miRNA* strands from miRNA duplexes, with the sorting determined by the structure of the small RNA and the preference of AGO2 for extensively matched targets (Okamura et al. 2009).

The role of Dicer, the enzyme responsible for processing pre-miRNAs into mature miRNAs, is central to this process. Dicer works in tandem with Argonaute proteins to facilitate the loading of small RNAs into RISCs, ensuring that the correct strand is selected for silencing. In the context of siRNA loading, Dicer and its associated factors, such as TRBP (TAR RNA-binding protein) and PACT, play pivotal roles in the processing and selection of the guide strand for RNAi. The Dicer complex biases the selection of the guide strand based on the thermodynamic properties of the siRNA duplex, particularly its asymmetry. This strand selection is critical for the efficient assembly of RISCs and for the subsequent degradation of target mRNA. In humans, TRBP has been shown to interact with siRNAs and to bind preferentially to the guide strand, further ensuring the accuracy of strand selection for RNAi. Additionally, studies suggest that Dicer can also influence miRNA strand selection through its RNA-binding properties, contributing to the assembly of miRNPs (miRNA ribonucleoproteins) and the miRNA-loading complex (Gredell et al. 2010; Noland et al. 2011). The loading of small RNAs into RISCs is a finely regulated process. In both Drosophila and mammals, Dicer and Argonaute proteins act in concert to ensure that small RNA duplexes are processed, sorted, and loaded efficiently into the appropriate RISC complexes. This process is not only dependent on the structural features of the small RNAs but also on the specific characteristics of the Argonaute proteins involved. For example, while AGO1 is primarily responsible for miRNA function in *Drosophila*, AGO2 is more specialized for siRNA-mediated RNAi. Furthermore, the formation of miRNA loading complexes (miRLCs) in mammals plays a critical role in guiding pre-miRNAs into RISCs for further processing (Jannot et al. 2008; Liu et al. 2012; Okamura et al. 2004).

#### 1.4.8.2 miRNA Duplex Unwinding Rules for Strand Selection

The core machinery of RNA interference involves the assembly of RISC, where Argonaute proteins play a pivotal role. In particular, AGO2, which is part of the RISC complex, is essential for the cleavage of target RNAs during RNAi. During miRNA and small interfering RNA (siRNA) processing, the duplex structure is crucial for guiding RISC to its target. For siRNAs, both strands of the duplex are initially loaded onto AGO2. However, only one strand, the guide strand, remains integrated into the active RISC, while the other, known as the passenger strand, is discarded. This process is mediated by the ATP-dependent unwinding of the siRNA duplex (Leuschner et al. 2006). A similar mechanism applies to miRNA duplexes, but the specificity of strand selection in miRNAs involves different regulatory features. Interestingly, the unwinding of the miRNA duplex and the removal of the passenger strand are not entirely random. It has been observed that the thermodynamic properties of the duplex, particularly at the 5′ ends, play a critical role in determining which strand is retained in the RISC. Studies have shown that miRNA precursors exhibit significant asymmetry in their duplex stability, especially at the 5′ end of the antisense strand. This thermodynamic asymmetry favors the selection of the guide strand and influences the efficiency of miRNA biogenesis (Khvorova et al. 2003). Furthermore, ATP is crucial for maintaining the integrity of the 5′ phosphate groups on the siRNA strands, which is important for the downstream steps of target RNA recognition and cleavage (Nykänen et al. 2001).

The process of strand selection is highly regulated and involves not only thermodynamic factors but also the action of specific cellular proteins. In *Drosophila*, for example, the 3′-to-5′ exoribonuclease Nibbler (CG9247) plays a crucial role in the 3′ end processing of miRNAs. Nibbler interacts with AGO1 and facilitates the trimming of miRNA 3′ ends, which is critical for the maturation of miRNAs and the regulation of target mRNA repression. This trimming process is essential for the generation of miRNA isoforms that can regulate different pools of target genes, suggesting that miRNA diversity is important for cellular function (Han et al. 2011). The functional diversity of miRNAs is not limited to the length of the mature miRNA but also involves variations in the processing of the precursor molecules.

Some miRNAs undergo multiple rounds of trimming, which may generate isoforms with slightly different 3′ ends and altered target gene specificity (Liu et al. 2011).

Another important aspect of miRNA biogenesis and strand selection involves the distinct roles of Dicer and Drosha, which process pri-miRNAs and pre-miRNAs, respectively. In some cases, such as with miR-142 in hematopoietic cells, alternative processing of pri-miRNA can lead to multiple distinct pre-miRNA isoforms, each of which is processed into a different miRNA variant. These miRNA variants may target distinct sets of genes, contributing to the regulation of a broader gene network. This type of alternative processing could represent a novel mechanism for diversifying the pool of target genes regulated by a single miRNA (Wu et al. 2009). The interplay between RNA helicases, such as the ATP-dependent RNA unwinding mechanism during RISC assembly, and the specificity of strand selection ensures that only one strand of the miRNA duplex enters the RISC to exert its regulatory function. It has been proposed that the helicase activity of certain proteins, including AGO2, is required to separate the two strands of the duplex before the guide strand is incorporated into the complex (Matranga et al. 2005). This mechanism is critical for the efficient silencing of target mRNAs by preventing the incorporation of the passenger strand, which would otherwise disrupt the silencing process (Rand et al. 2005). The regulation of miRNA function is not limited to strand selection but also involves intricate modulation of miRNA stability and target interaction. In some organisms, such as *Drosophila*, the presence of mismatched duplexes or specific modifications on the miRNA strands can affect the selection of the guide strand and its subsequent loading into RISC. These modifications can influence the strength of miRNA-target binding and thereby modulate the efficacy of target repression (Miyoshi et al. 2010). This complex regulation highlights the importance of miRNA duplex structure and the precise control of strand selection in determining the outcome of RNAi pathways.

#### 1.4.8.3 Regulation of the AGO Proteins

The regulation of Argonaute (AGO) proteins is critical for maintaining proper gene silencing pathways in eukaryotic organisms. AGO proteins, primarily recognized for their pivotal role in posttranscriptional gene silencing (PTGS), serve as effectors in the RNA-induced silencing complexes (RISCs), where they mediate small RNA-guided gene silencing. These proteins are integral for protecting the genome against exogenous elements like viruses and bacteria, as well as controlling the expression of endogenous genes, including repetitive elements and transposons, through both transcriptional gene silencing (TGS) and PTGS mechanisms (Mallory and Vaucheret 2010). AGO proteins are also essential in various biological processes, including development, stress responses, and homeostasis (Smibert et al. 2013). In addition to their canonical roles in the miRNA pathway, AGO proteins, particularly AGO1 and AGO2, are now understood to have noncanonical functions. For instance, in mouse embryonic stem cells, the deletion of AGO1 and AGO2 led to a reduction of the repressive histone mark H3K27me3, which is a hallmark of the Polycomb repressive complex 2 (PRC2). This effect suggests that AGOs may play a role in regulating chromatin states and gene expression independent of miRNA-mediated silencing (Müller et al. 2022). Moreover, AGO proteins are involved in regulating various other pathways, including alternative splicing and host defense mechanisms, indicating that their functions extend beyond simple RNA interference (Wu et al. 2020). Posttranslational modifications (PTMs) play a significant role in the regulation of AGO proteins. For example, AGO2, which contains a PIWI domain, is regulated by hydroxylation at proline 700 by the type I collagen prolyl-4-hydroxylase (C-P4H(I)) complex. This modification is crucial for AGO2 stability, as its mutation or depletion of C-P4H(I) components results in reduced AGO2 stability and impaired RNA interference (RNAi) activity (Qi et al. 2008).

Similarly, phosphorylation of AGO2 at serine 387 by the p38 MAPK pathway has been identified as a key regulatory mechanism. Phosphorylation at this site is induced under stress conditions and promotes the localization of AGO2 to processing bodies, which are cellular hubs for RNA silencing (Zeng et al. 2008). Moreover, phosphorylation at other sites, such as serine 387 by Akt3, regulates AGO2's functional switch between mRNA cleavage and translational repression. This phosphorylation alters AGO2's interactions with GW182 and its localization to cytoplasmic processing bodies (P-bodies), implicating it in the regulation of miRNA-targeted mRNAs (Horman et al. 2013). The role of protein chaperones like HSP90 in AGO function has also been noted, where HSP90 assists in AGO protein loading, thereby facilitating efficient RNA silencing (Johnston et al. 2010). The stability and abundance of AGO proteins are tightly regulated through homeostatic mechanisms. In both *Drosophila* and mammalian systems, miRNA biogenesis influences the stability of AGO proteins. For example, in *Drosophila*, disruptions in miRNA biogenesis led to the destabilization of AGO1, while ectopic miRNA transcription can induce AGO1 accumulation. Similarly, in mammals, AGO2 stability is dependent on the availability of miRNAs, and interference with the ubiquitin-proteasome system can rescue AGO2 levels, suggesting a feedback loop that links AGO protein stability to miRNA abundance. Additionally, AGO2 expression is regulated at the posttranscriptional level by miRNAs, with a feedback mechanism that couples AGO2 levels to small RNA abundance (Martinez and Gregory 2013; Smibert et al. 2013).

### 1.4.9 miRNA Intrinsic Regulation

The intrinsic regulation of miRNAs is a dynamic and multilayered process involving various molecular mechanisms that finely tune their expression and function in posttranscriptional gene silencing. It is now well established that single-nucleotide polymorphisms (SNPs) can influence miRNA biogenesis by altering critical sequences in miRNA genes or their processing components. These alterations can affect the interaction of pri-miRNAs with the Drosha–DGCR8 complex or impair Dicer-mediated cleavage, ultimately modulating miRNA abundance and potentially contributing to disease susceptibility, particularly in cancers (Kawahara et al. 2007; Ryan et al. 2010). It has also been shown that SNPs occurring within target sites of miRNAs in the 3′ untranslated regions (3′ UTRs) of mRNAs can disrupt miRNA binding, affecting translational repression or degradation of target transcripts (Hynes and Kakumani 2024). There is increasing evidence that RNA editing plays a crucial role in regulating miRNA biogenesis. It modifies specific adenosines to inosines (A-to-I editing) within pri- or pre-miRNAs, affecting their structure and recognition by the miRNA processing machinery. For example, editing of pri-miR-151 at defined positions impedes its cleavage by Dicer, resulting in the accumulation of unprocessed precursors and decreased levels of mature miRNA. This indicates that RNA editing serves as a posttranscriptional checkpoint in miRNA maturation, with implications for timing and precision in gene regulation (Kawahara et al. 2007).

It has also been recognized that miRNAs are not static molecules; their regulation extends beyond production to modifications at their termini. It is through a process known as tailing that miRNAs undergo 3′ end modifications, often by the addition of non-templated nucleotides. These modifications can alter miRNA stability and, in some cases, change target specificity. It appears that tailing, often mediated by terminal nucleotidyl transferases, contributes to the creation of distinct miRNA isoforms, termed isomiRs, which may possess divergent biological roles depending on the cell type or developmental stage (Ameres and Zamore 2013). Regulation of miRNA stability is another intrinsic mechanism that determines miRNA levels. Although miRNAs have traditionally been viewed as stable, recent findings have identified several pathways responsible for their active degradation. Specific exoribonucleases, such as small RNA-degrading nuclease (SDN) proteins in plants, degrade mature miRNAs in a sequence-independent manner, and loss of SDN function leads to increased miRNA levels and developmental defects (Ramachandran and Chen 2008). Similarly, in mammals, the exonuclease Eri1 has been shown to regulate miRNA abundance in lymphocytes. Its deficiency results in widespread miRNA accumulation and impaired immune cell function, indicating a conserved role in miRNA turnover across species (Thomas et al. 2012). There is compelling evidence that these intrinsic regulatory mechanisms are essential for orchestrating cell state transitions and immune functions. In embryonic stem cells, variation in miRNA expression contributes to heterogeneous gene expression states, facilitating developmental progression. In immune cells, particularly T cells, miRNAs regulate key transcriptional networks that define lineage specification and functional responses (Chakraborty et al. 2020; Cho et al. 2021).

## 1.5 Conclusion

MicroRNAs are pivotal regulators of gene expression, orchestrating diverse biological processes through finely tuned posttranscriptional mechanisms. Their biogenesis is a multistep process involving Drosha–DGCR8-mediated processing of primary transcripts (pri-miRNAs), Dicer-dependent cleavage of precursor miRNAs (pre-miRNAs), and precise strand selection during RNA-induced silencing complex (RISC) assembly. Argonaute (AGO) proteins are integral to this machinery, facilitating the incorporation of the guide strand and enabling target mRNA recognition and cleavage or translational repression. Thermodynamic asymmetry, helicase activity, and specific structural features of miRNA duplexes ensure accurate guide strand selection while RNA-binding proteins like TRBP and posttranslational modifications of AGO proteins modulate their stability, localization, and activity. Beyond canonical gene silencing, miRNAs also participate in noncanonical functions, including chromatin remodeling, alternative splicing regulation, and immune signaling modulation. Their functional versatility is further enhanced through intrinsic regulatory mechanisms such as RNA editing, 3′ end tailing, and sequence variations that generate miRNA isoforms (isomiRs), expanding the repertoire of regulatory targets. The stability and turnover of miRNAs are tightly controlled by exonucleases like Eri1 and SDNs, ensuring dynamic regulation in response to cellular needs.

Posttranscriptional and posttranslational regulatory layers highlight the complexity and adaptability of the miRNA machinery. Moreover, the interplay between miRNA expression and AGO protein abundance reveals feedback mechanisms that maintain homeostasis within the RNA silencing pathway. These insights underscore the evolutionary sophistication of miRNA-mediated regulation across species. Overall, the biogenesis, regulation, and diverse functional roles of miRNAs underscore their critical importance in gene expression control, cellular identity, and organismal development. Understanding these pathways not only provides insight into fundamental biological processes but also opens avenues for therapeutic interventions in diseases where miRNA pathways are dysregulated, including cancer, neurodegeneration, and immune disorders.

## References


Alarcón CR, Lee H, Goodarzi H, Halberg N, Tavazoie SF (2015) N6-methyladenosine marks primary microRNAs for processing. Nature 519(7544):482–485. https://doi.org/10.1038/nature14281

Ameres SL, Zamore PD (2013) Diversifying microRNA sequence and function. Nat Rev Mol Cell Biol 14(8):475–488. https://doi.org/10.1038/nrm3611

Amirouche A, Tadesse H, Miura P, Bélanger G, Lunde JA, Côté J, Jasmin BJ (2014) Converging pathways involving microRNA-206 and the RNA-binding protein KSRP control post-transcriptionally utrophin a expression in skeletal muscle. Nucleic Acids Res 42(6): 3982–3997. https://doi.org/10.1093/nar/gkt1350

Aravin AA, Hannon GJ, Brennecke J (2007) The Piwi-piRNA pathway provides an adaptive defense in the transposon arms race. Science 318(5851):761–764. https://doi.org/10.1126/science.1146484

Auyeung VC, Ulitsky I, McGeary SE, Bartel DP (2013) Beyond secondary structure: primary-sequence determinants license pri-miRNA hairpins for processing. Cell 152(4):844–858. https://doi.org/10.1016/j.cell.2013.01.031

Babiarz JE, Ruby JG, Wang Y, Bartel DP, Blelloch R (2008) Mouse ES cells express endogenous shRNAs, siRNAs, and other microprocessor-independent, Dicer-dependent small RNAs. Genes Dev 22(20):2773–2785. https://doi.org/10.1101/gad.1705308

Barr I, Smith AT, Chen Y, Senturia R, Burstyn JN, Guo F (2012) Ferric, not ferrous, heme activates RNA-binding protein DGCR8 for primary microRNA processing. Proc Natl Acad Sci USA 109(6):1919–1924. https://doi.org/10.1073/pnas.1114514109

Benoit MPMH, Imbert L, Palencia A, Pérard J, Ebel C, Boisbouvier J, Plevin MJ (2013) The RNA-binding region of human TRBP interacts with microRNA precursors through two independent domains. Nucleic Acids Res 41(7):4241–4252. https://doi.org/10.1093/nar/gkt086

Bernstein E, Caudy AA, Hammond SM, Hannon GJ (2001) Role for a bidentate ribonuclease in the initiation step of RNA interference. Nature 409(6818):363–366. https://doi.org/10.1038/35053110

Blahna MT, Hata A (2012) Smad-mediated regulation of microRNA biosynthesis. FEBS Lett 586(14):1906–1912. https://doi.org/10.1016/j.febslet.2012.01.041

Bohnsack MT, Czaplinski K, Gorlich D (2004) Exportin 5 is a RanGTP-dependent dsRNA-binding protein that mediates nuclear export of pre-miRNAs. RNA 10(2):185–191. https://doi.org/10.1261/rna.5167604

Brennecke J, Malone CD, Aravin AA, Sachidanandam R, Stark A, Hannon GJ (2008) An epigenetic role for maternally inherited piRNAs in transposon silencing. Science 322(5906):1387–1392. https://doi.org/10.1126/science.1165171

Cai X, Hagedorn CH, Cullen BR (2004) Human microRNAs are processed from capped, polyadenylated transcripts that can also function as mRNAs. RNA 10(12):1957–1966. https://doi.org/10.1261/rna.7135204

Chakraborty M, Hu S, Visness E, Del Giudice M, De Martino A, Bosia C, Sharp PA, Garg S (2020) MicroRNAs organize intrinsic variation into stem cell states. Proc Natl Acad Sci 117(12):6942–6950. https://doi.org/10.1073/pnas.1920695117

Chakravarthy S, Sternberg SH, Kellenberger CA, Doudna JA (2010) Substrate-specific kinetics of dicer-catalyzed RNA processing. J Mol Biol 404(3):392–402. https://doi.org/10.1016/j.jmb.2010.09.030

Cheloufi S, Dos Santos CO, Chong MMW, Hannon GJ (2010) A dicer-independent miRNA biogenesis pathway that requires ago catalysis. Nature 465(7298):584–589. https://doi.org/10.1038/nature09092

Chendrimada TP, Gregory RI, Kumaraswamy E, Norman J, Cooch N, Nishikura K, Shiekhattar R (2005) TRBP recruits the dicer complex to Ago2 for microRNA processing and gene silencing. Nature 436(7051):740–744. https://doi.org/10.1038/nature03868

Cheng T-L, Wang Z, Liao Q, Zhu Y, Zhou W-H, Xu W, Qiu Z (2014) MeCP2 suppresses nuclear microRNA processing and dendritic growth by regulating the DGCR8/Drosha complex. Dev Cell 28(5):547–560. https://doi.org/10.1016/j.devcel.2014.01.032

Chiang HR, Schoenfeld LW, Ruby JG, Auyeung VC, Spies N, Baek D, Johnston WK, Russ C, Luo S, Babiarz JE, Blelloch R, Schroth GP, Nusbaum C, Bartel DP (2010) Mammalian microRNAs: experimental evaluation of novel and previously annotated genes. Genes Dev 24(10):992–1009. https://doi.org/10.1101/gad.1884710

Cho S, Dong J, Lu L (2021) Cell-intrinsic and -extrinsic roles of miRNAs in regulating T cell immunity. Immunol Rev 304(1):126–140. https://doi.org/10.1111/imr.13029

Conrad T, Marsico A, Gehre M, Ørom UA (2014) Microprocessor activity controls differential miRNA biogenesis in vivo. Cell Rep 9(2):542–554. https://doi.org/10.1016/j.celrep.2014.09.007

Das AK, Carmichael GG (2007) ADAR editing wobbles the MicroRNA world. ACS Chem Biol 2(4):217–220. https://doi.org/10.1021/cb700064h

Davis BN, Hata A (2010) Mechanisms of control of microRNA biogenesis. J Biochem 148(4):381–392. https://doi.org/10.1093/jb/mvq096

Davis BN, Hilyard AC, Lagna G, Hata A (2008) SMAD proteins control DROSHA-mediated microRNA maturation. Nature 454(7200):56–61. https://doi.org/10.1038/nature07086

Davis BN, Hilyard AC, Nguyen PH, Lagna G, Hata A (2010) Smad proteins bind a conserved RNA sequence to promote microRNA maturation by Drosha. Mol Cell 39(3):373–384. https://doi.org/10.1016/j.molcel.2010.07.011

de Rie D, Abugessaisa I, Alam T, Arner E, Arner P, Ashoor H, Åström G, Babina M, Bertin N, Burroughs AM, Carlisle AJ, Daub CO, Detmar M, Deviatiiarov R, Fort A, Gebhard C, Goldowitz D, Guhl S, Ha TJ et al (2017) An integrated expression atlas of miRNAs and their promoters in human and mouse. Nat Biotechnol 35(9):872–878. https://doi.org/10.1038/nbt.3947

Denli AM, Tops BBJ, Plasterk RHA, Ketting RF, Hannon GJ (2004) Processing of primary microRNAs by the microprocessor complex. Nature 432(7014):231–235. https://doi.org/10.1038/nature03049

Di Carlo V, Grossi E, Laneve P, Morlando M, Dini Modigliani S, Ballarino M, Bozzoni I, Caffarelli E (2013) TDP-43 regulates the microprocessor complex activity during in vitro neuronal differentiation. Mol Neurobiol 48(3):952–963. https://doi.org/10.1007/s12035-013-8564-x

Faller M, Matsunaga M, Yin S, Loo JA, Guo F (2007) Heme is involved in microRNA processing. Nat Struct Mol Biol 14(1):23–29. https://doi.org/10.1038/nsmb1182

Faller M, Toso D, Matsunaga M, Atanasov I, Senturia R, Chen Y, Zhou ZH, Guo F (2010) DGCR8 recognizes primary transcripts of microRNAs through highly cooperative binding and formation of higher-order structures. RNA 16(8):1570–1583. https://doi.org/10.1261/rna.2111310

Fareh M, Yeom K-H, Haagsma AC, Chauhan S, Heo I, Joo C (2016) TRBP ensures efficient dicer processing of precursor microRNA in RNA-crowded environments. Nat Commun 7(1):13694. https://doi.org/10.1038/ncomms13694

Finnegan EF, Pasquinelli AE (2013) MicroRNA biogenesis: regulating the regulators. Crit Rev Biochem Mol Biol 48(1):51–68. https://doi.org/10.3109/10409238.2012.738643

Förstemann K, Tomari Y, Du T, Vagin VV, Denli AM, Bratu DP, Klattenhoff C, Theurkauf WE, Zamore PD (2005) Normal microRNA maturation and germ-line stem cell maintenance requires loquacious, a double-stranded RNA-binding domain protein. PLoS Biol 3(7):e236. https://doi.org/10.1371/journal.pbio.0030236

Förstemann K, Horwich MD, Wee L, Tomari Y, Zamore PD (2007) Drosophila microRNAs are sorted into functionally distinct Argonaute complexes after production by Dicer-1. Cell 130(2):287–297. https://doi.org/10.1016/j.cell.2007.05.056

Friedman RC, Farh KK-H, Burge CB, Bartel DP (2009) Most mammalian mRNAs are conserved targets of microRNAs. Genome Res 19(1):92–105. https://doi.org/10.1101/gr.082701.108

Fukunaga R, Han BW, Hung J-H, Xu J, Weng Z, Zamore PD (2012) Dicer partner proteins tune the length of mature miRNAs in flies and mammals. Cell 151(4):912. https://doi.org/10.1016/j.cell.2012.10.029

Georgakilas G, Vlachos IS, Paraskevopoulou MD, Yang P, Zhang Y, Economides AN, Hatzigeorgiou AG (2014) microTSS: accurate microRNA transcription start site identification reveals a significant number of divergent pri-miRNAs. Nat Commun 5(1):5700. https://doi.org/10.1038/ncomms6700

Gredell JA, Dittmer MJ, Wu M, Chan C, Walton SP (2010) Recognition of siRNA asymmetry by TAR RNA binding protein. Biochemistry 49(14):3148–3155. https://doi.org/10.1021/bi902189s

Gregory RI, Yan K, Amuthan G, Chendrimada T, Doratotaj B, Cooch N, Shiekhattar R (2004) The microprocessor complex mediates the genesis of microRNAs. Nature 432(7014):235–240. https://doi.org/10.1038/nature03120

Grishok A, Pasquinelli AE, Conte D, Li N, Parrish S, Ha I, Baillie DL, Fire A, Ruvkun G, Mello CC (2001) Genes and mechanisms related to RNA interference regulate expression of the small temporal RNAs that Control C. Elegans developmental timing. Cell 106(1):23–34. https://doi.org/10.1016/S0092-8674(01)00431-7

Gu S, Jin L, Zhang Y, Huang Y, Zhang F, Valdmanis PN, Kay MA (2012) The loop position of shRNAs and pre-miRNAs is critical for the accuracy of dicer processing in vivo. Cell 151(4):900–911. https://doi.org/10.1016/j.cell.2012.09.042

Guil S, Cáceres JF (2007) The multifunctional RNA-binding protein hnRNP A1 is required for processing of miR-18a. Nat Struct Mol Biol 14(7):591–596. https://doi.org/10.1038/nsmb1250

Gwizdek C, Ossareh-Nazari B, Brownawell AM, Doglio A, Bertrand E, Macara IG, Dargemont C (2003) Exportin-5 mediates nuclear export of minihelix-containing RNAs. J Biol Chem 278(8):5505–5508. https://doi.org/10.1074/jbc.C200668200

Ha M, Kim VN (2014) Regulation of microRNA biogenesis. Nat Rev Mol Cell Biol 15(8): 509–524. https://doi.org/10.1038/nrm3838

Haase AD, Jaskiewicz L, Zhang H, Lainé S, Sack R, Gatignol A, Filipowicz W (2005) TRBP, a regulator of cellular PKR and HIV-1 virus expression, interacts with dicer and functions in RNA silencing. EMBO Rep 6(10):961–967. https://doi.org/10.1038/sj.embor.7400509

Han J, Lee Y, Yeom K-H, Kim Y-K, Jin H, Kim VN (2004) The Drosha-DGCR8 complex in primary microRNA processing. Genes Dev 18(24):3016–3027. https://doi.org/10.1101/gad.1262504

Han J, Lee Y, Yeom K-H, Nam J-W, Heo I, Rhee J-K, Sohn SY, Cho Y, Zhang B-T, Kim VN (2006) Molecular basis for the recognition of primary microRNAs by the Drosha-DGCR8 complex. Cell 125(5):887–901. https://doi.org/10.1016/j.cell.2006.03.043

Han J, Pedersen JS, Kwon SC, Belair CD, Kim Y-K, Yeom K-H, Yang W-Y, Haussler D, Blelloch R, Kim VN (2009) Posttranscriptional crossregulation between Drosha and DGCR8. Cell 136(1):75–84. https://doi.org/10.1016/j.cell.2008.10.053

Han BW, Hung J-H, Weng Z, Zamore PD, Ameres SL (2011) The 3′-to-5′ exoribonuclease nibbler shapes the 3′ ends of MicroRNAs bound to Drosophila Argonaute1. Curr Biol 21(22): 1878–1887. https://doi.org/10.1016/j.cub.2011.09.034

Herbert KM, Pimienta G, DeGregorio SJ, Alexandrov A, Steitz JA (2013) Phosphorylation of DGCR8 increases its intracellular stability and induces a progrowth miRNA profile. Cell Rep 5(4):1070–1081. https://doi.org/10.1016/j.celrep.2013.10.017

Horman SR, Janas MM, Litterst C, Wang B, MacRae IJ, Sever MJ, Morrissey DV, Graves P, Luo B, Umesalma S, Qi HH, Miraglia LJ, Novina CD, Orth AP (2013) Akt-mediated phosphorylation of Argonaute 2 downregulates cleavage and upregulates translational repression of MicroRNA targets. Mol Cell 50(3):356–367. https://doi.org/10.1016/j.molcel.2013.03.015

Hutvágner G, McLachlan J, Pasquinelli AE, Bálint E, Tuschl T, Zamore PD (2001) A cellular function for the RNA-interference enzyme dicer in the maturation of the let-7 small temporal RNA. Science 293(5531):834–838. https://doi.org/10.1126/science.1062961

Hynes C, Kakumani PK (2024) Regulatory role of RNA-binding proteins in microRNA biogenesis. Front Mol Biosci 11:1374843. https://doi.org/10.3389/fmolb.2024.1374843

Iwasaki YW, Kiga K, Kayo H, Fukuda-Yuzawa Y, Weise J, Inada T, Tomita M, Ishihama Y, Fukao T (2013) Global microRNA elevation by inducible Exportin 5 regulates cell cycle entry. RNA 19(4):490–497. https://doi.org/10.1261/rna.036608.112

Jannot G, Boisvert M-EL, Banville IH, Simard MJ (2008) Two molecular features contribute to the Argonaute specificity for the microRNA and RNAi pathways in C. Elegans. RNA 14(5): 829–835. https://doi.org/10.1261/rna.901908

Jiang F, Ye X, Liu X, Fincher L, McKearin D, Liu Q (2005) Dicer-1 and R3D1-L catalyze microRNA maturation in Drosophila. Genes Dev 19(14):1674–1679. https://doi.org/10.1101/gad.1334005

Johnston M, Geoffroy M-C, Sobala A, Hay R, Hutvagner G (2010) HSP90 protein stabilizes unloaded Argonaute complexes and microscopic P-bodies in human cells. Mol Biol Cell 21(9):1462–1469. https://doi.org/10.1091/mbc.e09-10-0885

Kadener S, Rodriguez J, Abruzzi KC, Khodor YL, Sugino K, Marr MT, Nelson S, Rosbash M (2009) Genome-wide identification of targets of the drosha-pasha/DGCR8 complex. RNA 15(4):537–545. https://doi.org/10.1261/rna.1319309

Kawahara Y, Mieda-Sato A (2012) TDP-43 promotes microRNA biogenesis as a component of the Drosha and dicer complexes. Proc Natl Acad Sci USA 109(9):3347–3352. https://doi.org/10.1073/pnas.1112427109

Kawahara Y, Zinshteyn B, Chendrimada TP, Shiekhattar R, Nishikura K (2007) RNA editing of the microRNA-151 precursor blocks cleavage by the dicer-TRBP complex. EMBO Rep 8(8): 763–769. https://doi.org/10.1038/sj.embor.7401011

Ketting RF, Fischer SE, Bernstein E, Sijen T, Hannon GJ, Plasterk RH (2001) Dicer functions in RNA interference and in synthesis of small RNA involved in developmental timing in C. Elegans. Genes Dev 15(20):2654–2659. https://doi.org/10.1101/gad.927801

Khvorova A, Reynolds A, Jayasena SD (2003) Functional siRNAs and miRNAs exhibit Strand Bias. Cell 115(2):209–216. https://doi.org/10.1016/S0092-8674(03)00801-8

Kim VN, Han J, Siomi MC (2009) Biogenesis of small RNAs in animals. Nat Rev Mol Cell Biol 10(2):126–139. https://doi.org/10.1038/nrm2632

Kim Y-K, Wee G, Park J, Kim J, Baek D, Kim J-S, Kim VN (2013) TALEN-based knockout library for human microRNAs. Nat Struct Mol Biol 20(12):1458–1464. https://doi.org/10.1038/nsmb.2701

Knight SW, Bass BL (2001) A role for the RNase III enzyme DCR-1 in RNA interference and germ line development in Caenorhabditis elegans. Science 293(5538):2269–2271. https://doi.org/10.1126/science.1062039

Knuckles P, Vogt MA, Lugert S, Milo M, Chong MMW, Hautbergue GM, Wilson SA, Littman DR, Taylor V (2012) Drosha regulates neurogenesis by controlling neurogenin 2 expression independent of microRNAs. Nat Neurosci 15(7):962–969. https://doi.org/10.1038/nn.3139

Kozomara A, Birgaoanu M, Griffiths-Jones S (2019) miRBase: from microRNA sequences to function. Nucleic Acids Res 47(D1):D155–D162. https://doi.org/10.1093/nar/gky1141

Kumar MS, Pester RE, Chen CY, Lane K, Chin C, Lu J, Kirsch DG, Golub TR, Jacks T (2009) Dicer1 functions as a haploinsufficient tumor suppressor. Genes Dev 23(23):2700–2704. https://doi.org/10.1101/gad.1848209

Lau P-W, Potter CS, Carragher B, MacRae IJ (2009) Structure of the human dicer-TRBP complex by electron microscopy. Structure 17(10):1326–1332. https://doi.org/10.1016/j.str.2009.08.013

Lee HY, Doudna JA (2012) TRBP alters human precursor microRNA processing in vitro. RNA 18(11):2012–2019. https://doi.org/10.1261/rna.035501.112

Lee RC, Feinbaum RL, Ambros V (1993) The C. Elegans heterochronic gene lin-4 encodes small RNAs with antisense complementarity to lin-14. Cell 75(5):843–854. https://doi.org/10.1016/0092-8674(93)90529-y

Lee Y, Jeon K, Lee J-T, Kim S, Kim VN (2002) MicroRNA maturation: stepwise processing and subcellular localization. EMBO J 21(17):4663–4670. https://doi.org/10.1093/emboj/cdf476

Lee Y, Kim M, Han J, Yeom K-H, Lee S, Baek SH, Kim VN (2004) MicroRNA genes are transcribed by RNA polymerase II. EMBO J 23(20):4051–4060. https://doi.org/10.1038/sj.emboj.7600385

Lee Y, Hur I, Park S-Y, Kim Y-K, Suh MR, Kim VN (2006) The role of PACT in the RNA silencing pathway. EMBO J 25(3):522–532. https://doi.org/10.1038/sj.emboj.7600942

Leuschner PJF, Ameres SL, Kueng S, Martinez J (2006) Cleavage of the siRNA passenger strand during RISC assembly in human cells. EMBO Rep 7(3):314–320. https://doi.org/10.1038/sj.embor.7400637

Lin EA, Kong L, Bai X-H, Luan Y, Liu C (2009) miR-199a, a bone Morphogenic protein 2-responsive MicroRNA, regulates Chondrogenesis via direct targeting to Smad1. J Biol Chem 284(17):11326–11335. https://doi.org/10.1074/jbc.M807709200

Liu N, Abe M, Sabin LR, Hendriks G-J, Naqvi AS, Yu Z, Cherry S, Bonini NM (2011) The exoribonuclease nibbler controls 3′ end processing of microRNAs in Drosophila. Curr Biol 21(22):1888–1893. https://doi.org/10.1016/j.cub.2011.10.006

Liu X, Jin D-Y, McManus MT, Mourelatos Z (2012) Precursor MicroRNA-programmed silencing complex assembly pathways in mammals. Mol Cell 46(4):507–517. https://doi.org/10.1016/j.molcel.2012.03.010

Lund E, Güttinger S, Calado A, Dahlberg JE, Kutay U (2004) Nuclear export of microRNA precursors. Science 303(5654):95–98. https://doi.org/10.1126/science.1090599

Ma J-B, Yuan Y-R, Meister G, Pei Y, Tuschl T, Patel DJ (2005) Structural basis for 5′-end-specific recognition of guide RNA by the a. fulgidus Piwi protein. Nature 434(7033):666–670. https://doi.org/10.1038/nature03514

Ma H, Wu Y, Choi J-G, Wu H (2013) Lower and upper stem-single-stranded RNA junctions together determine the Drosha cleavage site. Proc Natl Acad Sci USA 110(51):20687–20692. https://doi.org/10.1073/pnas.1311639110

Macrae IJ, Zhou K, Li F, Repic A, Brooks AN, Cande WZ, Adams PD, Doudna JA (2006) Structural basis for double-stranded RNA processing by dicer. Science 311(5758):195–198. https://doi.org/10.1126/science.1121638

MacRae IJ, Zhou K, Doudna JA (2007) Structural determinants of RNA recognition and cleavage by dicer. Nat Struct Mol Biol 14(10):934–940. https://doi.org/10.1038/nsmb1293

Mallory A, Vaucheret H (2010) Form, function, and regulation of ARGONAUTE proteins. Plant Cell 22(12):3879–3889. https://doi.org/10.1105/tpc.110.080671

Martinez NJ, Gregory RI (2013) Argonaute2 expression is post-transcriptionally coupled to microRNA abundance. RNA 19(5):605–612. https://doi.org/10.1261/rna.036434.112

Matranga C, Tomari Y, Shin C, Bartel DP, Zamore PD (2005) Passenger-Strand cleavage facilitates assembly of siRNA into Ago2-containing RNAi enzyme complexes. Cell 123(4):607–620. https://doi.org/10.1016/j.cell.2005.08.044

McGeary SE, Lin KS, Shi CY, Pham TM, Bisaria N, Kelley GM, Bartel DP (2019) The biochemical basis of microRNA targeting efficacy. Science 366(6472):eaav1741. https://doi.org/10.1126/science.aav1741

Meijer HA, Smith EM, Bushell M (2014) Regulation of miRNA strand selection: follow the leader? Biochem Soc Trans 42(4):1135–1140. https://doi.org/10.1042/BST20140142

Melo SA, Moutinho C, Ropero S, Calin GA, Rossi S, Spizzo R, Fernandez AF, Davalos V, Villanueva A, Montoya G, Yamamoto H, Schwartz S, Esteller M (2010) A genetic defect in exportin-5 traps precursor microRNAs in the nucleus of cancer cells. Cancer Cell 18(4): 303–315. https://doi.org/10.1016/j.ccr.2010.09.007

Michlewski G, Guil S, Semple CA, Cáceres JF (2008) Posttranscriptional regulation of miRNAs harboring conserved terminal loops. Mol Cell 32(3):383–393. https://doi.org/10.1016/j.molcel.2008.10.013

Miyoshi K, Miyoshi T, Hartig JV, Siomi H, Siomi MC (2010) Molecular mechanisms that funnel RNA precursors into endogenous small-interfering RNA and microRNA biogenesis pathways in Drosophila. RNA 16(3):506–515. https://doi.org/10.1261/rna.1952110

Monteys AM, Spengler RM, Wan J, Tecedor L, Lennox KA, Xing Y, Davidson BL (2010) Structure and activity of putative intronic miRNA promoters. RNA 16(3):495–505. https://doi.org/10.1261/rna.1731910

Mori M, Triboulet R, Mohseni M, Schlegelmilch K, Shrestha K, Camargo FD, Gregory RI (2014) Hippo signaling regulates microprocessor and links cell-density-dependent miRNA biogenesis to cancer. Cell 156(5):893–906. https://doi.org/10.1016/j.cell.2013.12.043

Morlando M, Ballarino M, Gromak N, Pagano F, Bozzoni I, Proudfoot NJ (2008) Primary microRNA transcripts are processed co-transcriptionally. Nat Struct Mol Biol 15(9):902–909. https://doi.org/10.1038/nsmb.1475

Müller M, Schaefer M, Fäh T, Spies D, Hermes V, Ngondo RP, Peña-Hernández R, Santoro R, Ciaudo C (2022) Argonaute proteins regulate a specific network of genes through KLF4 in mouse embryonic stem cells. Stem Cell Rep 17(5):1070–1080. https://doi.org/10.1016/j.stemcr.2022.03.014

Noland CL, Ma E, Doudna JA (2011) siRNA repositioning for guide Strand selection by human dicer complexes. Mol Cell 43(1):110–121. https://doi.org/10.1016/j.molcel.2011.05.028

Nunes S, Bastos R, Marinho AI, Vieira R, Benício I, De Noronha MA, Lírio S, Brodskyn C, Tavares NM (2025) Recent advances in the development and clinical application of miRNAs in infectious diseases. Non-Coding RNA Res 10:41–54. https://doi.org/10.1016/j.ncrna.2024.09.005

Nykänen A, Haley B, Zamore PD (2001) ATP requirements and small interfering RNA structure in the RNA interference pathway. Cell 107(3):309–321. https://doi.org/10.1016/S0092-8674(01)00547-5

Okada C, Yamashita E, Lee SJ, Shibata S, Katahira J, Nakagawa A, Yoneda Y, Tsukihara T (2009) A high-resolution structure of the pre-microRNA nuclear export machinery. Science 326(5957): 1275–1279. https://doi.org/10.1126/science.1178705

Okamura K, Ishizuka A, Siomi H, Siomi MC (2004) Distinct roles for Argonaute proteins in small RNA-directed RNA cleavage pathways. Genes Dev 18(14):1655–1666. https://doi.org/10.1101/gad.1210204

Okamura K, Liu N, Lai EC (2009) Distinct mechanisms for MicroRNA Strand selection by Drosophila Argonautes. Mol Cell 36(3):431–444. https://doi.org/10.1016/j.molcel.2009.09.027

Ozsolak F, Poling LL, Wang Z, Liu H, Liu XS, Roeder RG, Zhang X, Song JS, Fisher DE (2008) Chromatin structure analyses identify miRNA promoters. Genes Dev 22(22):3172–3183. https://doi.org/10.1101/gad.1706508

Parker JS, Roe SM, Barford D (2005) Structural insights into mRNA recognition from a PIWI domain-siRNA guide complex. Nature 434(7033):663–666. https://doi.org/10.1038/nature03462

Pasquinelli AE (2012) MicroRNAs and their targets: recognition, regulation and an emerging reciprocal relationship. Nat Rev Genet 13(4):271–282. https://doi.org/10.1038/nrg3162

Pawlicki JM, Steitz JA (2008) Primary microRNA transcript retention at sites of transcription leads to enhanced microRNA production. J Cell Biol 182(1):61–76. https://doi.org/10.1083/jcb.200803111

Peláez N, Carthew RW (2012) Biological robustness and the role of microRNAs: a network perspective. Curr Top Dev Biol 99:237–255. https://doi.org/10.1016/B978-0-12-387038-4.00009-4

Piskounova E, Polytarchou C, Thornton JE, LaPierre RJ, Pothoulakis C, Hagan JP, Iliopoulos D, Gregory RI (2011) Lin28A and Lin28B inhibit let-7 microRNA biogenesis by distinct mechanisms. Cell 147(5):1066–1079. https://doi.org/10.1016/j.cell.2011.10.039

Qi HH, Ongusaha PP, Myllyharju J, Cheng D, Pakkanen O, Shi Y, Lee SW, Peng J, Shi Y (2008) Prolyl 4-hydroxylation regulates Argonaute 2 stability. Nature 455(7211):421–424. https://doi.org/10.1038/nature07186

Ramachandran V, Chen X (2008) Degradation of microRNAs by a family of exoribonucleases in Arabidopsis. Science 321(5895):1490–1492. https://doi.org/10.1126/science.1163728

Rand TA, Petersen S, Du F, Wang X (2005) Argonaute2 cleaves the anti-guide Strand of siRNA during RISC activation. Cell 123(4):621–629. https://doi.org/10.1016/j.cell.2005.10.020

Rodriguez BN, Huang H, Chia JJ, Hoffmann A (2024) The noncanonical NFκB pathway: regulatory mechanisms in health and disease. WIREs Mech Dis 16(6):e1646. https://doi.org/10.1002/wsbm.1646

Ruby JG, Jan CH, Bartel DP (2007) Intronic microRNA precursors that bypass Drosha processing. Nature 448(7149):83–86. https://doi.org/10.1038/nature05983

Ryan BM, Robles AI, Harris CC (2010) Genetic variation in microRNA networks: the implications for cancer research. Nat Rev Cancer 10(6):389–402. https://doi.org/10.1038/nrc2867

Saito K, Ishizuka A, Siomi H, Siomi MC (2005) Processing of pre-microRNAs by the Dicer-1-loquacious complex in Drosophila cells. PLoS Biol 3(7):e235. https://doi.org/10.1371/journal.pbio.0030235

Shiohama A, Sasaki T, Noda S, Minoshima S, Shimizu N (2003) Molecular cloning and expression analysis of a novel gene DGCR8 located in the DiGeorge syndrome chromosomal region. Biochem Biophys Res Commun 304(1):184–190. https://doi.org/10.1016/s0006-291x(03)00554-0

Shiohama A, Sasaki T, Noda S, Minoshima S, Shimizu N (2007) Nucleolar localization of DGCR8 and identification of eleven DGCR8-associated proteins. Exp Cell Res 313(20):4196–4207. https://doi.org/10.1016/j.yexcr.2007.07.020

Smibert P, Yang J-S, Azzam G, Liu J-L, Lai EC (2013) Homeostatic control of Argonaute stability by microRNA availability. Nat Struct Mol Biol 20(7):789–795. https://doi.org/10.1038/nsmb.2606

Sohn SY, Bae WJ, Kim JJ, Yeom K-H, Kim VN, Cho Y (2007) Crystal structure of human DGCR8 core. Nat Struct Mol Biol 14(9):847–853. https://doi.org/10.1038/nsmb1294

Song J-J, Smith SK, Hannon GJ, Joshua-Tor L (2004) Crystal structure of Argonaute and its implications for RISC slicer activity. Science 305(5689):1434–1437. https://doi.org/10.1126/science.1102514

Sundaram GM, Common JEA, Gopal FE, Srikanta S, Lakshman K, Lunny DP, Lim TC, Tanavde V, Lane EB, Sampath P (2013) "See-saw" expression of microRNA-198 and FSTL1 from a single transcript in wound healing. Nature 495(7439):103–106. https://doi.org/10.1038/nature11890

Suzuki HI, Yamagata K, Sugimoto K, Iwamoto T, Kato S, Miyazono K (2009) Modulation of microRNA processing by p53. Nature 460(7254):529–533. https://doi.org/10.1038/nature08199

Suzuki HI, Arase M, Matsuyama H, Choi YL, Ueno T, Mano H, Sugimoto K, Miyazono K (2011) MCPIP1 ribonuclease antagonizes dicer and terminates MicroRNA biogenesis through precursor MicroRNA degradation. Mol Cell 44(3):424–436. https://doi.org/10.1016/j.molcel.2011.09.012

Tang X, Zhang Y, Tucker L, Ramratnam B (2010) Phosphorylation of the RNase III enzyme Drosha at Serine300 or Serine302 is required for its nuclear localization. Nucleic Acids Res 38(19):6610–6619. https://doi.org/10.1093/nar/gkq547

Tang X, Li M, Tucker L, Ramratnam B (2011) Glycogen synthase kinase 3 beta (GSK3β) phosphorylates the RNAase III enzyme Drosha at S300 and S302. PLoS One 6(6):e20391. https://doi.org/10.1371/journal.pone.0020391

Tang X, Wen S, Zheng D, Tucker L, Cao L, Pantazatos D, Moss SF, Ramratnam B (2013) Acetylation of drosha on the N-terminus inhibits its degradation by ubiquitination. PLoS One 8(8):e72503. https://doi.org/10.1371/journal.pone.0072503

Thivierge C, Bellefeuille M, Diwan S-S, Dyakov BJA, Leventis R, Perron G, Najafabadi HS, Gravel S-P, Gingras A-C, Duchaine TF (2024) Paraspeckle-independent co-transcriptional regulation of nuclear microRNA biogenesis by SFPQ. Cell Rep 43(9):114695. https://doi.org/10.1016/j.celrep.2024.114695

Thomas MF, Abdul-Wajid S, Panduro M, Babiarz JE, Rajaram M, Woodruff P, Lanier LL, Heissmeyer V, Ansel KM (2012) Eri1 regulates microRNA homeostasis and mouse lymphocyte development and antiviral function. Blood 120(1):130–142. https://doi.org/10.1182/blood-2011-11-394072

Tian Y, Simanshu DK, Ma J-B, Park J-E, Heo I, Kim VN, Patel DJ (2014) A phosphate-binding pocket within the platform-PAZ-connector helix cassette of human dicer. Mol Cell 53(4):606–616. https://doi.org/10.1016/j.molcel.2014.01.003

Trabucchi M, Briata P, Garcia-Mayoral M, Haase AD, Filipowicz W, Ramos A, Gherzi R, Rosenfeld MG (2009) The RNA-binding protein KSRP promotes the biogenesis of a subset of microRNAs. Nature 459(7249):1010–1014. https://doi.org/10.1038/nature08025

Tsutsumi A, Kawamata T, Izumi N, Seitz H, Tomari Y (2011) Recognition of the pre-miRNA structure by Drosophila Dicer-1. Nat Struct Mol Biol 18(10):1153–1158. https://doi.org/10.1038/nsmb.2125

Ventura A, Young AG, Winslow MM, Lintault L, Meissner A, Erkeland SJ, Newman J, Bronson RT, Crowley D, Stone JR, Jaenisch R, Sharp PA, Jacks T (2008) Targeted deletion reveals essential and overlapping functions of the miR-17 through 92 family of miRNA clusters. Cell 132(5):875–886. https://doi.org/10.1016/j.cell.2008.02.019

Vermeulen A, Behlen L, Reynolds A, Wolfson A, Marshall WS, Karpilow J, Khvorova A (2005) The contributions of dsRNA structure to dicer specificity and efficiency. RNA 11(5):674–682. https://doi.org/10.1261/rna.7272305

Wada T, Kikuchi J, Furukawa Y (2012) Histone deacetylase 1 enhances microRNA processing via deacetylation of DGCR8. EMBO Rep 13(2):142–149. https://doi.org/10.1038/embor.2011.247

Wan G, Zhang X, Langley RR, Liu Y, Hu X, Han C, Peng G, Ellis LM, Jones SN, Lu X (2013) DNA-damage-induced nuclear export of precursor microRNAs is regulated by the ATM-AKT pathway. Cell Rep 3(6):2100–2112. https://doi.org/10.1016/j.celrep.2013.05.038

Wang Y, Juranek S, Li H, Sheng G, Tuschl T, Patel DJ (2008a) Structure of an argonaute silencing complex with a seed-containing guide DNA and target RNA duplex. Nature 456(7224): 921–926. https://doi.org/10.1038/nature07666

Wang Y, Sheng G, Juranek S, Tuschl T, Patel DJ (2008b) Structure of the guide-strand-containing argonaute silencing complex. Nature 456(7219):209–213. https://doi.org/10.1038/nature07315

Wang H-W, Noland C, Siridechadilok B, Taylor DW, Ma E, Felderer K, Doudna JA, Nogales E (2009) Structural insights into RNA processing by the human RISC-loading complex. Nat Struct Mol Biol 16(11):1148–1153. https://doi.org/10.1038/nsmb.1673

Weitz SH, Gong M, Barr I, Weiss S, Guo F (2014) Processing of microRNA primary transcripts requires heme in mammalian cells. Proc Natl Acad Sci USA 111(5):1861–1866. https://doi.org/10.1073/pnas.1309915111

Widmark A, Sagredo EA, Karlström V, Behm M, Biryukova I, Friedländer MR, Daniel C, Öhman M (2022) ADAR1- and ADAR2-mediated regulation of maturation and targeting of miR-376b to modulate GABA neurotransmitter catabolism. J Biol Chem 298(3):101682. https://doi.org/10.1016/j.jbc.2022.101682

Wu H, Ye C, Ramirez D, Manjunath N (2009) Alternative processing of primary microRNA transcripts by Drosha generates 5′ end variation of mature microRNA. PLoS One 4(10): e7566. https://doi.org/10.1371/journal.pone.0007566

Wu J, Yang J, Cho WC, Zheng Y (2020) Argonaute proteins: structural features, functions and emerging roles. J Adv Res 24:317–324. https://doi.org/10.1016/j.jare.2020.04.017

Xie M, Li M, Vilborg A, Lee N, Shu M-D, Yartseva V, Šestan N, Steitz JA (2013) Mammalian 5′-capped microRNA precursors that generate a single microRNA. Cell 155(7):1568–1580. https://doi.org/10.1016/j.cell.2013.11.027

Yang J-S, Lai EC (2011) Alternative miRNA biogenesis pathways and the interpretation of core miRNA pathway mutants. Mol Cell 43(6):892–903. https://doi.org/10.1016/j.molcel.2011.07.024

Yang J-S, Maurin T, Robine N, Rasmussen KD, Jeffrey KL, Chandwani R, Papapetrou EP, Sadelain M, O'Carroll D, Lai EC (2010) Conserved vertebrate mir-451 provides a platform for dicer-independent, Ago2-mediated microRNA biogenesis. Proc Natl Acad Sci 107(34): 15163–15168. https://doi.org/10.1073/pnas.1006432107

Yeom K-H, Lee Y, Han J, Suh MR, Kim VN (2006) Characterization of DGCR8/pasha, the essential cofactor for Drosha in primary miRNA processing. Nucleic Acids Res 34(16): 4622–4629. https://doi.org/10.1093/nar/gkl458

Yi R, Qin Y, Macara IG, Cullen BR (2003) Exportin-5 mediates the nuclear export of pre-microRNAs and short hairpin RNAs. Genes Dev 17(24):3011–3016. https://doi.org/10.1101/gad.1158803

Yigit E, Batista PJ, Bei Y, Pang KM, Chen C-CG, Tolia NH, Joshua-Tor L, Mitani S, Simard MJ, Mello CC (2006) Analysis of the C. Elegans Argonaute family reveals that distinct Argonautes act sequentially during RNAi. Cell 127(4):747–757. https://doi.org/10.1016/j.cell.2006.09.033

Yoda M, Kawamata T, Paroo Z, Ye X, Iwasaki S, Liu Q, Tomari Y (2010) ATP-dependent human RISC assembly pathways. Nat Struct Mol Biol 17(1):17–23. https://doi.org/10.1038/nsmb.1733

Yuan Y-R, Pei Y, Ma J-B, Kuryavyi V, Zhadina M, Meister G, Chen H-Y, Dauter Z, Tuschl T, Patel DJ (2005) Crystal structure of a. aeolicus Argonaute, a site-specific DNA-guided endoribonuclease, provides insights into RISC-mediated mRNA cleavage. Mol Cell 19(3): 405–419. https://doi.org/10.1016/j.molcel.2005.07.011

Zeng Y, Cullen BR (2004) Structural requirements for pre-microRNA binding and nuclear export by Exportin 5. Nucleic Acids Res 32(16):4776–4785. https://doi.org/10.1093/nar/gkh824

Zeng Y, Yi R, Cullen BR (2005) Recognition and cleavage of primary microRNA precursors by the nuclear processing enzyme Drosha. EMBO J 24(1):138–148. https://doi.org/10.1038/sj.emboj.7600491

Zeng Y, Sankala H, Zhang X, Graves PR (2008) Phosphorylation of Argonaute 2 at serine-387 facilitates its localization to processing bodies. Biochem J 413(3):429–436. https://doi.org/10.1042/BJ20080599

Zhang H, Kolb FA, Brondani V, Billy E, Filipowicz W (2002) Human dicer preferentially cleaves dsRNAs at their termini without a requirement for ATP. EMBO J 21(21):5875–5885. https://doi.org/10.1093/emboj/cdf582

Zhang H, Kolb FA, Jaskiewicz L, Westhof E, Filipowicz W (2004) Single processing center models for human dicer and bacterial RNase III. Cell 118(1):57–68. https://doi.org/10.1016/j.cell.2004.06.017

Zhang P, Frederick MI, Heinemann IU (2022) Terminal Uridylyltransferases TUT4/7 regulate microRNA and mRNA homeostasis. Cells 11(23):3742. https://doi.org/10.3390/cells11233742

# miRNA Deregulation and Its Role in Progression and Advancement of Cancer

# 2

Jugal R. Patil, Pranjal Agarwal, Amit K. Yadav, Dhiraj Bhatia, and Navneet Chaudhary

**Abstract**

MicroRNAs (miRNAs) are tiny noncoding RNAs that have a major regulatory impact on gene expression and are essential for cellular functions like apoptosis, differentiation, and proliferation. A thorough review of the ways that miRNA malfunction affects the development, spread, and metastasis of different cancer types is provided in this chapter. The chapter elucidates the mechanisms underlying miRNA deregulation, including genetic, epigenetic, and transcriptional alterations, and discusses their dualistic function as oncogenes or tumor suppressors. Recent research highlighting key deregulated miRNAs in major cancers is summarized, demonstrating their involvement in hallmark cancer pathways such as cell cycle regulation, apoptosis evasion, and metastasis promotion. Furthermore, the chapter explores the translational potential of targeting miRNA deregulation using novel therapeutics, underscoring the promise of miRNA modulation as a potent strategy in combinatorial cancer therapy. Overall, this chapter provides critical insights into the complex landscape of miRNA deregulation, laying a foundation for innovative, more effective approaches in cancer management.

**Keywords**

microRNA · Cancer progression · Deregulation · Oncogenes · Therapy

J. R. Patil · P. Agarwal · A. K. Yadav · D. Bhatia
Department of Biological Sciences and Engineering, Indian Institute of Technology Gandhinagar, Gandhinagar, Gujarat, India

N. Chaudhary (✉)
Nano for Energy, Department of Physics, Umeå University, Umeå, Sweden
e-mail: navneet.chaudhary@umu.se





S. Tiwari et al. (eds.), *microRNA Modulation: A Promising Combination Therapy for Cancer*, https://doi.org/10.1007/978-981-95-5989-3_2

**Learning Objectives**

- Understand the molecular mechanisms of miRNA deregulation.
- Analyze the functional role of deregulated miRNAs in cancer hallmarks.
- Assess the impact of miRNA deregulation on cancer progression and therapy resistance.
- Explore the clinical implications of deregulated miRNAs.

## 2.1 Introduction

Micro-RNA (miRNA) is a tiny, noncoding RNA molecule that is important for controlling and expressing genes. By targeting messenger RNA (mRNA), miRNA exerts significant indirect control on cellular signaling pathways and transcription (O'Brien et al. 2018). Crucially, miRNAs have widespread influence over the essential components of cellular life. By simultaneously fine-tuning hundreds of target mRNA with a dynamic regulatory network, they coordinate intricate programs of differentiation, apoptosis, proliferation, and stress response (Nunes et al. 2025). The entire signaling pathways can be inhibited by a single miRNA, which can direct stem cells toward specific destinies, repress oncogenes to stop unchecked divisions, or cause a death signal in injured cells (Choi 2013). miRNA quickly alters defense mechanisms in response to metabolic or environmental stress, guaranteeing cellular adaptability or programmed cell death (Li et al. 2025a). Because of this broad spread impact, miRNA is positioned not only as molecular switches but also as essential builders of cellular balance, whose dysregulation is the root cause of diseases. When this complex regulatory structure breaks down, miRNA dysregulation becomes an active cause of carcinogenesis rather than a passive effect. Aberrant miRNA directly speeds up the acquisition of malignant hallmarks by tampering with the mechanisms they evolved to regulate, including differentiation programs, apoptotic triggers, proliferation checkpoints, and stress resilience. To investigate the functional reconfiguration of cellular circuitry by certain mRNAs in order to provide signals for prolonged growth, to avoid growth suppressing agents, prevent apoptosis, allow for replicative immortality, promote angiogenesis, initiate metastasis, rewire metabolism, and avoid immune destruction.

The dysregulation of miRNAs has emerged as a pivotal factor in the advancement of cancer, redefining our understanding of molecular oncology (Vyas et al. 2025). Regulation of gene expression after transcription is orchestrated by miRNAs, which are tiny, noncoding RNA molecules that maintain cellular homeostasis and the accurate performance of critical biological functions. However, when their regulation is disrupted, miRNAs may become potent drivers of malignant transformation by either silencing tumor suppressor genes or enhancing the expression of oncogenes (Vyas et al. 2025). In this chapter, the complex field of miRNA deregulation is reviewed along with the several ways that cancer cells modify miRNA expression and function (Svoronos et al. 2016). It highlights the cumulative evidence linking aberrant miRNA profiles to cancer initiation, progression, metastatic spread, and therapy resistance (Kumar et al. 2025a). By examining recent discoveries in the field, the discussion underscores how miRNA deregulation not only contributes to hallmark capabilities of cancer cells but also opens new avenues for innovative diagnostic, prognostic, and therapeutic strategies. Through an integrated examination of molecular pathways, clinical correlations, and experimental findings, this chapter aims to provide readers with a comprehensive understanding of how miRNA deregulation underpins cancer complexity and why targeting these molecules offers considerable promise in developing combination therapies for improved cancer management.

### 2.1.1 miRNA Biogenesis

miRNA biogenesis is a multistep process that mostly takes place in the nucleus and cytoplasm, which guarantees the production of mature, functional miRNA. Transcription of the primary miRNA transcript initiates the biogenesis of miRNA, which is followed by nuclear processing of the miRNA precursor, export of primary miRNAs (pre-miRNA) from the nuclear compartment of the cell to the cytoplasmic compartment, cytoplasmic processing, and loading and maturation of the RNA-induced silencing complex (RISC) (Bofill-De Ros and Vang Ørom 2024). An essential step in miRNA synthesis is the transcription of primary microRNA (pri-miRNA) transcripts. RNA polymerase II (RNA pol-II) transcribes the majority of miRNA as lengthy, capped, and polyadenylated primary transcripts, which can range in length from a few kilobases to hundreds of kilobases (Ma et al. 2025a; Saini et al. 2007). These pre-miRNAs can be inserted into the exon or intron of protein-coding genes, or they can come from separate noncoding genes (Ma et al. 2025a; Saini et al. 2007). Tightly controlled transcription is frequently linked to co-transcriptional RNA processing activities such as capping, splicing, and polyadenylation, which occur (Yin et al. 2015). When mature miRNA sequences are detected in regions of pri-miRNA that fold into defective hairpin shapes, the microprocessor complex, which is a combination of the RNase III enzyme known as DROSHA and the protein that binds to RNA called DiGeorge Syndrome Critical Region 8 (DGCR-8), identifies and cleaves them (Chang et al. 2015). Because of quick and effective processing, these complex pre-miRNAs are found in cells in small amount, which makes it difficult to identify and analyze them (Chang et al. 2015). Figure 2.1 illustrates the stepwise miRNA biogenesis, beginning with their transcription in the nuclear compartment as pri-miRNAs.

### 2.1.2 Decay of miRNA

miRNA stability and degradation are essential mechanism that regulates their cellular abundance and, in turn, their regulatory roles in gene expression. The synthesis, biogenesis, and eventual destruction of mature miRNA regulate their levels. Some miRNAs have substantially shorter half-lives, degrading within a short time period in response to biological cues, cell cycle phases, but most miRNA are protected by

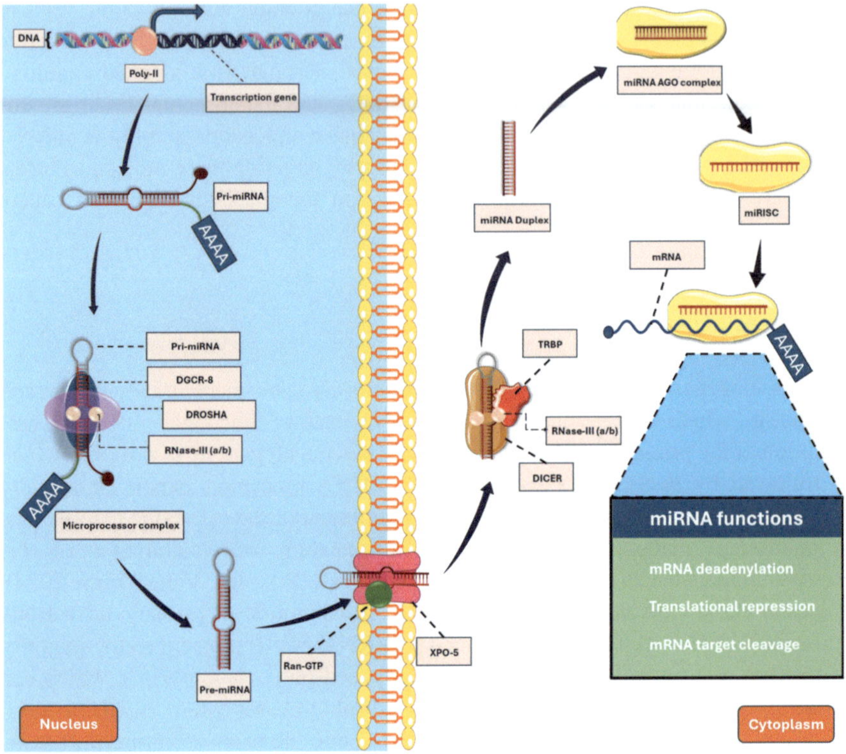


**Fig. 2.1** miRNAs sequential biogenesis and functions in gene regulation after transcription. miRNA transcription starts the process in the nucleus and continues through cytoplasmic maturation and functional target repression. At every stage, important molecular actors and complexes are emphasized

Argonaute (AGO) proteins, which can result in half-lives that last for days or even weeks in vivo. The significance of strict control over miRNA degradation for accurate gene regulation is highlighted by this dynamic variation in stability. Despite being well known for their function in silencing mRNAs and being taught as a stable entity, especially when protected within the AGO-bound RISC, miRNAs have a complex decay pathway that dynamically and precisely regulates their cellular lifetimes.

This controlled turnover is an active necessary process that guarantees the flexibility of gene expression network in response to environmental stressors, developmental signals, and cellular homeostasis. AGO association is the main defender of miRNA stability; loaded miRNA are shielded from general exonucleases (Yang et al. 2020a). But when needed, evolution has shaped a specialized method to circumvent this defence. Target-directed miRNA degradation (TDMD) is a well-known process. In this process, particularly, highly complementary target RNAs interact with the miRNA to cause conformational changes in the AGO protein rather than to silence it (Li et al. 2025b). The weak 3′ end of miRNA exposed by this remodeling. This exposure serves as a molecular beacon, causing nucleotides from the 3′ end to be removed ("trimming"), and non-template nucleotides, frequently adenines or uridines, to be added ("Tailing"). The Zinc finger SWIM domain-containing protein 8 (ZSWIM8E3) ubiquitin ligase complex, which identifies the remodeled AGO-miRNA complex, ubiquitinates AGO and targets it for proteasomal degradation, which is crucially involved in TDMD. After miRNA release from breakdown AGO, it becomes vulnerable to quick exonucleolytic decay, which eliminates it completely (Shi et al. 2020).

Sequence-specific post-maturation changes at the 3′ end of miRNA function as potent degradation signals that are independent of substantial target binding, complementing TDMD. The integration of non-templated nucleotide is catalyzed by terminal nucleotide transferases (TENTs), including terminal uridylyltransferases-4 (TUT4) and terminal uridylyltransferases-7 (TUT7), which causes uridylation, and terminal nucleotide transferases-2 (TENT2) and terminal nucleotide transferases-B4 (TENTB4), which result in adenylation (Yang et al. 2022a). This tailing, in particular, uridylation serves as a chemical tag that can flag the molecule for destruction and prevent miRNA action. Enzymes such as poly(A)-specific ribonuclease (PARN) break down poly(A) tailed miRNA, whereas DIS3-like 3′-5′ exoribonuclease 2 (DIS3L2), a 3′-5′ exonuclease that strongly prefers oligo-uridine tails which is major executor of uridylated miRNA degradation (Shukla et al. 2019). These tailing-trimming mechanisms, which are essential for both development and proliferation, offer a tenable approach for controlling miRNA abundance. Interestingly, it has been demonstrated that TENTB4-mediated adenylation of miR21 accelerates its degradation, underscoring the complex function of tailing in miRNA turnover (Boele et al. 2014).

In human diseases, particularly in cancer, the significance of accurate regulation of miRNA degradation is glaringly obvious. The buildup of aberrantly tailed, trimmed, or stabilized miRNA isoform is caused by dysregulation of decay pathways, which can be caused by aberrant TDMD, malfunctioning TENTs, or defective exonucleases. In contrast to their canonical forms, these mutated species may exhibit tumor-suppressive or carcinogenic actions. For example, shortened isoforms could acquire new targets or lose selectivity. As a result, miRNA decay is not a waste disposal mechanism by default. It is an essential mechanism that shapes functional miRNA landscape and is dynamic, finely controlled, and enzymatically driven. Cells exercise fine control over miRNA stability through processes such as sequence-specific tailing/trimming and TDMD, which enables rapid adaptation, precise developmental timing, and the maintenance of cellular health with profound implications when these pathways go awry.

## 2.2 The Roots of Dysregulation: How miRNAs Go Awry in Cancer

miRNAs are essential modulators of gene expression that impact vital cellular functions as death, differentiation, and proliferation. Dysregulation of miRNAs is a frequent occurrence in cancer and plays a role in the development, spread, and metastasis of tumors. Numerous processes, such as genomic changes, epigenetic silencing, transcriptional dysregulation, malfunctions in the machinery that processes miRNA, and environmental factors within the tumour microenvironment, contribute to this dysregulation. Deciphering these complicated causes of miRNA dysregulation is crucial to figuring out the intricacies of cancer pathogenesis and finding new targets for treatment. Figure 2.2 outlines the critical stages of miRNA biogenesis, starting from the transcription of pri-miRNAs in the nucleus to their maturation and functional incorporation into the RISC in the cytoplasm. Key

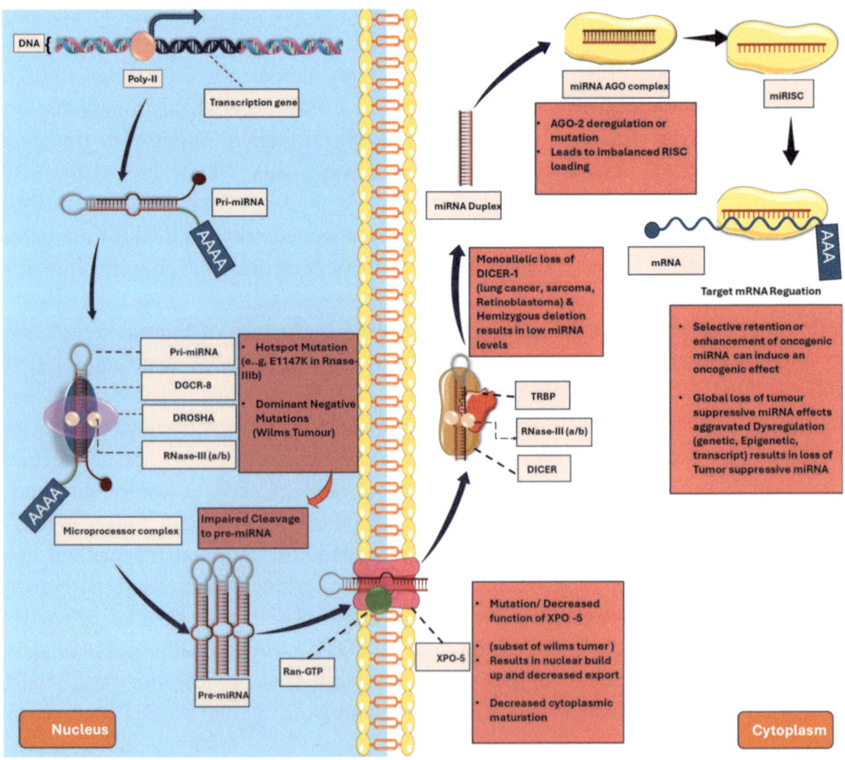


**Fig. 2.2** miRNA dysregulation in cancer, biogenesis. The image illustrates important stages in miRNA synthesis and shows how deregulation of processing components (e.g., DROSHA, DGCR8, DICER, AGO2, XPO-5) and genetic alterations affect miRNA maturation. These changes result in the development of cancer by either increasing oncogenic miRNAs or decreasing tumor-suppressive miRNAs

processing components, including DROSHA and DGCR8 in the nucleus, Exportin-5 (XPO-5) for nuclear export, and DICER, along with AGO in the cytoplasm, are highlighted. Dysregulation or genetic alterations affecting these processing factors can compromise miRNA maturation, resulting in aberrant expression profiles. Such perturbations may either elevate oncogenic miRNAs (oncomiRs) or reduce tumor-suppressive miRNAs, thereby promoting tumorigenesis and cancer progression by disrupting normal gene regulatory networks essential for cell homeostasis and growth control. Figure 2.2 emphasizes the pivotal role of miRNA biogenesis perturbations in cancer development.

### 2.2.1 Genomic Alterations

One of the most important processes causing the dysregulation of miRNA in cancer is genomic modifications, which radically change the transcriptome of cancerous cells. These changes can impact the structure and expression of miRNA genes, which can promote tumor development, progression, and resistance to treatment by either silencing tumor suppressive miRNA or aberrantly activating oncogenic miRNA. Gene amplification, in which some miRNA loci experience copy number increases, is one of the most common types of genomic modification. As a result, miRNAs that normally repress tumor suppressor genes are overexpressed. Amplification of miR-17-92 polycistronic cluster, which is found on chromosome 13q31.3 and is commonly seen in medulloblastomas, lung malignancies and B cell lymphomas, is a well-documented example (Fuziwara and Kimura 2015). By targeting several tumor suppressors, including Phosphatase and Tensin homolog (PTEN), Bcl-2-like protein 11 (BIM) and Transforming Growth Factor Beta Receptor (2TGFBR2), the overexpression of this cluster promotes cell proliferation, suppresses apoptosis, and aids in angiogenesis (Fuziwara and Kimura 2015).

The tumor suppressor function may be lost due to gene deletions. These deletions are linked to poor clinical outcomes and are frequently identified in delicate chromosomal regions. One well-known example is the deletion of miR-15a/16 1 cluster at chromosome 13q14, which is a region that is commonly eliminated in chronic lymphocytic leukaemia (CLL) (Klein et al. 2008). Anti-apoptotic proteins like BCL2 are upregulated when these miRNAs are lost, which increases cell survival and resistance to chemotherapy. Point mutation within miRNA genes or their regulatory components can also have a significant impact, in addition to large-scale deletions (Patel et al. 2015). The range of miRNA targets can be changed by mutation in seed region, the crucial 2–8 nucleotide sequence in charge of target recognition, which may turn a tumor suppressive miRNA into one with carcinogenic potential. Reduced quantities of mature miRNA can also result from changes in the flanking regions or stem-loop structure of primary miRNA transcript that affects DROSHA and DICER's ability to digest them (Patel et al. 2015).

Another type of genomic modification that has important ramifications for miRNA regulation is chromosomal translocation. These changes may reposition miRNA genes under the influence of heterologous enhancers or promoters, leading

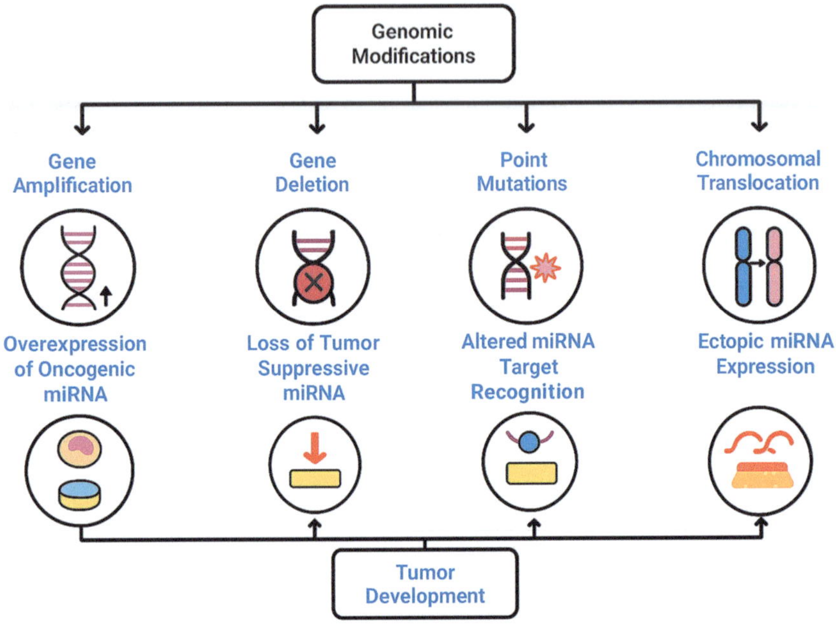


**Fig. 2.3** Genomic alterations driving miRNA dysregulation in cancer. Point mutations, chromosomal translocations, gene amplification, and deletion all result in abnormal miRNA expression or function, which can cause tumor growth by altering target identification, ectopic expression, loss of tumor-suppressive miRNAs, or overexpression of oncogenic miRNAs

to ectopic expression. For example, translocation involving miR-142 has been found in specific leukemia subtypes, where its misexpression targets genes involved in hematopoietic differentiation and proliferation, hence contributing to leukemogenesis (Huang et al. 2023). Although translocations are more frequently linked to genes that code for proteins, it is becoming more acknowledged that they can cause oncogenic reprogramming through their effect on miRNA. Figure 2.3 illustrates how diverse genomic alterations, including point mutations, chromosomal translocations, gene amplification, and gene deletions, disrupt normal miRNA expression and function in cancer.

### 2.2.2 Epigenetic Silencing

One important factor in the deregulation of gene expression during the development of cancer is the epigenetic silencing of miRNA. In contrast to genetic mutations, epigenetic modifications are heritable changes that impact gene function in different ways, which include DNA methylation, histone modification, and chromatin remodeling rather than altering the underlying DNA sequence. One common silencing mechanism in the context of miRNA is hypermethylation of cytosine and guanine nucleotides (CpG) islands inside promoter regions, which results in transcriptional inhibition of tumor suppressive miRNAs (Cui et al. 2023; Van Vlodrop et al. 2011). Histone deacetylation and restrictive histone markers like Histone H3 trimethylation at lysine 27 (H3K27me3) also contribute to chromatin compaction, which makes transcriptional access even more difficult (Ma et al. 2025b). The transcriptional machinery's inability to access miRNA genes is also a result of altered nucleosome placement caused by the chromatin remodeling complex. Oncogenesis is significantly impacted by this epigenetic suppression. By targeting oncogenes, many miRNAs suppress tumors; their silencing eliminates important inhibitory controls, which encourages unchecked cell division. For example, many tumors have the miR-34 family, which targets oncogenes such as MYC and BCL2, which are frequently silenced in many cancers (Li 2021). Hypomethylation, on the other hand, can result in overexpression of oncogenic miRNA, including miR-2, which suppresses tumor suppressors like programmed cell death protein 4 (PDCD4) and PTEN, which results in accelerating the growth of tumors (Wong 2024; Peralta-Zaragoza et al. 2016). Furthermore, epigenetic silencing of the genomic cluster of miRNA can decrease numerous tumor suppressive miRNAs at once, increasing the oncogenic consequences. The presence of feedback loops involving epigenetic machinery is particularly pernicious feature of this regulation. Some miRNAs, known as epi-miRNAs, directly influence key enzymes of epigenetic modification, such as DNA methyltransferases (DNMTs) and histone deacetylases (HDACs) (Arif et al. 2020). A self-reinforcing loop of epigenetic dysregulation is produced when their silence disrupts this regulatory equilibrium. When taken as a whole, these processes emphasize how important epigenetic miRNA silencing is to the development of cancer.

### 2.2.3 Transcriptional Dysregulation

Another characteristic of cancer is transcriptional dysregulation, which is often brought on by the loss of tumor suppressors like p53 or the aberrant activity of oncogenic transcription factors like c-Myc. These elements have a major impact on miRNA expression, which in turn affects the onset and spread of cancer. By attaching itself to the promoter of miRNA genes and either activating or repressing their transcription, the oncogenic transcription factor c-Myc is known to directly control miRNA production at the transcriptional level. For example, c-Myc suppresses the tumor-suppressive miRNA, such as the let-7 family miR-15a, miR-26, miR-29, and miR-30, which are essential for cell cycle regulation and apoptosis, while activating the transcription of the oncogenic miR17–92 cluster (Tao et al. 2014). Furthermore, via upregulating DROSHA, which is a crucial enzyme in miRNA processing, c-Myc regulates miRNA synthesis and promotes the maturation of certain miRNAs (Wang et al. 2013; Bui and Mendell 2010).

On the other hand, the tumor suppressor p53 plays both transcriptional and post-transcriptional activities as an aster regulator of miRNA expression. Numerous tumor suppressive miRNAs, such as miR-34 family (Chang et al. 2007), miR-145

(Ye et al. 2019), miR-107 (Yan et al. 2022), miR-192, and miR-215 (Braun et al. 2008), which mediates cell cycle arrest, apoptosis, and senescence. In addition to transcriptional regulation, p53 enhances the processing of initial miRNA transcripts in response to cellular stress or DNA damage by interacting with the DROSHA complex and promoting the development of growth suppressive miRNAs such as miR-16-1 and miR-145 (Ye et al. 2019; Calin et al. 2008). These mechanisms are upset when p53 is lost or mutated, which lowers the amounts of mature tumor-suppressive miRNA and promotes malignant transformation. Interestingly, mutant p53 can also disrupt the DROSHA complex, which worsens carcinogenic pathways and hinders miRNA production even more (Feng et al. 2011).

The intricacy of gene regulation network in cancer is highlighted by the complicated interaction between transcription factors and miRNA expression. In addition to contributing to the characteristics of cancer, dysregulation of miRNA expression due to changed transcription factor activity also identifies a possible therapeutic target for re-establishing regular cellular function.

### 2.2.4 Defects in miRNA Processing Machinery

The disruption of core miRNA biogenesis machinery, including DROSHA/DGCR8, Exportin-5 (XPO-5), DICER, and AGO protein, constitutes an additional and crucially important layer of miRNA dysregulation in cancer. The tumor-suppressive let-7 family and other mature miRNAs are widely lost in Wilms tumor due to recurrent somatic mutations that impede pri-miRNA processing (Rakheja et al. 2014). Especially the hotspot E1147K mutation in DROSHAs RNase-IIIb domain (Rakheja et al. 2014). These dominant negative mutations provide minimal levels necessary for tumor cell viability while impairing global miRNA maturation.

XPO-5 mediates the nuclear export of pre-miRNA after DROSHA's activity. A subset of Wilms tumors has a mutation or decreased XPO-5 function, which results in a decrease in cytoplasmic miRNA maturation and a nuclear buildup of pre-miRNAs. DICER converts the pre-miRNA into 22 nt duplex in the cytoplasm, as we discussed in the biogenesis of miRNA. But in lung cancer, sarcoma and retinoblastoma, monoallelic (haploinsufficiency) DICER loss speeds up tumor development, whereas biallelic loss is intolerable, suggesting that residual activity is necessary for tumor survivability (Kumar et al. 2009). Hemizygous deletion of DICER-1 in human malignancies is linked with a weak prognosis and lower levels of miRNA (Kumar et al. 2009).

Lastly, the active RISC is formed when processed miRNA is loaded into AGO proteins particularly AGO2 (Tan et al. 2011). The balance of RISC is shifted toward oncomiRs rather than tumor suppressive miRNA due to dysregulation of AGO proteins, which occur through changes expression or alternations, even though mutations are less common (Svoronos et al. 2016). These processing factors are significant biomarkers and potential therapeutic targets because, when combined, these flaws in the biogenesis of the miRNA pathway result in global loss of tumor suppressive miRNA and the selective retention or enhancement of oncogenic miRNAs, aggravating genetic, epigenetic, and transcriptional dysregulation in cancer.

### 2.2.5 Environmental Influences

The pathogenesis of cancer is further complicated by the tumor microenvironment, which exerts strong external forces that drastically alter miRNA expression profile in addition to inherent genetic, epigenetic, transcriptional, and processing abnormalities. A characteristic of solid tumors that arises from insufficient vascularization is hypoxia, which triggers the transcription factor Hypoxia-inducible factor 1-alpha (HIF-1$\alpha$) (Zhao et al. 2024). HIF-1$\alpha$ suppresses some miRNA, such as the tumor-suppressive miR-34a (Bouznad et al. 2023), while directly trans activating others, including the pro-angiogenic miR-210 (Huang et al. 2009). The transcription of some well-characterized oncomiRs, such as miR-21 and miR-155, is induced by (Nuclear factor kappa-light-chain-enhancer of activated B cells) NF–kB (Markopoulos et al. 2018), which targets key tumor suppressors (e.g., PTEN, PDCD4) (Meng et al. 2007; Zargar et al. 2019) and amplifies inflammatory signaling itself, creating a vicious cycle that fuels proliferation, survival, and immune evasion. On the other hand, NF–kB has the ability to inhibit specific miRNAs that suppress tumors. miRNA profiles are also reshaped by metabolic stress, which can result from nutritional storage (such as glucose or amino acids) or changes metabolite level in tumors that are growing quickly. The transcription and processing of miRNAs are influenced by stress-responsive pathways, including those involving AMP-activated protein kinase (AMPK) and (mammalian target of rapamycin) mTOR (Zhang et al. 2017; Sukumaran et al. 2020). For example, decreased expression of the glutamine transporter Solute Carrier Family 1 Member 5 (SLC1A5) and metabolic rewiring toward glutaminolysis can result from glucose starvation's downregulation of miR23-a/b (Dong et al. 2020). Moreover, components of the miRNA processing machinery might be directly impacted by metabolic enzymes.

## 2.3 miRNA Deregulation: Fueling the Hallmarks of Cancer Progression

Cellular homeostasis depends on the precise synthesis and controlled degradation of miRNAs, as explained in the sections above. But with cancer, this complex regulatory system is extremely susceptible. miRNA expression and function are disrupted by a combination of environmental stresses within the tumor microenvironment, transcriptional dysregulation, epigenetic silencing, genomic changes, and deficiencies in the core processing machinery. This widespread dysregulation is an active driving force that radically alters the biology of cancer cells rather than just being a passive result of cellular transformation. The consequences of miRNA disruption are far-reaching. When tumor-suppressive miRNAs are lost or oncogenic miRNAs (oncomiRs) are aberrantly overexpressed, the delicate balance controlling key cellular pathways is shattered. This imbalance directly fuels the acquisition and maintenance of the defining core capabilities that enable malignant growth, survival, invasion, and metastasis. The preceding exploration of the roots of Dysregulation established how miRNAs go awry. This section delves into the functional consequences, specifically discuss how miRNA deregulation acts as a potent accelerant for these hallmarks.

### 2.3.1 Enabling Replicative Immortality (Telomere Maintenance)

One important characteristic of cancer is the ability to induce replicative immortality, which is maintained by telomere length preservation mechanisms, as shown in Fig. 2.4. The alternate lengthening of telomeres (ALT) route or telomerase reactivation are the main components of these processes. The critical function of miRNAs as precise regulators of these telomere maintenance processes has been brought to light by recent studies. One of the best-studied instances is miR-138-5p, which suppresses tumors by specifically targeting the mRNA that codes for the catalytic core of telomerase, human telomerase reverse transcriptase (hTERT) (Wang et al. 2017). Overexpression of miR-138-5p in melanoma cells limits cell migration and proliferation by dramatically lowering telomerase activity and hTERT protein levels (Tarazón et al. 2021), which reduces telomerase activity and suppresses cell proliferation and migration, highlighting its tumor-suppressive role (Tarazón et al. 2021). Luciferase reporter assays have confirmed hTERT as a direct target of miR-138-5p, and functional studies demonstrate that miR-138-5p overexpression induces cell

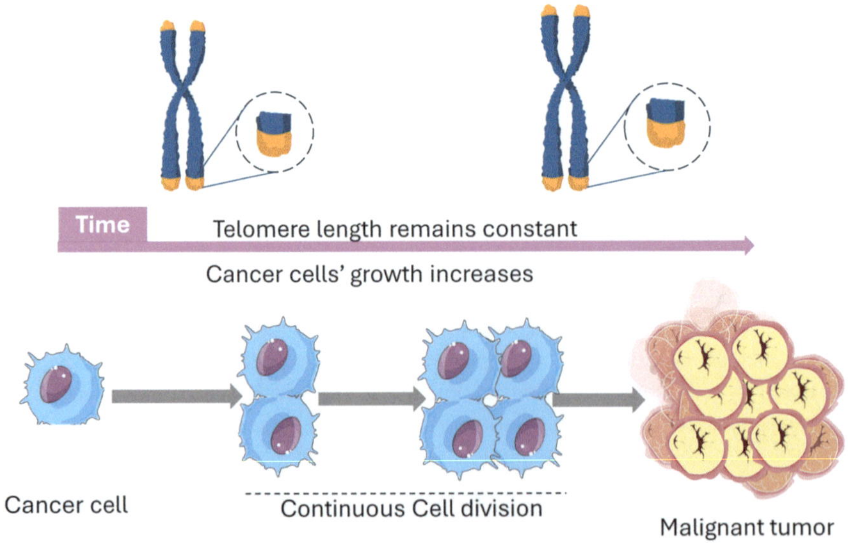


**Fig. 2.4** General ability of cancer cells to enable replicative immortality

cycle arrest and apoptosis in malignant melanoma cells (Li et al. 2017). Downregulation of these miRNAs can support cancer development.

The ALT route, a telomerase-independent mechanism common in some malignancies like glioblastomas and osteosarcomas, is similarly influenced by miRNAs. For example, a critical modulator that is differentially expressed in cancer cells that are ALT-positive versus those that are telomerase-positive is miR-708 (Kaul et al. 2021). miR-708, which is elevated in ALT cells, suppresses tumor growth by preventing angiogenesis, invasion, and cell migration. Together, these results highlight the dual function of miRNAs in regulating both alternative (ALT) and traditional (telomerase) telomere maintenance processes (Kaul et al. 2021). This complex regulatory layer presents encouraging opportunities for new therapeutic approaches meant to interfere with the replicative immortality that drives the development of cancer.

### 2.3.2 Reprogramming Energy Metabolism

To satisfy the needs of rapid multiplication, cancer cells undergo extensive metabolic reprogramming in addition to maintaining replicative immortality. Even when oxygen is present, this change often called the Warburg effect involves a preference for aerobic glycolysis over oxidative phosphorylation (Vander Heiden et al. 2009) as shown in Fig. 2.5. Crucial regulators of this metabolic reprogramming have been identified as miRNAs. Interestingly, hexokinase 2 (HK2), a rate-limiting enzyme in glycolysis that catalyzes the phosphorylation of glucose to glucose-6-phosphate, is specifically targeted by miR-143. MiR-143 expression is often downregulated in lung and breast cancer models, which results in increased glycolytic flux and HK2 protein levels (Ros and Schulze 2013). Restoring miR-143 expression highlights its tumor-suppressive function in metabolic control by lowering glucose uptake, lactate generation, and tumor growth (Miao et al. 2016).

Interestingly, miR-155 controls glycolysis via both transcriptional and post-transcriptional processes, functioning upstream in a regulatory cascade (Jiang et al. 2012). Pro-inflammatory cytokines cause miR-155 to suppress CCAAT/enhancer binding protein beta (C/EBPβ), a transcriptional activator of miR-143, while also activating Signal Transducer and Activator of Transcription 3 (STAT3) to enhance HK2 transcription (Jiang et al. 2012). Glycolytic activity in breast cancer cells is increased because of this dual effect, which suppresses miR-143 and increases HK2 expression (Jiang et al. 2012). An example of how inflammation-driven miRNA networks might rewire cancer metabolism is the miR-155/miR-143/HK2 axis (Jiang et al. 2012).

In addition to glycolysis, miRNAs affect mitochondrial activity, which is a crucial factor in maintaining cellular energy homeostasis. miR-378, which is encoded in the Peroxisome proliferator-activated receptor gamma coactivator 1-beta (Ppargc1b) gene's intron, targets important regulators of fatty acid oxidation and oxidative phosphorylation to modify mitochondrial metabolism. By suppressing Mediator complex subunit 13 (MED13) and carnitine acetyltransferase, it reduces

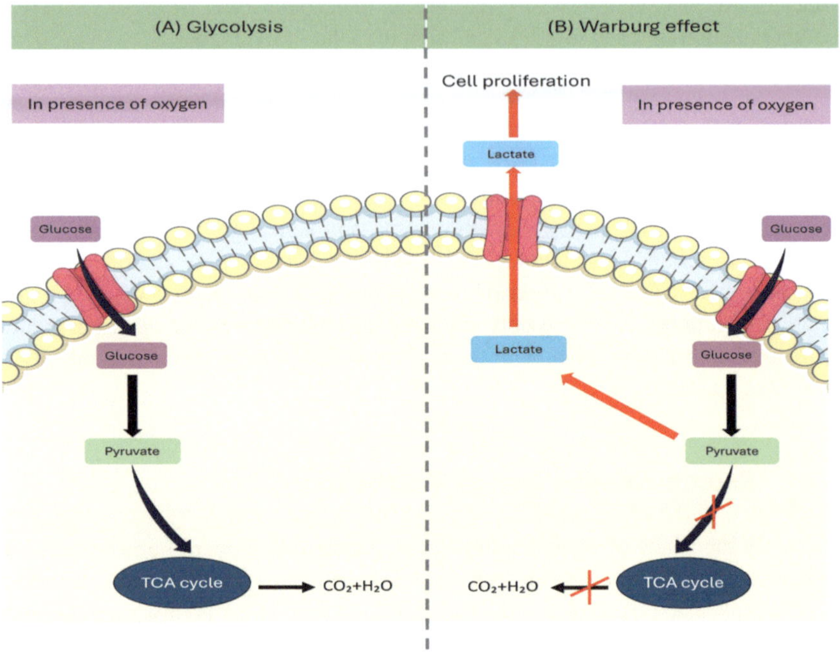


**Fig. 2.5** (**a**) Conventional oxidative metabolism where pyruvate enters the TCA cycle for complete oxidation to $CO_2 + H_2O$. (**b**) Warburg effect which is aerobic glycolysis, in this effect cancer cells convert glucose to lactate surprisingly in presence of oxygen

mitochondrial oxidative capability and encourages a glycolytic phenotype. MiR-378's function in systemic energy balance is highlighted by the increased mitochondrial respiration and resistance to diet-induced obesity shown in mice lacking it (Carrer et al. 2012). Moreover, miR-378 has been demonstrated to localize inside mitochondria and disrupt transcripts expressed by the mitochondria, indicating a direct function in regulating the expression of mitochondrial genes and bioenergetics.

These regulatory molecules emerge as master conductors of cancer hallmarks, building on the fundamental role of miRNAs in rewiring cancer metabolism, from influencing mitochondrial bioenergetics through miR-378 to enforcing glycolytic dependence via the miR-155/miR-143/HK2 axis.

### 2.3.3 Avoiding Immune Destruction

Beyond metabolic reprogramming, miRNA networks play a crucial role in helping tumors escapes immune destruction via a variety of strategies. These include reprogramming immune cells in the tumor microenvironment, interfering with antigen presentation, and directly suppressing immunological checkpoint molecules.

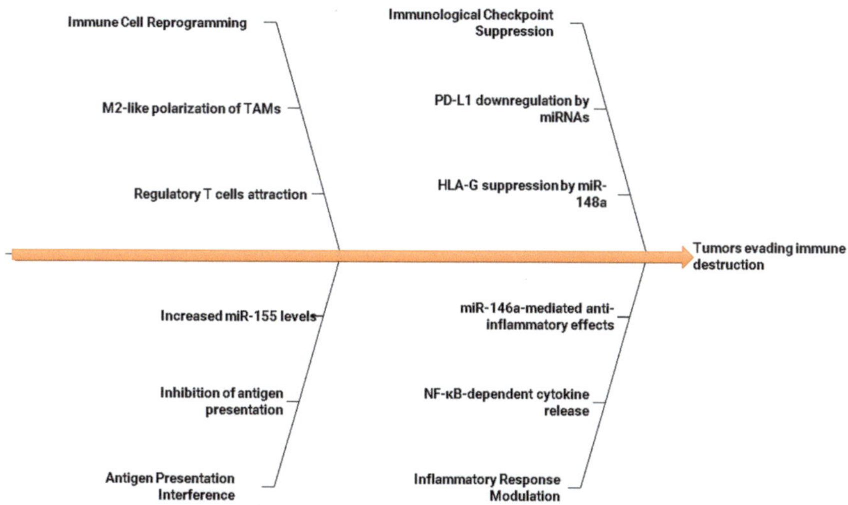


**Fig. 2.6** Various mechanisms of tumors to evade immune destruction regulated by various miRNA

A key immune evasion route, the PD-1/PD-L1 axis, is exploited by tumors (Lin et al. 2024). Here, miRNAs such as miR-138-5p and miR-424 directly bind the programmed death-ligand 1 (PD-L1) mRNA 3' Untranslated Region (3'UTR), causing posttranscriptional downregulation (Zhao et al. 2016; Chen et al. 2014). miRNAs are endogenous brakes on checkpoint-mediated T-cell fatigue since their restoration in preclinical models such as lung cancer and melanoma lowers PD-L1 surface expression, reactivates $CD8^{+}$ T-cell cytotoxicity, and inhibits tumor proliferation (Zabeti Touchaei and Vahidi 2024). The immunosuppressive ligand Human Leukocyte Antigen-G (HLA-G) is directly targeted by miR-148a (Zhang et al. 2019); ectopic expression in esophageal squamous cell carcinoma reduces HLA-G protein levels, which increases NK cell-mediated killing and tumor apoptosis, breaking down molecular cloaks that protect tumors from immunity. Immune cell behavior in the tumor stroma is regulated by miRNAs. Tumor-associated macrophages (TAMs) and myeloid-derived suppressor cells (MDSCs) have higher levels of miR-155 (Li et al. 2021; Chen et al. 2015), which promotes M2-like polarization, inhibits antigen presentation, and attracts regulatory T cells, causing NF–κB-dependent pro-inflammatory cytokine release. On the other hand, miR-146a functions as an anti-inflammatory checkpoint; pro-inflammatory cytokines are dampened when miR-146a mimics are delivered myeloid-specifically, suppressing NF–κB activators Interleukin-1 Receptor-Associated Kinase 1 (IRAK1) and TNF Receptor-Associated Factor 6 (TRAF6) (Han et al. 2023). This highlights its therapeutic potential in vivo by increasing T-cell activation and decreasing immunosuppressive cell infiltration to normalize myeloid function. Summary of the tumors evading immune destruction is described in Fig. 2.6.

## 2.4 Beyond Primary Tumors: miRNAs in Metastatic Cascades

Tumor cells must survive the harsh conditions of the bloodstream, where they are subject to shear stress, immune surveillance, and anoikis, during the metastatic cascade. Figure 2.7 illustrates the key stages of cancer metastasis, including local intravascular entry into the bloodstream, circulation as circulating tumor cells (CTCs) or via exosome-mediated transport, extravasation into distant tissues, and colonization of secondary organs. During this crucial stage, miRNAs are essential for improving the survival of CTCs. Notably, the miR-103/107 cluster and miR-21 have become important modulators of circulatory persistence and intravasation (Chen et al. 2012). One of the most widely expressed oncomiRs, miR-21, targets tumor suppressors such Phosphatase and TENsin homolog (PTEN), Programmed Cell Death 4 (PDCD4), Reversion-Inducing Cysteine-Rich Protein with Kazal Motifs (RECK), and Tissue Inhibitor of Metalloproteinases 3 (TIMP3) to increase cell motility and resistance to apoptosis and anoikis (Gabriely et al. 2008; Han et al. 2012). Both intravasation and the epithelial-to-mesenchymal transition (EMT) are significantly facilitated by this action (Han et al. 2012). Its upregulation in hypoxic settings through HIF-1$\alpha$ and its enrichment in exosomes imply that it plays a part in systemic dissemination, pre-metastatic niche formation, and intercellular communication (Yang et al. 2021). Additionally, miR-21 can alter immune cell function in the bloodstream and give recipient cells pro-metastatic characteristics by exosomal

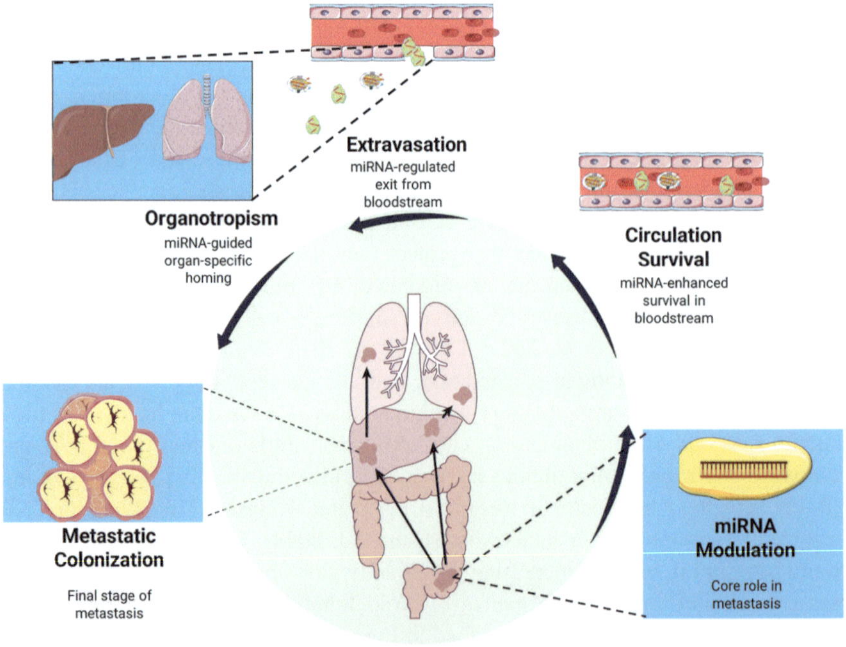


**Fig. 2.7** miRNA modulation in cancer metastasis

transfer. Likewise, miR-103 has been linked to endothelial junction disruption by targeting vascular endothelial cadherin (VE-cadherin) and DICER, which increases vascular permeability and tumor cell entrance into the bloodstream (Fang et al. 2018). After entering the bloodstream, CTCs have unique miRNA signatures that indicate their organotropism and potential for metastasis. miR-21, miR-10b, miR-200 family members (Han et al. 2012; Bahena-Ocampo et al. 2016; Schubert and Brabletz 2011), miR-141, and miR-375 are among the CTC-associated miRNAs that can be found in plasma and have been suggested as non-invasive indicators of prognosis, treatment response, and metastatic progression (Romeo et al. 2018; Yang et al. 2022b). According to recent research, platelet-coated CTCs that use transferred miRNAs can withstand shear stress and avoid immunological attack (Yang 2025).

Figure 2.7 depicts important stages of cancer metastasis, such as local invasion, the colonization of distant organs. Intravasation, CTCs or exosome mediated circulation, and extravasation. miRNAs are essential regulators at every step, impacting the growth of tumors and the development of metastases.

In order to colonize secondary sites after surviving in the circulation, CTCs must stop at distant arterial beds, attach to the endothelium, and then undergo extravasation processes that are actively regulated by miRNAs. At the same time, and even prior to the arrival of CTCs, primary tumor-derived miRNAs especially those transported by exosomes or extracellular vehicles (EVs) are essential for priming distant organs for colonization. Through inflammation, matrix remodeling, and immunosuppression, these tumor-secreted miRNAs, like miR-83 and miR-21, resident stromal cells like fibroblasts within the pre-metastatic niche, reprogramming the microenvironment to improve receptiveness (Sánchez et al. 2016). Importantly, miRNAs function as molecular compasses that actively regulate the exceptional specificity of metastatic dissemination, or organotropism. CTCs are pre-programmed by primary tumors with particular miRNA signatures that determine their propensity for the microenvironments of the brain, liver, lungs, or bones. For example, CTCs that target the liver often rely on miR-122 (Yang et al. 2020b) and lung-tropic CTCs use the miR-200 family (Dykxhoorn 2010). Brain metastasis involves miR-509 (promoting blood–brain barrier transmigration) (Xing et al. 2015) and exosomal miR-19 (promoting astrocyte-mediated vascular co-option) (Kaur et al. 2025). Certain miRNAs, such as miR-190 (Almog et al. 2013) and miR-23b (Ono et al. 2014), operate as molecular switches that control the crucial balance between cellular dormancy and cancer cell's destiny inside the metastatic niche after they have successfully extravasated and entered the parenchyma. Therefore, in addition to facilitating CTC extravasation and priming the distant soil, miRNA-mediated mechanisms also actively guide organ-specific homing and determine whether disseminated cells remain dormant or start lethal secondary growth. These organ-specific signatures have great potential as therapeutic targets and predictive biomarkers.

## 2.5 Conclusion and Future Directions

The hallmark capabilities of cancer, such as metabolic rewiring, immune evasion, replicative immortality, and metastatic dissemination, are driven by miRNAs, which are key, dynamic orchestrators of cancer progression and metastasis (Shao et al. 2018). These mechanisms range from transcriptional hijacking and environmental reprogramming to genomic changes and epigenetic silencing. Molecular biologists must decipher context-specific networks, clinicians must validate biomarkers and trial designs, and bioengineers must create targeted delivery systems (like nanoparticle carriers) (Kumar et al. 2025b), computational scientists must model regulatory circuits, and pharmacologists must optimize miRNA modulators (antagomiRs, miR-mimics) in order to realize their therapeutic potential. We can disrupt malignant circuits, resensitize tumors to treatment, and use liquid biopsies for early detection by precisely restoring tumor-suppressive miRNAs or silencing oncogenic ones. By combining immunotherapy, chemotherapy, and microenvironment reprogramming, we can fundamentally change the course of cancer toward better patient outcomes (Wolinsky et al. 2012).

In conclusion, miRNA deregulation has emerged as a fundamental mechanism driving the progression and advancement of cancer. Aberrant miRNA expression patterns resulting from genetic, epigenetic, and transcriptional alterations disrupt the finely tuned gene regulatory networks that maintain cellular homeostasis. miRNAs are both tumor suppressors and oncogenes that have a significant impact on pathways related to invasion, metastasis, resistance, apoptosis, proliferation, and interactions within the tumor microenvironment. Growing data demonstrate the potential of particular miRNAs to influence tumor behavior and treatment responses, making them useful as diagnostic, prognostic, and therapeutic biomarkers in a variety of cancers. Advances in delivery methods, improved understanding of the molecular determinants of miRNA deregulation, and integration of multi-omics data will facilitate personalized interventions targeting tumor-specific miRNA signatures. However, significant challenges remain, including off-target effects, delivery efficiency, and tumor heterogeneity. Ongoing research must address these obstacles by refining therapeutic strategies, identifying novel regulatory networks (including those involving noncoding RNAs such as lncRNAs and circRNAs), and validating emerging miRNA targets in clinical settings.

**Acknowledgments** D. B. thanks SERB, GoI for the Ramanujan Fellowship, DST-Nidhi Prayas for the start-up grant, and Gujcost-DST, GSBTM, BRNS-BARC, and HEFA-GoI, MoES for STARS for research grants. D. B. is a member of the Indian National Young Academy of Sciences (INYAS). A.K.Y expresses thankful to the Indian Institute of Technology Gandhinagar (IITGN) to provide financial support through the Early Career Fellowship (ECF).

**Conflicts of Interest** There are no conflicts to declare.

## References

Almog N et al (2013) Transcriptional changes induced by the tumor dormancy-associated microRNA-190. Transcription 4:177–191

Arif KMT, Elliott EK, Haupt LM, Griffiths LR (2020) Regulatory mechanisms of epigenetic miRNA relationships in human cancer and potential as therapeutic targets. Cancer 12:2922

Bahena-Ocampo I et al (2016) miR-10b expression in breast cancer stem cells supports self-renewal through negative PTEN regulation and sustained AKT activation. EMBO Rep 17:648–658

Boele J et al (2014) PAPD5-mediated 3′ adenylation and subsequent degradation of miR-21 is disrupted in proliferative disease. Proc Natl Acad Sci 111:11467–11472

Bofill-De Ros X, Vang Ørom UA (2024) Recent progress in miRNA biogenesis and decay. RNA Biol 21:36–43

Bouznad N, Rokavec M, Öner MG, Hermeking H (2023) miR-34a and IRE1A/XBP-1(S) form a double-negative feedback loop to regulate hypoxia-induced EMT, metastasis, Chemo-Resistance and Autophagy. Cancers (Basel) 15:1143

Braun CJ et al (2008) p53-responsive MicroRNAs 192 and 215 are capable of inducing cell cycle arrest. Cancer Res 68:10094–10104

Bui TV, Mendell JT (2010) Myc: maestro of MicroRNAs. Genes Cancer 1:568–575

Calin GA et al (2008) MiR-15a and miR-16-1 cluster functions in human leukemia. Proc Natl Acad Sci 105:5166–5171

Carrer M et al (2012) Control of mitochondrial metabolism and systemic energy homeostasis by microRNAs 378 and 378*. Proc Natl Acad Sci 109:15330–15335

Chang T-C et al (2007) Transactivation of miR-34a by p53 broadly influences gene expression and promotes apoptosis. Mol Cell 26:745–752

Chang TC, Pertea M, Lee S, Salzberg SL, Mendell JT (2015) Genome-wide annotation of microRNA primary transcript structures reveals novel regulatory mechanisms. Genome Res 25:1401–1409

Chen H-Y et al (2012) miR-103/107 promote metastasis of colorectal cancer by targeting the metastasis suppressors DAPK and KLF4. Cancer Res 72:3631–3641

Chen L et al (2014) Metastasis is regulated via microRNA-200/ZEB1 axis control of tumour cell PD-L1 expression and intratumoral immunosuppression. Nat Commun 5:5241

Chen S et al (2015) Host miR155 Promotes Tumor Growth through a Myeloid-Derived Suppressor Cell–Dependent Mechanism. Cancer Res 75:519–531

Choi E (2013) MicroRNAs as novel regulators of stem cell fate. World J Stem Cells 5:172

Cui G, Dong Q, Gai K, Qi S (2023) Chromatin dynamics: chromatin remodeler, epigenetic modification and diseases. In: Epigenetics – regulation and new perspectives. IntechOpen. https://doi.org/10.5772/intechopen.108385

Dong Y, Tu R, Liu H, Qing G (2020) Regulation of cancer cell metabolism: oncogenic MYC in the driver's seat. Signal Transduct Target Ther 5:124

Dykxhoorn DM (2010) MicroRNAs and metastasis: little RNAs go a long way. Cancer Res 70: 6401–6406

Fang J et al (2018) Hepatoma cell-secreted exosomal microRNA-103 increases vascular permeability and promotes metastasis by targeting junction proteins. Hepatology 68:1459–1475

Feng Z, Zhang C, Wu R, Hu W (2011) Tumor suppressor p53 meets microRNAs. J Mol Cell Biol 3: 44–50

Fuziwara CS, Kimura ET (2015) Insights into regulation of the miR-17-92 cluster of miRNAs in cancer. Front Med 2, 64

Gabriely G et al (2008) MicroRNA 21 promotes glioma invasion by targeting matrix metalloproteinase regulators. Mol Cell Biol 28:5369–5380

Han M et al (2012) Antagonism of miR-21 reverses epithelial-mesenchymal transition and cancer stem cell phenotype through AKT/ERK1/2 inactivation by targeting PTEN. PLoS One 7: e39520

Han R et al (2023) MicroRNA-146a negatively regulates inflammation via the IRAK1/TRAF6/NF-κB signaling pathway in dry eye. Sci Rep 13:11192
Huang X et al (2009) Hypoxia-inducible mir-210 regulates normoxic gene expression involved in tumor initiation. Mol Cell 35:856–867
Huang W, Paul D, Calin GA, Bayraktar R (2023) miR-142: a master regulator in hematological malignancies and therapeutic opportunities. Cells 13:84
Jiang S et al (2012) A novel miR-155/miR-143 cascade controls glycolysis by regulatinghexokinase 2in breast cancer cells: miR-155 links inflammation to the Warburg effect. EMBO J 31:1985–1998
Kaul Z et al (2021) Functional characterization of miR-708 microRNA in telomerase positive and negative human cancer cells. Sci Rep 11:17052
Kaur R, Gupta N, Dhawan A (2025) Astrocyte plasticity shapes brain metastasis progression through tumor-astrocyte crosstalk. Green Energy Sustain:1–15. https://doi.org/10.47248/chp2502010002
Klein U et al (2008) The DLEU2/Mir-15a/Mir-16-1 locus, commonly deleted in B-cell chronic lymphocytic leukemia (CLL), controls B-cell compartment expansion and its deletion leads to CLL in mice. Blood 112:25–25
Kumar MS et al (2009) Dicer1 functions as a haploinsufficient tumor suppressor. Genes Dev 23: 2700–2704
Kumar A et al (2025a) Clinical Applications of Targeted Nanomaterials. Pharmaceutics 17:379
Kumar A et al (2025b) Next-generation cancer Theragnostic: applications of carbon quantum dots. ChemNanoMat 11:e202500061
Li W (Jess) et al (2021) MicroRNA-34a: potent tumor suppressor, cancer stem cell inhibitor, and potential anticancer therapeutic. Front Cell Dev Biol 9:640587
Li X et al (2021) Tumor-associated macrophages secret exosomal miR-155 and miR-196a-5p to promote metastasis of non-small-cell lung cancer. Transl Lung Cancer Res 10:1338–1354
Li Y, Chen S, Rao H, Cui S, Chen G (2025a) MicroRNA gets a mighty award. Adv Sci 12:2414625
Li Z et al (2025b) Mechanistic insights into RNA cleavage by human Argonaute2–siRNA complex. Cell Res 35:453–464
Li C, Zhang Z, Gao T, Du M (2017) MicroRNA-138 suppresses cell proliferation of human malignant melanoma cells by targeting hTERT. Int J Clin Exp Med 10(4), 6517–6526
Lin X et al (2024) Regulatory mechanisms of PD-1/PD-L1 in cancers. Mol Cancer 23:108
Ma CP, Lo SJ, Chin-Ming Tan B (2025a) Good things come in small packages: the discovery of small RNAs in the smallest animal model. Biom J 48:100832
Ma J, Zhang Y, Li J, Dang Y, Hu D (2025b) Regulation of histone H3K27 methylation in inflammation and cancer. Mol Biomed 6:14
Markopoulos G et al (2018) Roles of NF-κB signaling in the regulation of miRNAs impacting on inflammation in cancer. Biomedicine 6:40
Meng F et al (2007) MicroRNA-21 regulates expression of the PTEN tumor suppressor gene in human hepatocellular cancer. Gastroenterology 133:647–658
Miao Y et al (2016) 18F-FDG PET/CT for monitoring the response of breast cancer to miR-143-based therapeutics by targeting tumor glycolysis. Mol Ther - Nucleic Acids 5:e357
Nunes S et al (2025) Recent advances in the development and clinical application of miRNAs in infectious diseases. Noncoding RNA Res 10:41–54
O'Brien J, Hayder H, Zayed Y, Peng C (2018) Overview of MicroRNA biogenesis, mechanisms of actions, and circulation. Front Endocrinol 9:402
Ono M et al (2014) Exosomes from bone marrow mesenchymal stem cells contain a microRNA that promotes dormancy in metastatic breast cancer cells. Sci Signal 7:ra63
Patel DJ, Am D, Bb T, Rh P, Li S, Patel DJ, Starega-Roslan J, Witkos T, Galka-Marciniak P, Krzyzosiak W (2015) Sequence features of Drosha and Dicer cleavage sites affect the complexity of IsomiRs. Int J Mol Sci 16:8110–8127
Peralta-Zaragoza O et al (2016) Relevance of miR-21 in regulation of tumor suppressor gene PTEN in human cervical cancer cells. BMC Cancer 16:215

Rakheja D et al (2014) Somatic mutations in DROSHA and DICER1 impair microRNA biogenesis through distinct mechanisms in Wilms tumours. Nat Commun 5:4802

Romeo P et al (2018) Circulating miR-375 as a novel prognostic marker for metastatic medullary thyroid cancer patients. Endocr Relat Cancer 25:217–231

Ros S, Schulze A (2013) Glycolysis Back in the limelight: systemic targeting of HK2 blocks tumor growth. Cancer Discov 3:1105–1107

Saini HK, Griffiths-Jones S, Enright AJ (2007) Genomic analysis of human microRNA transcripts. Proc Natl Acad Sci 104:17719–17724

Sánchez CA et al (2016) Exosomes from bulk and stem cells from human prostate cancer have a differential microRNA content that contributes cooperatively over local and pre-metastatic niche. Oncotarget. https://doi.org/10.18632/oncotarget.6540

Schubert J, Brabletz T (2011) p53 spreads out further: suppression of EMT and stemness by activating miR-200c expression. Cell Res 21:705–707

Shao Y et al (2018) Colorectal cancer-derived small extracellular vesicles establish an inflammatory premetastatic niche in liver metastasis. Carcinogenesis 39:1368–1379

Shi CY et al (2020) The ZSWIM8 ubiquitin ligase mediates target-directed microRNA degradation. Science 370:eabc9359

Shukla S, Bjerke GA, Muhlrad D, Yi R, Parker R (2019) The RNase PARN controls the levels of specific miRNAs that contribute to p53 regulation. Mol Cell 73:1204–1216.e4

Sukumaran A, Choi K, Dasgupta B (2020) Insight on transcriptional regulation of the energy sensing AMPK and biosynthetic mTOR pathway genes. Front Cell Dev Biol 8:671

Svoronos AA, Engelman DM, Slack FJ (2016) OncomiR or tumor suppressor? The duplicity of MicroRNAs in cancer. Cancer Res 76:3666–3670

Tan GS, Garchow BG, Liu X, Metzler D, Kiriakidou M (2011) Clarifying mammalian RISC assembly in vitro. BMC Mol Biol 12:19

Tao J, Zhao X, Tao J (2014) c-MYC–miRNA circuitry: A central regulator of aggressive B-cell malignancies. Cell Cycle 13:191–198

Tarazón E et al (2021) MiR-138-5p suppresses cell growth and migration in melanoma by targeting telomerase reverse transcriptase. Genes 12:1931

Van Vlodrop IJH et al (2011) Analysis of promoter CpG Island Hypermethylation in cancer: location, location, location! Clin Cancer Res 17:4225–4231

Vander Heiden MG, Cantley LC, Thompson CB (2009) Understanding the Warburg effect: the metabolic requirements of cell proliferation. Science 324:1029–1033

Vyas A et al (2025) Advancing the frontier of artificial intelligence on emerging technologies to redefine cancer diagnosis and care. Comput Biol Med 191:110178

Wang X, Zhao X, Gao P, Wu M (2013) C-Myc modulates microRNA processing via the transcriptional regulation of Drosha. Sci Rep 3:1942

Wang X et al (2017) miR-138-5p acts as a tumor suppressor by targeting hTERT in human colorectal cancer. Int J Clin Exp Pathol. 10(12):11516–11525

Wolinsky JB, Colson YL, Grinstaff MW (2012) Local drug delivery strategies for cancer treatment: gels, nanoparticles, polymeric films, rods, and wafers. J Control Release 159:14–26

Wong SR (2024) miR-21 expression and its correlation with demographics, subtypes, and tumour suppressor genes; PTEN and PDCD4 in breast cancer tissues in Malaysia. Eurasian J Med Oncol:59–73. https://doi.org/10.14744/ejmo.2024.24156

Xing F et al (2015) miR-509 suppresses brain metastasis of breast cancer cells by modulating RhoC and TNF-α. Oncogene 34:4890–4900

Yan J et al (2022) miR-107 inhibits the proliferation of gastric cancer cells in vivo and in vitro by targeting TRIAP1. Front Genet 13:855355

Yang J (2025) Advances and potentials in platelet-circulating tumor cell crosstalk. Am J Cancer Res 15:407–425

Yang A et al (2020a) AGO-bound mature miRNAs are oligouridylated by TUTs and subsequently degraded by DIS3L2. Nat Commun 11:2765

Yang E et al (2020b) Exosome-mediated metabolic reprogramming: the emerging role in tumor microenvironment remodeling and its influence on cancer progression. Signal Transduct Target Ther 5:242

Yang X et al (2021) The key role of exosomes on the pre-metastatic niche formation in tumors. Front Mol Biosci 8:703640

Yang A et al (2022a) TENT2, TUT4, and TUT7 selectively regulate miRNA sequence and abundance. Nat Commun 13:5260

Yang G et al (2022b) Circulating miR-141 as a potential biomarker for diagnosis, prognosis and therapeutic targets in gallbladder cancer. Sci Rep 12:10072

Ye D, Shen Z, Zhou S (2019) Function of microRNA-145 and mechanisms underlying its role in malignant tumor diagnosis and treatment. Cancer Manag Res 11:969–979

Yin S, Yu Y, Reed R (2015) Primary microRNA processing is functionally coupled to RNAP II transcription in vitro. Sci Rep 5:11992

Zabeti Touchaei A, Vahidi S (2024) Unraveling the interplay of CD8 + T cells and microRNA signaling in cancer: implications for immune dysfunction and therapeutic approaches. J Transl Med 22:1131

Zargar S et al (2019) A feedback loop between MicroRNA 155 (miR-155), programmed cell death 4, and activation protein 1 modulates the expression of miR-155 and tumorigenesis in tongue cancer. Mol Cell Biol 39:e00410-18

Zhang Y, Huang B, Wang H-Y, Chang A, Zheng XFS (2017) Emerging role of MicroRNAs in mTOR signaling. Cell Mol Life Sci 74:2613–2625

Zhang Y, Jin X, Wang J (2019) miR-148a modulates the viability, migration and invasion of oral squamous cell carcinoma cells by regulating HLA-G expression. Mol Med Rep. https://doi.org/10.3892/mmr.2019.10280

Zhao L et al (2016) The tumor suppressor miR-138-5p targets PD-L1 in colorectal cancer. Oncotarget 7:45370–45384

Zhao Y, Xing C, Deng Y, Ye C, Peng H (2024) HIF-1α signaling: essential roles in tumorigenesis and implications in targeted therapies. Genes Dis 11:234–251

# Significance of Altered miRNA Expression in Tumor and Therapeutic Scopes

3

Alisha Lalhall, Sagar Rana, Jogendra Singh Pawar, and Smita Kumari

**Abstract**

MicroRNAs (miRNAs) are small, noncoding RNAs that regulate gene expression and play crucial roles in cancer biology. Altered expression of miRNAseither through overexpression of oncogenic miRNAs or loss of tumor-suppressive miRNAsdisrupts signaling pathways controlling cell growth, apoptosis, angiogenesis, and metastasis. These deregulated profiles are tumor- and stage-specific, making miRNAs valuable candidates as biomarkers for early diagnosis, prognosis, and treatment monitoring. Therapeutically, strategies such as miRNA mimics to restore suppressed activity and antagomiRs to inhibit overactive miRNAs are being developed. Circulating miRNAs in body fluids further support noninvasive liquid biopsy approaches. Although challenges remain in delivery and specificity, advances in nanotechnology and molecular targeting are enhancing their clinical applicability. This chapter emphasizes the significance of altered miRNA

Authors Alisha Lalhall and Sagar Rana have equally contributed to this chapter.

A. Lalhall
Centre for Nanoscience and Nanotechnology (CNSNT), Panjab University, Chandigarh, India

S. Rana
Academy of Scientific and Innovative Research (AcSIR), Ghaziabad, India

CSIR-Central Scientific Instruments Organisation, Chandigarh, India

J. S. Pawar
College of Pharmacy, The Ohio State University, Columbus, OH, USA

The James Cancer Hospital and Solove Research Institute, Comprehensive Cancer Center, The Ohio State University, Columbus, OH, USA

S. Kumari (✉)
College of Pharmacy, The Ohio State University, Columbus, OH, USA
e-mail: kumari.45@osu.edu





S. Tiwari et al. (eds.), *microRNA Modulation: A Promising Combination Therapy for Cancer*, https://doi.org/10.1007/978-981-95-5989-3_3

expression in tumor progression and highlights their translational potential as biomarkers and therapeutic tools in precision oncology.

**Learning Objectives**

1. Explain the molecular mechanisms underlying altered miRNA expression in tumors.
2. Differentiate between oncogenic and tumor-suppressive miRNAs and their role in cancer progression.
3. Evaluate the potential of deregulated miRNAs as diagnostic and prognostic biomarkers in oncology.
4. Discuss therapeutic strategies targeting miRNAs, including current approaches, challenges, and future prospects.

## 3.1 Introduction to miRNAs in Cancer

The involvement of microRNAs (miRNAs) in cancer was first highlighted in studies of chronic lymphocytic leukemia, marking a pivotal discovery in cancer biology (Visone and Croce 2009). Since then, extensive research has established a strong link between abnormal miRNA expression and the fundamental characteristics of cancer. Multiple cellular processes, such as genetic alterations, abnormal DNA methylation, and changes in histone acetylation, have been implicated in the disruption of miRNA regulation in cancer cells. Furthermore, key cancer-associated transcription factors, including MYC and p53, are known to modulate miRNA expression. Regulatory mechanisms such as alternative splicing, polyadenylation, and genetic alterations in proteins involved in the miRNA biogenesis pathway can disrupt normal miRNA maturation. Disruptions in these processes can lead to abnormal expression levels of certain miRNAs, which have been linked to the onset, progression, spread, and treatment resistance of various cancers. The biological effects of miRNAs depend on their molecular targets, some suppress tumor growth, while others promote oncogenic activity, and their functional roles are frequently determined by the specific tissue (O'Brien et al. 2018). For instance, the let-7 family of miRNAs acts as tumor suppressors and is frequently downregulated in cancers such as those of the head and neck, lung, breast, ovary, and prostate. Let-7 miRNAs inhibit several oncogenes, thereby contributing to anticancer pathways. Similarly, miRNAs like the miR-34 family, miR-223, miR-143/145 cluster, and miR-204 are often found at reduced levels in different cancers, and restoring their expression has been shown to suppress tumor growth in experimental models. Interestingly, some miRNAs can exhibit dual roles depending on the cancer type. For example, miR-214 acts as an oncogene in osteosarcoma and nasopharyngeal cancer but serves as a tumor suppressor in glioma and colorectal cancer. Likewise, miR-125b promotes oncogenic behavior in colon and hematopoietic cancers but inhibits tumor growth in breast and liver cancers. Other miRNAs, including the miR-17/92 cluster, miR-21, miR-155, miR-221, miR-222, and miR-9, are commonly upregulated in various malignancies. Elevated levels of these miRNAs can activate oncogenic signaling pathways and contribute to tumorigenesis. For example, increased miR-17/92 levels enhance PI3K and NF-κB signaling in lymphomas, while miR-21 overexpression is linked to the development of pre-B cell lymphomas. Targeting miR-155, for instance, can restore tumor-suppressor gene expression and inhibit tumor growth in breast cancer. A comprehensive understanding of miRNAs in cancer requires insight into their molecular origins and the intricate processes underlying their synthesis and maturation. In the following, the molecular basis and biogenesis of microRNAs are explored, highlighting their regulatory roles in cellular homeostasis and tumorigenesis (Naidu et al. 2015).

### 3.1.1 Molecular Basis and Biogenesis of MicroRNAs (miRNAs)

The biogenesis of miRNAs begins with the transcription of miRNA genes, which can occur either post- or co-transcriptionally. miRNA genes are distributed throughout the genome, with approximately half being intragenic embedded within the introns (and less frequently, exons) of protein-coding genes while the remainder are intergenic and transcribed independently under the control of their own promoters (Fig. 3.1). In some cases, multiple miRNAs are transcribed together as long polycistronic clusters; these often share similar seed regions and are classified as miRNA families due to their related regulatory functions.

#### 3.1.1.1 Canonical Pathway of miRNA Biogenesis

The canonical pathway is the primary route for miRNA maturation. In this pathway, transcription is usually mediated by RNA polymerase II, resulting in the production of long primary miRNA transcripts (pri-miRNAs) that are 5′ capped and 3′ polyadenylated. Some miRNAs, however, can also be transcribed by RNA polymerase III. Within the nucleus, primary miRNA transcripts (pri-miRNAs) undergo processing by the microprocessor complex, which consists of the RNase III enzyme Drosha and the RNA-binding protein DGCR8; this complex recognizes and cleaves the pri-miRNAs. DGCR8 binds specific sequence motifs, such as N6-methyladenylated GGAC, and Drosha cleaves the stem-loop structure at the base, generating a precursor miRNA (pre-miRNA) with a characteristic 2-nucleotide 3′ overhang (Denli et al. 2004). The pre-miRNA is then exported from the nucleus to the cytoplasm by the Exportin-5/Ran-GTP complex. In the cytoplasm, the RNase III enzyme Dicer removes the terminal loop of the pre-miRNA, producing a ~ 20–22 nucleotide miRNA duplex. The two strands of this duplex are referred to as the 5p and 3p strands, originating from the respective ends of the pre-miRNA hairpin. Both strands can potentially be loaded into Argonaute (AGO) proteins (AGO1–4 in humans) in an ATP-dependent manner, but the proportion of each strand incorporated into AGO varies by cell type and context. The strand selection is influenced by the thermodynamic stability at the 5′ ends of the duplex and the presence of a 5′ uracil; generally, the strand with lower 5′ stability or a 5′ U is preferentially loaded as the guide strand, while the other (passenger strand) is

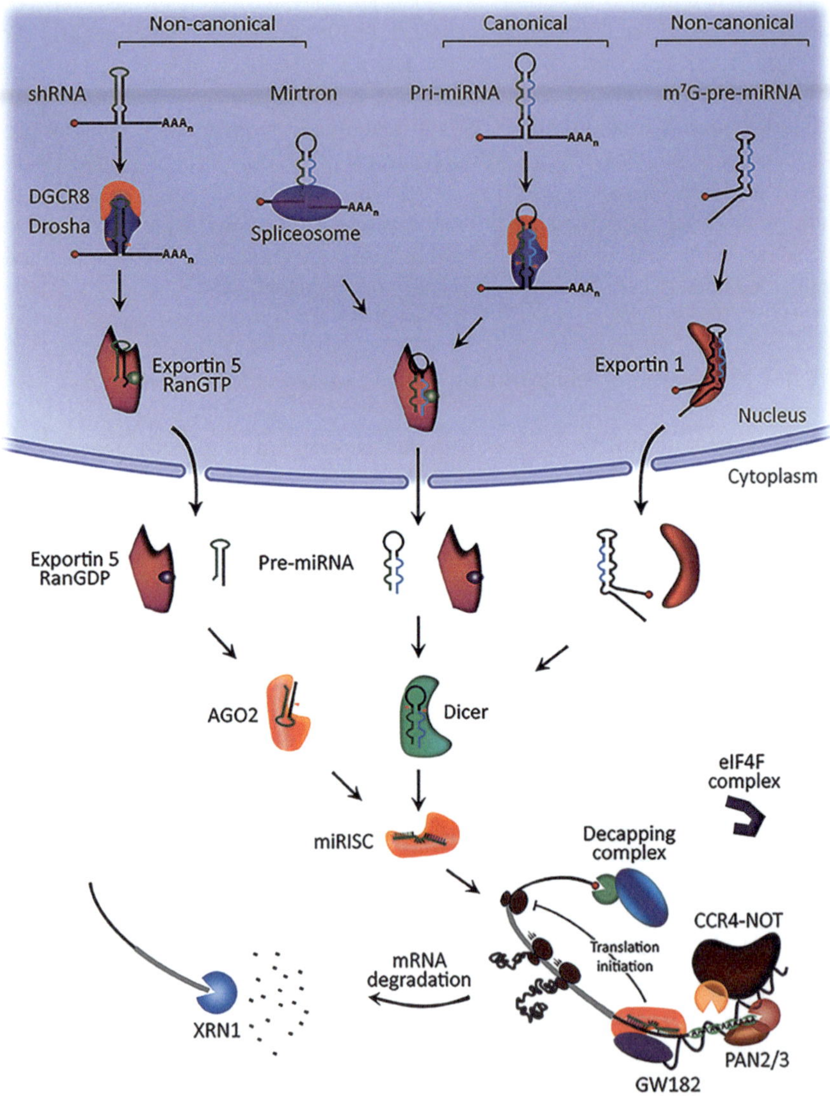


**Fig. 3.1** miRNA biogenesis and mechanism of action (O'Brien et al. 2018)

unwound and degraded. If the duplex contains perfect complementarity, the passenger strand may be cleaved by AGO2, resulting in a strong strand bias; otherwise, it is passively unwound and degraded. After being loaded into the RNA-induced silencing complex (RISC), the mature miRNA directs the complex to specific mRNAs by base pairing with complementary sites, most commonly located in the 3′ untranslated regions (3′ UTRs) of target transcripts. The seed region, a 6–8 nucleotide segment at the 5′ end of the miRNA, is usually critical for target recognition. While canonical seed-based interactions are most common, noncanonical binding modes have also been identified, expanding the regulatory potential of miRNAs.

#### 3.1.1.2 Noncanonical Pathways of miRNA Biogenesis

Beyond the canonical pathway, several noncanonical routes for miRNA processing have been described. These pathways often utilize alternative combinations of the core miRNA machinery (Drosha, Dicer, Exportin-5, and AGO2) and can be classified as either Drosha/DGCR8-independent or Dicer-independent (O'Brien et al. 2018).

(i) *Drosha-/DGCR8-Independent Pathways:*
Mirtrons are a notable example, originating from the splicing of introns within protein-coding genes. After splicing and debranching, these intronic sequences fold into hairpin structures that resemble pre-miRNAs and are exported to the cytoplasm, bypassing Drosha processing (Ruby et al. 2007). Another example involves pre-miRNAs capped with 7-methylguanosine (m7G), which can bypass Drosha-mediated cleavage and are exported from the nucleus via Exportin-1. These pre-miRNAs commonly show a preference for 3p strand incorporation into Argonaute proteins, likely due to the m7G cap hindering 5p strand loading.

(ii) *Dicer-Independent Pathways:*
Some miRNAs, such as those derived from endogenous short hairpin RNA (shRNA) transcripts, are processed by Drosha but are too short to serve as Dicer substrates. These pre-miRNAs are directly loaded into AGO2, which cleaves the 3p strand, and subsequent exonucleolytic trimming of the 5p strand completes maturation.

### 3.1.2 Physiological Roles of miRNAs in Cellular Homeostasis

miRNAs are essential regulators of cellular homeostasis, exerting control over gene expression networks that govern cell cycle progression, apoptosis, and differentiation. By fine-tuning the expression of key regulatory genes, miRNAs contribute to tissue-specific gene expression patterns and orchestrate developmental processes. For example, distinct miRNA expression profiles are critical during embryogenesis, organ development, and maintenance of tissue identity. Abnormalities in miRNA regulation are strongly associated with the development and advancement of cancer, as changes in miRNA expression can drive tumor formation and promote malignancy (García-López et al. 2013).

### 3.1.3 Regulation and Additional Functions of miRNAs

The biogenesis and function of miRNAs are tightly regulated at multiple levels, including transcription, processing, export, strand selection, RISC loading, and decay. Regulatory proteins such as DEAD-box RNA helicases and SMADs modulate Drosha activity, while the KH-type splicing regulatory protein (KSRP) participates in both Drosha and Dicer complexes, influencing the maturation of specific miRNAs. Methyltransferase-like 3 (METTL3) methylates pri-miRNAs, enhancing their recognition by DGCR8 and promoting efficient processing. Beyond their classical gene-silencing role, miRNAs can also function as signaling molecules. For instance, tumor-derived miR-21 and miR-29a can bind to Toll-like receptors on immune cells, triggering pro-inflammatory responses that may promote tumor progression and metastasis. Some miRNAs also interact with pathways such as nuclear factor κB (NF–κB) in natural killer cells, further expanding their functional repertoire (Finnegan and Pasquinelli 2013).

## 3.2 Molecular Mechanisms Underlying Aberrant miRNA Expression in Neoplasia

Altered miRNA expression is commonly observed in human cancers and plays a vital role in cancer development, advancement, spread, and resistance to treatment. Depending on their functional roles and expression patterns, miRNAs can be broadly classified into oncomiRs, which promote tumorigenesis by inhibiting tumor-suppressor genes, and tumor-suppressive miRNAs, which inhibit oncogenic pathways. Perturbations in the expression of either class can result from multiple molecular mechanisms, including epigenetic alterations, genetic abnormalities, transcription factor dysregulation, and defects in miRNA biogenesis (Visone and Croce 2009).

### 3.2.1 Epigenetic Modifications

Epigenetic changes significantly influence miRNA dysregulation by modifying chromatin organization and altering the accessibility of miRNA gene promoter regions. DNA methylation involving excessive methylation of CpG islands in the promoter regions of tumor-suppressor miRNAs may cause these genes to become transcriptionally inactive. Conversely, hypomethylation may activate the expression of oncogenic miRNAs (Weber et al. 2007).

(i) *Histone Modifications*: Histone acetylation typically enhances miRNA transcription, whereas histone deacetylation or methylation can repress miRNA gene expression by inducing a closed chromatin conformation.
(ii) *Epigenetic Silencing*: Tumor-suppressor miRNAs are frequently silenced in cancer via combined DNA and histone modifications, contributing to uncontrolled cell growth, formation of new blood vessels, epithelial–mesenchymal transition (EMT), and the spread of cancer cells to distant sites.

### 3.2.2 Genetic Alterations

Genetic changes at miRNA loci or their regulatory elements can disrupt miRNA biogenesis or alter their functional targets.

(i) *Mutations*: Point mutations within the miRNA sequence or regulatory regions (e.g., promoters or seed regions) can disrupt maturation, expression, or target binding.
(ii) *Copy Number Variations (CNVs):* Genomic amplification of regions containing oncomiRs leads to increased expression and tumor-promoting activity, while deletions involving tumor-suppressor miRNAs reduce their regulatory capacity.
(iii) *Chromosomal Rearrangements*: Structural alterations such as translocations can reposition miRNA genes under abnormal promoters or disrupt their sequences, resulting in inappropriate levels of expression.

### 3.2.3 Dysregulated Transcription Factor Activity

Aberrant activity of key transcription factors in cancer (e.g., MYC, TP53, STAT3, NF–κB, ZEB1, and RREB1) profoundly influences miRNA expression patterns:

- c-MYC upregulates oncogenic miRNAs (e.g., miR-155, miR-17–92 cluster) while repressing tumor suppressors like let-7, miR-23b, and miR-29a (Chang et al. 2008).
- TP53 directly activates tumor-suppressor miRNAs, including miR-34a, miR-145, and miR-605, thereby promoting apoptosis and cell cycle arrest.
- STAT3 and NF–κB are involved in upregulating miR-21, a well-established oncomiR, by binding to its promoter region (Iliopoulos et al. 2010).
- ZEB1 represses miR-200c and miR-141, promoting EMT and metastatic potential.

### 3.2.4 Impaired miRNA Biogenesis

Defects in the core components of the miRNA biogenesis machinery: Drosha, DGCR8, Dicer, and TRBP can lead to global downregulation of mature miRNAs, thereby contributing to oncogenesis. Cancer-associated mutations or dysregulated expression of these molecules disrupt miRNA maturation and reduce the abundance of functional miRNA molecules.

### 3.2.5 Deregulation of MetastamiRs

Specific miRNAs involved in regulating metastasis, termed metastamiRs, control key steps in the metastatic cascade, including EMT, stemness, adhesion, migration, and invasion. Oncogenic metastamiRs such as miR-10b, miR-21, and miR-31 promote invasion and dissemination, while tumor-suppressive metastamiRs like miR-200c, miR-34a, and miR-205 inhibit metastatic traits (Iorio and Croce 2012).

### 3.2.6 Impact of miRNA Dysregulation on Therapeutic Resistance

Dysregulated miRNAs contribute significantly to therapeutic resistance in cancer by modulating key genes involved in cell cycle control (e.g., CDKs), apoptosis (e.g., PTEN, BCL-2), DNA repair, and drug metabolism (e.g., cytochrome P450 enzymes). The overexpression of miR-21, miR-27a, and miR-451 enhances the expression of the drug-efflux transporter P-glycoprotein, thereby promoting chemoresistance. Additionally, miRNAs such as miR-122, miR-132, and miR-320 regulate the metabolism and clearance of anticancer agents, influencing drug efficacy and toxicity (Ali Syeda et al. 2020).

(i) *Alterations in miRNA Processing Enzymes.*
Genetic alterations or reduced levels of essential elements involved in the miRNA biogenesis process, including Drosha, Dicer, DGCR8, and Exportin-5, can impair the maturation of primary miRNAs (pri-miRNAs) into their functional, mature forms. These alterations result in broad miRNA expression changes, often favoring oncogenic pathways and resistance phenotypes.

(ii) *Disruption of Regulatory Proteins.*
Dysfunction in regulatory proteins involved in miRNA processing such as SMADs, KSRP, and METTL3 further contributes to abnormal miRNA expression. These proteins are crucial for pri-miRNA recognition, methylation, and recruitment of processing complexes, and their disruption exacerbates global miRNA misregulation in cancer.

### 3.2.7 Global Alterations in miRNA Expression Profiles in Malignant Tissues

Global changes in miRNA expression represent a defining feature of numerous malignancies. Most tumors exhibit overall downregulation of miRNAs compared to adjacent normal tissues. This pervasive suppression stems from a combination of genetic (e.g., CNVs, mutations, and chromosomal rearrangements) and epigenetic changes (e.g., promoter methylation, histone modification), as well as biogenesis defects. Collectively, these alterations disrupt the balance between tumor-suppressive and oncogenic miRNAs, leading to upregulation of oncogenes and inhibition of tumor-suppressor pathways. These expression patterns not only drive

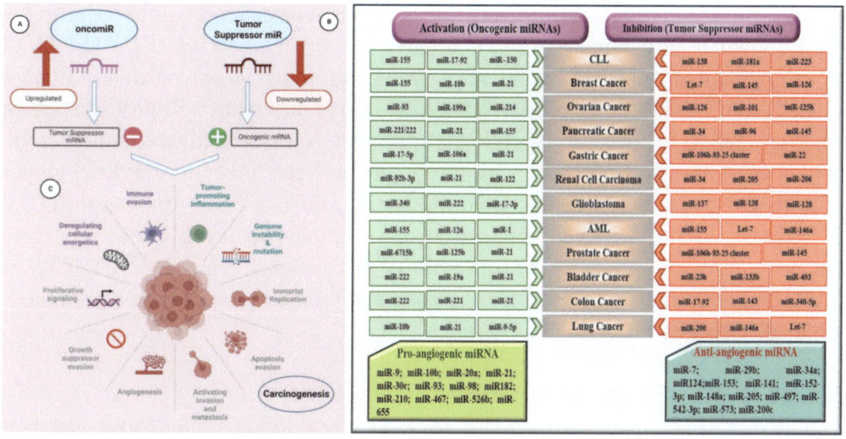


**Fig. 3.2** Classification of miRNAs in cancer: (**a**) oncomiRs facilitate tumor development, (**b**) tumor-suppressor miRNAs inhibit cancer progression, and (**c**) both contribute to the hallmarks of carcinogenesis (Chakrabortty et al. 2023a; Menon et al. 2022)

tumorigenesis and metastasis but also serve as potential biomarkers for diagnosis, prognosis, and therapy response (Fig. 3.2).

### 3.2.7.1 OncomiRs

OncomiRs are a subset of miRNAs that are often overexpressed in different types of cancer and play a role in tumor development by suppressing tumor-suppressor genes. Their overexpression facilitates uncontrolled cell proliferation, promotes metastasis, and supports other hallmarks of cancer (Menon et al. 2022). One well-characterized group is the miR-17-92 cluster, which includes miR-17, miR-18a, miR-19a, miR-20a, miR-19b, and miR-92a. This cluster exerts oncogenic effects by targeting and suppressing several key tumor-suppressors and regulatory pathways, such as PTEN (phosphatase and tensin homolog), E2F transcription factors, the TGF-β signaling pathway, BCL2L11 (BIM), and thrombospondin-1 (TSP-1). Elevated levels of the miR-17-92 group have been observed in various malignancies such as pulmonary small-cell carcinoma, hepatoma, thyroid neoplasm, colorectal adenomas, and renal carcinoma, and are linked to accelerated tumor development and disease progression. miR-21 is another prominent oncomiR, widely reported to be elevated in a range of malignancies including breast, ovarian, colon, lung, prostate, pancreatic, and gastric cancers. miR-21 promotes oncogenesis by downregulating tumor-suppressor genes such as PTEN, tropomyosin 1 (TPM1), and programmed cell death 4 (PDCD4). Elevated miR-21 levels are detectable in both tumor tissues and the blood serum or plasma of cancer patients, suggesting its potential use as a biomarker for both diagnostic and prognostic applications. Experimental downregulation of miR-21 has been shown to inhibit tumor cell proliferation and reverse drug resistance in cancers such as pancreatic, ovarian, and breast cancer (Si et al. 2007). miR-181(Gupta et al. 2017) is also recognized for its oncogenic activity and is upregulated in several cancer types. For example, miR-181a-5p has been shown to drive breast cancer progression by modulating the PTEN/AKT signaling pathway through NDRG2, and by inhibiting targets such as SPRY4, PHLPP2, and INPP4B. In prostate cancer, miR-181 promotes cell proliferation by targeting DAX-1. Elevated miR-181 expression is associated with unfavorable outcomes in oral squamous cell carcinoma and has been linked to the development of drug resistance in melanoma. Other oncomiRs, such as miR-146a and miR-221/222, are also frequently overexpressed in various malignancies. Elevated miR-146a has been observed in the plasma of breast cancer patients, while miR-221/222 are associated with liver, breast, colon, and pancreatic cancers, where they contribute to tumorigenesis and progression.

#### 3.2.7.2 Tumor-Suppressor miRNAs (TS-miRNAs)

Tumor-suppressor microRNAs (TS-miRNAs) are a subset of miRNAs that help prevent cancer initiation and advancement by suppressing oncogenes as well as genes involved in cell proliferation and survival (Inoue and Inazawa 2021). The loss or reduced expression of these miRNAs is a critical event in tumorigenesis, as it can lead to the unchecked activation of oncogenic pathways. TS-miRNAs are frequently situated within genomic loci that exhibit a high susceptibility to instability, such as fragile sites or cancer-associated loci, making them more susceptible to genetic alterations and mutations. Their downregulation can result from several factors, such as mutations, epigenetic silencing, or defects in the miRNA processing machinery. Disruption of key components involved in miRNA biogenesis, such as Drosha, DGCR8, and Dicer, has been shown to globally decrease miRNA levels, thereby promoting cellular transformation and tumorigenesis. A well-studied example is miR-16, whose diminished expression is linked to the progression of chronic lymphocytic leukemia, as well as gastric, prostate, and pancreatic cancers. The let-7 family of miRNAs also acts as potent tumor suppressors by targeting oncogenes such as RAS and MYC. The restoration of let-7 expression in lung cancer cells leads to increased apoptosis and reduced tumor growth. The let-7 miRNAs have also been shown to suppress other oncogenes, including HMGA2 and MYCN, and inhibit estrogen receptor alpha (ERα)-mediated signaling in breast cancer, thereby impeding malignant transformation (Zhou et al. 2014). Other notable TS-miRNAs include miR-29 and miR-34. Lower expression of these miRNAs is linked to increased tumor growth, invasiveness, and metastatic potential across multiple cancers, including breast, lung, neuroblastoma, glioblastoma, colon, stomach, ovarian, bladder, cervical, cholangiocarcinoma, melanoma, and prostate cancers. Reduced levels of miR-29 have been linked with heightened cellular growth and chemoresistance, such as cisplatin resistance in ovarian cancer and aggressive behavior in osteosarcoma (Liu et al. 2019). Similarly, loss of miR-34 expression is associated with enhanced cellular growth and poor prognosis in pancreatic, lung, colorectal, gastric, and ovarian cancers (Gao et al. 2024). miR-362-3p has been shown to suppress tumor growth and disrupt cell cycle progression, contributing to improved outcomes in colorectal cancer patients. miR-193b acts as a tumor suppressor in both breast and pancreatic cancers; its upregulation leads to decreased fatty

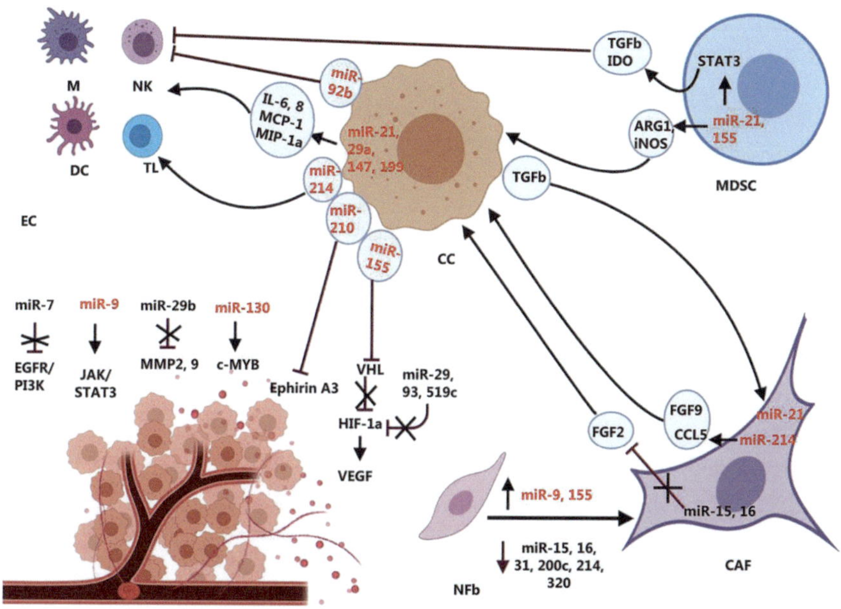


**Fig. 3.3** Role of dysregulated miRNAs in tumor microenvironment crosstalk. OncomiRs (red) and tumor-suppressor miRNAs (black) influence interactions between tumor cells and surrounding stromal and immune cells (Arghiani and Shah 2021)

acid synthase (FASN) expression, sensitizing triple-negative breast cancer cells to metformin, and its loss is associated with increased proliferation, migration, and invasiveness in pancreatic cancer cells. Genetic polymorphisms in miRNA genes can also influence cancer susceptibility. For example, single-nucleotide polymorphisms (SNPs) such as rs3746444 in miR-499 and rs4919510 in miR-608 have been linked to an increased risk of lung and prostate cancers (Hu et al. 2014). Additionally, long noncoding RNAs, such as XIST, can modulate the expression of TS-miRNAs, further influencing tumor progression, as seen in ovarian cancer through the regulation of the miR-355/BCL2L2 axis. Overall, these results highlight the crucial function of TS-miRNAs in maintaining cellular homeostasis and preventing malignant transformation. Their downregulation not only facilitates tumor progression but also represents a potential target for therapeutic intervention (Fig. 3.3).

## 3.3 Functional Consequences of Altered miRNA Expression

Alterations in miRNA expression patterns are characteristic of numerous human cancers and contribute directly to the onset, growth, and spread of tumors (Croce 2009). Acting as posttranscriptional regulators, miRNAs help maintain cellular

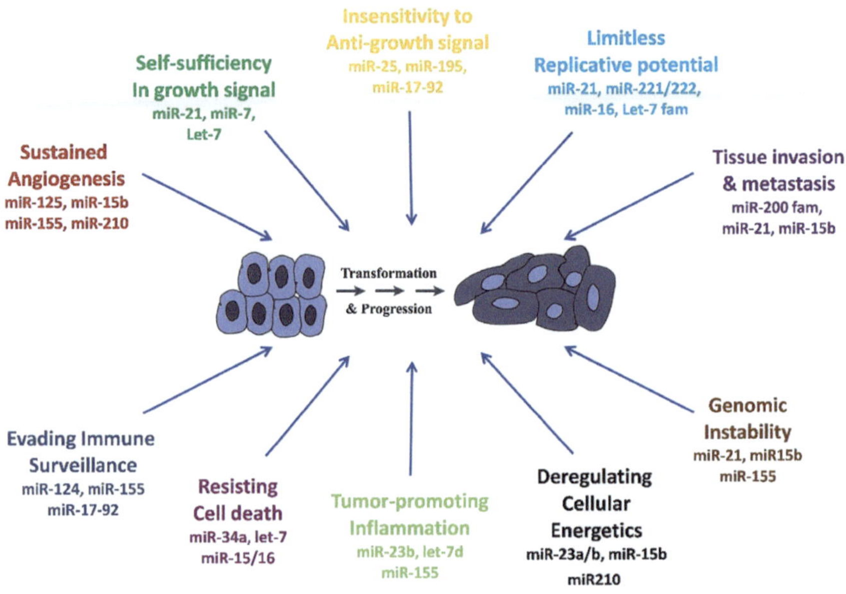


**Fig. 3.4** Altered expression of microRNAs regulates key cancer hallmarks, driving tumor transformation and progression (Shah et al. 2016)

equilibrium; when their levels become unbalanced, it can result in the activation of cancer-promoting pathways or the impairment of tumor suppression (Fig. 3.4).

### 3.3.1 Oncogenic Functions of Dysregulated miRNAs

OncomiRs, or oncogenic miRNAs, are commonly found at elevated levels in various cancers and are critically involved in tumorigenesis through the suppression of tumor-suppressor gene expression. This repression facilitates neoplastic transformation, enhances cellular proliferation, and promotes survival. A prominent example is the miR-17-92 cluster, recognized as the first miRNA cluster, identified as acting as a mammalian oncogene (Hill and Tran 2021). This cluster is located at the 13q31.3 locus, a region often amplified in lymphomas and solid tumors. In experimental models, such as the Eμ-Myc transgenic mouse model of B-cell lymphoma, overexpression of the miR-17-92 cluster accelerates disease onset and progression. The transcription of this cluster is directly driven by the c-Myc oncogene, establishing a regulatory feedback loop that amplifies oncogenic signaling. Additionally, the miR-17-92 cluster and its related cluster miR-106b-25 are tightly linked to the E2F family of transcription factors, targeting E2F1 and being activated by E2F, forming a negative feedback loop that further influences cell cycle regulation. The oncogenic miRNAs can modulate several key cancer hallmarks such as autonomous growth signaling, avoidance of programmed cell death, continuous formation of new blood vessels, and increased invasion and spread of cancer cells. For instance, miR-21 is a well-characterized oncomiR that targets crucial tumor suppressors such as PTEN and PDCD4. Overexpression of miR-21 leads to increased cell proliferation, resistance to apoptosis, and enhanced metastatic potential. Other oncomiRs, like miR-155, have been shown to drive the development of lymphoproliferative diseases and lymphomas in transgenic mouse models, underscoring the role of miRNA deregulation as an early event in oncogenesis. miRNAs further promote tumor development by influencing the production of proteins participating in the regulation of cell cycle advancement and apoptosis. For example, miR-221/222 target the cell cycle inhibitors p27Kip1 and p57Kip2, promoting cell cycle progression. The miR-17-92 and miR-106b-25 clusters further promote proliferation by targeting p21Waf1/Cip1, while also inhibiting pro-apoptotic factors such as Bim. Additionally, miRNAs like miR-21 and miR-10b have been implicated in tumor progression and metastasis. miR-21, for example, regulates genes involved in cell migration and invasion, such as TPM1, PDCD4, and maspin, and its suppression in metastatic cancer cells reduces invasion and metastasis. Increased levels of miR-10b, driven by the transcription factor Twist, enhance metastatic potential by suppressing HOXD10 and elevating RHOC expression, thereby facilitating greater cancer cell movement.

### 3.3.2 Tumor-Suppressive Functions of miRNAs

In contrast, TS-miRNAs inhibit tumor growth and metastasis by targeting and suppressing oncogenes as well as genes involved in cell division, movement, and the formation of new blood vessels (angiogenesis). The reduction or absence of these miRNAs in cancer cells results in unregulated activation of their target oncogenes, promoting tumor development and advancement. For example, the miR-15a/16–1 cluster targets the anti-apoptotic gene BCL2, and their loss is associated with increased cell survival and resistance to apoptosis, contributing to cancer development. These miRNAs also regulate cell cycle progression by targeting cyclins and CDKs, such as cyclin D1, cyclin D3, cyclin E1, and CDK6, and their downregulation results in the accumulation of cell cycle-promoting genes. The let-7 family of miRNAs targets key oncogenes like RAS and MYC, and reduced expression of let-7 is linked to enhanced tumorigenesis and poor prognosis in various cancers. Similarly, miR-122, which is downregulated in hepatocellular carcinoma, targets cyclin G1, and its loss leads to increased tumor incidence. The restoration or upregulation of tumor-suppressive miRNAs in cancer models has been shown to reduce tumorigenicity, inhibit metastasis, and sensitize cancer cells to therapy (Otmani and Lewalle 2021).

## 3.4 miRNAs in Tumor Invasion and Metastatic Cascade

Metastasis refers to cancer cells' capacity to move from the original tumor site to other parts of the body and establish secondary tumors in other locations within the body. This process can involve infiltration into adjacent tissues or transport through the circulatory and lymphatic networks. This phenomenon accounts for most fatalities associated with cancer. miRNAs have emerged as critical regulators of the metastatic cascade, influencing multiple steps involved in tumor cell invasion, migration, and colonization of distant organs. Among the key miRNAs implicated in metastasis, miR-10b has been widely investigated for its involvement in facilitating breast cancer metastasis, especially to the lungs. Elevated miR-10b expression is also observed in high-grade gliomas, where its inhibition suppresses tumor growth, invasion, and angiogenesis without affecting normal neural cells, indicating a degree of tumor specificity (Teplyuk et al. 2016). In pancreatic cancer, elevated levels of miR-320a are linked to enhanced cell proliferation, invasiveness, metastatic potential, and drug resistance, partly through the induction of epithelial–mesenchymal transition (EMT) and downregulation of the tumor-suppressor PDCD4. Interestingly, in breast cancer, miR-320a acts as a metastasis suppressor: its upregulation inhibits invasion and migration by targeting the oncogene metadherin, and xenograft studies confirm its ability to reduce metastatic spread in vivo (Yu et al. 2016). miR-335 functions as a metastasis suppressor in neuroblastoma and other cancers. Loss of miR-335 results in the onset of the transforming growth factor-β (TGF-β) signaling pathway and its downstream effectors, such as ROCK1, MAPK1, and LRG1, resulting in enhanced cell invasion and migration (Lynch et al. 2012). The miR-200 family and miR-506 play key roles in regulating EMT, a process where epithelial cells gain mesenchymal characteristics, aiding in the spread of cancer cells to distant sites. The miR-200 family sustains the epithelial phenotype by suppressing ZEB1 and ZEB2, transcriptional inhibitors of E-cadherin, thereby inhibiting EMT and reducing metastatic potential in various cancers (Park et al. 2008). Similarly, miR-506 suppresses EMT and metastasis by targeting key EMT drivers and is known to inhibit tumor progression in ovarian and breast cancers (Sun et al. 2014). Clinically, altered expression of EMT-regulating miRNAs such as the miR-200 family and miR-506 correlates with poor prognosis, increased metastatic risk, and reduced overall survival in multiple cancer types. Thus, EMT-associated miRNA signatures are being explored as prognostic biomarkers and potential therapeutic targets in oncology (Humphries and Yang 2015).

## 3.5 miRNAs as Diagnostic and Prognostic Biomarkers in Oncology

### 3.5.1 Clinical Utility of Circulating miRNAs in Noninvasive Cancer Diagnostics

A significant hurdle in leveraging miRNAs as diagnostic biomarkers is achieving sensitive and specific detection of target miRNAs in complex biological samples (Sundarbose et al. 2013). Recent technological advances, such as the development of

DNA nanomachines, have addressed these limitations. For example, researchers have developed a DNA nanomachine based on the zeolite imidazole framework-8 (ZIF-8) metal–organic framework (MOF) in combination with fluorescently labeled hairpin probes. This system is designed for the specific detection of miR-21 within cancer cells (Xie et al. 2022). Upon exposure to acidic tumor microenvironments, the MOF disassembles, releasing the probes, which then interact with miR-21 to trigger catalytic hairpin assembly (CHA) and amplify the detection signal. This system enables the visualization of miR-21 at concentrations as low as 27 pM, demonstrating the potential for highly sensitive cancer. Several studies have revealed the presence of miRNAs in different bodily fluids such as plasma, urine, and saliva which underscores their potential as noninvasive biomarkers for cancer screening. For example, elevated levels of miR-129-1-3p and miR-566 in urine have been associated with early-stage colorectal cancer, while increased urinary miR-21, miR-141, and miR-375 have been linked to prostate cancer. Similarly, elevated levels of miR-21, miR-125b, miR-126, miR-141, let-7, miR-205, and miR-375 have been detected in the plasma of individuals with prostate cancer. In colon cancer, exosomal miR-23a and miR-1246 are upregulated in early disease and can distinguish malignant from healthy tissues. Salivary miRNA profiles, such as the overexpression of miR-144, miR-10b, miR-451, and miR-21 in esophageal cancer, and miR-21, miR-23, miR-181a, miR-181b, and miR-196a in pancreatic cancer, further underscore the diagnostic versatility of miRNAs across cancer types.

### 3.5.2 miRNAs as Biomarkers of Early Cancer Development

While a standardized miRNA panel for early cancer detection has yet to be developed, increasing research highlights the usefulness of miRNA expression analysis in detecting initial cancer-related alterations. Alterations in miRNA patterns serve as molecular signatures, enabling the detection of both primary and metastatic tumors and providing insights into the tissue of origin. These signatures often outperform mRNA-based classifications, particularly for poorly differentiated or advanced cancers (Lan et al. 2015). For instance, a panel of approximately 200 miRNAs has demonstrated high accuracy in classifying tumors by tissue of origin, surpassing traditional mRNA profiling. Large-scale clinical validation studies have reinforced the diagnostic power of miRNA panels. For example, a three-miRNA panel (miR-15b, miR-21, and miR-31) has achieved high sensitivity and specificity for colorectal cancer detection in independent cohorts. Similarly, a five-miRNA panel has been shown to reliably distinguish brain malignancies from non-cancer controls (Peng et al. 2020; Sur et al. 2022). Importantly, combining miRNA signatures with established protein biomarkers, such as PSA for prostate cancer or CA19–9 for pancreatic cancer, can further improve diagnostic and prognostic accuracy. In addition to their diagnostic value, miRNA profiles support pathological classification and offer prognostic information regarding cancer development and progression. These analyses can be performed on tissue samples, including those preserved as formalin-fixed paraffin-embedded (FFPE) specimens, and are increasingly accessible through minimally invasive techniques such as liquid biopsies. Liquid biopsies using peripheral blood, saliva, urine, or tears allow for the detection of circulating miRNAs, many of which are encapsulated in exosomes or other extracellular vesicles and reflect changes occurring within the tumor microenvironment. Despite these advances, several challenges remain before miRNA-based diagnostics can be widely adopted in clinical practice. There is a gap in correlating tissue-derived miRNA profiles with those found in circulating exosomes, which limits the immediate translation of miRNA-based diagnostics into clinical practice. The interpretation of miRNA data also relies on robust bioinformatics tools, further underscoring the need for standardized methodologies in sample collection, miRNA quantification, normalization, and data analysis. Considering the wide range of genetic variations and cancer subtypes, it is improbable that one miRNA panel will be effective across all cases. Continued research is essential to tailor miRNA biomarkers to specific populations and cancer types, as well as to different stages of disease. As miRNAs are capable of crossing the blood–brain barrier, their measurement may also extend to the detection of central nervous system disorders, providing molecular insights that complement conventional imaging techniques.

### 3.5.3 Prognostic Relevance of miRNA Expression in Cancer Outcomes

The prognostic value of miRNA expression profiles is increasingly recognized in oncology. Altered levels of specific miRNAs can provide insights into disease progression, recurrence, and patient survival. For example, in lung cancer, low levels of the let-7 family and increased expression of miR-155 are linked to worse patient outcomes, whereas higher amounts of miR-221 and let-7a are associated with improved prognosis (Shen and Jiang 2012; Pop-Bica et al. 2020). Similarly, increased miR-21 in colon cancer correlates with poor therapeutic response and unfavorable outcomes. In ovarian cancer, high levels of circulating cell-free miRNAs, such as miR-92a, miR-200c, miR-320b, miR-320c, miR-335, miR-375, and miR-486, are linked to adverse clinical features, supporting their role as independent prognostic markers.

### 3.5.4 miRNAs in Tumor Staging and Disease Monitoring

miRNA expression can also aid in differentiating between localized and advanced disease stages. For instance, overexpression of miR-141, miR-200, and miR-375 has been linked to more aggressive forms of prostate cancer. In metastatic colon cancer, increased plasma concentrations of miR-122 and members of the miR-200 family are commonly observed (Chakrabortty et al. 2023b; Sapre and Selth 2013). Moreover, exosomal miRNAs such as let-7 g, miR-15b, miR-155, and miR-328 have been associated with predicting both cancer relapse and the emergence of liver metastases.

Increased expression of miR-155, miR-200c, and miR-210 are also connected to local recurrence, metastatic spread, and unfavorable outcomes in advanced cancer cases.

### 3.5.5 Predicting and Monitoring Treatment Resistance with miRNAs

miRNAs also serve as important indicators for predicting and tracking treatment resistance (Angerilli et al. 2021). The elevated levels of miR-34a, miR-205, and miR-31 are associated with enhanced responsiveness to taxane chemotherapy, while increased miR-106b is linked to resistance against radiotherapy. In prostate cancer, variations in the expression of miR-125b-2, miR-708, miR-375, miR-141, miR-143, miR-362-5p, miR-214, let-7i, and miR-545 can differentiate between castration-sensitive and castration-resistant forms of the disease (Zedan et al. 2020). The downregulation of miR-452 is associated with doxorubicin resistance, while its upregulation can restore drug sensitivity by targeting IGF-1R. In breast cancer, loss of miR-200c increases sensitivity to doxorubicin by reducing MDR1 gene expression, and miR-542-3p downregulation may contribute to trastuzumab resistance. High miR-205 levels predict better response to docetaxel, doxorubicin, and cyclophosphamide (TAC) therapy by suppressing VEGFA and FGF2. miR-21 is implicated in gemcitabine resistance via PTEN downregulation and EMT induction (Chakrabortty et al. 2023a; Chauhan et al. 2020). In triple-negative breast cancer, increased miR-181a and miR-182 promote metastasis and chemoresistance, while high miR-621 levels enhance chemosensitivity by promoting apoptosis through FBXO11 suppression and p53 activation. Overexpression of miR-340 inhibits oncogenic transcription factors and cell cycle regulators, offering potential therapeutic benefit (Fig. 3.5).

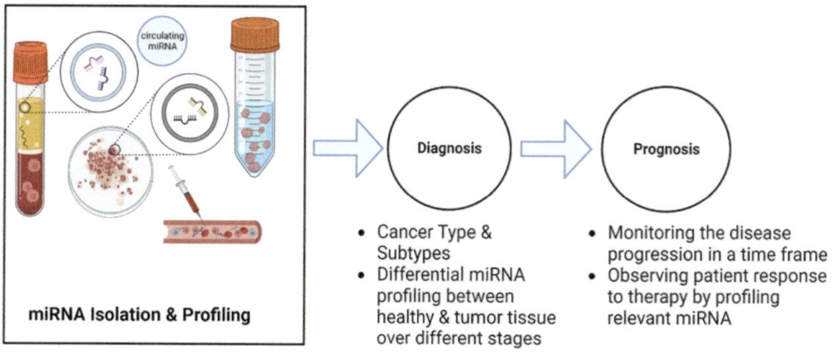


**Fig. 3.5** Clinical applications of microRNAs (miRNAs) as cancer biomarkers. miRNAs can be isolated from various biological samples, including tumor tissues, blood, plasma, serum, and saliva. Profiling of miRNA expression enables cancer diagnosis, staging, and prognosis (Chakrabortty et al. 2023a)

## 3.6 Therapeutic Modulation of miRNAs in Cancer Treatment

Recent advances in cancer therapeutics have underscored the therapeutic potential of modulating miRNAs to regulate gene expression and disrupt oncogenic signaling networks (Inoue and Inazawa 2021). miRNA-based approaches leverage the ability of these small noncoding RNAs to simultaneously target multiple genes involved in tumor development, progression, metastasis, and therapy resistance. Two primary therapeutic strategies have emerged: inhibition of oncomiRs and restoration of TS-miRNAs.

OncomiRs are miRNAs that are frequently overexpressed in various cancer types and function by repressing tumor-suppressor genes, facilitating cellular transformation, proliferation, migration, angiogenesis, and resistance to therapy (Fig. 3.6). Therapeutically targeting oncomiRs seeks to reverse these processes and can be achieved through several platforms:

(i) *AntimiRs (Antisense Oligonucleotides):*
Chemically modified antisense oligonucleotides (ASOs), known as antimiRs, are designed to bind to specific oncomiRs and inhibit their function by preventing interaction with target mRNAs. Structural modifications such as locked nucleic acids (LNAs), phosphorothioate backbones, and 2′-O-methyl substitutions enhance nuclease resistance, binding affinity, and in vivo stability (Visone and Croce 2009).

(ii) *miRNA Sponges:*
These are synthetic RNA transcript decoys containing tandem repeats of miRNA complementary binding sites. They act as competitive inhibitors, sequestering oncomiRs within the cytoplasm and suppressing their activity across multiple targets (Avendaño-Portugal et al. 2025).

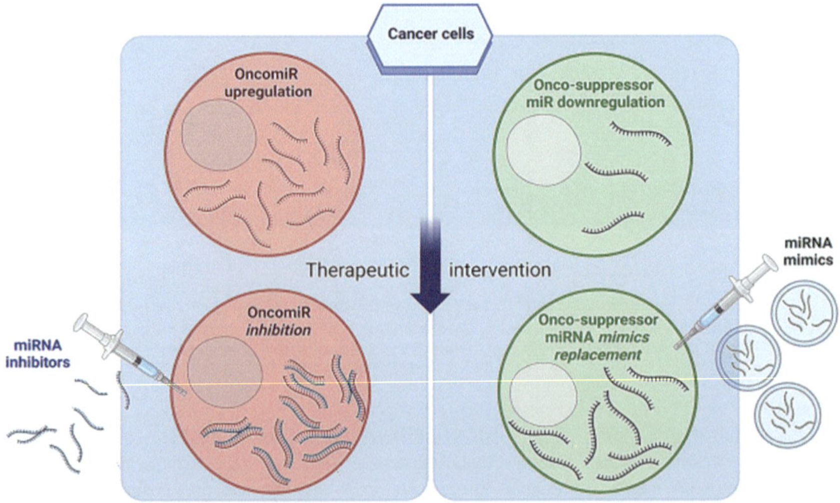


**Fig. 3.6** miRNA as a therapeutic target in cancer (Martino et al. 2025)

(iii) *Genome Editing:*
Genome editing technologies, particularly CRISPR/Cas9, offer a precise approach to directly disrupt, delete, or inactivate oncogenic miRNA (oncomiR) genes at the genomic level, enabling sustained suppression of aberrant miRNA activity. This strategy holds promise for long-term therapeutic benefit; however, concerns regarding off-target effects, immune responses, and efficient in vivo delivery remain significant challenges under active investigation (Rabaan et al. 2023).

### 3.6.1 Restoration of Tumor-Suppressor miRNAs via miRNA Mimics

Unlike oncomiRs, TS-miRNAs are often downregulated or lost in cancer. These miRNAs normally act to inhibit oncogene expression and maintain cellular homeostasis. Their therapeutic reintroduction aims to restore normal regulatory function and suppress malignant behavior (Otmani and Lewalle 2021).

(i) *Synthetic miRNA Mimics*:
These are chemically synthesized, double-stranded RNA molecules designed to replicate the function of natural TS-miRNAs. Upon cellular uptake, they integrate into the RNA-induced silencing complex (RISC) and direct the suppression of cancer-related gene transcripts (Martino et al. 2025). Delivery of miRNA mimics in preclinical cancer models has successfully reduced tumor growth, metastasis, stemness, and resistance to chemotherapy and radiotherapy (Fig. 3.6).

(ii) *Preclinical and Clinical Development:*
Several miRNA mimics have progressed into early-phase clinical trials. A key example is MRX34, a liposomal formulation of miR-34a, which demonstrated preliminary efficacy (e.g., tumor regression in hepatocellular carcinoma and melanoma) in Phase I trials. However, progress was discontinued because of immune-related side effects, underscoring the importance of thoroughly optimizing delivery methods, dosing, and overall safety. Ongoing research is focused on improving the stability, targeting precision, and tolerability of miRNA-based drugs. Combinatorial approaches integrating miRNA mimics with existing chemotherapies or immunotherapies are also under investigation, offering a multifaceted strategy to enhance clinical outcomes.

## 3.7 Challenges and Prospects in miRNA-Based Therapeutics

### 3.7.1 Obstacles to Clinical Translation of miRNA Therapeutics

miRNAs are now firmly established as pivotal regulators in cancer biology, with their dysregulation implicated in virtually all tumor types. Their ability to modulate multiple gene targets and entire signaling networks, coupled with their unique expression profiles and remarkable stability in biological fluids, has positioned them as attractive candidates for diagnostic and therapeutic applications. Therapeutic strategies typically focus on either inhibiting overexpressed oncomiRs or restoring the function of downregulated tumor-suppressor miRNAs (Reda El Sayed et al. 2021a). Numerous preclinical and in vivo studies have demonstrated that miRNA-based modulation used alone or in combination with conventional therapies can yield significant antitumor effects. For example, the targeted delivery of miR-34a along with K-ras siRNA in mouse models of lung cancer led to substantial tumor regression and reduced expression of genes such as Ccnd1, Sirt1, Cdk6, and Ccne2 (Xue et al. 2014). Co-delivery of miR-34 and let-7 has been shown to inhibit major tumor drivers like Lin28b, c-Met, and Myc, significantly inhibiting tumor progression in models of non-small cell lung cancer. Similarly, restoration of miR-200c sensitized lung tumors to radiation by targeting key genes involved in DNA repair and oxidative stress (RAD51, Nrf2, and SESN1). The therapeutic promise of miRNAs extends to gastrointestinal and other solid tumors as well. The administration of miR-520e and miR-375 has demonstrated the ability to suppress liver tumor growth in xenograft models, and the re-expression of miR-217 has limited progression in pancreatic ductal adenocarcinoma. Inhibition of miR-21, especially when combined with chemotherapy (e.g., gemcitabine), has shown encouraging results in pancreatic cancer (Tesfaye et al. 2019). Anti-metastatic effects were observed in colon cancer and glioma following modulation of miR-182 and miR-106b-5p, respectively, while miR-1 and miR-142-3p have demonstrated anti-angiogenic and anti-invasive potential. In breast and ovarian cancer models, knockdown of miR-20b and in vivo delivery of miR-31 have shown tumor-suppressive and chemosensitizing effects. Despite these promising findings, clinical translation has proven challenging. Various key challenges include (Brillante et al. 2024):

- *Establishing specificity and selectivity*: Avoiding off-target interactions, as miRNAs often regulate multiple genes simultaneously.
- *Mitigating immune responses*: Addressing immune-related adverse effects caused by miRNA analogs and some delivery systems.
- *Improving delivery efficiency*: Enabling reliable, tumor-specific delivery across biological barriers such as dense stroma.
- *Optimizing dosing strategies*: Accounting for stability, biodistribution, and discrepancies between animal models and human physiology.

To date, only a limited number of miRNA-based therapies, including MRX34 (a liposomal miR-34a mimic), have entered clinical trials. However, none have progressed beyond Phase II, with multiple trials being discontinued early because of safety issues, especially immune-related adverse effects, underscoring the complexity of using miRNAs that target multiple genes and pathways simultaneously. The delivery of miRNAs remains a major hurdle as miRNAs are inherently unstable in circulation, prone to enzymatic degradation, and must traverse biological barriers like dense tumor stroma. The off-target effects, resulting from miRNA promiscuity, pose toxicity risks to healthy cells. Although viral vectors offer high delivery

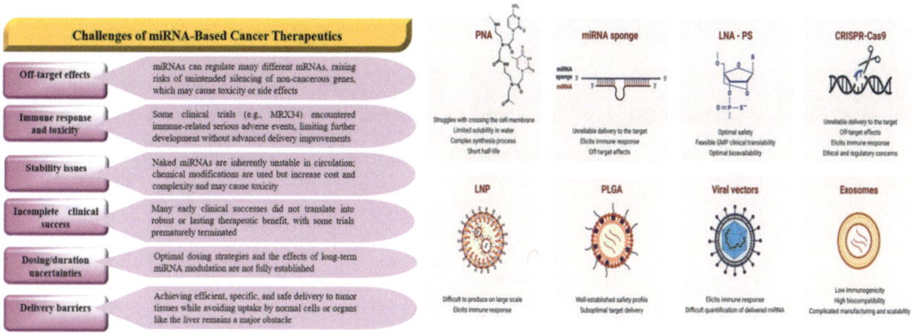


**Fig. 3.7** Challenges associated with miRNA-based therapeutic strategies (Martino et al. 2025)

efficiency, they can trigger immune responses or carry the risk of insertional mutagenesis. Nonviral systems, such as lipid nanoparticles, polymers, and exosomes, have emerged as safer alternatives, yet face limitations including cytotoxicity, limited tissue specificity, and scalability issues for mass production (Fig. 3.7). Furthermore, complex pharmacokinetics, dosing constraints, and inadequate oral delivery options compound the challenges. Existing miRNA mimics/antimiRs often require chemical modifications to enhance stability, yet these can unpredictably alter efficacy, biodistribution, and safety. Precise dosing is further complicated by interspecies differences between animal studies and human conditions, emphasizing the need for rigorous PK/ADME studies in larger animal models prior to human trials (Martino et al. 2025; Seyhan 2024).

In addition, regulatory and manufacturing challenges remain a significant bottleneck. Unlike siRNA-based drugs, of which over 50 have entered clinical trials and at least two have been approved by the FDA, miRNA therapeutics still lack clear regulatory guidance for quality control, formulation standards, and large-scale manufacturing. Ultimately, while miRNA therapeutics hold substantial promise, especially for malignancies resistant to standard treatments, their successful translation will require a multifaceted approach: optimization of delivery systems (potentially using nanotechnology and targeting ligands), development of patient-specific dosing algorithms, improved miRNA target validation, and supportive regulatory frameworks (Fig. 3.8). A deeper understanding of miRNA biology, coupled with advances in precision delivery technologies, will be crucial for unlocking the complete potential of miRNA-based treatments in tailored cancer care.

### 3.7.2 Innovative Strategies and Future Directions

To address these obstacles, innovative strategies are being developed that leverage advances in chemistry and nanotechnology. Chemical modifications, such as the incorporation of locked nucleic acids (LNAs) or other backbone alterations, have been shown to enhance the stability and affinity of synthetic miRNAs, protecting

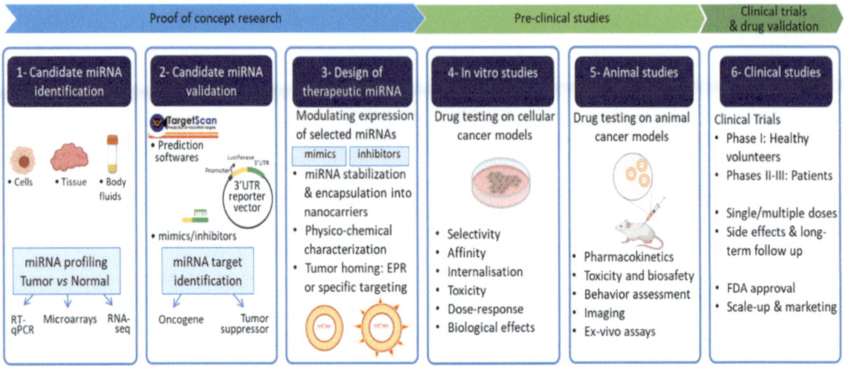


**Fig. 3.8** Translating miRNA therapeutics from bench to market in cancer (Reda El Sayed et al. 2021a)

them from enzymatic degradation and improving their pharmacokinetic profiles (Diener et al. 2022; Herkt and Thum 2021). A variety of delivery strategies are under active development. Viral vectors, including lentiviral, adenoviral, and adeno-associated viruses (AAVs), have demonstrated effectiveness in delivering miRNA antagonists or mimics, with advances in capsid engineering enhancing cell-specific targeting (Forterre et al. 2020; Geisler and Fechner 2016). Non-integrating AAVs and exosome-based delivery systems offer additional avenues for miRNA replacement therapy. The development of miRNA sponges' vectors expressing tandem repeats of miRNA antisense sequences enables simultaneous targeting of miRNA families sharing common seed sequences.

Nanotechnology-based delivery vehicles are at the forefront of these innovations. Engineered nanoparticles, including liposomes, polymeric nanoparticles, gold, carbon, and silica-based carriers, can encapsulate and protect miRNAs, facilitate their transport through biological barriers, and enable controlled release at the tumor site. For example, antibody-coated silica nanoparticles loaded with miR-34a have achieved significant tumor inhibition in neuroblastoma models, while PEI-based miR-145 nanoparticles have effectively eliminated metastatic lung tumors in combination with radiation and chemotherapy. The advent of "nanotheranostics," which integrate diagnostic and therapeutic functions into a single nanoparticle, represents an innovative platform for real-time monitoring and personalized cancer therapy.

The integration of miRNA profiling into precision oncology represents a promising direction for personalized cancer therapy. By analyzing the unique miRNA expression signatures of individual tumors, clinicians can tailor therapeutic interventions to target the specific molecular drivers of each patient's cancer. This approach enhances therapeutic efficacy while concurrently minimizing the incidence and severity of adverse effects. In addition, integrating miRNA-based therapies with traditional methods like chemotherapy, radiotherapy, or immunotherapy has demonstrated synergistic benefits in preclinical studies, further supporting their potential in multidimensional cancer care. Despite these advances, challenges related to immunogenicity, toxicity, scalability, and regulatory hurdles remain. Ongoing research into biodegradable, biocompatible nanomaterials and improved targeting strategies is expected to overcome many of these hurdles, facilitating the expanded clinical translation and implementation of miRNA-based therapeutic modalities.

#### 3.7.2.1 The Expanding Role of Nanotechnology in miRNA Therapeutics

Nanotechnology has become a pivotal approach for addressing the key challenges associated with miRNA-based cancer treatments, particularly those related to stability, targeted delivery, and bioavailability (Lee et al. 2019). Delivery systems based on nanoparticles, such as liposomes, polymeric nanoparticles, and inorganic nanocarriers, have been designed to protect miRNAs from enzymatic breakdown, which improves their stability and prolongs their circulation time. The functionalization with targeting ligands allows for selective delivery to tumor cells, increasing therapeutic efficacy while minimizing off-target effects and toxicity. Recent innovations enable nanoparticles to release their miRNA payload activated by specific stimuli within the tumor microenvironment, including variations in pH or enzymatic activity, thereby achieving targeted and regulated therapeutic effects. Both natural and synthetic polymers have been utilized as miRNA carriers. PEI/miR-145 complexes have significantly reduced colon tumor proliferation and increased apoptosis. PEI/miR-145 complexes have significantly reduced colon tumor proliferation and increased apoptosis. In animal model, PEI/miR-145 complexes have significantly reduced colon tumor proliferation and increased apoptosis (Yu et al. 2015). Natural polymers like chitosan and dextran have also shown promise, though modifications are often needed to improve drug release.

Inorganic nanoparticles, such as gold (AuNPs) and mesoporous silica nanoparticles (MSNs), offer tunable size, surface properties, and multifunctionality (Al-Thani et al. 2024). For example, AuNPs functionalized with thiol groups or PEG have delivered miR-145 to prostate and breast cancer cells, while GD2-targeted MSNs have enabled efficient miR-34 delivery in neuroblastoma models. Lipid-based nanoparticles (LNPs), including liposomes and solid lipid nanoparticles (SLNs), are particularly attractive due to their biocompatibility and efficient cellular uptake (Brillante et al. 2024). These can be engineered with cationic and neutral lipids, PEG, and targeting ligands for enhanced stability and tumor-specific delivery. Anti-EGFR immunoliposomes loaded with miR-135a have inhibited gallbladder carcinoma growth, and PD-L1 antibody-conjugated liposomes co-delivering miR-130a and oxaliplatin have shown potent antitumor effects in gastric cancer models (Reda El Sayed et al. 2021b). SLNs offer the added benefit of encapsulating both hydrophilic and hydrophobic drugs, providing controlled release and synergistic effects, as seen with co-delivery of miR-34a and paclitaxel in melanoma. Aptamers are short nucleic acid ligands that enable receptor-specific miRNA delivery to tumor cells, enhancing targeting precision. Hydrogel-based systems (Sharma et al. 2023) offer localized, sustained miRNA release within the tumor microenvironment, supporting controlled therapeutic delivery. Other advanced delivery systems include micelles, which offer good tissue penetration, and bacterially derived nanocells (EDVs), which have enabled targeted delivery of miR-16 to mesothelioma in both preclinical and clinical settings (Reda El Sayed et al. 2021b; Ghezzi et al. 2021). Exosome-mimetic nanosystems (EMNs) are also being developed to combine the advantages of natural exosomes with efficient miRNA loading. Therefore, these diverse nanotechnology platforms are propelling the development of next-generation miRNA therapeutics, offering improved stability, targeted delivery, and the potential for combination therapies.

### 3.7.3 Limitations and Future Perspectives

Despite considerable advances, each delivery platform for miRNA-based therapeutics presents a few limitations. Viral vectors though efficient in gene transfer can provoke immune responses and pose a risk of insertional mutagenesis. Nanoparticle-based systems also face several challenges, including cytotoxicity, off-target effects, limited tumor penetration in solid cancers, and scalability or production cost-related limitations. Nevertheless, functionalization strategies such as the incorporation of targeting ligands like aptamers are being actively explored to improve tumor selectivity, reduce systemic toxicity, and enhance delivery efficiency.

The complex and heterogeneous nature of cancer further underscores the necessity for advanced and multifaceted treatment approaches. In this regard, miRNAs are promising therapeutic candidates because of their capacity to simultaneously modulate several pathways involved in cancer progression (Fig. 3.9).

Accumulating evidence substantiates the critical involvement of miRNAs in cancer diagnostics, prognostic assessment, and therapeutic interventions. The therapeutic strategies focusing on restoring tumor-suppressive miRNAs (e.g., let-7,

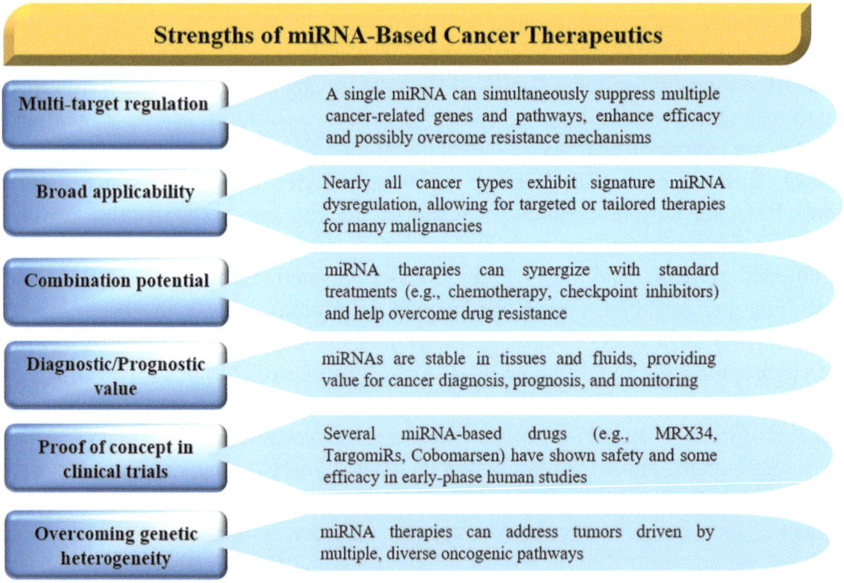


**Fig. 3.9** Strengths of miRNA-based therapeutic strategies

miR-16, and miR-31) or inhibiting oncogenic miRNAs have consistently shown antitumor effects in preclinical models. One key example is MRX34, a liposome-formulated miR-34 mimic, which advanced to Phase I clinical trials, representing a significant milestone in translating miRNA-based therapy into the clinic (Hong et al. 2020). However, challenges such as molecular instability, immune reactions, and delivery inefficiencies have tempered progress (Baumann and Winkler 2014). To overcome these challenges, chemical modifications such as 2'-O-methylation, locked nucleic acids (LNAs), and phosphorothioate backbones are being employed to enhance the stability, affinity, and pharmacokinetic profiles of therapeutic miRNAs (Martino et al. 2025). Alongside chemical innovations, nanotechnology-based delivery systems have demonstrated great potential in improving miRNA bioavailability, cellular uptake, and targeting specificity. Smart nanocarriers, including lipid nanoparticles, polymeric micelles, exosome-mimetic systems (EMNs), and responsive nanovehicles, are presently engineered to facilitate controlled release of their therapeutic payload triggered by tumor-specific microenvironmental cues, including pH or enzymatic activity, ensuring controlled and localized action (Liao et al. 2020). These advancements contribute significantly to improving both therapeutic index and safety. Moreover, the integration of miRNA profiling into precision oncology offers a transformative approach to personalized medicine. By analyzing individual miRNA expression signatures, clinicians can develop tailored treatment strategies targeting the specific molecular alterations of a patient's tumor. This enhances therapeutic efficacy and minimizes off-target toxicity. Finally, combining miRNA-based therapies with conventional treatments, including chemotherapy, radiotherapy, and immunotherapy, has demonstrated synergistic effects in preclinical models, further expanding their potential utility in comprehensive cancer care. Although technical, biological, and regulatory hurdles remain, miRNA therapeutics represent a promising frontier in oncology. The continued innovation in delivery technologies, chemical engineering, and molecular profiling will be critical to advancing miRNA-based therapies from the bench to bedside and realizing their full potential in the era of precision cancer medicine (Table 3.1).

**Table 3.1** Clinical challenges in the therapeutic application of microRNAs and corresponding strategies to overcome translational barriers (Tong and Nemunaitis 2008)

| Clinical challenge | Potential strategies |
|---|---|
| Reduced levels of tumor-suppressive miRNAs | Supplementation with miRNA precursors |
| | Application of antagomirs to inhibit harmful miRNAs |
| Limited therapeutic effectiveness | Multiple dosing schedules |
| | Use of chemically enhanced oligonucleotides (e.g., locked nucleic acids) |
| | Delivery via coding vectors instead of oligonucleotides |
| | Selection of miRNAs with diverse regulatory effects |
| | Improved knowledge of cellular bottlenecks in miRNA processing |

(continued)

**Table 3.1** (continued)

| Clinical challenge | Potential strategies |
|---|---|
| Broad target range of miRNA candidates | Computational (*in silico*) screening |
| | Experimental validation of expression and function |
| Challenges in delivery within living organisms | Controlled expression using oncolytic viral vectors |
| | Delivery via immunoliposomes or nanoparticles |
| Potential for toxic side effects | Employing delivery systems with high specificity for target cells |

## 3.8 Case Studies and Recent Advances in miRNA Research

### 3.8.1 Translational Milestones in miRNA-Based Cancer Therapies

Over the past two decades, miRNAs have become recognized as key regulators of gene expression in cancer, acting either as oncogenes (oncomiRs) or tumor suppressors based on the specific cellular environment. Their widespread dysregulation in solid tumors and hematological malignancies has made them compelling candidates for therapeutic development and biomarker-based diagnostics. Therapeutic strategies generally rely either on silencing oncogenic miRNAs using antisense oligonucleotides (antimiRs), sponges, or decoys, or on restoring tumor-suppressive miRNAs via synthetic mimics or gene delivery platforms. Several miRNA-targeted agents have progressed from preclinical studies to clinical trials, offering critical insights into efficacy, delivery, safety, and resistance mechanisms (Table 3.2).

One of the earliest and most notable examples is MRX34, a liposomal mimic of miR-34a. As a tumor-suppressor miRNA regulated by p53, miR-34a targets multiple oncogenes. MRX34 was evaluated in a Phase I trial for advanced solid tumors, including hepatocellular carcinoma, NSCLC, renal cell carcinoma, and melanoma (Hong et al. 2020; Li et al. 2021). The trial was discontinued early due to severe immune-related adverse effects, including treatment-related deaths, despite evidence of target gene knockdown, highlighting the crucial need for safer, more targeted delivery systems. Similarly, Cobomarsen (MRG-106), an LNA-modified antimiR targeting miR-155, showed promising early clinical outcomes in cutaneous T-cell lymphoma and other hematologic malignancies before development was paused for strategic reasons (Brillante et al. 2024). Other miRNA drugs, though originally developed for non-oncology indications, have laid the foundation for RNA-targeting therapeutics. Miravirsen, a β-D-oxy-LNA-modified antimiR targeting liver-specific miR-122, entered Phase II trials for hepatitis C virus (HCV) infection (Ottosen et al. 2015). Although it successfully lowered HCV RNA levels in about half of treated individuals, viral rebound due to mutations in the miR-122 binding site underscored the challenge of acquired resistance. A related candidate, RG-101 used GalNAc-conjugation to enhance hepatocyte uptake and showed long-

**Table 3.2** Key miRNA-based therapeutics in cancer clinical trials (Martino et al. 2025)

| Drug name | miRNA target | Drug type | Delivery platform | Phase | Cancer type(s) | Status/Outcome |
|---|---|---|---|---|---|---|
| MRX34 | miR-34a | Double-stranded mimic | Liposome nanoparticle | Phase 1 | Liver, SCLC, lymphoma, melanoma, RCC, NSCLC | Terminated (immune toxicity) |
| MRX34 | miR-34a | Double-stranded mimic | Liposome + Dexamethasone | Phase 1 | Melanoma | Withdrawn |
| TargomiR | miR-16 | Synthetic mimic | Bacterial minicell (EDV) | Phase 1 | Mesothelioma, NSCLC | Completed |
| MRG-106 (Cobomarsen) | miR-155 | LNA-based inhibitor | N/A | Phase 1 | CTCL, CLL, DLBCL, ATLL | Completed |
| MRG-106 (Cobomarsen) | miR-155 | LNA-based inhibitor | N/A | Phase 2 | CTCL | Early Termination |
| INT-1B3 | miR-193a-3p | Synthetic mimic | Lipid nanoparticle (LNP) | Phase 1 | TNBC, NSCLC, melanoma, colon, HCC | Recruiting/ongoing |
| TTX-MC138 | miR-10b | Antisense inhibitor | Dextran-coated iron oxide NP | Preclinical | Met. breast, glioblastoma, pancreatic ca, SCLC, OS | Preclinical |
| RGLS5579 | miR-10b | Antisense inhibitor | Unknown | Preclinical | Glioblastoma | Preclinical |
| miR-506 mimics | miR-506 | Synthetic mimic | Not specified | Preclinical | Ovarian, gastric, colon, NSCLC | Preclinical |

*ATLL* Adult T-cell leukemia/lymphoma, *RCC* Renal cell carcinoma, *CTCL* Cutaneous T-cell lymphoma, *CLL* Chronic lymphocytic leukemia, *DLBCL* Diffuse Large B-cell lymphoma, *TNBC* Triple negative breast cancer, *HCC* Hepatocellular carcinoma, *OS* Osteosarcoma, *EDV* EnGeneIC Dream Vector

lasting HCV suppression with minimal toxicity. While both drugs targeted a viral mechanism, they demonstrated the clinical potential of miRNA inhibition platforms, setting essential precedents for oncology applications.

In oncology-focused efforts, the MesomiR-1 trial evaluated TargomiR, a miR-16 mimic delivered via EGFR-targeted bacterial nanocells (EnGeneIC Dream Vectors, EDVs), in patients with relapsed malignant pleural carcinoma and NSCLC (Kim and Croce 2023; van Zandwijk et al. 2017). miR-16, a tumor suppressor downregulated in these cancers, is co-located with miR-15a at the frequently deleted 13q14.3 locus in CLL. Although the response rate was modest, the delivery system was well tolerated, indicating platform viability for future development. Another noteworthy approach involves TTX-MC138, a miR-10b inhibitor, the therapeutic for the treatment of metastatic cancers (Halim et al. 2025). The preclinical models of breast and pancreatic cancer showed significant regression of metastases. In a recent Phase 0 microdose trial, radiolabeled TTX-MC138 demonstrated target engagement and accumulation at metastatic sites, with a 66% reduction in circulating miR-10b levels observed post-administration. FDA clearance for a Phase I/II dose-escalation study marks a step forward in translating miRNA-targeted nanotherapies for aggressive solid tumors.

Beyond these clinical efforts, numerous preclinical studies continue to validate the therapeutic benefits of modulating miRNA levels. miR-101 and miR-184 are downregulated in colorectal, breast, lung, and endometrial cancers, and their reintroduction using mimics inhibits tumor growth and metastasis. miR-193b, involved in hematopoietic homeostasis, is reduced in acute myeloid leukemia and shows antiproliferative effects when restored (Liu et al. 2022; Wu et al. 2017). The lentiviral delivery of miR-15a/16 in CLL models resulted in sustained reduction of malignant B-1 cells and decreased tumor burden. The miR-181 family, notably a/b and c/d isoforms, plays dual roles depending on tissue context, blocking melanoma growth or enhancing chemotherapy sensitivity in biliary tract cancers. Conversely, the oncomiRs miR-21 and miR-22 have been effectively targeted by RNA decoys and LNA-antimiRs to suppress EMT and metastasis in breast and other cancers (Bautista-Sánchez et al. 2020; Song et al. 2013). Emerging candidates such as miR-506 mimics are showing potential in sensitizing ovarian, gastric, and colorectal cancers to chemotherapy. These collective findings reflect the ability of miRNA-directed therapeutics to modulate pathways linked to apoptosis, EMT, angiogenesis, metastasis, and stemness. Despite exciting progress, key challenges remain, including immune toxicity, off-target effects, and variability in delivery efficiency. Future success will rely on optimizing delivery vectors such as exosomes, lipid nanoparticles, EDVs, and DNA nanostructures and pairing miRNA therapeutics with precise molecular profiling to ensure effective, targeted intervention.

### 3.8.2 Emergence of miRNA Profiling in Precision Medicine

The introduction of high-throughput sequencing and sophisticated bioinformatics has made miRNA profiling a crucial aspect of precision oncology. The techniques such as next-generation sequencing (NGS), microarrays, and quantitative RT-PCR have facilitated the identification of new miRNAs and supported the development of diagnostic and prognostic signatures for various cancers. Circulating miRNAs, including miR-21, miR-92a, miR-122, and miR-210, have been detected in peripheral blood, saliva, and urine, enabling their application as noninvasive liquid biopsy indicators for early detection and ongoing disease surveillance (Wang et al. 2018; Kosaka et al. 2010). In gastric cancer, combined analysis of miRNAs such as miR-21, miR-106b, and miR-196a has demonstrated greater sensitivity and specificity than traditional biomarkers, offering a promising adjunct to tissue biopsy for disease staging and prognosis (Wu et al. 2020). Beyond detection, miRNA expression profiles are now being used to predict treatment response. For example, elevated miR-31 levels in colorectal cancer have been associated with resistance to anti-EGFR therapy, while low expression of the let-7 family is associated with adverse clinical outcomes and reduced chemotherapy responsiveness in NSCLC. Similarly, miR-200c has been implicated in radio sensitivity modulation in both lung and breast cancers.

Recent developments in computational platforms, such as CAMIRADA (Cancer microRNA Association Discovery Algorithm), leverage artificial intelligence to integrate miRNA expression data with clinical and genomic information for disease-specific target discovery, patient stratification, and therapeutic prioritization (Shamsizadeh et al. 2019). The integration of miRNA profiling with other "omics" platforms, including mRNA transcriptomics, epigenomics, and proteomics, holds significant promise for the development of multidimensional, data-driven models in personalized cancer care. Together, these advances underscore the growing role of miRNA-based strategies in both therapy and diagnostics. As the field progresses, the convergence of miRNA therapeutics and miRNA-guided molecular stratification may redefine current paradigms in oncology, shaping the future of personalized medicine across diverse cancer types.

Current and upcoming research aims to refine targeted delivery methods, reduce unintended effects, and address resistance mechanisms in order to enhance the effectiveness of treatments. Integrating miRNA regulation with advanced immunotherapeutic approaches, including immune checkpoint blockade and CAR-T cell treatments, also presents a compelling frontier in immuno-oncology (Youssef et al. 2025; Sareen et al. 2025). With continued innovation and interdisciplinary collaboration, miRNA-based therapeutics are anticipated to constitute essential elements of advanced oncology treatment paradigms, providing a multifactorial and precision medicine approach to enhance clinical outcomes, including in malignancies exhibiting resistance to conventional therapies.

## 3.9 Conclusion

Since the discovery of miRNAs, the understanding of their biogenesis, regulatory roles, and significance in cancer has expanded tremendously. miRNAs function as potent modulators of gene expression, participating in transcriptional and posttranscriptional networks that govern cell fate. Their deregulation through altered gene dosage, epigenetic changes, aberrant transcription factor activity, and defects in biogenesis plays a pivotal role in driving core oncogenic processes, including sustained proliferative signaling, resistance to programmed cell death, enhanced invasive capacity, and metastatic dissemination. miRNAs may act as tumor suppressors or oncogenes by targeting multiple signaling pathways, but future research must clarify the contributions of specific miRNA-target interactions to malignant transformation. The technological advances have accelerated miRNA profiling, demonstrating the close association between miRNA signatures and tumor type, grade, and prognosis. These findings establish miRNAs, including those found in exosomes, as promising biomarkers for early diagnosis and therapy stratification. Likewise, therapeutic delivery systems ranging from nanoparticles and nanomaterials to novel topical formulations have shown promise in overcoming challenges related to in vivo stability and targeting specificity. However, translating these advances to the clinic requires rigorous validation, robust detection technologies, and large-scale cohort studies. miRNAs' roles extend beyond molecular diagnostics; they support synergistic immunotherapies and personalized treatments, leveraging their capacity to modulate multiple genes and signaling pathways simultaneously. The interdisciplinary research spanning bioinformatics, materials science, and molecular medicine is pivotal for improving detection sensitivity, understanding the complexity of miRNA-mediated regulation, and developing innovative therapeutics. Despite technical and biological challenges, ongoing research on miRNA function, delivery, and detection is driving the integration of miRNAs into precision oncology, holding great potential to enhance clinical diagnosis, prognosis, and treatment of cancer.

## References

Ali Syeda Z, Langden SSS, Munkhzul C, Lee M, Song SJ (2020) Regulatory mechanism of MicroRNA expression in cancer. Int J Mol Sci 21(5). https://doi.org/10.3390/ijms21051723

Al-Thani AN, Jan AG, Abbas M, Geetha M, Sadasivuni KK (2024) Nanoparticles in cancer theragnostic and drug delivery: A comprehensive review. Life Sci 352:122899. https://doi.org/10.1016/j.lfs.2024.122899

Angerilli V, Galuppini F, Businello G, Dal Santo L, Savarino E, Realdon S et al (2021) MicroRNAs as predictive biomarkers of resistance to targeted therapies in gastrointestinal tumors. Biomedicine 9(3). https://doi.org/10.3390/biomedicines9030318

Arghiani N, Shah K (2021) Modulating microRNAs in cancer: next-generation therapies. Cancer Biol Med 19(3):289–304. https://doi.org/10.20892/j.issn.2095-3941.2021.0294

Avendaño-Portugal C, Montaño-Samaniego M, Guttman-Bazbaz R, Bravo-Estupiñan DM, Ibáñez-Hernández M (2025) Therapeutic applications of poly-miRNAs and miRNA sponges. Int J Mol Sci 26(10):4535

Baumann V, Winkler J (2014) miRNA-based therapies: strategies and delivery platforms for oligonucleotide and non-oligonucleotide agents. Future Med Chem 6(17):1967–1984. https://doi.org/10.4155/fmc.14.116

Bautista-Sánchez D, Arriaga-Canon C, Pedroza-Torres A, De La Rosa-Velázquez IA, González-Barrios R, Contreras-Espinosa L et al (2020) The promising role of miR-21 as a cancer

biomarker and its importance in RNA-based therapeutics. Mol Ther Nucleic Acids 20:409–420. https://doi.org/10.1016/j.omtn.2020.03.003
Brillante S, Volpe M, Indrieri A (2024) Advances in MicroRNA therapeutics: from preclinical to clinical studies. Hum Gene Ther 35(17–18):628–648. https://doi.org/10.1089/hum.2024.113
Chakrabortty A, Patton DJ, Smith BF, Agarwal P (2023a) miRNAs: potential as biomarkers and therapeutic targets for cancer. Genes 14(7):1375
Chakrabortty A, Patton DJ, Smith BF, Agarwal P (2023b) miRNAs: potential as biomarkers and therapeutic targets for cancer. Genes (Basel) 14(7). https://doi.org/10.3390/genes14071375
Chang T-C, Yu D, Lee Y-S, Wentzel EA, Arking DE, West KM et al (2008) Widespread microRNA repression by Myc contributes to tumorigenesis. Nat Genet 40(1):43–50. https://doi.org/10.1038/ng.2007.30
Chauhan N, Dhasmana A, Jaggi M, Chauhan SC, Yallapu MM (2020) miR-205: A potential biomedicine for cancer therapy. Cells 9(9). https://doi.org/10.3390/cells9091957
Croce CM (2009) Causes and consequences of microRNA dysregulation in cancer. Nat Rev Genet 10(10):704–714. https://doi.org/10.1038/nrg2634
Denli AM, Tops BB, Plasterk RH, Ketting RF, Hannon GJ (2004) Processing of primary microRNAs by the microprocessor complex. Nature 432(7014):231–235. https://doi.org/10.1038/nature03049
Diener C, Keller A, Meese E (2022) Emerging concepts of miRNA therapeutics: from cells to clinic. Trends Genet 38(6):613–626. https://doi.org/10.1016/j.tig.2022.02.006
Finnegan EF, Pasquinelli AE (2013) MicroRNA biogenesis: regulating the regulators. Crit Rev Biochem Mol Biol 48(1):51–68. https://doi.org/10.3109/10409238.2012.738643
Forterre A, Komuro H, Aminova S, Harada M (2020) A comprehensive review of cancer MicroRNA therapeutic delivery strategies. Cancers (Basel) 12(7). https://doi.org/10.3390/cancers12071852
Gao W, Zhou J, Morshedi M (2024) MicroRNA-34 and gastrointestinal cancers: a player with big functions. Cancer Cell Int 24(1):163. https://doi.org/10.1186/s12935-024-03338-w
García-López J, Brieño-Enríquez MA, Del Mazo J (2013) MicroRNA biogenesis and variability. Biomol Concepts 4(4):367–380. https://doi.org/10.1515/bmc-2013-0015
Geisler A, Fechner H (2016) MicroRNA-regulated viral vectors for gene therapy. World J Exp Med 6(2):37–54. https://doi.org/10.5493/wjem.v6.i2.37
Ghezzi M, Pescina S, Padula C, Santi P, Del Favero E, Cantù L et al (2021) Polymeric micelles in drug delivery: an insight of the techniques for their characterization and assessment in biorelevant conditions. J Control Release 332:312–336. https://doi.org/10.1016/j.jconrel.2021.02.031
Gupta P, Bhattacharjee S, Sharma AR, Sharma G, Lee SS, Chakraborty C (2017) miRNAs in Alzheimer disease—A therapeutic perspective. Curr Alzheimer Res 14(11):1198–1206. https://doi.org/10.2174/1567205014666170829101016
Halim A, Kim B, Kenyon E, Moore A (2025) miR-10b as a clinical marker and a therapeutic target for metastatic breast cancer. Technol Cancer Res Treat 24:15330338251339256. https://doi.org/10.1177/15330338251339256
Herkt M, Thum T (2021) Pharmacokinetics and proceedings in clinical application of nucleic acid therapeutics. Mol Ther 29(2):521–539. https://doi.org/10.1016/j.ymthe.2020.11.008
Hill M, Tran N (2021) miRNA interplay: mechanisms and consequences in cancer. Dis Model Mech 14(4). https://doi.org/10.1242/dmm.047662
Hong DS, Kang YK, Borad M, Sachdev J, Ejadi S, Lim HY et al (2020) Phase 1 study of MRX34, a liposomal miR-34a mimic, in patients with advanced solid tumours. Br J Cancer 122(11):1630–1637. https://doi.org/10.1038/s41416-020-0802-1
Hu Y, Yu C-Y, Wang J-L, Guan J, Chen H-Y, Fang J-Y (2014) MicroRNA sequence polymorphisms and the risk of different types of cancer. Sci Rep 4(1):3648. https://doi.org/10.1038/srep03648
Humphries B, Yang C (2015) The microRNA-200 family: small molecules with novel roles in cancer development, progression and therapy. Oncotarget 6(9):6472

Iliopoulos D, Jaeger SA, Hirsch HA, Bulyk ML, Struhl K (2010) STAT3 activation of miR-21 and miR-181b-1 via PTEN and CYLD are part of the epigenetic switch linking inflammation to cancer. Mol Cell 39(4):493–506. https://doi.org/10.1016/j.molcel.2010.07.023

Inoue J, Inazawa J (2021) Cancer-associated miRNAs and their therapeutic potential. J Hum Genet 66(9):937–945. https://doi.org/10.1038/s10038-021-00938-6

Iorio MV, Croce CM (2012) Causes and consequences of microRNA dysregulation. Cancer J 18(3): 215–222. https://doi.org/10.1097/PPO.0b013e318250c001

Kim T, Croce CM (2023) MicroRNA: trends in clinical trials of cancer diagnosis and therapy strategies. Exp Mol Med 55(7):1314–1321. https://doi.org/10.1038/s12276-023-01050-9

Kosaka N, Iguchi H, Ochiya T (2010) Circulating microRNA in body fluid: a new potential biomarker for cancer diagnosis and prognosis. Cancer Sci 101(10):2087–2092. https://doi.org/10.1111/j.1349-7006.2010.01650.x

Lan H, Lu H, Wang X, Jin H (2015) MicroRNAs as potential biomarkers in cancer: opportunities and challenges. Biomed Res Int 2015:125094. https://doi.org/10.1155/2015/125094

Lee SWL, Paoletti C, Campisi M, Osaki T, Adriani G, Kamm RD et al (2019) MicroRNA delivery through nanoparticles. J Control Release 313:80–95. https://doi.org/10.1016/j.jconrel.2019.10.007

Li W, Wang Y, Liu R, Kasinski AL, Shen H, Slack FJ et al (2021) MicroRNA-34a: potent tumor suppressor, cancer stem cell inhibitor, and potential anticancer therapeutic (Review). Front Cell Dev Biol 9. https://doi.org/10.3389/fcell.2021.640587

Liao Z, Wong SW, Yeo HL, Zhao Y (2020) Smart nanocarriers for cancer treatment: clinical impact and safety. NanoImpact 20:100253. https://doi.org/10.1016/j.impact.2020.100253

Liu Q, Geng P, Shi L, Wang Q, Wang P (2019) miR-29 promotes osteosarcoma cell proliferation and migration by targeting PTEN. Oncol Lett 17(1):883–890. https://doi.org/10.3892/ol.2018.9646

Liu N, Yang C, Gao A, Sun M, Lv D (2022) MiR-101: an important regulator of gene expression and tumor ecosystem. Cancers (Basel) 14(23). https://doi.org/10.3390/cancers14235861

Lynch J, Fay J, Meehan M, Bryan K, Watters KM, Murphy DM et al (2012) MiRNA-335 suppresses neuroblastoma cell invasiveness by direct targeting of multiple genes from the non-canonical TGF-β signalling pathway. Carcinogenesis 33(5):976–985. https://doi.org/10.1093/carcin/bgs114

Martino MTD, Tagliaferri P, Tassone P (2025) MicroRNA in cancer therapy: breakthroughs and challenges in early clinical applications. J Exp Clin Cancer Res 44(1):126. https://doi.org/10.1186/s13046-025-03391-x

Menon A, Abd-Aziz N, Khalid K, Poh CL, Naidu R (2022) miRNA: A promising therapeutic target in cancer. Int J Mol Sci 23(19):11502

Naidu S, Magee P, Garofalo M (2015) MiRNA-based therapeutic intervention of cancer. J Hematol Oncol 8(1):68. https://doi.org/10.1186/s13045-015-0162-0

O'Brien J, Hayder H, Zayed Y, Peng C (2018) Overview of MicroRNA biogenesis, mechanisms of actions, and circulaton (Review). Front Endocrinol 9. https://doi.org/10.3389/fendo.2018.00402

Otmani K, Lewalle P (2021) Tumor suppressor miRNA in cancer cells and the tumor microenvironment: mechanism of deregulation and clinical implications. Front Oncol 11:708765. https://doi.org/10.3389/fonc.2021.708765

Ottosen S, Parsley TB, Yang L, Zeh K, van Doorn LJ, van der Veer E et al (2015) In vitro antiviral activity and preclinical and clinical resistance profile of miravirsen, a novel anti-hepatitis C virus therapeutic targeting the human factor miR-122. Antimicrob Agents Chemother 59(1):599–608. https://doi.org/10.1128/aac.04220-14

Park SM, Gaur AB, Lengyel E, Peter ME (2008) The miR-200 family determines the epithelial phenotype of cancer cells by targeting the E-cadherin repressors ZEB1 and ZEB2. Genes Dev 22(7):894–907. https://doi.org/10.1101/gad.1640608

Peng X, Wang J, Zhang C, Liu K, Zhao L, Chen X et al (2020) A three-miRNA panel in serum as a noninvasive biomarker for colorectal cancer detection. Int J Biol Markers 35(3):74–82. https://doi.org/10.1177/1724600820950740

Pop-Bica C, Pintea S, Magdo L, Cojocneanu R, Gulei D, Ferracin M et al (2020) The clinical utility of miR-21 and let-7 in non-small cell lung cancer (NSCLC). A systematic review and meta-analysis. Front Oncol 10:516850. https://doi.org/10.3389/fonc.2020.516850

Rabaan AA, AlSaihati H, Bukhamsin R, Bakhrebah MA, Nassar MS, Alsaleh AA et al (2023) Application of CRISPR/Cas9 Technology in Cancer Treatment: A future direction. Curr Oncol 30(2):1954–1976. https://doi.org/10.3390/curroncol30020152

Reda El Sayed S, Cristante J, Guyon L, Denis J, Chabre O, Cherradi N (2021a) MicroRNA therapeutics in cancer: current advances and challenges. Cancer 13(11):2680

Reda El Sayed S, Cristante J, Guyon L, Denis J, Chabre O, Cherradi N (2021b) MicroRNA therapeutics in cancer: current advances and challenges. Cancers (Basel) 13(11). https://doi.org/10.3390/cancers13112680

Ruby JG, Jan CH, Bartel DP (2007) Intronic microRNA precursors that bypass Drosha processing. Nature 448(7149):83–86. https://doi.org/10.1038/nature05983

Sapre N, Selth LA (2013) Circulating MicroRNAs as biomarkers of prostate cancer: the state of play. Prostate Cancer 2013:539680. https://doi.org/10.1155/2013/539680

Sareen G, Mohan M, Mannan A, Dua K, Singh TG (2025) A new era of cancer immunotherapy: vaccines and miRNAs. Cancer Immunol Immunother 74(5):163. https://doi.org/10.1007/s00262-025-04011-5

Seyhan AA (2024) Trials and tribulations of MicroRNA therapeutics. Int J Mol Sci 25(3):1469

Shah MY, Ferrajoli A, Sood AK, Lopez-Berestein G, Calin GA (2016) microRNA therapeutics in cancer—an emerging concept. EBioMedicine 12:34–42. https://doi.org/10.1016/j.ebiom.2016.09.017

Shamsizadeh S, Goliaei S, Razaghi Moghadam Z (2019) CAMIRADA: cancer microRNA association discovery algorithm, a case study on breast cancer. J Biomed Inform 94:103180. https://doi.org/10.1016/j.jbi.2019.103180

Sharma R, Lalhall A, Puri S, Wangoo N (2023) Design of Fmoc-Phenylalanine Nanofibrillar Hydrogel and Mechanistic Studies of its antimicrobial action against both gram-positive and gram-negative bacteria. ACS Appl Bio Mater 6(2):494–506. https://doi.org/10.1021/acsabm.2c00767

Shen J, Jiang F (2012) Applications of MicroRNAs in the diagnosis and prognosis of lung cancer. Expert Opin Med Diagn 6(3):197–207. https://doi.org/10.1517/17530059.2012.672970

Si ML, Zhu S, Wu H, Lu Z, Wu F, Mo YY (2007) miR-21-mediated tumor growth. Oncogene 26(19):2799–2803. https://doi.org/10.1038/sj.onc.1210083

Song SJ, Poliseno L, Song MS, Ala U, Webster K, Ng C et al (2013) MicroRNA-antagonism regulates breast cancer Stemness and metastasis via TET-family-dependent chromatin remodeling. Cell 154(2):311–324. https://doi.org/10.1016/j.cell.2013.06.026

Sun Y, Mezzanzanica D, Zhang W (2014) MiR-506: A multitasker in suppression of the epithelial-to-mesenchymal transition. RNA Dis 1(1):e447. https://doi.org/10.14800/rd.447

Sundarbose K, Kartha RV, Subramanian S (2013) MicroRNAs as biomarkers in cancer. Diagnostics (Basel) 3(1):84–104. https://doi.org/10.3390/diagnostics3010084

Sur D, Advani S, Braithwaite D (2022) MicroRNA panels as diagnostic biomarkers for colorectal cancer: A systematic review and meta-analysis. Front Med (Lausanne) 9:915226. https://doi.org/10.3389/fmed.2022.915226

Teplyuk NM, Uhlmann EJ, Gabriely G, Volfovsky N, Wang Y, Teng J et al (2016) Therapeutic potential of targeting microRNA-10b in established intracranial glioblastoma: first steps toward the clinic. EMBO Mol Med 8(3):268–287. https://doi.org/10.15252/emmm.201505495

Tesfaye AA, Azmi AS, Philip PA (2019) miRNA and gene expression in pancreatic ductal adenocarcinoma. Am J Pathol 189(1):58–70. https://doi.org/10.1016/j.ajpath.2018.10.005

Tong AW, Nemunaitis J (2008) Modulation of miRNA activity in human cancer: a new paradigm for cancer gene therapy? Cancer Gene Ther 15(6):341–355. https://doi.org/10.1038/cgt.2008.8

van Zandwijk N, Pavlakis N, Kao SC, Linton A, Boyer MJ, Clarke S et al (2017) Safety and activity of microRNA-loaded minicells in patients with recurrent malignant pleural mesothelioma: a first-in-man, phase 1, open-label, dose-escalation study. Lancet Oncol 18(10):1386–1396. https://doi.org/10.1016/s1470-2045(17)30621-6

Visone R, Croce CM (2009) MiRNAs and cancer. Am J Pathol 174(4):1131–1138. https://doi.org/10.2353/ajpath.2009.080794

Wang H, Peng R, Wang J, Qin Z, Xue L (2018) Circulating microRNAs as potential cancer biomarkers: the advantage and disadvantage. Clin Epigenetics 10(1):59. https://doi.org/10.1186/s13148-018-0492-1

Weber B, Stresemann C, Brueckner B, Lyko F (2007) Methylation of human microRNA genes in normal and neoplastic cells. Cell Cycle 6(9):1001–1005. https://doi.org/10.4161/cc.6.9.4209

Wu G, Liu J, Wu Z, Wu X, Yao X (2017) MicroRNA-184 inhibits cell proliferation and metastasis in human colorectal cancer by directly targeting IGF-1R. Oncol Lett 14(3):3215–3222. https://doi.org/10.3892/ol.2017.6499

Wu SR, Wu Q, Shi YQ (2020) Recent advances of miRNAs in the development and clinical application of gastric cancer. Chin Med J 133(15):1856–1867. https://doi.org/10.1097/cm9.0000000000000921

Xie H, Liu X, Huang Z, Xu L, Bai R, He F et al (2022) Nanoscale Zeolitic Imidazolate framework (ZIF)–8 in cancer Theranostics: current challenges and prospects. Cancer 14(16):3935

Xue W, Dahlman JE, Tammela T, Khan OF, Sood S, Dave A et al (2014) Small RNA combination therapy for lung cancer. Proc Natl Acad Sci USA 111(34):E3553–E3561. https://doi.org/10.1073/pnas.1412686111

Youssef E, Fletcher B, Palmer D (2025) Enhancing precision in cancer treatment: the role of gene therapy and immune modulation in oncology (Review). Front Med 11. https://doi.org/10.3389/fmed.2024.1527600

Yu Y, Nangia-Makker P, Farhana L, G. Rajendra S, Levi E, Majumdar APN (2015) miR-21 and miR-145 cooperation in regulation of colon cancer stem cells. Mol Cancer 14(1):98. https://doi.org/10.1186/s12943-015-0372-7

Yu J, Wang JG, Zhang L, Yang HP, Wang L, Ding D et al (2016) MicroRNA-320a inhibits breast cancer metastasis by targeting metadherin. Oncotarget 7(25):38612–38625. https://doi.org/10.18632/oncotarget.9572

Zedan AH, Osther PJS, Assenholt J, Madsen JS, Hansen TF (2020) Circulating miR-141 and miR-375 are associated with treatment outcome in metastatic castration resistant prostate cancer. Sci Rep 10(1):227. https://doi.org/10.1038/s41598-019-57101-7

Zhou Z, Qiao JX, Shetty A, Wu G, Huang Y, Davidson NE et al (2014) Regulation of estrogen receptor signaling in breast carcinogenesis and breast cancer therapy. Cell Mol Life Sci 71(8):1549. https://doi.org/10.1007/s00018-013-1376-3

# Oncogenic miRNAs as Therapeutic Targets: Opportunities and Challenges

4

Prashanth Kotthapalli and Sumana Kumar

**Abstract**

MicroRNAs (miRNAs) play a pivotal role in cancer pathogenesis, acting as oncogenes or tumor suppressors depending on their targets and cellular context. Oncogenic miRNAs, or oncomiRs, promote cancer by inhibiting tumor suppressor genes, while tumor suppressor miRNAs inhibit oncogenes. This dual functionality presents challenges and opportunities for therapeutic intervention. Recent advances highlight miRNAs' potential as therapeutic targets. Targeting oncomiRs with inhibitors can suppress tumor growth and enhance chemotherapy sensitivity. Conversely, mimicking tumor suppressor miRNAs can restore normal gene expression patterns, inhibiting cancer progression. For instance, miR-10b is an oncomiRs in glioblastoma, and its inhibition reduces cancer cell proliferation and invasion. Similarly, miR-21 is associated with therapy resistance in ovarian cancer, and its downregulation can induce apoptosis and reverse drug resistance. The therapeutic potential of miRNA is enhanced by its role in modulating the tumor microenvironment and immune response. For example, miR-155 can sensitize cancer cells to radiation therapy by suppressing the DNA repair mechanism. However, the complex interactions between miRNAs and other cellular components necessitate a holistic approach to miRNA-based therapies. This includes using immunocompetent mouse models and systems biology to fully understand miRNA networks and their implications for cancer treatment. In conclusion, targeting oncogenic miRNAs offers a promising therapeutic strategy for cancer management. By understanding miRNAs' multifaceted roles and developing effective delivery systems, researchers can harness their potential to

P. Kotthapalli · S. Kumar (✉)
Department of Microbiology, School of Life Sciences, JSS Academy of Higher Education and Research, Shivarathreeshwara Nagar, Bannimantap, Mysuru, Karnataka, India
e-mail: sumana.k@jssuni.edu.in



S. Tiwari et al. (eds.), *microRNA Modulation: A Promising Combination Therapy for Cancer*, https://doi.org/10.1007/978-981-95-5989-3_4

improve treatment outcomes and patient survival. Future studies should focus on integrating miRNA therapies with existing treatments to maximize their therapeutic benefits.

**Learning Objectives**

- *Describe* the biological functions and regulatory pathways of oncogenic miRNAs in different cancer types.
- *Identify* therapeutic approaches aimed at silencing or modulating oncogenic miRNAs and their molecular basis.
- *Critically assess* the preclinical and clinical evidence supporting oncomiR-targeted interventions.
- *Examine* ethical, regulatory, and translational challenges associated with developing miRNA-based therapeutics for oncology.

## 4.1 Introduction to Oncogenic miRNAs

Oncogenic microRNAs (oncomiRs) are microRNAs whose aberrant expression promotes cancer by downregulating tumor suppressor genes or activating oncogenic pathways. OncomiRs promote tumorigenesis by downregulating tumor suppressor genes through mechanisms such as translational repression and mRNA destabilization, leading to reduced protein expression essential for controlling cell growth. Their dysregulation leads to increased cell proliferation, invasion, metastasis through processes like epithelial-to-mesenchymal transition (EMT), and resistance to cell death, making them crucial contributors to malignant transformation and tumor progression. OncomiRs are often overexpressed in cancerous tissues and can serve as biomarkers for diagnosis, prognosis, and potential therapeutic targets in various cancers, such as the lung, prostate, brain, and colon cancers. OncomiRs have been shown to act as regulators in the tumor microenvironment (TME), including immune modulation and the response to therapies, making them potential clinical biomarkers and therapeutic targets. OncomiRs can repress translation or destabilize the mRNAs of tumor suppressor genes by binding to specific sequences, leading to reduced protein expression that normally maintains cell homeostasis. Their expression may be modulated by genetic, epigenetic, and environmental factors, such as changes in DNA methylation or chromatin structure can alter miRNA levels in tumors. These miRNAs regulate multiple cellular pathways, including PI3K/AKT/mTOR, MAPK, and cell cycle regulators, by targeting key genes involved in cell proliferation, apoptosis, angiogenesis, and immune evasion.

## 4.2 Therapeutic Targeting Strategies

### 4.2.1 Anti-miRNA Oligonucleotides (AMOs): Chemistry, Mechanisms, and Clinical Progress

Anti-miRNA oligonucleotides (AMOs) are synthetic, short strands of nucleic acids, specifically designed to bind to microRNAs through complementary base pairing, thereby inhibiting their function in gene regulation. By targeting and neutralizing the activity of specific miRNAs, anti-miRNA oligonucleotides effectively prevent these miRNAs from interacting with their mRNA targets and thus derepress the expression of genes that would otherwise be inhibited. The design of AMOs involves incorporating chemical modifications to enhance their stability, binding affinity, and resistance to degradation by nucleases, which is crucial for their effectiveness both in vitro and in vivo. Common chemical modifications include 2′-O-methyl RNA nucleotides, locked nucleic acid (LNA) analogs, and phosphorothioate backbone linkages, which together improve the durability of molecules and affinity for their target miRNA seed regions. AMOs work primarily through a steric blocking mechanism, forming highly stable duplexes with their miRNA targets and preventing miRNA-induced silencing complexes (miRISCs) from repressing mRNA. This precise and potent interference is highly sequence-specific, with even single-nucleotide mismatches greatly reducing AMO activity, reflecting their strict targeting requirements. Clinically, AMOs show promise for treating diseases related to miRNA dysregulation, including cancers, fibrosis, viral infections, and genetic disorders, where aberrant miRNA expression leads to pathological gene expression. In a study, anti-miRNA oligonucleotides (AMOs) inhibited the proliferation of A549 lung cancer cells primarily by inducing apoptosis and cell cycle arrest, targeting specific oncogenic microRNAs (miRNAs) such as miR-21, miR-16, miR-181a, and miR-150. These AMOs suppress proliferation by downregulating these miRNAs, which regulate genes involved in cancer cell survival and proliferation (Fei et al. 2008). The miR-150-5p plays a significant oncogenic role in colon adenocarcinoma by directly repressing the tumor suppressor gene TP53. Studies have shown that miR-150-5p levels are significantly elevated in colon cancer tissues and cell lines, correlating with reduced TP53 expression. Conversely, overexpression of miR-150-5p promotes cancer cell proliferation by downregulating TP53, revealing a critical miR-150-50p-TP53 signaling pathway that regulates tumor growth. This pathway represents a potential target for diagnostic approaches in colorectal cancer (F. Liu and Di Wang 2019). In another study, intratumoral delivery of anti-miR-150 expression vectors in mice resulted in decreased tumor volume and weight by suppressing miR-150 and increasing the expression of the tumor suppressor p53, a critical tumor suppressor protein that regulates apoptosis and cell cycle checkpoints. This indicates that targeting miR-150 not only inhibits tumor growth but also triggers natural tumor-suppressive pathways in vivo (Li et al. 2012). Conversely, antisense oligonucleotides (ASOs), another class of synthetic nucleic acids, were used to target oncogenic miRNAs such as miR-20, miR-106, and miR-150, which can induce apoptosis and sensitize A549 cells to chemotherapeutic agents like cisplatin and experimental anticancer compounds like demethylcantharidin, thereby enhancing the efficacy of cancer treatment. These miRNAs typically promote cancer cell survival by regulating genes such as BCL-2, E2F1, E2F3, RB1, and P53. The combined treatment of ASOs, particularly against miR-106 and miR-150, with anticancer drugs demonstrated a greater reduction in viable A549 cells than either treatment alone. Tumor suppressor miRNAs such as miR-16, miR-17, miR-34a-c, and miR-125 were also identified; however, the ASO effect predominantly involved the inhibition of oncogenic miRNAs (Wang et al. 2010). MicroRNA, i.e., miR-155, has been strongly implicated in the pathogenesis of cutaneous T-cell lymphoma due to its role in promoting tumor cell survival and proliferation. A first-in-human dose-escalation study evaluated the safety, tolerability, and efficacy of MRG-106, a synthetic oligonucleotide inhibitor targeting miR-155, administered intratumorally, subcutaneously, or intravenously in patients with stage I-III mycosis fungoides (MF) presenting with plaques and tumor lesions. MRG-106 was well-tolerated, with only minor local injection site reactions observed, and no dose-limiting toxicities were identified, as the maximum tolerated dose was not reached (Foss et al. 2017). In addition to Cobomarsen, other AMOs targeting miR-221, miR-17, and miR-21 are currently undergoing advanced clinical trials, highlighting the therapeutic potential of miRNA modulation. Although several AMO candidates have advanced to early-stage clinical trials, challenges such as efficient delivery to target tissues, minimizing off-target effects, and avoiding immune responses continue to impede broader therapeutic application (Di Martino et al. 2022). miR-21 is among the most extensively studied OncomiRs, recognized for its overexpression in a wide array of cancers and its role in promoting tumor proliferation and resistance to chemotherapy. Lademirsen, an antisense oligonucleotide targeting miR-21, is currently being investigated in phase II trials for kidney disease (Alport syndrome) and has demonstrated favorable safety profiles, although efficacy outcomes are still under evaluation (Gale et al. 2024). Recent advancements have led to the development of innovative antisense morpholino oligomer (AMO) designs, which are distinguished by increased specificity, potency, and reduced toxicity. This marks a rapidly evolving area in RNA-based therapeutic innovation. The progress in AMO development has been substantial, with novel chemical modifications playing a pivotal role in enhancing both their efficacy and safety. A recent innovation involves the incorporation of a non-nucleotide modifier known as ZEN (N, N-diethyl-4-(4-nitronaphthalen-1-ylazo)-phenylamine), which, when combined with a 2′-O-methyl RNA backbone, substantially enhances the binding affinity of AMOs to target microRNAs (miRNAs) and significantly increases their resistance to exonuclease degradation. This advancement yields oligonucleotides that exhibit high potency at low nanomolar concentrations, superior specificity, and minimal toxicity in vitro. These attributes effectively address the primary limitations of earlier generation AMOs, which utilized phosphorothioate linkages or locked nucleic acid (LNA) modifications and occasionally encountered toxicity issues (Lennox et al. 2013; Meng et al. 2025; Saw and Song 2024). AMOs offer a robust platform for the posttranscriptional modulation of gene expression by selectively inhibiting microRNAs (miRNAs), presenting significant potential implications for both research and clinical therapy.

### 4.2.2 miRNA Sponges: Design, Competitive Inhibition, and Therapeutic Applications

miRNA sponges are RNA molecules designed to competitively inhibit microRNAs by comprising multiple binding sites complementary to target miRNAs. They act by "sponging" or "sequestering" miRNAs, preventing them from interacting with their natural mRNA targets, thereby modulating gene expression. This competitive inhibition can be long-lasting when sponges are expressed from transgenes, making them valuable tools for both functional studies and therapeutic applications. miRNA sponges typically consist of transcripts with multiple tandem miRNA-binding sites (MBSs). The binding sites are either perfectly complementary or contain bulged sequences at central positions to avoid degradation by the RNA cleavage mechanism. Multiple copies of MBS are included to enhance sponge efficacy. Spacers of 4–6 nucleotides separate the binding sites to optimize binding and reduce RNA secondary structure formation. Sponge constructs are often delivered via vectors or transgenes with strong promoters for robust expression. These design features allow miRNA sponges to inhibit entire miRNA seed families or clusters effectively. Designs with 4–12 repeated binding sites are common, with the exact number balanced to maximize inhibition while avoiding risks like RNA degradation or recombination.

miRNA sponges function by binding active miRNA–RISC complexes, preventing miRNAs from recognising and cleaving their natural mRNA targets. This sequestration mimics a loss-of-function phenotype for the inhibited miRNAs but does so in a reversible and tunable manner. Because sponges target the miRNA seed region, they can inhibit multiple miRNAs sharing that seed sequence, thus allowing broad regulation of miRNA families. Their expression from stable transgenes allows long-term miRNA inhibition, unlike transit antisense oligonucleotide. miRNA sponges have been developed to target disease-associated miRNAs in various conditions showing promise as therapeutic agents. miRNA Sponges targeting oncogenic miRNAs such as miR-21 and miR-155 have been used to inhibit tumor growth and hematological malignancies. miR-155-5p acts mainly as an oncogenic miRNA in breast cancer and is frequently overexpressed, especially in aggressive subtypes like triple-negative and HER2-enriched tumors. High levels are linked to poor prognosis, loss of hormone receptors, increased proliferation, and homologous recombination deficiency (HRD) (Li et al. 2021; Pasculli et al. 2020). Conversely, miR-21-5p is an oncogene widely overexpressed in breast cancer and other cancers. It promotes migration and angiogenesis. It can be transferred via exosomes from cancer cells to endothelial cells, where it suppresses Krev interaction trapped protein 1 (KRIT1) and activates beta-catenin signaling to induce angiogenesis and vascular permeability (Degheidy et al. 2025; He et al. 2021; Liu et al. 2021). Studies that engineered exosomes to carry a miR-21-sponge construct have demonstrated significant therapeutic potential in a rat model of glioblastoma. By effectively delivering this miRNA sponge, the overexpressed oncogenic miR-21 in glioblastoma cells is inhibited, leading to the restoration of tumor suppressor genes such as PDCD4 and RECK. This inhibition results in reduced tumor cell proliferation and promotes apoptosis, culminating in a marked decrease in tumor volume in treated rats. The utilization of engineered exosomes as delivery vehicles is particularly advantageous due to their capacity to traverse the blood–brain barrier and specifically target brain tumor cells. This approach represents a novel targeted therapeutic strategy against glioblastoma, with the potential to enhance treatment outcomes for this aggressive cancer (Gaur et al. 2011; Monfared et al. 2019). Silencing miR-21 markedly enhances the cytotoxic efficacy of the antiangiogenic agent sunitinib in glioblastoma. Upon silencing miR-21 in glioma cells, there is an upregulation of tumor suppressor levels, which leads to increased caspase activation and heightened apoptosis. Notably, the combination of miR-21 silencing with sunitinib treatment yields a synergistic antitumor effect, resulting in a more pronounced reduction in tumor cell viability compared to either treatment alone. This effect appears to be specific to sunitinib, as a similar enhancement is not observed with conventional chemotherapeutic agents such as temozolomide (TMZ) (Costa et al. 2013; Zhou and Gallo 2009). Conversely, miR-92a sponges facilitate vascular repair and angiogenesis following injury. The application of miR-92a inhibitors augments endothelial progenitor cell (EPC)-mediated regeneration of damaged arteries, primarily by enhancing endothelial cell proliferation, migration, and angiogenic function, while mitigating injury-induced apoptosis and oxidative stress. miR-92a is upregulated in endothelial cells and EPCs under pathological conditions such as vascular injury, hypoxia, and hyperglycemic environments, which impairs their regenerative capabilities. Inhibition of miR-92a accelerates re-endothelialization by promoting EPC and endothelial cell migration, proliferation, and angiogenesis, which are essential for repairing the damaged vascular lining of arteries. miR-92a inhibition de-represses target genes such as integrin a5 and sirtuin1, thereby supporting endothelial recovery by enhancing cell survival and function (Dos Santos and Ruivo 2018; Chamorro-Jorganes et al. 2013; Daniel et al. 2014). Additionally, miR-92a inhibitors decrease reactive oxygen species (ROS) production and apoptosis in EPCs under stress conditions like hypoxia, thus preserving and enhancing EPC viability and function. The beneficial effects involve the regulation of key signaling pathways, including SMAD2/3/FAK/AKT/eNOS, activated through the modulation of growth factors such as GDF11, leading to improved nitric oxide production and EPC-mediated vascular repair (Huang et al. 2019). Systemic administration of locked nucleic acid (LNA)-based miR-92a inhibitors in animal models accelerated endothelial recovery after arterial injury, reduced neointimal lesion formation, and decreased leukocyte accumulation, which are critical for arterial regeneration. Elevated miR-92a expression is found in the aortic endothelium of diabetic mice and in renal arteries from diabetic patients, particularly under conditions of advanced glycation end products (AGEs) and oxidized low-density lipoprotein (LDL) exposure. Overexpression of miR-92a impairs endothelium-dependent relaxations (EDRs) and suppresses HO-1 (Heme oxygenase-1) expression. Inhibiting miR-92a significantly increases HO-1 expression in human endothelial cells and mouse aortas, which leads to reduced ROS production and oxidative stress in the vascular endothelium. Treatment with miR-92a inhibitors improves endothelium function by enhancing EDRs and normalizing ROS levels in mouse aortas. The protective effects of miR-92a inhibition are reversed when HO-1 is inhibited or knocked down, indicating that HO-1 is a critical mediator of these vascular benefits (Gou et al. 2018; Scioli et al. 2020). A notable study targeted four oncogenic miRNAs, such as miR-21, miR-155, and miR-221/222, common in breast and pancreatic cancers by designing a multi-potent miRNA sponge with perfect and bulged MBS sequences to ensure high affinity and stable binding. The sponge was shown to effectively inhibit these four miRNAs simultaneously, confirmed by luciferase reporter assays and PCR. The expression of the multi-potent sponge increased the protein levels of genes normally suppressed by these miRNAs, sensitized cancer cells to chemotherapy drugs like doxorubicin, and reduced cell migration and proliferation. The bulged version of the sponge showed stronger functional inhibition, likely due to better interaction with the miRNA machinery (Jung et al. 2015).

### 4.2.3 Small Molecule Inhibitors and miRNA Masking: Molecular Mechanisms and Validation

Small molecule inhibitors (SMIs) are chemical compounds designed to target specific molecules involved in cancer development. Their mechanisms include inhibiting signaling pathways, modulating immune checkpoints, such as CTLA-4, PD-1/PD-L1, interfering with gene transcription, DNA repair, epigenetic modification, or protein function critical for tumor growth and immune evasion. These inhibitors can enhance immune response against tumors and disrupt tumor-supporting pathways. A study utilizing a panel of human breast cancer cell lines demonstrated that basal-type and post-epithelial-to-mesenchymal transition (post-EMT) breast cancer cells exhibited the highest sensitivity to dasatinib treatment, resulting in significant growth inhibition. Dasatinib is an orally active small-molecule inhibitor that targets SRC and ABL kinases. Specifically, it is a potent dual inhibitor that targets the active and inactive conformations of these kinases with high affinity, which differentiates it from other inhibitors like imatinib that only bind one conformation. ABL kinase, particularly the BCR–ABL fusion protein tyrosine kinase, is a key driver in certain leukemias such as chronic myeloid leukemia (CML). Dasatinib inhibits this kinase, suppressing the abnormal signaling that promotes cancer cell growth. SRC family kinases are non-receptor tyrosine kinases involved in various cellular processes, including proliferation, survival, and migration (Sun et al. 2024). Dasatinib selectively inhibits the proliferation of basal-type or "triple-negative" breast cancer (TNBC) cell lines in vitro. Moreover, dasatinib shows promise as part of combination therapy with agents such as paclitaxel to overcome chemotherapy resistance in TNBC cells by targeting breast cancer stem cells (BCSCs) and reducing tumor growth in preclinical models (Finn et al. 2007; Tian et al. 2018; Tryfonopoulos et al. 2011). Breast cancer cells can evade apoptosis-mediated selective killing by SMI targeting the BCL-2 protein family, which are key regulators of apoptosis. SMI of BCL-2, such as YC137, have been designed to block the binding of pro-apoptotic proteins to BCL-2, aiming to restore apoptosis selectively in cancer cells that overexpress BCL-2 (Real et al. 2004). However, challenges include drug resistance mechanisms such as target mutation, drug efflux, and metabolic inactivation. Recent advances have also focused on small-molecule inhibitors targeting oncogenic miRNAs, offering potential for improved cellular uptake and stability compared to oligonucleotide therapies. These inhibitors can bind to pre-miRNAs to inhibit their biogenesis or function, representing an emerging cancer treatment modality. Small molecule inhibitors have been shown to overcome tumor immune evasion and drug resistance, while miRNA masking and small RNA zippers have demonstrated the capability to rescue gene expression and sensitize tumors to chemotherapeutics like doxorubicin in breast cancer cell lines (Zangouei et al. 2021) (Table 4.1).

miRNA masking involves the use of synthetic oligonucleotides, called miR-masks, which are complementary to the 3′ untranslated region (3′ UTR) of specific target mRNAs. These miR-masks competitively block endogenous miRNAs from binding to their target mRNAs, thus preventing the miRNA-mediated repression of gene expression. This approach allows selective inhibition of miRNA activity on specific mRNA targets without affecting global miRNA levels. By rescuing the expression of tumor suppressor genes or silencing oncogenes, miRNA masking holds promise in cancer therapeutics. Additionally, "small RNA zippers" are a novel type of miRNA inhibitor that locks miRNA molecules by connecting them end to end with high-affinity DNA–RNA binding, effectively knocking down miRNA levels and reversing oncogenic effects, demonstrated in breast cancer models with miR-221 as a target.

**Table 4.1** MicroRNAs modulating doxorubicin resistance mechanisms in breast cancer

| Authors | MicroRNA | Role in DOX Resistance | Target/Pathway | Effect on breast tumor cells |
|---|---|---|---|---|
| Dai et al. (2019) | miR-222 | Promotes DOX resistance | Targets BIM (Bcl-2 interacting mediator), affecting apoptosis via Bim-caspase pathway | Upregulates in resistant cells, suppresses apoptosis by downregulating BIM, leading to increased proliferation and migration, reducing DOX sensitivity. Targeting miR-222 can reverse resistance via BIM-caspase pathway activation |
| Du et al. (2019) | miR-137 | Alleviates DOX resistance | Represses chemoresistance development | Enhances DOX sensitivity |
| Chen et al. (2012) | miR-200c | Negatively regulates P-glycoprotein (P-gp) | ABC transporter p-gp (ABCB1) | Downregulates p-gp, reducing drug efflux, thereby reversing DOX resistance. P-gp overexpression is a major contributor to drug resistance by pumping DOX out of cells |

(continued)

**Table 4.1** (continued)

| Authors | MicroRNA | Role in DOX Resistance | Target/Pathway | Effect on breast tumor cells |
|---|---|---|---|---|
| Armada et al. (2019) | miR-203 | Negatively regulates P-gp | ABC transporter p-gp (ABCB1) | Sensitizes cells to DOX |
| Bao et al. (2012) | miR-298 | Suppresses P-gp expression | Binds 3' UTR of P-gp mRNA | Direct regulation of P-gp expression reducing the efflux of DOX, facilitating drug retention and reversing resistance |
| H. Yi et al. (2016) | miR-129-5p | Suppresses P-gp and CDK6 | Inhibits P-gp and CDK6 (cell cycle kinase) | Inhibits P-gp, increasing intracellular DOX accumulation; suppresses CDK6 causing G2 phase cell cycle arrest, enhancing DOX sensitivity |
| M. Gao et al. (2016), Wuxiao et al. (2020) | miR-145 | Enhances DOX sensitivity | Targets MRP1 (multidrug resistance-associated protein 1) | Reduces MRP-1-mediated drug efflux, increasing intracellular drug concentration, leading to increased apoptosis and reduced tumor growth |
| Lu et al. (2015) | miR-134 | Reverses DOX resistance | Suppresses MRP1 | Inhibits proliferation, triggers apoptosis, and enhances DOX sensitivity by targeting MRP1 transporter |
| Lu et al. (2022) | miR-186-5p | Overcomes DOX resistance | Targets MRP1 | Targets MRP1 to reverse drug resistance by reducing efflux and promoting apoptosis |
| Chang et al. (2018) | miR-199a | Regulates DOX resistance | MRP1 axis | Contributing to DOX resistance via modulating MRP1 expression, impacting drug efflux |
| Kim et al. (2023) | miR-221-3p | Tumor suppressor, reverses DOX resistance | Targets P-gp and interacts with lncRNA GAS5/Wnt/beta-catenin signaling | GAS5 acts as competing endogenous RNA (ceRNA) modulating miR-221 and P-gp, regulating resistance. Its upregulation represses P-gp and Wnt signaling, reversing resistance |

(continued)

**Table 4.1** (continued)

| Authors | MicroRNA | Role in DOX Resistance | Target/Pathway | Effect on breast tumor cells |
|---|---|---|---|---|
| Bourguignon et al. (2009) | miR-21 | Indirectly regulates P-gp & modulates DOX resistance | Hyaluronan-CD44-PKC pathway | Promotes DOX resistance by mediating P-gp expression via the pathways, enhancing drug efflux, and tumor survival |
| Zhao et al. (2016) | miR-302 family | Sensitizes to DOX | Suppresses P-gp via MEKK1 inhibition | Inhibits MAP/ERK kinase 1 (MEKK1), decreasing P-gp levels and enhancing DOX cytotoxicity in resistant cells |
| Tormo et al. (2019) | miR-449 family | Mediates DOX resistance | Cell cycle regulation involved | Overexpression leads to cell cycle regulation alterations that confer resistance; modulation can affect DOX sensitivity in triple-negative breast cancer |
| Sun et al. (2018) | miR-574 | Enhances DOX resistance | Downregulates SMAD4 | Promotes drug resistance by targeting SMAD4, which influences pathways related to cell proliferation and survival |

### 4.2.4 Locked Nucleic Acids (LNAs): Stability Enhancement, Modifications, and in Vivo Applications

Locked nucleic acids (LNAs) are chemically modified RNA nucleotides with a methylene bridge locking the ribose ring in a rigid 3′-endo conformation, greatly enhancing duplex stability, binding affinity, and specificity. This chemical modification improves thermal stability, resistance to enzymatic degradation, and mismatch discrimination, making LNAs powerful tools in nucleic acid therapeutics and diagnostics. The methylene bridge in LNAs locks the sugar conformation, which increases thermal duplex stability and enhances hybridization specificity. LNAs have increased melting temperature (Tm) by several degrees per modification compared to unmodified DNA or RNA, and they show high resistance to nucleases, thus offering improved biostability both in vitro and in vivo (Elmén et al. 2005; Gao et al. 2024; Kauppinen et al. 2005; Vester and Wengel 2004). LNAs have been explored extensively for ASO therapies targeting cancer-related genes. Full LNA oligonucleotides exhibit remarkable stability and potent gene-silencing activity in vivo, with demonstrated tumor growth inhibition in mouse xenograft cancer models. LNAs targeting genes like POLR2A have shown sequence-specific tumor suppression without significant toxicity at therapeutically relevant doses. Thie superior pharmacokinetics make them promising candidates for cancer therapeutics (Fluiter et al. 2003).

## 4.3 Delivery Systems and Technologies

### 4.3.1 Nanoparticles and Lipoplexes: Targeted Delivery and Efficiency

Nanoparticles and lipoplexes are advanced delivery systems used in cancer therapy for targeted delivery, improved efficiency, and reduction of side effects. They work by specifically targeting cancer cells, delivering drugs or genes effectively while sparing normal cells, and have shown clinical promise across various cancers. Nanoparticles (NPs) are engineered to deliver chemotherapeutic drugs, nucleic acids or immunotherapeutic agents directly to tumor sites. Their small size, surface properties, and functionalization with targeting ligands, like antibodies, peptides and folic acid, enable precise targeting of tumor cells through enhanced permeability and retention effects., this targeting increases drug concentration in tumor tissue and reduces exposure to normal cells, minimizing side effects such as cardiotoxicity observed with free drugs such as doxorubicin-loaded PEGylated liposomes. NPs can also bypass drug resistance mechanisms in cancer cells and improve pharmacokinetics by extending drug half-life. Conversely, lipoplexes are lipid-based complexes primarily used for delivering nucleic acids such as tumor suppressor genes to cancer cells. Their often-cationic charge favors interaction with the negatively charged surface of cancer cells, enhancing selective uptake via endocytosis. Functionalization with polyethylene glycol (PEG) and targeting molecules like folic acid can further improve stability, selectivity, and efficiency of gene transfer to cancer cells. The Oncoprex delivery system utilizes lipoplexes to deliver tumor suppressor genes intravenously, showing promising anticancer activity in trails by selectively targeting tumor cells without immunogenicity. In breast cancer, nanoparticle formulations like PEGylated liposomes and nanoparticle albumin-bound paclitaxel (Nab-paclitaxel) have demonstrated reduced toxicity and higher tolerated doses compared to conventional chemotherapy. PEGylated liposomal doxorubicin (Doxil) has a much longer half-life (about 74 h) compared to free doxorubicin (about 5 min). This leads to less peak plasma concentration and reduced cardiac toxicity, a common side effect of conventional doxorubicin (Fiegl et al. 2011; Li et al. 2019; Tang et al. 2017). On the other hand, Nab-paclitaxel allows higher doses to be administered safely, leading to improved clinical benefit and prolonged survival in metastatic breast cancer patients (Singh et al. 2014). In ovarian cancer, the folate receptor (FR) alpha is significantly overexpressed, rendering it an ideal target for receptor-mediated delivery systems such as folate-targeted lipoplexes (FTLs). These lipoplexes are lipid-based nanoparticles (NPs) engineered to deliver therapeutic genes specifically to FR-positive ovarian cancer cells. The targeted delivery of therapeutic genes, driven by tumor-specific promoters such as hTERT, has resulted in enhanced transfection efficiency and gene expression in FR-positive ovarian cancer cells, as well as cell cycle arrest and increased apoptosis in treated cancer cells (He et al. 2013a, 2016). A study designed folate-linked lipoplexes to deliver short hairpin RNA (shRNA) targeting claudin-3 (CDLN3), which has demonstrated promising antitumor effects in ovarian cancer xenograft models. Both FR alpha and claudin-3 are highly overexpressed in ovarian cancers and are associated with tumor progression and poor patient prognosis. The F-P-LP/CDLN3 complex formulation selectively targets ovarian cancer cells via FR alpha and delivers shRNA to downregulate CDLN3 expression (He et al. 2013b). In glioblastoma, nanoparticles conjugated with tumor-penetrating peptides (e.g., iGRD) facilitate brain tumor targeting, improving drug delivery and efficacy in glioblastoma models (Rahman et al. 2025). The iGRD peptide is known for its ability to penetrate tumors efficiently by binding to specific receptors like neuropilin-1 on tumor cells and tumor vasculature. This peptide facilitates deep penetration of nanoparticles beyond the vascular endothelium into tumor tissues, overcoming physical barriers like the blood–brain barrier and dense tumor extracellular matrix (Kang et al. 2020; Zuo 2019). Studies have shown that combining iGRD with proapoptotic peptide-bearing NPs improves the internalization of these nanocarriers into tumor cells and tumor endothelial cells, effectively inhibiting tumor growth and significantly prolonging survival in murine glioblastoma models. Compared to conventional peptides, iGRD improves the intratumoral accumulation and distribution of drug-loaded nanoparticles, enhancing drug absorption by tumor cells in glioblastoma (Agemy et al. 2011; Gu et al. 2013).

### 4.3.2 Viral Vector-Based Delivery for miRNA Mimics or Inhibitors

Viral vector-mediated delivery of microRNA (miRNA) mimics or inhibitors represents a highly effective approach for the introduction or suppression of miRNAs within target cells. Commonly employed viral vectors for this purpose include retroviruses, lentiviruses, adenoviruses, and adeno-associated viruses (AAVs), owing to their superior transfection efficiency and capacity to facilitate long-term or stable expression of miRNA mimics or inhibitors. The miRNA pathways can be significantly altered in response to noncoding RNA delivery via viral vector-based gene therapy. This alteration can be caused both by the viral vectors themselves and by the delivered noncoding RNA genetic material. These vectors can disrupt cellular miRNA expression and processing pathways, leading to change sin gene regulation and cellular metabolism. Viral vectors can induce immune responses and cytotoxicity affecting miRNA pathways, which play crucial roles in transcription regulation.

Adenovirus synthesizes virus-associated RNAs (VA RNAs), predominantly VA RNAI and VA RNAII, which compete with endogenous pre-miRNAs for nuclear export (Aparicio et al. 2009; Kamel et al. 2013). This competition saturates the Exportin 5 (Exp-5) protein, a RanGTP-dependent transporter responsible for translocating pre-miRNAs and certain mRNAs into the cytoplasm, thereby diminishing the export and processing of cellular pre-miRNAs by Dicer (Bohnsack et al. 2004; Yi et al. 2003). VA RNAs are processed into mivaRNAs, also referred to as small viral RNAs or svaRNAs, which integrate into the RNA-induced silencing complex (RISC) and can interfere with miRNA activity by repressing host gene expression. The cumulative effect of Exportin-5 saturation, Dicer inhibition, and mivaRNAs-mediated gene silencing results in the disruption of the host miRNA pathway and the downregulation of critical host gene expressions. This likely facilitates viral evasion of host antiviral responses and optimizes the cellular environment for viral replication (Savenkova et al. 2022; Kumar et al. 2024). AAV infections (e.g., AAV2. AAV3) can downregulate specific miRNAs such as hsa-miR-4488, altering gene expression. Hsa-miR-4488 targets important genes like APC, which is a tumor suppressor involved in regulating the Wnt signaling pathway, cell adhesion, and cell cycle, and CERS2, involved in lipid synthesis. The decrease in miR-4488 upon AAV infection can therefore influence these pathways and cellular processes (Yin et al. 2020). Additionally, research shows that AAV infections consistently reduce hsa-miR-4488 levels in infected cell lines such as Huh-7 and ARPE-19, leading to changes in gene expression important for virus-host interactions (Arumugam et al. 2020). Vector-derived artificial microRNAs (amiRNAs) can effectively inhibit West Nile virus (WNV) replication and protein expression. Studies have shown that amiRNAs targeting conserved genomic regions of WNV, such as the NS5 region, significantly reduce viral replication and protein levels in infected cells. Dual amiRNA expression cassettes targeting multiple viral gene regions have shown even better efficacy in blocking WNV infection and replication (Karothia et al. 2020). Lentivirus and HIV-1 infection impact multiple miRNA pathways related to immune response and neuropathogenesis. Some studies report that carefully designed pri-amiRNA delivery by AAV does not adversely affect endogenous miRNA profiles, exemplified in therapies for Huntington's disease (Chinniah et al. 2022; Gupta and Saksena 2013; White et al. 1999). The advantages of utilizing viral vector delivery encompass high transduction efficiency, the capability to deliver miRNA mimics or inhibitors across diverse tissue types, and stable expression upon integration. Additionally, tissue-specific targeting can be achieved using various viral serotypes or miRNA-regulated expression systems, rendering this method effective in both miRNA replacement therapy (via miRNA mimics) and miRNA inhibition therapy (via inhibitors such as antagomirs and sponges). However, challenges persist, including potential immunogenicity, particularly with adenoviral vectors, which may constrain repeated administrations, as well as possible genetic instability or insertional mutagenesis associated with integrating viruses like retroviruses. Strategies to confine transgene expression to specific cell types using miRNA target sites for regulation are under exploration (Khare et al. 2011; Lee et al. 2017). Nonviral methods, such as polyethylenimine (PEI) nanoparticle delivery, are also being investigated due to safety concerns associated with viral vectors, although they may exhibit lower transfection efficiency (Lungwitz et al. 2005; Mellott et al. 2012; Zakeri et al. 2018) (Fig. 4.1).

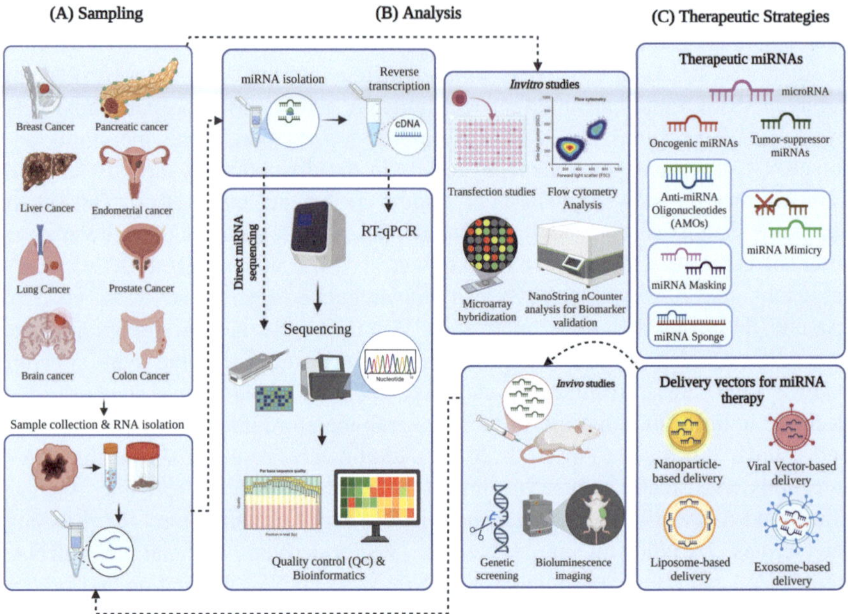


**Fig. 4.1** Integrated Workflow of MicroRNA-Based Cancer Diagnostics and Therapeutics. The schematic illustrates (**A**) tumor sampling & RNA isolation, (**B**) miRNA processing & analysis, and downstream quantification (dotted arrows indicate sequential progress). (C) Right panel depicts in vitro and/or in vivo evaluation, therapeutic strategies, and delivery vectors for miRNA therapy

## 4.4 Future Directions and Conclusion

Recent advancements in technology have sought to modulate miRNA activity to either inhibit tumor growth or restore tumor-suppressive miRNAs. Oncogenic miRNAs function as biomarkers for cancer diagnosis and prognosis within the realm of personalized medicine, facilitating the customization of treatments according to individual molecular profiles. Emerging therapeutic strategies emphasize the combination of miRNA replacement with gene silencing or chemotherapy, the enhancement of delivery methods, and the integration of computational approaches to elucidate miRNA interactions and identify novel targets. The CRISPR/Cas9 genome-editing technology holds promise for the precise targeting of miRNA genes, although further development is necessary for clinical application. miRNA profiling from tissue samples or liquid biopsies aids in patient stratification and the personalization of therapy based on molecular characteristics. Single-nucleotide polymorphisms that affect miRNAs or their binding sites can influence cancer risk and drug response, thereby supporting personalized treatment approaches. The complexity of miRNA regulatory networks and the incomplete understanding of miRNA–mRNA interactions pose challenges to precise targeting.

The combination of miRNA therapies with conventional chemotherapies enhances therapeutic efficacy and helps overcome resistance. Computational and systems biology approaches are employed to identify novel miRNA regulators and therapeutic targets. The development of more effective and safer delivery platforms is crucial for achieving clinical success.

## References


Agemy L, Friedmann-Morvinski D, Kotamraju VR, Roth L, Sugahara KN, Girard OM, Mattrey RF, Verma IM, Ruoslahti E (2011) Targeted nanoparticle enhanced proapoptotic peptide as potential therapy for glioblastoma. Proc Natl Acad Sci USA 108(42):17450–17455. https://doi.org/10.1073/PNAS.1114518108/SUPPL_FILE/PNAS.201114518SI.PDF

Aparicio O, Carnero E, Abad X, Razquin N, Guruceaga E, Segura V, Fortes P (2009) Adenovirus VA RNA-derived miRNAs target cellular genes involved in cell growth, gene expression and DNA repair. Nucleic Acids Res 38(3):750. https://doi.org/10.1093/NAR/GKP1028

Armada A, Gomes BC, Viveiros M, Rueff J, Rodrigues AS (2019) Regulation of ABCB1 activity by microRNA-200c and microRNA-203a in breast cancer cells: the quest for microRNAs' involvement in cancer drug resistance. Cancer Drug Resistance 2(3):897. https://doi.org/10.20517/CDR.2019.24

Arumugam S, Mary B, Kumar M, Jayandharan GR (2020) Analysis of hepatic and retinal cell microRNAome during AAV infection reveals their diverse impact on viral transduction and cellular physiology. Gene 724:144157. https://doi.org/10.1016/J.GENE.2019.144157

Bao L, Hazari S, Mehra S, Kaushal D, Moroz K, Dash S (2012) Increased expression of P-glycoprotein and doxorubicin Chemoresistance of metastatic breast cancer is regulated by miR-298. Am J Pathol 180(6):2490–2503. https://doi.org/10.1016/J.AJPATH.2012.02.024

Bohnsack MT, Czaplinski K, Görlich D (2004) Exportin 5 is a RanGTP-dependent dsRNA-binding protein that mediates nuclear export of pre-miRNAs. RNA 10(2):185. https://doi.org/10.1261/RNA.5167604

Bourguignon LYW, Spevak CC, Wong G, Xia W, Gilad E (2009) Hyaluronan-CD44 interaction with protein kinase Cε promotes oncogenic signaling by the stem cell marker nanog and the production of microRNA-21, leading to down-regulation of the tumor suppressor protein PDCD4, anti-apoptosis, and chemotherapy resistance in breast tumor cells. J Biol Chem 284(39):26533–26546. https://doi.org/10.1074/JBC.M109.027466

Chamorro-Jorganes A, Araldi E, Suárez Y (2013) MicroRNAs as pharmacological targets in endothelial cell function and dysfunction. Pharmacol Res 75:15–27. https://doi.org/10.1016/J.PHRS.2013.04.002

Chang L, Hu Z, Zhou Z, Zhang H (2018) Linc00518 contributes to multidrug resistance through regulating the MiR-199a/MRP1 Axis in breast cancer. Cell Physiol Biochem 48(1):16–28. https://doi.org/10.1159/000491659

Chen J, Tian W, Cai H, He H, Deng Y (2012) Down-regulation of microRNA-200c is associated with drug resistance in human breast cancer. Med Oncol 29(4):2527–2534. https://doi.org/10.1007/S12032-011-0117-4/FIGURES/4

Chinniah R, Adimulam T, Nandlal L, Arumugam T, Ramsuran V (2022) The effect of miRNA gene regulation on HIV disease. Front Genet 13:862642. https://doi.org/10.3389/FGENE.2022.862642

Costa PM, Cardoso AL, Nóbrega C, Pereira de Almeida LF, Bruce JN, Canoll P, Pedroso de Lima MC (2013) MicroRNA-21 silencing enhances the cytotoxic effect of the antiangiogenic drug sunitinib in glioblastoma. Hum Mol Genet 22(5):904–918. https://doi.org/10.1093/HMG/DDS496

Dai H, Xu LY, Qian Q, Zhu QW, Chen WX (2019) MicroRNA-222 promotes drug resistance to doxorubicin in breast cancer via regulation of miR-222/bim pathway. Biosci Rep 39(7): BSR20190650. https://doi.org/10.1042/BSR20190650

Daniel JM, Penzkofer D, Teske R, Dutzmann J, Koch A, Bielenberg W, Bonauer A, Boon RA, Fischer A, Bauersachs J, Van Rooij E, Dimmeler S, Sedding DG (2014) Inhibition of miR-92a improves re-endothelialization and prevents neointima formation following vascular injury. Cardiovasc Res 103(4):564. https://doi.org/10.1093/CVR/CVU162

Degheidy MS, Abou-Elalla AA, Kamel MM, Abdel-Ghany S, Arneth B, Sabit H (2025) Regulatory roles of miR-155-5p, miR-21-5p, miR-93-5p, and miR-140-5p in breast cancer progression. Curr Issues Mol Biol 47:377. https://doi.org/10.3390/CIMB47050377

Di Martino MT, Arbitrio M, Caracciolo D, Cordua A, Cuomo O, Grillone K, Riillo C, Caridà G, Scionti F, Labanca C, Romeo C, Siciliano MA, D'Apolito M, Napoli C, Montesano M, Farenza V, Uppolo V, Tafuni M, Falcone F et al (2022) miR-221/222 as biomarkers and targets for therapeutic intervention on cancer and other diseases: A systematic review. Mol Ther Nucleic Acids 27:1191. https://doi.org/10.1016/J.OMTN.2022.02.005

Dos Santos A, Ruivo JM (2018) The use of miR-92a inhibitor to enhance endothelial progenitor cell-mediated regeneration of injured arteries. Doctoral Thesis, UCL (University College London)

Du F, Yu L, Wu Y, Wang S, Yao J, Zheng X, Xie S, Zhang S, Lu X, Liu Y, Chen W (2019) miR-137 alleviates doxorubicin resistance in breast cancer through inhibition of epithelial-mesenchymal transition by targeting DUSP4. Cell Death Dis 10(12):1–10. https://doi.org/10.1038/s41419-019-2164-2

Elmén J, Thonberg H, Ljungberg K, Frieden M, Westergaard M, Xu Y, Wahren B, Liang Z, Ørum H, Koch T, Wahlestedt C (2005) Locked nucleic acid (LNA) mediated improvements in siRNA stability and functionality. Nucleic Acids Res 33(1):439. https://doi.org/10.1093/NAR/GKI193

Fei J, Lan F, Guo M, Li Y, Liu Y (2008) Inhibitory effects of anti-miRNA oligonucleotides (AMOs) on A549 cell growth. J Drug Target 16(9):688–693. https://doi.org/10.1080/10611860802295946

Fiegl M, Mlineritsch B, Hubalek M, Bartsch R, Pluschnig U, Steger GG (2011) Single-agent pegylated liposomal doxorubicin (PLD) in the treatment of metastatic breast cancer: results of an Austrian observational trial. BMC Cancer 11:373. https://doi.org/10.1186/1471-2407-11-373

Finn RS, Dering J, Ginther C, Wilson CA, Glaspy P, Tchekmedyian N, Slamon DJ (2007) Dasatinib, an orally active small molecule inhibitor of both the src and abl kinases, selectively inhibits growth of basal-type/"triple- negative" breast cancer cell lines growing in vitro. Breast Cancer Res Treat 105(3):319–326. https://doi.org/10.1007/S10549-006-9463-X

Fluiter K, ten Asbroek ALMA, de Wissel MB, Jakobs ME, Wissenbach M, Olsson H, Olsen O, Oerum H, Baas F (2003) In vivo tumor growth inhibition and biodistribution studies of locked nucleic acid (LNA) antisense oligonucleotides. Nucleic Acids Res 31(3):953. https://doi.org/10.1093/NAR/GKG185

Foss FM, Querfeld C, Porcu P, Kim YH, Pacheco T, Halwani AS, DeSimone J, William BM, Seto AG, Ruckman J, Landry ML, Jackson AL, Pestano LA, Dickinson BA, Sanseverino M, Rodman DM, Rubin P, Gordon GS, Marshall WS (2017) Phase 1 trial evaluating MRG-106, a synthetic inhibitor of microRNA-155, in patients with cutaneous t-cell lymphoma (CTCL). J Clin Oncol 35(15_suppl):7564–7564. https://doi.org/10.1200/JCO.2017.35.15_SUPPL.7564

Gale DP, Gross O, Wang F, Esteban De La Rosa RJ, Hall M, Sayer JA, Appel G, Hariri A, Liu S, Maski M, Shen Y, Zhang Q, Iqbal S, Kowthalam MU, Lin J, Ding J, Chakera A, Chauveau D, Soler Romeo MJ et al (2024) A randomized controlled clinical trial testing effects of Lademirsen on kidney function decline in adults with Alport syndrome. Clin J Am Soc Nephrol 19(8): 995–1004. https://doi.org/10.2215/CJN.0000000000000458

Gao M, Miao L, Liu M, Li C, Yu C, Yan H, Yin Y, Wang Y, Qi X, Ren J (2016) miR-145 sensitizes breast cancer to doxorubicin by targeting multidrug resistance-associated protein-1. Oncotarget 7(37):59714. https://doi.org/10.18632/ONCOTARGET.10845

Gao J, Nutan B, Gargouri D, Pisal ND, Do V, Zubair M, Alanzi H, Wang H, Lee D, Joshi N, Ullah A (2024) Unlocking the potential of chemically modified nucleic acid therapeutics. Adv Ther 7(11):2400231. https://doi.org/10.1002/ADTP.202400231;SUBPAGE:STRING:FULL

Gaur AB, Holbeck SL, Colburn NH, Israel MA (2011) Downregulation of Pdcd4 by mir-21 facilitates glioblastoma proliferation in vivo. Neuro-Oncology 13(6):580–590. https://doi.org/10.1093/NEUONC/NOR033

Gou L, Zhao L, Song W, Wang L, Liu J, Zhang H, Lau CW, Yao X, Tian XY, Wong WT, Luo JY, Huang Y, Huang Y (2018) Inhibition of miR-92a suppresses oxidative stress and improves endothelial function by upregulating Heme Oxygenase-1 in db/db mice. Antioxid Redox Signal 28(5):358–370. https://doi.org/10.1089/ARS.2017.7005/ASSET/IMAGES/LARGE/FIGURE7.JPEG

Gu G, Gao X, Hu Q, Kang T, Liu Z, Jiang M, Miao D, Song Q, Yao L, Tu Y, Pang Z, Chen H, Jiang X, Chen J (2013) The influence of the penetrating peptide iRGD on the effect of paclitaxel-loaded MT1-AF7p-conjugated nanoparticles on glioma cells. Biomaterials 34(21): 5138–5148. https://doi.org/10.1016/j.biomaterials.2013.03.036

Gupta P, Saksena NK (2013) MiRNAs: small molecules with a big impact on HIV infection and pathogenesis. Future Virol 8(8):769–781. https://doi.org/10.2217/FVL.13.59;WGROUP:STRING:PUBLICATION

He ZY, Wei XW, Luo M, Luo ST, Yang Y, Yu YY, Chen Y, Ma CC, Liang X, Guo FC, Ye TH, Shi HS, Shen GB, Wang W, Gong FM, He G, Yang L, Zhao X, Song XR, Wei YQ (2013a) Folate-linked lipoplexes for short hairpin RNA targeting claudin-3 delivery in ovarian cancer xenografts. J Control Release 172(3):679–689. https://doi.org/10.1016/j.jconrel.2013.10.015

He Z, Yu Y, Zhang Y, Yan Y, Zheng Y, He J, Xie Y, He G, Wei Y, Song X (2013b) Gene delivery with active targeting to ovarian cancer cells mediated by folate receptor α. J Biomed Nanotechnol 9(5):833–844. https://doi.org/10.1166/JBN.2013.1587

He ZY, Deng F, Wei XW, Ma CC, Luo M, Zhang P, Sang YX, Liang X, Liu L, Qin HX, Shen YL, Liu T, Liu YT, Wang W, Wen YJ, Zhao X, Zhang XN, Qian ZY, Wei YQ (2016) Ovarian cancer treatment with a tumor-targeting and gene expression-controllable lipoplex. Sci Rep 6(1):1–13. https://doi.org/10.1038/SREP23764

He Q, Ye A, Ye W, Liao X, Qin G, Xu Y, Yin Y, Luo H, Yi M, Xian L, Zhang S, Qin X, Zhu W, Li Y (2021) Cancer-secreted exosomal miR-21-5p induces angiogenesis and vascular permeability by targeting KRIT1. Cell Death Dis 12(6):1–14. https://doi.org/10.1038/s41419-021-03803-8

Huang H-T, Liu Z-C, Wu K-Q, Gu S-R, Lu T-C, Zhong C-J, Zhou Y-X (2019) MiR-92a regulates endothelial progenitor cells (EPCs) by targeting GDF11 via activate SMAD2/3/FAK/Akt/eNOS pathway. Ann Transl Med 7(20):563. https://doi.org/10.21037/ATM.2019.09.35

Jung J, Yeom C, Choi YS, Kim S, Lee EJ, Park MJ, Kang SW, Kim SB, Chang S (2015) Simultaneous inhibition of multiple oncogenic miRNAs by a multi-potent microRNA sponge. Oncotarget 6(24):20370. https://doi.org/10.18632/ONCOTARGET.4827

Kamel W, Segerman B, Öberg D, Punga T, Akusjärvi G (2013) The adenovirus VA RNA-derived miRNAs are not essential for lytic virus growth in tissue culture cells. Nucleic Acids Res 41(9): 4802. https://doi.org/10.1093/NAR/GKT172

Kang S, Lee S, Park S (2020) iRGD peptide as a tumor-penetrating enhancer for tumor-targeted drug delivery. Polymers 12(9):1906. https://doi.org/10.3390/POLYM12091906

Karothia D, Kumar Dash P, Parida M, Bhagyawant SS, Kumar JS (2020) Vector derived artificial miRNA mediated inhibition of West Nile virus replication and protein expression. Gene 729: 144300. https://doi.org/10.1016/J.GENE.2019.144300

Kauppinen S, Vester B, Wengel J (2005) Locked nucleic acid (LNA): high affinity targeting of RNA for diagnostics and therapeutics. Drug Discov Today Technol 2(3):287. https://doi.org/10.1016/J.DDTEC.2005.08.012

Khare R, Chen YC, Weaver EA, Barry MA (2011) Advances and future challenges in adenoviral vector pharmacology and targeting. Curr Gene Ther 11(4):241. https://doi.org/10.2174/156652311796150363

Kim J-Y, Jung EJ, Kim J-M, Son Y, Lee HS, Kwag S-J, Park J-H, Cho J-K, Kim H-G, Park T, Jeong S-H, Jeong C-Y, Ju Y-T (2023) MiR-221 and miR-222 regulate cell cycle progression and affect chemosensitivity in breast cancer by targeting ANXA3. Exp Ther Med 25(3):127. https://doi.org/10.3892/ETM.2023.11826

Kumar, A., Arya, N., Tiwari, S., Khan, R. (Eds.). (2024). Diagnostic, prognostic, and therapeutic role of MicroRNAs in head and neck cancer. Elsevier

Lee CS, Bishop ES, Zhang R, Yu X, Farina EM, Yan S, Zhao C, Zeng Z, Shu Y, Wu X, Lei J, Li Y, Zhang W, Yang C, Wu K, Wu Y, Ho S, Athiviraham A, Lee MJ et al (2017) Adenovirus-mediated gene delivery: potential applications for gene and cell-based therapies in the new era of personalized medicine. Genes Dis 4(2):43. https://doi.org/10.1016/J.GENDIS.2017.04.001

Lennox KA, Owczarzy R, Thomas DM, Walder JA, Behlke MA (2013) Improved performance of anti-miRNA oligonucleotides using a novel non-nucleotide modifier. Mol Ther Nucleic Acids 2 (AUG):e117. https://doi.org/10.1038/MTNA.2013.46/ATTACHMENT/6A71DB22-7669-48D2-AC8C-42242D3F9E97/MMC1.PDF

Li YJ, Zhang YX, Wang PY, Chi YL, Zhang C, Ma Y, Lv CJ, Xie SY (2012) Regression of A549 lung cancer tumors by anti-miR-150 vector. Oncol Rep 27(1):129–134. https://doi.org/10.3892/OR.2011.1466

Li R, Tian F, Qi Y, Ma L, Zhou T, Li Y, Hui T, Zhang L, Wang S, Song Z (2019) Pegylated liposomal doxorubicin plus cyclophosphamide followed by docetaxel as neoadjuvant chemotherapy in locally advanced breast cancer (registration number: ChiCTR1900023052). Sci Rep 9(1):1–8. https://doi.org/10.1038/S41598-019-54387-5

Li Y, Zhang L, Dong Z, Xu H, Yan L, Wang W, Yang Q, Chen C (2021) MicroRNA-155-5p promotes tumor progression and contributes to paclitaxel resistance via TP53INP1 in human breast cancer. Pathol – Res Pract 220:153405. https://doi.org/10.1016/J.PRP.2021.153405

Liu F, Di Wang X (2019) miR-150-5p represses TP53 tumor suppressor gene to promote proliferation of colon adenocarcinoma. Sci Rep 9(1):6740. https://doi.org/10.1038/S41598-019-43231-5

Liu M, Mo F, Song X, He Y, Yuan Y, Yan J, Yang Y, Huang J, Zhang S (2021) Exosomal hsa-miR-21-5p is a biomarker for breast cancer diagnosis. PeerJ 9:e12147. https://doi.org/10.7717/PEERJ.12147/SUPP-9

Lu L, Ju F, Zhao H, Ma X (2015) MicroRNA-134 modulates resistance to doxorubicin in human breast cancer cells by downregulating ABCC1. Biotechnol Lett 37(12):2387–2394. https://doi.org/10.1007/S10529-015-1941-Y

Lu Q, Chen W, Ji Y, Liu Y, Xue X (2022) Ursolic acid enhances cytotoxicity of doxorubicin-resistant triple-negative breast cancer cells via ZEB1-AS1/miR-186-5p/ABCC1 Axis. Cancer Biother Radiopharm 37(8):673–683. https://doi.org/10.1089/CBR.2020.4147

Lungwitz U, Breunig M, Blunk T, Göpferich A (2005) Polyethylenimine-based non-viral gene delivery systems. Eur J Pharm Biopharm 60(2):247–266. https://doi.org/10.1016/J.EJPB.2004.11.011

Mellott AJ, Forrest ML, Detamore MS (2012) Physical non-viral gene delivery methods for tissue engineering. Ann Biomed Eng 41(3):446. https://doi.org/10.1007/S10439-012-0678-1

Meng Y, Serio VB, Tong Y, Huang T, Frullanti E, Palmieri M (2025) Editorial: recent advancements in RNA-based and targeted therapeutics. Front Genet 16:1603498. https://doi.org/10.3389/FGENE.2025.1603498

Monfared H, Jahangard Y, Nikkhah M, Mirnajafi-Zadeh J, Mowla SJ (2019) Potential therapeutic effects of exosomes packed with a miR-21-sponge construct in a rat model of glioblastoma. Front Oncol 9(AUG):443866. https://doi.org/10.3389/FONC.2019.00782/BIBTEX

Pasculli B, Barbano R, Fontana A, Biagini T, Di Viesti MP, Rendina M, Valori VM, Morritti M, Bravaccini S, Ravaioli S, Maiello E, Graziano P, Murgo R, Copetti M, Mazza T, Fazio VM, Esteller M, Parrella P (2020) Hsa-miR-155-5p up-regulation in breast cancer and its relevance for treatment with poly[ADP-ribose] polymerase 1 (PARP-1) inhibitors. Front Oncol 10:545842. https://doi.org/10.3389/FONC.2020.01415/BIBTEX

Rahman MA, Jalouli M, Yadab MK, Al-Zharani M (2025) Progress in drug delivery systems based on nanoparticles for improved glioblastoma therapy: addressing challenges and investigating opportunities. Cancer 17(4):701. https://doi.org/10.3390/CANCERS17040701

Real PJ, Cao Y, Wang R, Nikolovska-Coleska Z, Sanz-Ortiz J, Wang S, Fernandez-Luna JL (2004) Breast cancer cells can evade apoptosis-mediated selective killing by a novel small molecule inhibitor of Bcl-2. Cancer Res 64(21):7947–7953. https://doi.org/10.1158/0008-5472.CAN-04-0945

Savenkova DA, Makarova ALA, Shalik IK, Yudkin DV (2022) miRNA pathway alteration in response to non-coding RNA delivery in viral vector-based gene therapy. Int J Mol Sci 23: 14954. https://doi.org/10.3390/IJMS232314954

Saw PE, Song E (2024) Advancements in clinical RNA therapeutics: present developments and prospective outlooks. Cell Rep Med 5(5):101555. https://doi.org/10.1016/J.XCRM.2024.101555

Scioli MG, Storti G, D'amico F, Guzmán RR, Centofanti F, Doldo E, Miranda EMC, Orlandi A (2020) Oxidative stress and new Pathogenetic mechanisms in endothelial dysfunction: potential diagnostic biomarkers and therapeutic targets. J Clin Med 9(6):1995. https://doi.org/10.3390/JCM9061995

Singh RK, Singh RK, Pankaj S, Kumar S, Rajkota V (2014) A retrospective study of efficacy and safety of albumin-bound paclitaxel in metastatic breast cancer. World J Oncol 5(5):204–209. https://doi.org/10.4021/wjon.v5i5.865

Sun FD, Wang PC, Luan RL, Zou SH, Du X (2018) MicroRNA-574 enhances doxorubicin resistance through down-regulating SMAD4 in breast cancer cells. Eur Rev Med Pharmacol Sci 22(5):1342–1350. https://doi.org/10.26355/EURREV_201803_14476

Sun L, Yang PC, Luan L, Sun JF, Wang YT (2024) Harmonizing the craft of crafting clinically endorsed small-molecule BCR-ABL tyrosine kinase inhibitors for the treatment of hematological malignancies. Eur J Pharm Sci 193:106678. https://doi.org/10.1016/J.EJPS.2023.106678

Tang X, Loc WS, Dong C, Matters GL, Butler PJ, Kester M, Meyers C, Jiang Y, Adair JH (2017) The use of nanoparticulates to treat breast cancer. Nanomedicine 12(19):2367. https://doi.org/10.2217/NNM-2017-0202

Tian J, Al Raffa F, Dai M, Moamer A, Khadang B, Hachim IY, Bakdounes K, Ali S, Jean-Claude B, Lebrun JJ (2018) Dasatinib sensitises triple negative breast cancer cells to chemotherapy by targeting breast cancer stem cells. Br J Cancer 119(12):1495–1507. https://doi.org/10.1038/S41416-018-0287-3

Tormo E, Ballester S, Adam-Artigues A, Burgués O, Alonso E, Bermejo B, Menéndez S, Zazo S, Madoz-Gúrpide J, Rovira A, Albanell J, Rojo F, Lluch A, Eroles P (2019) The miRNA-449 family mediates doxorubicin resistance in triple-negative breast cancer by regulating cell cycle factors. Sci Rep 9(1):1–14. https://doi.org/10.1038/s41598-019-41472-y

Tryfonopoulos D, Walsh S, Collins DM, Flanagan L, Quinn C, Corkery B, McDermott EW, Evoy D, Pierce A, O'Donovan N, Crown J, Duffy MJ (2011) Src: a potential target for the treatment of triple-negative breast cancer. Ann Oncol 22(10):2234–2240. https://doi.org/10.1093/ANNONC/MDQ757

Vester B, Wengel J (2004) LNA (locked nucleic acid): high-affinity targeting of complementary RNA and DNA. Biochemistry 43(42):13233–13241. https://doi.org/10.1021/bi0485732

Wang PY, Li YJ, Zhang S, Li ZL, Yue Z, Xie N, Xie SY (2010) Regulating A549 cells growth by ASO inhibiting miRNA expression. Mol Cell Biochem 339(1–2):163–171. https://doi.org/10.1007/S11010-009-0380-2

White SM, Renda M, Nam N-Y, Klimatcheva E, Zhu Y, Fisk J, Halterman M, Rimel BJ, Federoff H, Pandya S, Rosenblatt JD, Planelles V (1999) Lentivirus vectors using human and simian immunodeficiency virus elements. J Virol 73(4):2832. https://doi.org/10.1128/JVI.73.4.2832-2840.1999

Wuxiao Z, Wang H, Su Q, Zhou H, Hu M, Tao S, Xu L, Chen Y, Hao X (2020) MicroRNA-145 promotes the apoptosis of leukemic stem cells and enhances drug-resistant K562/ADM cell sensitivity to adriamycin via the regulation of ABCE1. Int J Mol Med 46(4):1289. https://doi.org/10.3892/IJMM.2020.4675

Yi R, Qin Y, Macara IG, Cullen BR (2003) Exportin-5 mediates the nuclear export of pre-microRNAs and short hairpin RNAs. Genes Dev 17(24):3011. https://doi.org/10.1101/GAD.1158803

Yi H, Liu L, Sheng N, Li P, Pan H, Cai L, Ma Y (2016) Synergistic therapy of doxorubicin and miR-129-5p with self-cross-linked bioreducible polypeptide nanoparticles reverses multidrug resistance in cancer cells. Biomacromolecules 17(5):1737–1747. https://doi.org/10.1021/ACS.BIOMAC.6B00141

Yin L, Keeler GD, Zhang Y, Hoffman BE, Ling C, Qing K, Srivastava A (2020) AAV3-miRNA vectors for growth suppression of human hepatocellular carcinoma cells in vitro and human liver tumors in a murine xenograft model in vivo. Gene Ther 28(7–8):422. https://doi.org/10.1038/S41434-020-0140-1

Zakeri A, Kouhbanani MAJ, Beheshtkhoo N, Beigi V, Mousavi SM, Hashemi SAR, Zade AK, Amani AM, Savardashtaki A, Mirzaei E, Jahandideh S, Movahedpour A (2018) Polyethylenimine-based nanocarriers in co-delivery of drug and gene: a developing horizon. Nano Rev Exp 9(1):1488497. https://doi.org/10.1080/20022727.2018.1488497

Zangouei AS, Alimardani M, Moghbeli M (2021) MicroRNAs as the critical regulators of doxorubicin resistance in breast tumor cells. Cancer Cell Int 21(1):213. https://doi.org/10.1186/S12935-021-01873-4

Zhao L, Wang Y, Jiang L, He M, Bai X, Yu L, Wei M (2016) MIR-302a/b/c/d cooperatively sensitizes breast cancer cells to adriamycin via suppressing P-glycoprotein(P-gp) by targeting MAP/ERK kinase kinase 1 (MEKK1). J Exp Clin Cancer Res 35(1). https://doi.org/10.1186/S13046-016-0300-8

Zhou Q, Gallo JM (2009) Differential effect of sunitinib on the distribution of temozolomide in an orthotopic glioma model. Neuro-Oncology 11(3):301. https://doi.org/10.1215/15228517-2008-088

Zuo H (2019) iRGD: A promising peptide for cancer imaging and a potential therapeutic agent for various cancers. J Oncol 2019:9367845. https://doi.org/10.1155/2019/9367845

# Therapeutic Approach with Tumor Suppressor miRNA

# 5

Archana Joon and Rekha Jalandra

**Abstract**

miRNAs are small noncoding RNAs that regulate gene expression and play crucial roles in the activation and inhibition of various cellular processes like proliferation, cell division, and apoptosis. miRNAs have emerged as a promising therapeutic tool for targeting oncogenes and tumor-suppressive pathways. This chapter examines the therapeutic approaches where miRNAs are being used in inhibition strategies in various kinds of cancers alongside viral and nonviral delivery systems. There are several preclinical and clinical trials exploring the use of miRNA mimics and inhibitors to target oncogenic miRNAs. Both miRNA mimics and inhibitors struggle to reach the cancer site because of their stability, off-target effects, and toxicity. We discussed more specific and stable innovative delivery systems, such as miRNA-loaded nanoparticles and exosomes. Future research is more likely to focus on minimizing off-target effects, toxicity, increased target specificity, and expanding clinical applications of tumor suppressor miRNA therapies.

**Keywords**

miRNA · Tumor · Cancer · Immunotherapy

A. Joon
Department of Pediatrics and Department of Microbiology & Immunology, Albert Einstein College of Medicine, New York, NY, USA

R. Jalandra (✉)
Department of Pharmacology & Immunology, Medical University of South Carolina, Charleston, SC, USA



S. Tiwari et al. (eds.), *microRNA Modulation: A Promising Combination Therapy for Cancer*, https://doi.org/10.1007/978-981-95-5989-3_5

**Learning Objectives**

- Explain the role of tumor suppressor miRNAs in regulating key oncogenic pathways and maintaining cellular homeostasis.
- Discuss mechanisms leading to the loss or downregulation of tumor suppressor miRNAs in different cancers.
- Evaluate therapeutic strategies to restore tumor suppressor miRNA function, including synthetic mimics, viral vectors, and nanoparticle-based delivery systems.
- Assess the challenges and future prospects of tumor suppressor miRNA-based therapies in clinical oncology, such as delivery efficiency, stability, and specificity.

## 5.1 Introduction

Tumor suppressor microRNAs (miRNAs) are small, noncoding RNAs that regulate gene expression by silencing oncogenes or genes involved in tumor progression. Restoring or mimicking their function offers promising anticancer strategies (Otmani and Lewalle 2021). Cancer is an uncontrolled cell division. Tumor suppressor miRNA can inhibit cell proliferation, leading to increased apoptosis, thus preventing uncontrolled cell growth. miRNAs can also interfere in the crucial process of spreading cancer to other parts of the body, which is epithelial–mesenchymal transition (EMT). EMT is a process of losing cell-to-cell attachment and polarity, acquiring mesenchymal properties, which allows them to move and invade other parts of the body (Roche 2018). miRNA can prevent cancer cells from acquiring these properties, restricting cancer cells' movement. miRNA can also target genes involved in angiogenesis, which is the formation of new blood vessels (Dong et al. 2021). Angiogenesis is crucial for the growth of tumors as they are a source of nutrients for cancer cells. There are several other ways in which miRNA target cancer cell growth, proliferation, and migration (Landskroner-Eiger et al. 2013). We have discussed some of them in the following sections.

## 5.2 miRNA Replacement Therapy

miRNA replacement therapy has emerged as a promising therapeutic strategy for cancer, particularly by restoring the expression of tumor suppressor miRNAs that are frequently downregulated in malignancies (Calin and Croce 2006; Rupaimoole and Slack 2017). Replacement therapy involves the delivery of synthetic miRNA mimics, typically encapsulated in nanoparticles, liposomes, or viral vectors, to reintroduce these miRNAs into tumor cells. These miRNAs bind to complementary mRNA and induce changes in them posttranslationally, leading to repression, silence, or degradation of tumor mRNA (Fig. 5.1) (Lin and Gregory 2015).

Loss of tumor suppressor miRNAs such as miR-34a, the let-7 family, and miR-145 has been implicated in the initiation and progression of various cancers

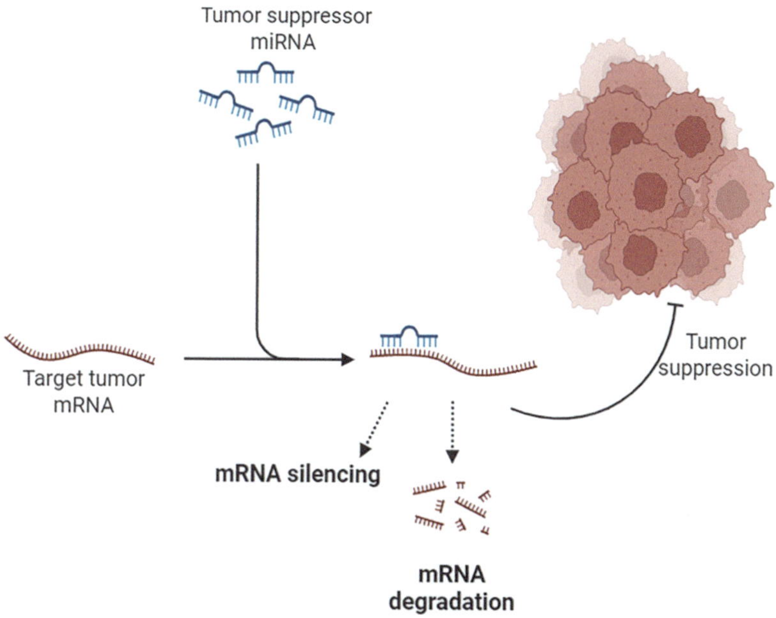


**Fig. 5.1** Mechanism of tumor suppressor miRNA-mediated tumor suppression. Tumor suppressor miRNAs bind to complementary sequences on target tumor mRNAs, leading to either translational silencing or degradation of the target mRNA. This process inhibits the expression of oncogenic proteins, ultimately resulting in tumor growth suppression

(Esquela-Kerscher and Slack 2006). Reintroduces synthetic miRNA mimics or vectors encoding tumor suppressor miRNAs to restore their lost function in cancer cells. MRX34, a liposomal miR-34a mimic, which entered phase I clinical trials for patients with advanced solid tumors and showed promising biological activity, though development was halted due to immune-related adverse events (Beg et al. 2017). MRX34 is a first-in-human miRNA therapy trial. Preclinical studies showed that miR-34a has significant antitumor effects in hepatocellular carcinoma and non-small cell lung cancer models (Wiggins et al. 2010). Additionally, let-7 mimics have demonstrated tumor-suppressive effects in preclinical lung, breast, and ovarian cancer models (Trang et al. 2011). Table 5.1 summarizes miRNAs used in replacement therapy, such as miR-145, miR-16, miR-200 family, miR-29, and miR-101.

miRNA replacement therapy represents a promising strategy for cancer treatment by restoring the expression of tumor-suppressive miRNAs lost or downregulated in malignancies (Matsui and Corey 2017). Preclinical studies and early-phase clinical trials have demonstrated the therapeutic potential of several miRNAs, including miR-34a, let-7, and miR-16, in suppressing tumor growth, metastasis, and chemoresistance. Advances in delivery methods, such as nanoparticles and bacterial minicells, have further enhanced the suitability of miRNA-based therapies (Nana-Sinkam and Croce 2013). Table 5.1 summarizes tumor suppressor miRNAs, their molecular targets, types of cancer, and key findings. While challenges related to

**Table 5.1** Tumor suppressor miRNAs in miRNA therapy

| miRNA | Key targets | Cancer types | Study highlights | References |
|---|---|---|---|---|
| miR-34a | BCL2, CDK6, MET | Solid tumors, hematological malignancies | Phase I clinical trial (MRX34); preclinical efficacy in liver, lung | Beg et al. (2017), Wiggins et al. (2010) |
| let-7 family | RAS, HMGA2 | Lung, breast, ovarian | Systemic let-7 mimic reduced lung tumor growth in mice | Trang et al. (2011) |
| miR-145 | FSCN1, SOX2, c-Myc | Colorectal, bladder | Mimic inhibited tumor growth in xenografts | Sachdeva and Mo (2010) |
| miR-16 | BCL2, cyclins | Mesothelioma | Phase I trial using bacterial minicell-delivered TargomiRs | van Zandwijk et al. (2017) |
| miR-200 family | ZEB1/2 | Breast, ovarian | Restoration suppressed EMT and metastasis in preclinical models | Korpal and Kang (2008) |
| miR-29 family | DNMT3A/B, ECM proteins | Lung, liver | Replacement therapy reduced fibrosis and tumor growth | Fabbri et al. (2007) |
| miR-101 | EZH2, COX-2 | Prostate, liver | Mimic delivery inhibited tumor growth in animal studies | Su et al. (2009) |

stability, targeted delivery, and immune responses remain, the encouraging outcomes from ongoing research underscore the translational potential of miRNA replacement strategies as adjunct or standalone cancer therapeutics.

## 5.3 Delivery Methods of miRNA

Tumor suppressor miRNAs play a crucial role in regulating gene expression and suppressing tumor growth. However, their instability and difficulty in reaching target cells within the tumor microenvironment limit their therapeutic applications. To overcome these limitations, various delivery methods have been developed, broadly categorized into viral and nonviral systems (Segal and Slack 2020). Viral delivery systems genetically modified viruses used to deliver tumor suppressor miRNAs by incorporating the miRNA gene into the viral genome. Nonviral delivery methods employ nonviral carriers to encapsulate and transport tumor suppressor miRNAs, offering potential benefits such as lower immunogenicity and toxicity compared to viral vectors (Yang 2015). Some common examples are miRNA within nanoparticles such as lipid nanoparticles, extracellular vehicles, and exosomes. We have discussed these approaches in detail in the following sections.

### 5.3.1 Viral Vector Delivery

Viral vector delivery is the most effective approach for delivering these miRNAs into cancer cells via viral vectors, which offer high transduction efficiency, stable expression, and the potential for systemic administration. Adenoviral, lentiviral, and adeno-associated viral (AAV) vectors are among the most widely used systems for miRNA replacement therapy. These vectors can efficiently introduce synthetic or plasmid-based tumor suppressor miRNAs directly into tumor tissues or systemically, leading to sustained expression and therapeutic effects.

**Adenoviral Vectors**

Adenoviral vectors are widely utilized for delivering tumor suppressor miRNAs in cancer therapy due to their high transduction efficiency, capacity for large genetic payloads, and ability to infect both dividing and nondividing cells (Wang et al. 2013). These vectors enable robust, transient expression of tumor suppressor miRNAs, which can downregulate oncogenic pathways, induce apoptosis, and inhibit tumor progression. miR-34a, one of the most extensively studied tumor suppressor miRNAs, has been delivered via adenoviral vectors in multiple preclinical models (Wiggins et al. 2010). Li et al. (2021) showed that restoring miR-34a expression in hepatocellular carcinoma stem cells led to decreased stemness properties, reduced CD44 expression, and sensitized cells to chemotherapy (Li et al. 2021). Similarly, adenoviral delivery of miR-145 has been shown to inhibit tumor proliferation and invasion in colorectal and breast cancer by targeting oncogenes involved in cell cycle regulation and metastasis (Iorio and Croce 2012; Sachdeva and Mo 2010). Adenoviral vectors' high transduction rates facilitate efficient miRNA delivery to tumor cells, overcoming some limitations of nonviral methods such as low uptake and poor stability. Additionally, the transient nature of adenoviral expression reduces risks associated with permanent genetic modifications. However, immune responses against adenoviruses remain a challenge, which researchers are addressing by engineering less immunogenic vectors or using local delivery strategies.

Overall, adenoviral vector-mediated delivery of tumor suppressor miRNAs holds great promise for cancer therapeutics, with multiple preclinical studies supporting their efficacy in tumor growth inhibition and reversal of malignant phenotypes.

**Lentiviral Vectors**

Lentiviral vectors have emerged as powerful tools for delivering tumor suppressor miRNAs in cancer gene therapy owing to their ability to integrate into the host genome, ensuring long-term and stable transgene expression. Unlike adenoviral vectors, lentiviruses can efficiently transduce both dividing and nondividing cells, making them particularly valuable for targeting solid tumors and cancer stem cells that contribute to tumor recurrence and resistance (Ju 2025). miR-34a is also delivered through lentiviral vectors, which inhibit proliferation, induce apoptosis, and suppress metastasis in hepatocellular carcinoma and prostate cancer models (Xu et al. 2015; Yan et al. 2019). Another example of lentiviral delivery is miR-145, which significantly reduces tumor growth and invasion in bladder and colorectal cancer models by targeting key oncogenic pathways involved in cell cycle progression and metastasis (Chiyomaru et al. 2010; Spizzo et al. 2010). lentiviral vector-mediated delivery of tumor suppressor miRNAs represents a promising avenue for durable, targeted cancer gene therapy, with encouraging preclinical outcomes.

**Adeno-Associated Virus (AAV) Vectors**

AAV vectors have become attractive tools for gene and miRNA-based cancer therapies due to their favorable safety profile, low immunogenicity, and capability to sustain long-term gene expression in both proliferating and nondividing cells. Unlike integrating lentiviruses, AAV vectors predominantly exist as episomes within the host cell, reducing the risk of insertional mutagenesis while providing persistent transgene expression—an essential feature for chronic diseases like cancer (Xie et al. 2015). In the context of tumor suppressor miRNA therapy, AAV vectors have shown promise in delivering therapeutic miRNAs such as miR-26a, miR-34a, and miR-124. For example, Kota et al. (2009) demonstrated that systemic delivery of AAV-expressing miR-26a in a murine hepatocellular carcinoma model led to significant tumor regression by inducing cell cycle arrest and apoptosis without noticeable toxicity (Kota et al. 2009). Similarly, another study utilized AAV-mediated miR-34a delivery in glioblastoma models, achieving effective tumor growth suppression by downregulating oncogenic targets like BCL2 and MET (Li et al. 2009).

Moreover, AAV-mediated miR-124 expression has been reported to suppress glioma proliferation and invasion by targeting CDK6 and STAT3 pathways (Fang and Hong 2021). The tissue tropism flexibility of various AAV serotypes, such as AAV8 for liver and AAV9 for CNS and systemic delivery, allows tailored approaches for different tumor types (Ju 2025).

While the limited packaging capacity (~4.7 kb) of AAV vectors poses challenges, miRNAs' small size makes them ideal candidates for AAV-based therapies. Table 5.2 summarizes the use of various viral vectors in miRNA-based therapy. Current research focuses on optimizing vector design, serotype selection, and co-delivery strategies to enhance tumor targeting and therapeutic efficacy.

### 5.3.2 Nonviral Delivery of miRNA

**Nanoparticle-Based Delivery**

Nanoparticle-based delivery systems have emerged as a powerful platform for miRNA therapeutics in cancer due to their biocompatibility, tunable surface modifications, and ability to protect miRNAs from nuclease degradation. These systems enable targeted delivery of tumor-suppressor miRNAs to tumor tissues while minimizing off-target effects. Various nanocarriers, including lipid-based nanoparticles (LNPs), polymeric nanoparticles, and inorganic nanomaterials (such as gold and silica nanoparticles), have been engineered for miRNA delivery (He et al. 2025; Hou et al. 2021).

**Table 5.2** Comparison of viral vectors for miRNA-based tumor suppression

| Feature | Adeno-associated virus (AAV) | Lentivirus | Adenovirus |
|---|---|---|---|
| Genome type | Single-stranded DNA | RNA (converted to DNA, integrates into host) | Double-stranded DNA |
| Integration | Mostly episomal (low integration risk) | Integrates into the host genome | Episomal (non-integrating) |
| Packaging capacity | ~4.7 kb | ~8 kb | ~36 kb |
| Tropism/ target cells | Broad (serotype-dependent) | Broad (dividing and non-dividing cells) | Broad (dividing and non-dividing cells) |
| Immune response | Low | Moderate (risk of insertional mutagenesis) | High |
| Duration of expression | Long-term (non-dividing cells especially) | Long-term (stable integration) | Transient |
| Clinical safety profile | Excellent (used in FDA-approved gene therapies) | Moderate (integration-related oncogenesis risk) | Limited (immune-related toxicities) |
| Use in miRNA cancer therapy | miR-26a, miR-34a, miR-124 in HCC, glioma, etc. | miR-34a, let-7, miR-200 family in breast, lung, glioma | miR-145, miR-34a in lung, prostate cancers |
| Key references | Fang and Hong (2021), Kota et al. (2009), Xie et al. (2015) | Chiyomaru et al. (2010), Xu et al. (2015) | Esquela-Kerscher and Slack (2006), Scholz et al. (2022) |

For instance, miR-34a-loaded lipid nanoparticles demonstrated significant tumor growth inhibition in a mouse model of non-small cell lung cancer (Trang et al. 2011). Similarly, miR-200c encapsulated in polymeric nanoparticles effectively suppressed breast cancer metastasis (Liu et al. 2016). Gold nanoparticles conjugated with miR-205 have shown enhanced tumor accumulation and apoptosis induction in prostate cancer cells (Jeyaram and Jay 2017). These studies highlight the versatility and effectiveness of nanoparticle-mediated miRNA delivery in restoring tumor-suppressive functions. Nanogels, with their highly hydrophilic, biocompatible nature, have shown promise for the co-delivery of miRNAs and chemotherapeutic drugs, offering controlled, pH-responsive release properties. A recent study demonstrated the use of polymeric micelles for the co-delivery of doxorubicin, showing effects in suppressing cancer growth (Lee et al. 2005). Chitosan-based polymeric nanogels and hydrogels carrying tumor suppressor miRNA have shown increased uptake and tumor inhibition in colorectal cancer (Piotrowska and Orzechowska 2024).

Importantly, lipid nanoparticle-based miRNA therapy (MRX34), a synthetic miR-34a mimic, became the first miRNA-based therapy to enter clinical trials for liver cancer, though it was later halted due to immune-related adverse events (Beg et al. 2017). Beyond miR-34a, other tumor-suppressor miRNAs such as miR-29b, miR-145, and let-7 have also been successfully delivered using nanoparticles. For instance, dextran-coated iron oxide nanoparticles have been used to deliver miR-29a in breast cancer models, resulting in reduced tumor growth (Yalcin 2019). Despite this, ongoing advancements in nanoparticle engineering, surface modification, and controlled-release systems continue to refine this approach for safer and more precise cancer therapies.

**miRNA-Loaded Exosomes Mediated Delivery**
Exosomes are 30–150 nm vesicles secreted by almost all cells, carrying molecular cargo such as miRNAs, mRNAs, proteins, and lipids. They facilitate intercellular communication and can be engineered to deliver therapeutic miRNAs with low immunogenicity. Their ability to cross physiological barriers, including the blood–brain barrier, makes them attractive for treating tumors in difficult locations such as glioblastomas (D. Zhang et al. 2016). Exosomes are isolated and administered systemically or locally. Upon reaching the tumor microenvironment, exosomes are taken up by cancer cells, releasing miRNAs that suppress oncogenic pathways by targeting mRNA transcripts, leading to reduced tumor growth and metastasis. Surface modifications on exosomes (e.g., tumor-specific ligands) enhance target efficiency and reduce off-target effects. Kamerkar et al. (2017) developed engineered exosomes loaded with siRNA targeting the KRAS oncogene, demonstrating tumor regression in pancreatic cancer models. This strategy established a precedent for miRNA-based exosome therapeutics (Kamerkar et al. 2017). Another study showed that exosomes from human bone marrow mesenchymal stem cells carrying miR-199a-3p suppress HCC progression by downregulating mTOR signaling, highlighting exosome-mediated miRNA delivery's therapeutic potential (Lou et al. 2017). Mesenchymal stem cell-derived exosomes loaded with tumor-suppressor miR-146b were shown to reduce glioma growth in rodents (Katakowski et al. 2013). Delivery of miR-34a via exosomes inhibits breast cancer progression by targeting anti-apoptotic genes and EMT pathways (Sun et al. 2025).

Du et al. (2020) reported that exosomes carrying miR-139-5p inhibited colorectal cancer metastasis by downregulating Wnt and Ras signaling pathways (Du et al. 2020). Another study demonstrated that exosome miR-199a-5p suppressed ovarian cancer cell proliferation and invasion by targeting the oncogene HOXA10 (Lian et al. 2020).

## 5.4 Combination Therapy

Combining tumor suppressor miRNA with conventional cancer therapies such as chemotherapy, radiotherapy, or targeted therapies is a promising strategy to enhance therapeutic outcomes. Chemotherapy combined with miRNA therapy can modulate pathways driven by aberrant expression of oncogenes, anti-apoptotic proteins, drug efflux pumps, and components of DNA repair pathways. It can downregulate genes associated with drug resistance (e.g., MDR1, BCL2, ABC transporters), enhance apoptosis, and cell cycle arrest. For example, miR-15/16 + Gemcitabine targets anti-apoptotic genes (BCL2, MCL1) and enhances gemcitabine-induced cell death in pancreatic cancer cells (Sun et al. 2014). miR-200c reverses EMT and sensitizes cells to paclitaxel. miR-200c + Paclitaxel therapy enhances paclitaxel sensitivity in ovarian cancer cells by targeting class III β-tubulin and ZEB1 (Leskelä et al. 2011). Another therapy combined chemotherapy (5-FU) with miR-21 and reported increased sensitivity to drugs in CRC models (Li et al. 2013). Yang et al. (2021) demonstrated that miR-34a mimics, along with doxorubicin, inhibit breast cancer growth in vivo, lowering the required chemotherapeutic dose while maintaining efficacy (Yang et al. 2021).

miRNA therapy can be combined with immunotherapy, such as immune checkpoint inhibitors (ICIs) targeting PD-1, PD-L1, and CTLA-4 has emerged as a promising strategy to enhance antitumor immune responses and overcome resistance to ICIs (Cortez et al. 2019). Tumor suppressor miRNAs can directly downregulate immunosuppressive factors (e.g., PD-L1, IDO1) and oncogenic pathways (e.g., PI3K/AKT, STAT3, WNT/β-catenin) that contribute to immune evasion. Additionally, miRNAs modulate the infiltration and activity of cytotoxic T lymphocytes (CTLs), dendritic cells, and natural killer (NK) cells within the TME, enhancing antitumor immunity (Xing et al. 2021). The following are some examples of miRNA therapy combined with immunotherapy.

miR-200 and PD-L1 Regulation: Zhang et al. (2023) demonstrated that miR-200 family members suppress PD-L1 expression in non-small cell lung cancer (NSCLC) cells, enhancing T-cell-mediated killing. Combining miR-200 delivery with anti-PD-1 therapy improved tumor control in mouse models (Zhang et al. 2023).

miR-34a and Immune Checkpoints: miR-34a has been shown to target PD-L1 in various cancers, including NSCLC and hepatocellular carcinoma. In vivo studies combining miR-34a mimics with anti-PD-1 antibodies resulted in synergistic tumor regression and prolonged survival (Cortez et al. 2015).

miR-155 and Immunotherapy: miR-155 enhances the activation of T cells and dendritic cells by targeting SOCS1, a negative regulator of immune responses. Combining miR-155 overexpression or delivery with ICIs boosts immunotherapy efficacy, especially in immunologically "cold" tumors (Ji et al. 2015).

Exosome-Mediated miRNA Delivery with ICIs: Engineered exosomes carrying tumor suppressor miRNAs like miR-124 and miR-34a, when combined with anti-PD-L1 therapy, have demonstrated enhanced tumor shrinkage and immune cell infiltration in glioma and melanoma models (Roshani Asl et al. 2021; Yueh et al. 2025).

While clinical trials directly combining miRNA therapeutics with immunotherapy are limited, ongoing phase I/II studies evaluating miRNA mimics (e.g., MRX34, a miR-34a mimic) have laid the groundwork for such strategies.

## 5.5 Conclusion and Future Perspectives

In recent years, tumor suppressor miRNAs have emerged as pivotal regulators of oncogenic processes, offering novel avenues for cancer diagnosis, prognosis, and therapy. These small, noncoding RNAs orchestrate complex gene regulatory networks, often acting as critical antagonists of tumor-promoting pathways. The therapeutic restoration of downregulated tumor suppressor miRNAs holds promise in reversing malignant phenotypes, sensitizing tumors to conventional treatments, and improving overall clinical outcomes (Menon et al. 2022).

Multiple preclinical studies have demonstrated the potential of miRNA-based therapies in suppressing tumor growth, inhibiting metastasis, and modulating the tumor microenvironment. Delivery strategies utilizing nanoparticles, liposomes, viral vectors, and exosomes have shown considerable advancement in overcoming challenges related to stability, specificity, and targeted delivery (Chen et al. 2015; Dasgupta and Chatterjee 2021). Moreover, combination approaches integrating tumor suppressor miRNAs with chemotherapy, radiotherapy, immunotherapy, or targeted agents are gaining traction as synergistic strategies to enhance therapeutic efficacy and counteract drug resistance (Seo et al. 2019).

Despite these encouraging developments, several translational hurdles remain. Issues surrounding off-target effects, immunogenicity, optimal dosing, and long-term safety profiles need to be meticulously addressed before miRNA-based therapeutics can be seamlessly incorporated into routine clinical practice (Momin et al. 2021). Additionally, the heterogeneity of tumor biology, patient-specific genetic variations, and the dynamic nature of miRNA expression profiles necessitate a more personalized and adaptable therapeutic framework.

Looking ahead, the integration of high-throughput sequencing, bioinformatics, and AI-driven predictive models will be instrumental in identifying novel tumor suppressor miRNAs and tailoring individualized therapeutic regimens (Kausar 2025). The development of multifunctional, tumor-targeted delivery systems capable of co-delivering miRNAs with other therapeutic agents offers exciting prospects for next-generation combination therapies. Furthermore, expanding our understanding of miRNA interactions within the tumor immune microenvironment may unlock new opportunities to synergize miRNA therapy with immunotherapeutic interventions.

In conclusion, while miRNA-based tumor suppressor therapeutics are still in the early stages of clinical translation, their multifaceted role in cancer biology positions them as a compelling and versatile component of future oncology treatment paradigms. Continued interdisciplinary research and carefully designed clinical trials will be pivotal in realizing the full therapeutic potential of tumor suppressor miRNAs in precision cancer medicine.

## References


Beg MS, Brenner AJ, Sachdev J, Borad M, Kang Y-K, Stoudemire J, Smith S, Bader AG, Kim S, Hong DS (2017) Phase I study of MRX34, a liposomal miR-34a mimic, administered twice weekly in patients with advanced solid tumors. Investig New Drugs 35(2):180–188. https://doi.org/10.1007/s10637-016-0407-y

Calin GA, Croce CM (2006) MicroRNA signatures in human cancers. Nat Rev Cancer 6(11): 857–866. https://doi.org/10.1038/nrc1997

Chen Y, Gao D-Y, Huang L (2015) In vivo delivery of miRNAs for cancer therapy: challenges and strategies. Adv Drug Deliv Rev 81:128–141. https://doi.org/10.1016/j.addr.2014.05.009

Chiyomaru T, Enokida H, Tatarano S, Kawahara K, Uchida Y, Nishiyama K, Fujimura L, Kikkawa N, Seki N, Nakagawa M (2010) miR-145 and miR-133a function as tumour suppressors and directly regulate FSCN1 expression in bladder cancer. Br J Cancer 102(5): 883–891. https://doi.org/10.1038/sj.bjc.6605570

Cortez MA, Ivan C, Valdecanas D, Wang X, Peltier HJ, Ye Y, Araujo L, Carbone DP, Shilo K, Giri DK, Kelnar K, Martin D, Komaki R, Gomez DR, Krishnan S, Calin GA, Bader AG, Welsh JW (2015) PDL1 regulation by p53 via miR-34. JNCI. J Natl Cancer Inst 108(1):djv303. https://doi.org/10.1093/jnci/djv303

Cortez MA, Anfossi S, Ramapriyan R, Menon H, Atalar SC, Aliru M, Welsh J, Calin GA (2019) Role of miRNAs in immune responses and immunotherapy in cancer. Genes Chromosomes Cancer 58(4):244. https://doi.org/10.1002/gcc.22725

Dasgupta I, Chatterjee A (2021) Recent advances in miRNA delivery systems. Methods Protocols 4(1):Article 1. https://doi.org/10.3390/mps4010010

Dong B, Li S, Zhu S, Yi M, Luo S, Wu K (2021) MiRNA-mediated EMT and CSCs in cancer chemoresistance. Exp Hematol Oncol 10(1):12. https://doi.org/10.1186/s40164-021-00206-5

Du F, Cao T, Xie H, Li T, Sun L, Liu H, Guo H, Wang X, Liu Q, Kim T, Franklin JL, Graves-Deal R, Han W, Tian Z, Ge M, Nie Y, Fan D, Coffey RJ, Lu Y, Zhao X (2020) KRAS mutation-responsive miR-139-5p inhibits colorectal cancer progression and is repressed by Wnt signaling. Theranostics 10(16):7335. https://doi.org/10.7150/thno.45971

Esquela-Kerscher A, Slack FJ (2006) Oncomirs—microRNAs with a role in cancer. Nat Rev Cancer 6(4):259–269. https://doi.org/10.1038/nrc1840

Fabbri M, Garzon R, Cimmino A, Liu Z, Zanesi N, Callegari E, Liu S, Alder H, Costinean S, Fernandez-Cymering C, Volinia S, Guler G, Morrison CD, Chan KK, Marcucci G, Calin GA, Huebner K, Croce CM (2007) MicroRNA-29 family reverts aberrant methylation in lung cancer by targeting DNA methyltransferases 3A and 3B. Proc Natl Acad Sci USA 104(40): 15805–15810. https://doi.org/10.1073/pnas.0707628104

Fang Y, Hong X (2021) miR-124-3p inhibits microglial secondary inflammation after basal ganglia hemorrhage by targeting TRAF6 and repressing the activation of NLRP3 Inflammasome. Front Neurol 12:653321. https://doi.org/10.3389/fneur.2021.653321

He W, Zhang M, Zhong Y, Gao Y, Fan D, Lu X (2025) Diverse nanoparticles deliver mRNA to enhance tumor immunotherapy. BMB Rep 58(3):124–132. https://doi.org/10.5483/BMBRep.2024-0165

Hou X, Zaks T, Langer R, Dong Y (2021) Lipid nanoparticles for mRNA delivery. Nat Rev Mater 6(12):1078–1094. https://doi.org/10.1038/s41578-021-00358-0

Iorio MV, Croce CM (2012) MicroRNA dysregulation in cancer: diagnostics, monitoring and therapeutics. A comprehensive review. EMBO Mol Med 4(3):143–159. https://doi.org/10.1002/emmm.201100209

Jeyaram A, Jay SM (2017) Preservation and storage stability of extracellular vesicles for therapeutic applications. AAPS J 20(1):1. https://doi.org/10.1208/s12248-017-0160-y

Ji Y, Wrzesinski C, Yu Z, Hu J, Gautam S, Hawk NV, Telford WG, Palmer DC, Franco Z, Sukumar M, Roychoudhuri R, Clever D, Klebanoff CA, Surh CD, Waldmann TA, Restifo NP, Gattinoni L (2015) miR-155 augments CD8+ T-cell antitumor activity in lymphoreplete hosts by enhancing responsiveness to homeostatic γc cytokines. Proc Natl Acad Sci 112(2): 476–481. https://doi.org/10.1073/pnas.1422916112

Ju J (2025) Challenges and opportunities in microRNA-based cancer therapeutics. Cell Rep Med 6(4):102057. https://doi.org/10.1016/j.xcrm.2025.102057

Kamerkar S, LeBleu VS, Sugimoto H, Yang S, Ruivo CF, Melo SA, Lee JJ, Kalluri R (2017) Exosomes facilitate therapeutic targeting of oncogenic KRAS in pancreatic cancer. Nature 546(7659):498–503. https://doi.org/10.1038/nature22341

Katakowski M, Buller B, Zheng X, Lu Y, Rogers T, Osobamiro O, Shu W, Jiang F, Chopp M (2013) Exosomes from marrow stromal cells expressing miR-146b inhibit glioma growth. Cancer Lett 335(1):201–204. https://doi.org/10.1016/j.canlet.2013.02.019

Kausar N (2025) Machine learning and explainable artificial intelligence reveals the MicroRNAs associated with survival of head and neck squamous cell carcinoma patients. Comput Biol Chem 118:108503. https://doi.org/10.1016/j.compbiolchem.2025.108503

Korpal M, Kang Y (2008) The emerging role of miR-200 family of MicroRNAs in epithelial-mesenchymal transition and cancer metastasis. RNA Biol 5(3):115–119. https://doi.org/10.4161/rna.5.3.6558

Kota J, Chivukula RR, O'Donnell KA, Wentzel EA, Montgomery CL, Hwang H-W, Chang T-C, Vivekanandan P, Torbenson M, Clark KR, Mendell JR, Mendell JT (2009) Therapeutic microRNA delivery suppresses tumorigenesis in a murine liver cancer model. Cell 137(6): 1005–1017. https://doi.org/10.1016/j.cell.2009.04.021

Landskroner-Eiger S, Moneke I, Sessa WC (2013) miRNAs as modulators of angiogenesis. Cold Spring Harb Perspect Med 3(2):a006643. https://doi.org/10.1101/cshperspect.a006643

Lee ES, Na K, Bae YH (2005) Doxorubicin loaded pH-sensitive polymeric micelles for reversal of resistant MCF-7 tumor. J Control Release 103(2):405–418. https://doi.org/10.1016/j.jconrel.2004.12.018

Leskelä S, Leandro-García LJ, Mendiola M, Barriuso J, Inglada-Pérez L, Muñoz I, Martínez-Delgado B, Redondo A, de Santiago J, Robledo M, Hardisson D, Rodríguez-Antona C (2011) The miR-200 family controls beta-tubulin III expression and is associated with paclitaxel-based treatment response and progression-free survival in ovarian cancer patients. Endocr Relat Cancer 18(1):85–95. https://doi.org/10.1677/ERC-10-0148

Li Y, Guessous F, Zhang Y, DiPierro C, Kefas B, Johnson E, Marcinkiewicz L, Jiang J, Yang Y, Schmittgen TD, Lopes B, Schiff D, Purow B, Abounader R (2009) microRNA-34a inhibits glioblastoma growth by targeting multiple oncogenes. Cancer Res 69(19):7569. https://doi.org/10.1158/0008-5472.CAN-09-0529

Li T, Leong MH, Harms B, Kennedy G, Chen L (2013) MicroRNA-21 as a potential colon and rectal cancer biomarker. World J Gastroenterol 19(34):5615. https://doi.org/10.3748/wjg.v19.i34.5615

Li W, Wang Y, Liu R, Kasinski AL, Shen H, Slack FJ, Tang DG (2021) MicroRNA-34a: potent tumor suppressor, cancer stem cell inhibitor, and potential anticancer therapeutic. Front Cell Dev Biol 9:640587

Li, W. (Jess), Wang, Y., Liu, R., Kasinski, A. L., Shen, H., Slack, F. J., & Tang, D. G. (2021). MicroRNA-34a: potent tumor suppressor, cancer stem cell inhibitor, and potential anticancer therapeutic. Front Cell Dev Biol, 9, 640587. https://doi.org/10.3389/fcell.2021.640587

Lian X-Y, Zhang H, Liu Q, Lu X, Zhou P, He S-Q, Tang R-X, Cui J (2020) Ovarian cancer-excreted exosomal miR-199a-5p suppresses tumor metastasis by targeting hypoxia-inducible factor-2α in hypoxia microenvironment. Cancer Commun 40(8):380–385. https://doi.org/10.1002/cac2.12034

Lin S, Gregory RI (2015) MicroRNA biogenesis pathways in cancer. Nat Rev Cancer 15(6):321. https://doi.org/10.1038/nrc3932

Liu J, Meng T, Yuan M, Wen L, Cheng B, Liu N, Huang X, Hong Y, Yuan H, Hu F (2016) MicroRNA-200c delivered by solid lipid nanoparticles enhances the effect of paclitaxel on breast cancer stem cell. Int J Nanomedicine 11:6713. https://doi.org/10.2147/IJN.S111647

Lou G, Chen Z, Zheng M, Liu Y (2017) Mesenchymal stem cell-derived exosomes as a new therapeutic strategy for liver diseases. Exp Mol Med 49(6):e346. https://doi.org/10.1038/emm.2017.63

Matsui M, Corey DR (2017) Non-coding RNAs as drug targets. Nat Rev Drug Discov 16(3): 167–179. https://doi.org/10.1038/nrd.2016.117

Menon A, Abd-Aziz N, Khalid K, Poh CL, Naidu R (2022) miRNA: A promising therapeutic target in cancer. Int J Mol Sci 23(19):11502. https://doi.org/10.3390/ijms231911502

Momin MY, Gaddam RR, Kravitz M, Gupta A, Vikram A (2021) The challenges and opportunities in the development of MicroRNA therapeutics: A multidisciplinary viewpoint. Cells 10(11): 3097. https://doi.org/10.3390/cells10113097

Nana-Sinkam SP, Croce CM (2013) Clinical applications for microRNAs in cancer. Clin Pharmacol Ther 93(1):98–104. https://doi.org/10.1038/clpt.2012.192

Otmani K, Lewalle P (2021) Tumor suppressor miRNA in cancer cells and the tumor microenvironment: mechanism of deregulation and clinical implications. Front Oncol 11:708765. https://doi.org/10.3389/fonc.2021.708765

Piotrowska U, Orzechowska K (2024) Advances in chitosan-based smart hydrogels for colorectal cancer treatment. Pharmaceuticals 17(10):Article 10. https://doi.org/10.3390/ph17101260

Roche J (2018) The Epithelial-to-Mesenchymal Transition in Cancer. Cancers (Basel) 10(2):52. https://doi.org/10.3390/cancers10020052

Roshani Asl E, Rasmi Y, Baradaran B (2021) MicroRNA-124-3p suppresses PD-L1 expression and inhibits tumorigenesis of colorectal cancer cells via modulating STAT3 signaling. J Cell Physiol 236(10):7071–7087. https://doi.org/10.1002/jcp.30378

Rupaimoole R, Slack FJ (2017) MicroRNA therapeutics: towards a new era for the management of cancer and other diseases. Nat Rev Drug Discov 16(3):203–222. https://doi.org/10.1038/nrd.2016.246

Sachdeva M, Mo Y-Y (2010) MicroRNA-145 suppresses cell invasion and metastasis by directly targeting mucin 1. Cancer Res 70(1):378–387. https://doi.org/10.1158/0008-5472.CAN-09-2021

Scholz J, Weil PP, Pembaur D, Koukou G, Aydin M, Hauert D, Postberg J, Kreppel F, Hagedorn C (2022) An adenoviral vector as a versatile tool for delivery and expression of miRNAs. Viruses 14(9):1952. https://doi.org/10.3390/v14091952

Segal M, Slack FJ (2020) Challenges identifying efficacious miRNA therapeutics for cancer. Expert Opin Drug Discov 15(9):987. https://doi.org/10.1080/17460441.2020.1765770

Seo HA, Moeng S, Sim S, Kuh HJ, Choi SY, Park JK (2019) MicroRNA-based combinatorial cancer therapy: effects of MicroRNAs on the efficacy of anti-cancer therapies. Cells 9(1):29. https://doi.org/10.3390/cells9010029

Spizzo R, Nicoloso MS, Lupini L, Lu Y, Fogarty J, Rossi S, Zagatti B, Fabbri M, Veronese A, Liu X, Davuluri R, Croce CM, Mills G, Negrini M, Calin GA (2010) miR-145 participates with TP53 in a death-promoting regulatory loop and targets estrogen receptor-alpha in human breast cancer cells. Cell Death Differ 17(2):246–254. https://doi.org/10.1038/cdd.2009.117

Su H, Yang J-R, Xu T, Huang J, Xu L, Yuan Y, Zhuang S-M (2009) MicroRNA-101, Down-regulated in hepatocellular carcinoma, promotes apoptosis and suppresses Tumorigenicity. Cancer Res 69(3):1135–1142. https://doi.org/10.1158/0008-5472.CAN-08-2886

Sun T, Kong X, Du Y, Li Z (2014) Aberrant MicroRNAs in pancreatic cancer: researches and clinical implications. Gastroenterol Res Pract 2014:386561. https://doi.org/10.1155/2014/386561

Sun WY, Lee D-S, Park JH, Kim O-H, Choi HJ, Kim S-J (2025) Utilizing miR-34a-loaded HER2-targeting exosomes to improve breast cancer treatment: insights from an animal model. J Breast Cancer 28. https://doi.org/10.4048/jbc.2024.0262

Trang P, Wiggins JF, Daige CL, Cho C, Omotola M, Brown D, Weidhaas JB, Bader AG, Slack FJ (2011) Systemic delivery of tumor suppressor microRNA mimics using a neutral lipid emulsion inhibits lung tumors in mice. Mol Ther 19(6):1116–1122. https://doi.org/10.1038/mt.2011.48

van Zandwijk N, Pavlakis N, Kao SC, Linton A, Boyer MJ, Clarke S, Huynh Y, Chrzanowska A, Fulham MJ, Bailey DL, Cooper WA, Kritharides L, Ridley L, Pattison ST, MacDiarmid J, Brahmbhatt H, Reid G (2017) Safety and activity of microRNA-loaded minicells in patients with recurrent malignant pleural mesothelioma: a first-in-man, phase 1, open-label, dose-escalation study. Lancet Oncol 18(10):1386–1396. https://doi.org/10.1016/S1470-2045(17)30621-6

Wang Y, Huang C, Chintagari NR, Bhaskaran M, Weng T, Guo Y, Xiao X, Liu L (2013) miR-375 regulates rat alveolar epithelial cell trans-differentiation by inhibiting Wnt/β-catenin pathway. Nucleic Acids Res 41(6):3833. https://doi.org/10.1093/nar/gks1460

Wiggins JF, Ruffino L, Kelnar K, Omotola M, Patrawala L, Brown D, Bader AG (2010) Development of a lung cancer therapeutic based on the tumor suppressor microRNA-34. Cancer Res 70(14):5923–5930. https://doi.org/10.1158/0008-5472.CAN-10-0655

Xie J, Burt DR, Gao G (2015) AAV-mediated miRNA delivery and therapeutics. Semin Liver Dis 35(1):81. https://doi.org/10.1055/s-0034-1397352

Xing Y, Wang Z, Lu Z, Xia J, Xie Z, Jiao M, Liu R, Chu Y (2021) MicroRNAs: immune modulators in cancer immunotherapy. Immunotherapy Advances 1(1):ltab006. https://doi.org/10.1093/immadv/ltab006

Xu X, Chen W, Miao R, Zhou Y, Wang Z, Zhang L, Wan Y, Dong Y, Qu K, Liu C (2015) miR-34a induces cellular senescence via modulation of telomerase activity in human hepatocellular carcinoma by targeting FoxM1/c-Myc pathway. Oncotarget 6(6):3988–4004. https://doi.org/10.18632/oncotarget.2905

Yalcin S (2019) Dextran-coated iron oxide nanoparticle for delivery of miR-29a to breast cancer cell line. Pharm Dev Technol 24(8):1032–1037. https://doi.org/10.1080/10837450.2019.1623252

Yan X, Tang B, Chen B, Shan Y, Yang H, Reproducibility Project: Cancer Biology (2019) Replication study: the microRNA miR-34a inhibits prostate cancer stem cells and metastasis by directly repressing CD44. eLife 8:e43511. https://doi.org/10.7554/eLife.43511

Yang N (2015) An overview of viral and nonviral delivery systems for microRNA. Int J Pharm Invest 5(4):179. https://doi.org/10.4103/2230-973X.167646

Yang X, Shang P, Yu B, Jin Q, Liao J, Wang L, Ji J, Guo X (2021) Combination therapy with miR34a and doxorubicin synergistically inhibits dox-resistant breast cancer progression via down-regulation of snail through suppressing notch/NF-κB and RAS/RAF/MEK/ERK signaling pathway. Acta Pharm Sin B 11(9):2819–2834. https://doi.org/10.1016/j.apsb.2021.06.003

Yueh P-F, Chiang I-T, Weng Y-S, Liu Y-C, Wong RCB, Chen C-Y, Hsu JB-K, Jeng L-B, Shyu W-C, Hsu F-T (2025) Innovative dual-gene delivery platform using miR-124 and PD-1 via umbilical cord mesenchymal stem cells and exosome for glioblastoma therapy. J Exp Clin Cancer Res 44(1):107. https://doi.org/10.1186/s13046-025-03336-4

Zhang D, Lee H, Zhu Z, Minhas JK, Jin Y (2016) Enrichment of selective miRNAs in exosomes and delivery of exosomal miRNAs in vitro and in vivo. Am J Physiol Lung Cell Mol Physiol 312(1):L110. https://doi.org/10.1152/ajplung.00423.2016

Zhang Q, Pan J, Xiong D, Zheng J, McPherson KN, Lee S, Huang M, Xu Y, Chen S-H, Wang Y, Hildebrandt Ruiz L, You M (2023) Aerosolized miR-138-5p and miR-200c targets PD-L1 for lung cancer prevention. Front Immunol 14:1166951. https://doi.org/10.3389/fimmu.2023.1166951

# MiRNA in the Regulation of DNA Damage Repair, Epithelial–Mesenchymal Transition, and Cancer Therapy Resistance

6

Joanna Szczepanek

**Abstract**

MicroRNAs are posttranscriptional regulators of gene expression that influence biological processes, including cell cycle control, apoptosis, and differentiation. In recent years, their role in cancer biology has become increasingly apparent, especially in DNA damage response (DDR), epithelial–mesenchymal transition (EMT), and resistance to cancer therapies. This chapter provides an overview of current knowledge on how miRNAs regulate these three interconnected processes, with a focus on their mechanisms and clinical implications.

MiRNAs affect the DNA damage repair machinery by targeting key components involved in homologous recombination (HR) and non-homologous end joining (NHEJ). Dysregulation of these molecules can compromise genomic stability and contribute to tumor progression or resistance to therapy. Moreover, specific microRNAs act as key regulators of epithelial–mesenchymal transition (EMT), a reversible process that enhances the migratory and invasive potential of cancer cells. By targeting transcription factors such as ZEB1, Snail, and Twist, miRNAs can either suppress or promote EMT, depending on their expression levels. Significantly, miRNAs also influence cancer therapy resistance by modulating cell survival pathways, DNA repair mechanisms, and the tumor microenvironment. Both oncomiRs and tumor-suppressor miRNAs are involved in resistance to chemotherapy, radiotherapy, and targeted treatments, underscoring their potential as predictive biomarkers and therapeutic targets.

This chapter discusses key miRNA families involved in DDR, EMT, and resistance (e.g., miR-34, miR-200, let-7, and miR-21), elucidates their molecular interactions, and examines their clinical relevance in oncology. The therapeutic

J. Szczepanek (✉)
Centre for Modern Interdisciplinary Technologies, Nicolaus Copernicus University, Torun, Poland
e-mail: szczepanekj@umk.pl



S. Tiwari et al. (eds.), *microRNA Modulation: A Promising Combination Therapy for Cancer*, https://doi.org/10.1007/978-981-95-5989-3_6

modulation of miRNAs—through the use of mimics or inhibitors—offers a promising approach to sensitize tumors to treatment and overcome resistance. Understanding miRNA-driven regulatory networks is therefore essential for the development of innovative, personalized cancer therapies.

**Keywords**

miRNA · DDR · EMT · Therapy resistance · Chemoresistance · Tumor plasticity · Biomarkers · Therapeutic miRNA targeting

**Learning Objectives**

- *Explain* how specific miRNAs regulate DNA damage response and repair pathways, influencing genomic stability in cancer cells.
- *Analyze* the role of miRNAs in modulating epithelial–mesenchymal transition (EMT) and their contribution to tumor invasion and metastasis.
- *Evaluate* the impact of deregulated miRNAs on cancer therapy resistance, including chemotherapy, radiotherapy, and targeted therapies.
- *Assess* the potential of targeting miRNAs involved in DNA repair, EMT, and resistance as therapeutic strategies in precision oncology.

## 6.1 Fundamental Cellular Mechanisms in Cancer—DNA Repair, EMT, and Therapy Resistance

### 6.1.1 DNA Damage Response in Genomic Instability

Cancer cells accumulate genomic alterations through defects in DNA damage response (DDR) pathways. In a typical cell, DDR signaling—through components such as ATM and ATR kinases or γH2AX formation—detects DNA damage and activates repair processes that safeguard the genome's stability (Moyret-Lalle et al. 2022). But when these DDR pathways falter, genomic instability can occur. This instability fuels clonal evolution, resulting in a patchwork of distinct cells within a tumor. Interestingly, the epithelial–mesenchymal transition (EMT) itself can cause chromosomal instability. Recent in vivo studies have shown that EMT facilitates chromosomal rearrangements, including extensive chromosome fragmentation and reassembly events, thereby accelerating the generation of heterogeneous tumor cell clones (Carstens and Lovisa 2025). Moreover, the overexpression of Twist1, a key regulator of EMT, tends to exacerbate genome instability, thereby contributing to tumor cell diversity and potentially to resistance to chemotherapy (Khot et al. 2020; De Las Rivas et al. 2021). These findings highlight how losing genome stability—whether through faulty DDR or EMT-linked changes—produces a broad variety of cancer cells, some of which might survive treatment (Khot et al. 2020; De Las Rivas et al. 2021).

### 6.1.2 EMT as a Driver of Phenotypic Heterogeneity

EMT is a flexible and reversible program in which epithelial cancer cells lose their polarity and adhesion properties, instead gaining mesenchymal features, such as enhanced movement and invasiveness. In cancer, EMT plays a significant role in tumor initiation, growth, spread, and the development of stem-like traits (Debaugnies et al. 2023a; Nieto et al. 2016; Shibue and Weinberg 2017). Transcription factors central to EMT, such as ZEB1/2, Snail, and Twist, suppress epithelial genes and activate mesenchymal ones, enabling cells to migrate and adapt to various environments. What's worth noting is that EMT isn't simply an on–off event; instead, tumors often contain cells in intermediate, "hybrid" epithelial/mesenchymal states, which broadens the range of cell behaviors and increases plasticity. This diversity inside tumors helps cancer cells cope with stresses—including therapies (Khot et al. 2020; De Las Rivas et al. 2021). Those cells stuck in partial EMT states often show the highest tolerance to drugs and other cytotoxic challenges. So, the EMT program grants specific cancer cells a kind of flexibility that can be exploited to survive the pressure of treatments (Khot et al. 2020; De Las Rivas et al. 2021; Shibue and Weinberg 2017; Zhang et al. 2021; Chaffer et al. 2016).

### 6.1.3 Crosstalk between DDR, EMT, and Drug Resistance

Emerging evidence reveals tight molecular links between DDR pathways and EMT regulators. For example, the DNA damage kinase ATM directly phosphorylates the EMT factor ZEB1 in response to DNA breaks (Moyret-Lalle et al. 2022; Zhang et al. 2014, 2018b). Phospho-ZEB1 then binds the deubiquitinase USP7, stabilizing the checkpoint kinase CHK1 and enhancing homologous recombination repair (Moyret-Lalle et al. 2022; Zhang et al. 2014). Moreover, ZEB1 forms a complex with p300/pCAF to activate the ATM gene promoter, creating a positive feedback loop that boosts DNA repair capacity after genotoxic stress (Moyret-Lalle et al. 2022). Similarly, RNF8-mediated ubiquitination of TWIST1 promotes its nuclear localization and upregulates homologous repair genes (Zhou et al. 2019; Cao et al. 2022; Lee et al. 2016). Conversely, inhibition of key DDR enzymes can induce EMT: for instance, PARP1 loss triggers ZEB1 expression and an EMT phenotype. Together, these studies suggest that EMT transcription factors can regulate DNA repair choices and that the status of DNA repair can impact EMT plasticity. In cancer models, EMT-driven alterations in DNA repair have concrete effects on therapy response. In colorectal cancer, ZEB2-driven EMT upregulates the nucleotide excision repair (NER) gene ERCC1, thereby increasing the repair of platinum-induced DNA lesions and promoting cell survival (Moyret-Lalle et al. 2022; Sreekumar et al. 2021). Thus, EMT in this setting actually enhances genomic stability in response to chemotherapeutic insult. Consistent with this, high ZEB1 expression is frequently observed in tumors resistant to radiotherapy or chemotherapy (Moyret-Lalle et al. 2022). EMT not only reprograms cell phenotype but can also promote DNA repair mechanisms, enabling cancer cells to withstand DNA damage from therapy.

### 6.1.4 Phenotypic Heterogeneity and Therapy Resistance

Tumor heterogeneity is a key driver of both intrinsic (preexisting) and acquired drug resistance (Roszkowska 2024; Crucitta et al. 2022; Nussinov et al. 2025; Zhang et al. 2022; Lim and Ma 2019). Within a heterogeneous tumor, some clones possess innate resistance mechanisms, while others develop resistance during treatment. EMT contributes to this heterogeneity, and EMT-type cells display inherent resistance to cytotoxic therapies that selectively eliminate epithelial-like cancer cells, reflecting their altered signaling, survival, and repair mechanisms. For example, in a mouse skin squamous cell carcinoma model, EPCAM$^{-}$ (EMT) tumor cells were profoundly resistant to cisplatin/5-FU chemotherapy in vivo, whereas EPCAM$^{+}$ epithelial cells were sensitive (Debaugnies et al. 2023b). In vitro, EMT cells survived multiple genotoxic stresses (e.g., cisplatin, topoisomerase inhibitors, radiation) that efficiently induced apoptosis in epithelial cells. Mechanistically, EMT-associated cells upregulate drug efflux pumps, anti-apoptotic factors, and DNA repair regulators (such as the RHOJ GTPase) that enhance DNA replication stress tolerance and repair (Debaugnies et al. 2023b). Consequently, EMT promotes both tumor cell plasticity and survival during therapy. Essentially, the interaction of genomic instability, EMT-driven phenotypic plasticity, and DNA damage response remodeling creates a constantly changing tumor cell microenvironment. Genome instability generates diverse clones, EMT processes encourage phenotypic switching and invasion, and coordinated DDR changes support the survival of these variants under stress. The interplay of genome instability, EMT-driven plasticity, and DDR alterations accelerates tumor evolution, leading to treatment failure (Carstens and Lovisa 2025; Gu et al. 2024; Hosea et al. 2024).

## 6.2 MicroRNA-Mediated Integration of DNA Repair, Epithelial–Mesenchymal Transition, and Therapy Resistance

In cancer cells, miRNAs act as key regulators, integrating the DNA damage response, epithelial–mesenchymal transition, and therapy resistance into a coherent regulatory network (Ha and Kim 2014). By targeting detectors and mediators of genomic stress (e.g, *ATM*, *BRCA1*, *RAD51*) (Valenti et al. 2019), EMT transcription factors (e.g., *ZEB1/2*, *Snail*, *Twist*) (Siemens et al. 2014; Brabletz and Brabletz 2010), and apoptotic regulators (e.g., *Bcl*-2 family, *PTEN*) (Meng et al. 2007), miRNAs regulate the balance between proliferative stability and adaptive plasticity, enabling tumor cells to respond to therapeutic challenges in real time. Dysregulation of this posttranscriptional layer leads to unchecked genomic instability and the development of hybrid EMT states, which promote the selection of subclones with enhanced survival and migratory abilities (Brabletz and Brabletz 2010). Consequently, the tumor becomes heterogeneous—subpopulations with impaired DDR accumulate mutations, while those with moderate EMT develop optimal resistance to chemotherapy and radiotherapy (Gregory et al. 2008). Aberrant miRNA-mediated

DDR control can both trigger mutational cascades and enhance the regenerative potential of cancer stem cell-like populations posttreatment (Iorio and Croce 2012).

### 6.2.1 The Role of microRNAs in the Regulation of DNA Repair Genes and Resistance to Anticancer Therapy

DNA damage repair is essential for maintaining genomic stability in cells. Dysregulation of DNA repair pathways leads to the accumulation of mutations and genomic instability, and contributes to carcinogenesis (Peraza-Vega et al. 2022; He et al. 2016). Long, uninterrupted DNA fragments are primarily restored through homologous recombination (HR) and non-homologous end joining (NHEJ). In contrast, minor lesions are repaired via mechanisms such as base excision repair (BER) and mismatch repair (MMR) (Dexheimer 2013; Chatterjee and Walker 2017). MicroRNAs constitute a critical posttranscriptional regulatory layer of genes involved in DNA repair, and their dysregulation plays a significant role in shaping tumor resistance to genotoxic therapies. This regulation is crucial for cancer cell sensitivity to genotoxic treatments; for example, modulation of miRNA expression alters the capacity for double-strand break (DSB) repair, thereby impacting the efficacy of chemotherapy and radiotherapy (Table 6.1).

Under physiological conditions, the MRN complex (MRE11–RAD50–NBS1) initiates the detection of DNA double-strand breaks and activates ATM/ATR kinases, triggering cell cycle arrest and recruitment of repair proteins (Espinosa-Diez et al. 2018; Williams et al. 2007; Lee et al. 2013; Hopfner 2014). miR-494 and miR-24, frequently overexpressed in various cancers, directly suppress MRE11 and NBS1, resulting in inhibited phosphorylation of H2AX (γH2AX) and a shortened G2/M phase, thereby accelerating the progression of damaged cells through the cell cycle, promoting chromosomal aberrations, and increasing resistance to chemotherapy and radiotherapy (Peraza-Vega et al. 2022; Espinosa-Diez et al. 2018).

In homologous recombination, the recombinase RAD51 and BRCA1/2 proteins play a pivotal role by forming filaments that can search for the sister chromatid template (Prakash et al. 2015; Spies et al. 2021). MiR-182 and miR-155, often overexpressed in breast cancer and lymphomas, reduce RAD51 levels, initially enhancing sensitivity to platinum-based drugs and PARP inhibitors due to HR deficiency. However, prolonged RAD51 downregulation selects clones that employ alternative repair mechanisms, leading to the development of secondary resistance (Li et al. 2022; Gasparini et al. 2014; Moskwa et al. 2011). Conversely, tumor-suppressive miRNAs, such as let-7e and miR-506, increase the efficacy of cisplatin chemotherapy in ovarian cancer models by repressing RAD51 and BRCA1, highlighting their clinical potential as adjuvants for synthetic lethality-based therapies (Gasparini et al. 2014; Moskwa et al. 2011; Xiao et al. 2017; Xiao et al. 2020; Mao et al. 2025; Liu et al. 2015; Setton et al. 2021). Regulation of PARP1 by miRNAs is also crucial for the development of drug resistance. For instance, miR-7-5p targets the 3′ untranslated region (3′ UTR) of PARP1, thereby significantly reducing its expression. In small cell lung cancer (SCLC) cell models, miR-7-5p

**Table 6.1** Summary of key DNA repair pathways, associated genes, regulatory miRNAs, and their relevance to therapy resistance in cancer

| DNA repair mechanism | Description | Target genes | Regulatory miRNAs | Significance for cancer therapy |
|---|---|---|---|---|
| BER (base excision repair) | Removal of damaged bases (e.g., oxidized, methylated) | UNG2, DNA polymerase β | miR-16, miR-34c, miR-199a, miR-499 | Resistance to alkylating agents (e.g., temozolomide, melphalan), cisplatin, and radiotherapy |
| NER (nucleotide excision repair) | Removal of lesions caused by UV radiation and chemical agents | ERCC3 (XPB), ERCC4 (XPF), RAD23B | miR-192, miR-373 | Involved in cancer development associated with HBV infection and hypoxia, resistance to cisplatin (ovarian and lung cancer) |
| MMR (mismatch repair) | Correction of replication errors (e.g., insertions, deletions) | MSH2, MSH6, MLH1 | miR-155, miR-21 | Associated with microsatellite instability (MSI) and resistance to 5-FU, resistance to alkylating agents (e.g., temozolomide), sensitivity to immunotherapy in MSI-high tumors; potential predictive markers. |
| TLS (translesion synthesis) | DNA damage tolerance via specialized polymerases during replication blockage | RAD18, REV1 | miR-145, miR-630, miR-96, miR-155, miR-21 | Overcoming resistance to 5-FU and PARP inhibitors, replication protection under stress, resistance to cisplatin, carboplatin, and other cytostatics; mechanism associated with mutagenesis and acquired resistance |
| *DSB repair by precise (HR) or error-prone (NHEJ) mechanisms* | | | | |
| HR (homologous recombination) | Accurate DSB repair using a sister chromatid as a template | BRCA1, BRCA2, RAD51, RAD52, | miR-182, miR-125b, miR-193b, miR-148b, miR-210, | Increased sensitivity to PARP inhibitors (e.g., olaparib) and cisplatin; |

(continued)

**Table 6.1** (continued)

| DNA repair mechanism | Description | Target genes | Regulatory miRNAs | Significance for cancer therapy |
|---|---|---|---|---|
| | | FANCD2, FANCG | miR-96, miR-302, miR-239 | modulation of sensitivity to radiotherapy and chemotherapy |
| NHEJ (non-homologous end joining) | Template-independent, fast but error-prone DSB repair | DNA-PKcs, 53BP1 | miR-101, miR-34a | Potential enhancement of radiotherapy efficacy, associated with resistance to radiation and cytostatics (e.g., etoposide) |

overexpression inhibits RAD51 and BRCA1 induction following doxorubicin treatment, attenuating HR and restoring drug sensitivity (Lai et al. 2019). In other words, miR-7-5p overabundance reverses doxorubicin resistance by inducing PARP1 deficiency and compromising DNA repair. Similar miRNA-dependent regulatory networks affecting DNA repair components warrant further investigation.

NHEJ is a rapid but error-prone DSB repair pathway active throughout the cell cycle, especially in G1 phase. It initiates with the Ku70/Ku80 heterodimer binding to DNA ends, the activation of DNA–PKcs kinase, and the recruitment of the Lig4–XRCC4 complex (and XLF) to ligate the DNA ends (Gasparini et al. 2014; Moskwa et al. 2011; Xiao et al. 2017, 2020; Mao et al. 2025; Liu et al. 2015; Tan et al. 2023; Stinson and Loparo 2021). Recent studies demonstrate that microRNAs actively modulate the classical NHEJ pathway, critically influencing tumor sensitivity to radiotherapy and genotoxic chemotherapy. miR-145 binds to the 3′ UTR of PRKDC (encoding DNA-PKcs), thereby reducing DNA–PKcs levels in breast cancer cells. Consequently, ionizing radiation results in prolonged γH2AX foci persistence, impaired XRCC4–Lig4 recruitment, and diminished DSB repair efficiency, thereby increasing apoptosis and tumor radiosensitization in vivo (Sudhanva et al. 2022). Moreover, miR-145 acts synergistically with miR-125b and miR-24, which are co-overexpressed in glioma and lung cancer cells, regulating DNA–PKcs and modulating Cdk2 activity during the G1/S phase (He et al. 2016). This interaction shifts the repair balance toward error-prone NHEJ under HR-compromised conditions, promoting faulty end-joining and the accumulation of small deletions at DSB sites, which correlates clinically with resistance to low-dose radiation (2–4 Gy) and necessitates higher doses for tumor control (He et al. 2016). The miRNA network regulating NHEJ extends beyond miR-145. OncomiRs miR-24 and miR-151a target Ku70/Ku80 expression, destabilizing NHEJ initiation and impairing rapid DSB repair. Reduced Ku70/80 availability triggers partial switching to alternative end-joining pathways (alt-NHEJ/MMEJ), characterized by increased nonhomologous recombination and the formation of gene rearrangements that promote tumor progression (Tessitore et al. 2014). Additionally, miR-1246 and miR-1587 regulate XRCC4 and Lig4, thereby influencing the end ligation of DSBs. Deregulation of these miRNAs in breast and lung cancers associates with poorer treatment outcomes, as excessive yet unstable repair enhances selection of genotoxic therapy-resistant clones (Tessitore et al. 2014). Kinetic analysis of DSB repair under miRNA modulation shows that reducing DNA–PKcs and Ku70/80 extends G2 arrest. This, after temporary inhibition of HR, leads to more faulty NHEJ. As a result, genomic diversity increases; patients with high levels of miR-145 or miR-24 have significantly worse overall survival and shorter progression-free periods after radiotherapy. These results indicate that delivering anti–miR–145 or inhibiting miR-24 could be new adjuvant strategies to improve the effectiveness of genotoxic therapy.

Base excision repair (BER) removes small DNA lesions such as oxidation, deamination, alkylation, methylation, and uracil incorporation arising from replication errors or cytosine deamination (Natarajan 2016; Wallace et al. 2012). BER dysfunction leads to the accumulation of mutations and genomic instability, which are features often seen in many cancers. A well-known example is miR-16, which binds to the 3′ UTR of UNG2 mRNA, encoding uracil-DNA glycosylase responsible for removing uracil from DNA. The miR-16-driven downregulation of UNG2 decreases BER activity, potentially increasing DNA damage risk during the G2/M phase of the cell cycle (Hegre et al. 2013). Similar effects are seen with miR-34c, a member of the p53-regulated miRNA family that links genotoxic stress response to BER regulation. miR-199a, which also targets UNG2, affects how cancer cells respond to DNA-damaging treatments, with roles that vary depending on the context. Additionally, miR-499 controls the expression of DNA polymerase β, which is essential for gap filling after base excision. Studies on esophageal cancer cells show that overexpressing miR-499 lowers pol β levels, making cells more sensitive to cisplatin. (Wang et al. 2015). This suggests that miR-499 could act as a potential enhancer of platinum therapy. Notably, many of these miRNAs, including miR-16 and miR-34c, also function as tumor suppressors in other biological contexts by regulating the cell cycle, apoptosis, and signaling pathways, highlighting their therapeutic significance (Zhang et al. 2019; Slabáková et al. 2017; Sadeghi et al. 2025; Fu et al. 2023).

The mismatch repair (MMR) system corrects replication errors, with major proteins including MSH2–MSH6 and MLH1–PMS2 (Friker et al. 2025; Pecina-Slaus et al. 2020). OncomiR-21 suppresses MSH2 and MLH1 in colorectal cancer cells, causing microsatellite instability (MSI) and resistance to fluorouracil and other DNA synthesis inhibitors (Valeri et al. 2010a). Similarly, miR-155 reduces MSH2/MSH6 levels, which promotes the accumulation of mutations and diminishes the response to genotoxic therapy (Ding et al. 2018; Valeri et al. 2010b). Furthermore, miR-138 influences the availability of phosphorylated H2AX, thereby impacting the quantity and stability of repair foci; its overexpression diminishes the radiation response, thereby expediting the selection of resistant tumor clones (Majidinia and Yousefi 2016). miR-363-3p enhances JNK pathway activity and DSB repair via the DUSP10–JNK axis, contributing to resistance against R-CHOP chemotherapy in diffuse large B-cell lymphoma (DLBCL), suggesting anti–miR-363-3p as a potential chemosensitizer (Zhou et al. 2022).

### 6.2.2 The Role of microRNAs in the Regulation of Genes Associated with Epithelial–Mesenchymal Transition

Previous research on the role of microRNAs in the context of the EMT has provided valuable insights into the complexity of genetic regulation during carcinogenesis. However, there is limited data on this topic in the context of various types of tissue cancers. Previous work has primarily focused on a kind of cancer, which represents a significant omission in understanding the universal and specific regulatory mechanisms between microRNAs and EMT.

The process of transition from epithelial cells to mesenchymal cells, known as epithelial-mesenchymal transition, plays a key role in the ability of cancer cells to migrate and invade. There are three subtypes of EMT: developmental EMT, wound healing EMT, and cancer-related EMT. Of these, "type 3" is associated with cancer development, also referred to as oncogenic EMT. In this process, epithelial cells transform into mesenchymal cells, which leads to increased invasive and migratory abilities (Gupta and Maitra 2016; Demirkan 2013). EMT is a complex molecular process in which specific genes, especially transcription factors, play a significant role. SNAIL, Slug, ZEB1, ZEB2, and Twist are key regulators of EMT-related gene expression. Their action is to repress epithelial genes such as E-cadherin, which leads to the cells transitioning to a mesenchymal state. ZEB factors (ZEB1 and ZEB2) play a crucial role in inhibiting the expression of adhesion proteins, thereby destabilizing the cellular structure and facilitating migration. Similarly, SNAIL and Slug regulate the EMT process by controlling gene expression, and their actions are particularly related to the loss of epithelial characteristics. Additionally, the Twist gene acts as a regulator of EMT by influencing the expression of genes associated with the mesenchymal phenotype. Its role is particularly noticeable in the context of cells' ability to invade and migrate (Zhang et al. 2018a; Wang et al. 2018; You et al. 2017). Additionally, molecular signals play an essential role in the induction of EMT. Growth factors, such as TGF-β, cytokines, and signaling pathways, including the Wnt pathway, influence this process, initiating appropriate changes in gene expression and promoting the transition of cancer cells into the mesenchymal state. In the context of the TGF-β pathway, this is a key signaling mechanism regulating EMT. Transforming growth factor beta (TGF-β) activates a receptor on the cell surface, which initiates a cascade of intracellular signals. These signals, particularly through Smad transcription factors, regulate the expression of genes associated with EMT. TGF-β is therefore a key player in controlling the balance between epithelial and mesenchymal features in the context of cancer (Wang et al. 2017; Ding et al. 2013).

The regulation of EMT is also modulated by microRNAs, which play a crucial role in controlling the expression of genes involved in this process. They significantly influence the EMT process in cancer, regulating cell proliferation, migration, invasion, the ability to create stem cells, drug resistance, angiogenesis, tumor growth, and even metastasis (Hill and Tran 2021; Tang et al. 2016; Berindan-Neagoe and Calin 2014; D'Amato et al. 2013). Changes in cell adhesion are another essential element of the EMT process. The expression of genes related to cell adhesion, such as E-cadherin, is modified, a characteristic of the transition of cancer cells from the epithelial to the mesenchymal phase. These changes in cell adhesion contribute to the migration and invasion of cancer cells. EMT has significant clinical implications, particularly in the context of tumor aggressiveness, metastasis, and resistance to therapies. The genetic basis of EMT in cancer is multifaceted, encompassing the interactions of various transcription factors, molecular signals, microRNAs, and alterations in cell adhesion. They are still insufficiently characterized. Understanding these underlying genetic mechanisms is essential from a therapeutic perspective. Research on these genes opens up the possibility of developing therapies targeting EMT, which may be promising in the context of inhibiting invasion and metastasis, as well as increasing the effectiveness of anticancer treatment. Therefore, understanding the molecular basis of EMT is a key step towards developing more effective cancer treatment strategies.

## 6.3 Characterization of the Role of miRNAs in Selected Types of Cancer

### 6.3.1 Breast Cancer

Breast cancer (BC) is considered one of the most common causes of cancer death among women around the world. The EMT is a complex program crucial for the acquisition of migratory and invasive capabilities by carcinoma cells in breast cancer metastasis. TGF-β1-induced EMT plays a significant role in this process, and its regulation involves various microRNAs. One such miRNA, miR-23a, is upregulated in breast cancer cells, especially in metastatic samples, and is induced by TGF-β1. The Smad proteins directly bind to the RNA Smad-binding element in miR-23a, mediating TGF-β1-associated regulation. Inhibition of miR-23a expression suppresses TGF-β1-induced EMT, leading to reduced migration, invasiveness, and metastatic potential in breast cancer cells. miR-23a directly targets CDH1, a key EMT modulator, and its suppression activates the Wnt/β-catenin signalling cascade, thereby enhancing BC metastasis (Ma et al. 2017). Another upregulated miRNA, miR-27a, contributes to the migratory potential of breast cancer cells by inducing EMT. It targets FBXW7, and the overexpression of FBXW7 inhibits EMT and the migratory abilities of breast cancer cells, highlighting the regulatory role of miR-27a in breast cancer metastasis. It has been found that restoring FBXW7 expression can at least partially reverse the oncogenic effect (Jiang et al. 2018). Additionally, miR-29a, upregulated in breast cancer samples, correlates with distant metastasis and poor clinical outcomes. Silencing miR-29a suppresses proliferation and migration, with its target, Ten Eleven Translocation 1 (TET1), identified as a key regulator. Upregulation of TET1 attenuates proliferation and migration, indicating that miR-29a-mediated downregulation of TET1 enhances EMT (Pei et al. 2016). Similarly, miR-122-5p had a promoting effect on triple-negative breast cancer cells (TNBC), which led to the EMT process and activation of the MAPK pathway (Wang and Wang 2020). Several other miRNAs, both upregulated (miR-93, miR-125b, miR-199a-3p, miR-221) and downregulated (miR-34a, miR-92b, miR-124, miR-138-5p, miR-153, miR-516a-3p, miR-524-5p), impact the EMT process in breast cancer (Du et al. 2020; Yang et al. 2015; Wu et al. 2017; Lu et al. 2021). These miRNAs exert their influence by targeting receptors that convey signals from EMT inducers or various EMT components, collectively illustrating the intricately regulated network of EMT in breast cancer.

### 6.3.2 Ovarian Cancer

Ovarian cancers are among the most aggressive female malignancies. The functions of microRNAs in the process of ovarian cancer (EOC) metastasis are complex and diverse. Direct and indirect regulation occurs through influencing the transition from an epithelial to a mesenchymal state, by targeted interactions with E-cadherin and E-cadherin repressors, such as ZEB, SNAIL, Slug, and Twist, along with associated signaling pathways. MiR-9 directly targets E-cadherin, activating EMT, while miR-25, miR-101-3p, miR-130b, miR-145, miR-150, and miR-1236-3b directly inhibit the expression of E-cadherin repressors, including ZEB1, SNAI2, and TWIST1 (Nguyen et al. 2020). Numerous miRNAs, including miR-18b, miR-19b, miR-205, miR-216b, and miR-552, promote EMT by targeting PTEN, thereby activating the PI3K/AKT pathway. Additionally, the upregulation of miR-590-3p, miR-130a, miR-181a, and miR-520 h induces the Wnt, mTOR, and TGF-β signalling pathways, which are known to promote EMT (Lei et al. 2020). In contrast, miR-936, miR-375-5p, miR-509-3p, miR-340, miR-377, miR-219-5p, miR-25, and miR-203 inhibit the PI3K/AKT, Hippo, Wnt, and TGF-β signalling pathways concurrently, thereby weakening cell migration, invasion, and EMT (Nguyen et al. 2020). Examples of identified miRNAs also include miR-506, miR-130a, miR-200, and miR-429, which function both as oncomiRs and TsmiRs (Wang et al. 2014; Liu et al. 2014; Prahm et al. 2021). MiR-506, acting as a TsmiR, has been identified as an inhibitor of EMT, restraining ovarian cancer cells during the invasion process (Liu et al. 2014). MiR-130a and miR-200 act as suppressors of EMT, limiting the invasive ability of ovarian cancer cells (Prahm et al. 2021). Meanwhile, miR-429 plays a dual role as both an oncomiR and TsmiR, depending on the context, emphasizing the complexity of EMT regulation by miRNAs in ovarian tumors (Chen et al. 2011).

### 6.3.3 Lung Cancer

Lung cancer is one of the most common and deadliest cancers in the world. One notable family of EMT–TFs is the SNAIL gene family, which includes SNAI1

(SNAIL), SNAI2 (SLUG), and SNAI3 (SMUC). MiRNAs intricately regulate EMT by directly targeting members of the SNAIL family. The p53/miR-34 axis, for instance, modulates SNAIL expression in lung cancer cells, with p53 knockdown leading to SNAIL upregulation by down-regulating miR-34 expression (Kang et al. 2013; Kawami et al. 2021). Additionally, miR-126 exhibits suppressive effects on lung cancer cell invasion and migration by directly targeting the PI3K/AKT/SNAIL signaling axis. Other miRNAs, such as miR-346, miR-22, and miR-30a, further regulate EMT in lung cancer by directly targeting SNAIL, influencing cellular invasion and migration, and modulating the expression levels of key EMT markers, including E-cadherin and N-cadherin (Wang et al. 2017; Cao et al. 2012). Another set of EMT–TFs, ZEB1/ZEB2, extensively modulates cell plasticity, metastasis, and treatment resistance. The miR-200 family emerges as a crucial regulator of this axis, forming a feedback loop wherein ZEB1/ZEB2 inhibits miR-200 family transcription, and, reciprocally, the miR-200 family directly suppresses ZEB1. This intricate interplay influences lung cancer cell invasiveness, tumor progression, and therapeutic resistance (Liu et al. 2022; Zhao et al. 2018; Zhang et al. 2017). Furthermore, miR-205 acts as a suppressor of EMT by directly targeting the 3' UTR of ZEB1 and ZEB2, underscoring its role in reducing the metastatic potential of lung cancer cells. Downstream of ZEB1/ZEB2, miR-489 emerges as a regulator that inhibits EMT by targeting SUZ12, and miR-33b suppresses Wnt/β-β-catenin/ZEB1 signaling, thereby affecting lung cancer cell growth, invasion, and EMT progression (You et al. 2017). Twist, another key EMT–TF, is regulated by miRNAs such as miR-98 and miR-92b, which inhibit Twist-induced EMT, highlighting their potential role as modulators of lung cancer cell invasion and migration (Li and Wu 2016; Nakashima et al. 2012). Beyond directly targeting EMT–TFs, miRNAs also influence key signaling pathways involved in EMT. For example, miR-155 targets Smad2 in the TGF-β pathway, and miR-148b affects EMT by targeting ROCK1 in the Rho/ROCK pathway. These interactions demonstrate the complex regulatory network driven by miRNAs in lung cancer EMT (Du et al. 2019; Luo and Liang 2018).

### 6.3.4 Gastric Cancer

Gastric cancer is a serious health issue, and its link to the EMT process has been well established. One of the main pathways targeted by miRNAs in EMT regulation is the Wnt/β-catenin pathway. MiR-143, miR-145, miR-204, and miR-491-5p are identified as suppressors of EMT by activating E-cadherin, a key component of the Wnt/β-catenin pathway, while also inhibiting direct targets such as MYO6 and SNAIL. On the other hand, overexpression of miR-93 and miR-106a is associated with lymph node metastasis and vascular invasion in GC. These miRNAs target TIMP2, which promotes E-cadherin expression and inhibits EMT, thus facilitating metastasis development. The complex interaction between miRNAs and the Wnt/β-catenin pathway highlights their role in regulating EMT during GC metastasis (Yu et al. 2018; Lei et al. 2017; Guan et al. 2017). Beyond the Wnt/β-catenin pathway, miRNAs exert regulatory control over the PI3K/AKT/mTOR signalling cascade, influencing processes such as invasion, migration, and EMT. MiR-340, miR-379, miR-520a-3p, and miR-1254 have been shown to inhibit the PI3K/AKT/mTOR pathway by targeting SPP1, FAK, WEE1, and SMURF1 (Song et al. 2019; Xu et al. 2017). Notably, increased expression of these miRNAs correlates with decreased AKT phosphorylation, suggesting a suppressive effect on this signaling pathway. Additionally, miR-142-5p, miR-216, miR-491, miR-28, and miR-197 directly target components of the PI3K/AKT/mTOR pathway, further emphasizing their role in modulating EMT-related events in GC metastasis (Song et al. 2019; Kipkeeva et al. 2020; Li et al. 2018). Furthermore, miRNAs such as miR-7, miR-9, and miR-508-3p function as tumor suppressors by inhibiting the NF–κB pathway, while the dual role of miR-216a adds complexity to its involvement in GC metastasis. While some studies suggest its oncogenic role by activating the NF–κB pathway, others propose its tumor-suppressive function by suppressing JAK2 in the JAK/STAT pathway, thereby preventing the development of metastasis. The intricately regulatory network orchestrated by these miRNAs highlights the multifaceted nature of EMT regulation in gastric cancer metastasis (Kipkeeva et al. 2020; Huang et al. 2016; Wu et al. 2018).

### 6.3.5 Colon Cancer

Colorectal cancer is a common health problem. Additionally, miRNAs that play an essential role in regulating EMT have been identified. In the context of colon cancer, exposure to TGF-β induces the expression of miR-21 and miR-31, which contribute to heightened cellular motility and invasiveness. Notably, the increased levels of these miRNAs have been linked to lymph node positivity and the development of distant metastases in colorectal cancer patients (Slaby et al. 2007). In the progression of colorectal cancer, both miR-21 and miR-31 play a significant role in promoting the TGF-β-induced epithelial–mesenchymal transition (EMT). This occurs through the repression of TIAM1 (T-cell lymphoma invasion and metastasis 1), a guanine nucleotide exchange factor involved in the activation of the Rac GTPase (Cottonham et al. 2010). Coincidentally, in established metastases, the activation of miR-31 demonstrates a dual role, inducing the regression of metastasis and extending patient survival. Moreover, its upregulation has been associated with a reduction in the metastatic potential of cancer cells, achieved through the targeting of RhoA (Valastyan et al. 2011). This multifaceted regulatory mechanism highlights the intricate role of miR-31 in modulating both the initiation and regression phases of metastasis in colorectal cancer, offering insights into potential therapeutic strategies. In colorectal cancer cells, miR-34 acts as a TsmiR, inhibiting EMT by regulating the p53 pathway. MiR-34 regulates the expression of EMT-related genes, including SNAIL, SLUG, ZEB1, and ZEB2 (Wang et al. 2018; Gu et al. 2018). As in gastric cancer, miRNAs from the miR-200 family play a key role in regulating EMT in colorectal cancer. miR-200 expression maintains the epithelial phenotype, limiting the invasive capacity of cancer cells (Shang et al. 2017; O'Brien et al. 2021). Again, miR-21, a type of oncomiR, is involved in the EMT process in colorectal cancer. Its excessive expression leads to increased invasiveness of cancer cells and promotes disease progression (Slaby et al. 2007; Cottonham et al. 2010).

## 6.4 Therapeutic Using Mimics and Inhibitors

For the treatment of cancers, particularly those associated with the EMT process, research is focusing on the development of miRNA mimics and inhibitors. miRNA mimics are synthetic analogues of miRNAs that are designed to mimic the functions of endogenous miRNAs. In turn, miRNA inhibitors are molecules designed to inhibit the function of specific miRNAs that are overexpressed in cancer cells. The use of miRNA mimics, which mimic endogenous miRNAs, may lead to the inhibition of the EMT process, thereby limiting the invasive ability of cancer cells. In turn, miRNA inhibitors can block the action of oncomiRs, which contributes to stopping the EMT process and reducing the invasiveness of cancer cells (Szczepanek et al. 2022; Romano et al. 2021).

Therapy using microRNA mimics and inhibitors is an innovative approach to controlling gene expression at the posttranscriptional level, which is essential in the context of diseases, especially in oncological therapy. The advantages of this type of therapy include precise regulation of gene expression, which allows the therapy to be tailored to the patient's specific genetic profile. miRNA mimics can be used to increase the level of suppressive miRNAs, leading to the inhibition of cell proliferation, induction of apoptosis, or inhibition of invasive and metastatic processes in cancer cells. In turn, miRNA inhibitors enable the exclusion of oncogenic miRNAs, which may contribute to limiting cancer progression (Takahashi et al. 2019; Ors-Kumoglu et al. 2019).

The prospects for therapy using miRNA mimics and inhibitors are promising because they enable nearly surgical precision in regulating gene expression, which may lead to reduced therapeutic toxicity to healthy cells. Moreover, the use of these therapies can be flexible, allowing for adaptation to the tumor's changing genetic profile during treatment. Potential applications of miRNA therapy and inhibitor therapy include the treatment or support of various types of cancer, especially those where changes in miRNA expression are a key factor. In the context of clinical application, therapy with miRNA mimics and inhibitors may represent a novel option for patients, especially in cases where conventional treatments are limited or ineffective. It can be used alone or in combination with existing therapies, potentially increasing the effectiveness of treatment (Takahashi et al. 2019; Ors-Kumoglu et al. 2019; Reda El Sayed et al. 2021).

Currently used solutions in oncological therapy, based on miRNA mimics and inhibitors, are in the clinical trial phase, which proves current scientific efforts toward the practical application of these methods. Some miRNA-based solutions already in use include the use of the miRNA-34a mimic in the treatment of advanced lung cancer. Moreover, research on miRNA-122 inhibitors in the context of hepatocellular carcinoma brings promising results (Wu et al. 2019; Diener et al. 2022; Yang et al. 2023). Examples of studies include evaluating the effectiveness of miRNA therapies in treating various types of cancer, such as identifying predictive biomarkers and assessing potential adverse effects.

MicroRNAs play a key role in regulating the EMT process in various types of cancer, such as breast, ovarian, and lung cancer. Identified miRNAs acting as both oncomiRs and TsmiRs constitute potential therapeutic targets. The use of miRNA mimics and inhibitors may open new therapeutic possibilities, allowing for precise control of the EMT process and limiting the invasive capacity of cancer cells. As research on miRNAs and their role in cancer continues to progress, further development of innovative therapies targeting the EMT process can be anticipated. Despite the multitude of studies conducted to identify microRNA markers crucial for the epithelial-to-mesenchymal transition process, a significant gap persists in translating these findings into effective therapeutic interventions. Despite extensive research efforts, the development of miRNA-based therapeutics capable of halting or reversing EMT remains an unmet challenge. While numerous studies have successfully identified miRNAs associated with EMT, the translation of this knowledge into clinically applicable treatments has been limited. The existing therapeutic landscape lacks agents specifically designed to modulate miRNA expression effectively in a way that curbs the progression of EMT in various cancer types.

The necessity for this research arises from the pressing need to bridge the gap between the wealth of knowledge accumulated regarding EMT-associated miRNAs and their tangible application in therapeutic contexts. Despite the identification of candidate miRNAs, there is a notable absence of successful translation of these discoveries into therapeutic strategies that can be employed in clinical settings. Consequently, this research project aims to address this critical translational gap by systematically investigating the potential of miRNA mimics and antagomirs as therapeutic tools. By delving into the complexities of miRNA regulation during EMT across diverse tumor types, the project aspires to unravel the elusive mechanisms that hindered previous efforts. Through the strategic use of specialized biological databases, high-throughput sequencing techniques, and functional assays, the goal is to identify not only key miRNA regulators but also to elucidate their downstream targets and pathways, providing a comprehensive understanding of the intricacies involved. The anticipated outcomes of this research extend beyond the identification of miRNAs with therapeutic potential. The project aims to elucidate the underlying mechanisms and functional implications, thereby paving the way for the development of innovative interventions in the treatment of various types of cancer. Ultimately, the significance of this research lies in its potential to contribute to the realization of effective miRNA-based therapeutics, bringing us closer to addressing the longstanding challenge of controlling EMT progression in a clinical setting.

## 6.5 Conclusion

MicroRNAs play a crucial role in regulating the DNA damage response and epithelial–mesenchymal transition, particularly in the context of cancer therapy resistance. By linking genomic stability with phenotypic flexibility, miRNAs enable tumor cells to adapt to therapeutic stress. These small non-coding RNAs can either promote or inhibit DNA repair pathways, thereby directly influencing the sensitivity of cells to chemotherapeutic drugs and ionizing radiation. Simultaneously, miRNAs control EMT, a process that enables tumor cell invasion, migration, and metastasis. Notably, the same miRNAs can function as tumor suppressors or oncogenes, depending on the biological context and tumor type.

The clinical application of miRNAs will require further investigation, especially in establishing stable and specific circulating biomarkers for DDR and EMT activity. To accomplish this, advanced bioinformatics methods and multi-omics datasets will be essential, considering the differences among tumors and disease-specific expression patterns. Regarding treatment options, significant progress is being made in developing targeted miRNA modulation therapies using mimics, inhibitors, and delivery systems such as lipid nanoparticles. These therapies should ideally be tailored to the molecular characteristics of an individual's tumors and their microenvironment, and tested in conjunction with conventional treatment regimens.

Nevertheless, miRNAs hold substantial promise as both therapeutic targets and predictive or prognostic biomarkers. Their context-dependent, multifaceted functions present challenges but also open opportunities for advancing precision oncology and overcoming treatment resistance. Realizing this potential will require close collaboration among clinicians, molecular biologists, and bioinformaticians, along with the integration of insights from molecular biology, computational sciences, and translational medicine.

## References


Berindan-Neagoe I, Calin GA (2014) Molecular pathways: microRNAs, cancer cells, and microenvironment. Clin Cancer Res 20(24):6247–6253

Brabletz S, Brabletz T (2010) The ZEB/miR-200 feedback loop--a motor of cellular plasticity in development and cancer? EMBO Rep 11(9):670–677

Cao M, Seike M, Soeno C et al (2012) MiR-23a regulates TGF-beta-induced epithelial-mesenchymal transition by targeting E-cadherin in lung cancer cells. Int J Oncol 41(3):869–875

Cao L, Liu X, Zheng B et al (2022) Role of K63-linked ubiquitination in cancer. Cell Death Discov 8(1):410

Carstens JL, Lovisa S (2025) Epithelial-to-mesenchymal transition drives cancer genomic instability. J Exp Clin Cancer Res 44(1):135

Chaffer CL, San Juan BP, Lim E et al (2016) EMT, cell plasticity and metastasis. Cancer Metastasis Rev 35(4):645–654

Chatterjee N, Walker GC (2017) Mechanisms of DNA damage, repair, and mutagenesis. Environ Mol Mutagen 58(5):235–263

Chen J, Wang L, Matyunina LV et al (2011) Overexpression of miR-429 induces mesenchymal-to-epithelial transition (MET) in metastatic ovarian cancer cells. Gynecol Oncol 121(1):200–205

Cottonham CL, Kaneko S, Xu L (2010) miR-21 and miR-31 converge on TIAM1 to regulate migration and invasion of colon carcinoma cells. J Biol Chem 285(46):35293–35302
Crucitta S, Cucchiara F, Mathijssen R et al (2022) Treatment-driven tumour heterogeneity and drug resistance: lessons from solid tumours. Cancer Treat Rev 104:102340
D'Amato NC, Howe EN, Richer JK (2013) MicroRNA regulation of epithelial plasticity in cancer. Cancer Lett 341(1):46–55
De Las Rivas J, Brozovic A, Izraely S et al (2021) Cancer drug resistance induced by EMT: novel therapeutic strategies. Arch Toxicol 95(7):2279–2297
Debaugnies M, Rodriguez-Acebes S, Blondeau J et al (2023a) RHOJ controls EMT-associated resistance to chemotherapy. Nature 616(7955):168–175
Debaugnies M, Rodríguez-Acebes S, Blondeau J et al (2023b) RHOJ controls EMT-associated resistance to chemotherapy. Nature 616(7955):168–175
Demirkan B (2013) The roles of epithelial-to-mesenchymal transition (EMT) and mesenchymal-to-epithelial transition (MET) in breast cancer bone metastasis: potential targets for prevention and treatment. J Clin Med 2(4):264–282
Dexheimer TS (2013) DNA repair pathways and mechanisms. In: DNA repair of cancer stem cells, pp 19–32
Diener C, Keller A, Meese E (2022) Emerging concepts of miRNA therapeutics: from cells to clinic. Trends Genet 38(6):613–626
Ding X, Park SI, McCauley LK et al (2013) Signaling between transforming growth factor beta (TGF-beta) and transcription factor SNAI2 represses expression of microRNA miR-203 to promote epithelial-mesenchymal transition and tumor metastasis. J Biol Chem 288(15): 10241–10253
Ding L, Lan Z, Xiong X et al (2018) The dual role of MicroRNAs in colorectal cancer progression. Int J Mol Sci 19(9):2791
Du W, Tang H, Lei Z et al (2019) miR-335-5p inhibits TGF-beta1-induced epithelial-mesenchymal transition in non-small cell lung cancer via ROCK1. Respir Res 20(1):225
Du C, Wang Y, Zhang Y et al (2020) LncRNA DLX6-AS1 contributes to epithelial-mesenchymal transition and cisplatin resistance in triple-negative breast cancer via modulating Mir-199b-5p/ Paxillin Axis. Cell Transplant 29:963689720929983
Espinosa-Diez C, Wilson R, Chatterjee N et al (2018) MicroRNA regulation of the MRN complex impacts DNA damage, cellular senescence, and angiogenic signaling. Cell Death Dis 9(6):632
Friker LL, Perwein T, Waha A et al (2025) MSH2, MSH6, MLH1, and PMS2 immunohistochemistry as highly sensitive screening method for DNA mismatch repair deficiency syndromes in pediatric high-grade glioma. Acta Neuropathol 149(1):11
Fu J, Imani S, Wu M-Y et al (2023) MicroRNA-34 family in cancers: role, mechanism, and therapeutic potential. Cancer 15(19):4723
Gasparini P, Lovat F, Fassan M et al (2014) Protective role of miR-155 in breast cancer through RAD51 targeting impairs homologous recombination after irradiation. Proc Natl Acad Sci USA 111(12):4536–4541
Gregory PA, Bert AG, Paterson EL et al (2008) The miR-200 family and miR-205 regulate epithelial to mesenchymal transition by targeting ZEB1 and SIP1. Nat Cell Biol 10(5):593–601
Gu J, Wang G, Liu H et al (2018) SATB2 targeted by methylated miR-34c-5p suppresses proliferation and metastasis attenuating the epithelial-mesenchymal transition in colorectal cancer. Cell Prolif 51(4):e12455
Gu X, Wei S, Lv X (2024) Circulating tumor cells: from new biological insights to clinical practice. Signal Transduct Target Ther 9(1):226
Guan H, Li W, Li Y et al (2017) MicroRNA-93 promotes proliferation and metastasis of gastric cancer via targeting TIMP2. PLoS One 12(12):e0189490
Gupta S, Maitra A (2016) EMT: matter of life or death? Cell 164(5):840–842
Ha M, Kim VN (2014) Regulation of microRNA biogenesis. Nat Rev Mol Cell Biol 15(8):509–524
He M, Zhou W, Li C et al (2016) MicroRNAs, DNA damage response, and cancer treatment. Int J Mol Sci 17(12):2087

Hegre SA, Saetrom P, Aas PA et al (2013) Multiple microRNAs may regulate the DNA repair enzyme uracil-DNA glycosylase. DNA Repair (Amst) 12(1):80–86
Hill M, Tran N (2021) miRNA interplay: mechanisms and consequences in cancer. Dis Model Mech 14(4):dmm047662
Hopfner KP (2014) ATP puts the brake on DNA double-strand break repair: a new study shows that ATP switches the Mre11-Rad50-Nbs1 repair factor between signaling and processing of DNA ends. BioEssays 36(12):1170–1178
Hosea R, Hillary S, Naqvi S et al (2024) The two sides of chromosomal instability: drivers and brakes in cancer. Signal Transduct Target Ther 9(1):75
Huang T, Kang W, Zhang B et al (2016) miR-508-3p concordantly silences NFKB1 and RELA to inactivate canonical NF-kappaB signaling in gastric carcinogenesis. Mol Cancer 15:9
Iorio MV, Croce CM (2012) microRNA involvement in human cancer. Carcinogenesis 33(6): 1126–1133
Jiang G, Shi W, Fang H et al (2018) miR-27a promotes human breast cancer cell migration by inducing EMT in a FBXW7-dependent manner. Mol Med Rep 18(6):5417–5426
Kang J, Kim E, Kim W et al (2013) Rhamnetin and cirsiliol induce radiosensitization and inhibition of epithelial-mesenchymal transition (EMT) by miR-34a-mediated suppression of Notch-1 expression in non-small cell lung cancer cell lines. J Biol Chem 288(38):27343–27357
Kawami M, Takenaka S, Akai M et al (2021) Characterization of miR-34a-induced epithelial-mesenchymal transition in non-small lung cancer cells focusing on p53. Biomolecules 11(12): 1853
Khot M, Sreekumar D, Jahagirdar S et al (2020) Twist1 induces chromosomal instability (CIN) in colorectal cancer cells. Hum Mol Genet 29(10):1673–1688
Kipkeeva F, Muzaffarova T, Korotaeva A et al (2020) MicroRNA in gastric cancer development: mechanisms and biomarkers. Diagnostics (Basel) 10(11):891
Lai J, Yang H, Zhu Y et al (2019) MiR-7-5p-mediated downregulation of PARP1 impacts DNA homologous recombination repair and resistance to doxorubicin in small cell lung cancer. BMC Cancer 19(1):602
Lee JH, Mand MR, Deshpande RA et al (2013) Ataxia telangiectasia-mutated (ATM) kinase activity is regulated by ATP-driven conformational changes in the Mre11/Rad50/Nbs1 (MRN) complex. J Biol Chem 288(18):12840–12851
Lee HJ, Li CF, Ruan D et al (2016) The DNA damage transducer RNF8 facilitates cancer Chemoresistance and progression through twist activation. Mol Cell 63(6):1021–1033
Lei C, Du F, Sun L et al (2017) miR-143 and miR-145 inhibit gastric cancer cell migration and metastasis by suppressing MYO6. Cell Death Dis 8(10):e3101
Lei Y, Chen L, Zhang G et al (2020) MicroRNAs target the Wnt/β-catenin signaling pathway to regulate epithelial-mesenchymal transition in cancer (Review). Oncol Rep. https://doi.org/10.3892/or.2020.7703
Li L, Wu D (2016) miR-32 inhibits proliferation, epithelial-mesenchymal transition, and metastasis by targeting TWIST1 in non-small-cell lung cancer cells. Onco Targets Ther 9:1489–1498
Li L, Zhu X, Shou T et al (2018) MicroRNA-28 promotes cell proliferation and invasion in gastric cancer via the PTEN/PI3K/AKT signalling pathway. Mol Med Rep 17(3):4003–4010
Li Y, Tong Y, Liu J et al (2022) The role of MicroRNA in DNA damage response. Front Genet 13: 850038
Lim Z-F, Ma PC (2019) Emerging insights of tumor heterogeneity and drug resistance mechanisms in lung cancer targeted therapy. J Hematol Oncol 12(1):134
Liu G, Sun Y, Ji P et al (2014) MiR-506 suppresses proliferation and induces senescence by directly targeting the CDK4/6–FOXM1 axis in ovarian cancer. J Pathol 233(3):308–318
Liu G, Yang D, Rupaimoole R et al (2015) Augmentation of response to chemotherapy by microRNA-506 through regulation of RAD51 in serous ovarian cancers. J Natl Cancer Inst 107(7):djv108

Liu J, Cao L, Li Y et al (2022) Pirfenidone promotes the levels of exosomal miR-200 to down-regulate ZEB1 and represses the epithelial-mesenchymal transition of non-small cell lung cancer cells. Hum Cell 35(6):1813–1823

Lu Y, Hu X, Yang X (2021) miR-934 promotes breast cancer metastasis by regulation of PTEN and epithelial-mesenchymal transition. Tissue Cell 71:101581

Luo H, Liang C (2018) MicroRNA-148b inhibits proliferation and the epithelial-mesenchymal transition and increases radiosensitivity in non-small cell lung carcinomas by regulating ROCK1. Exp Ther Med 15(4):3609–3616

Ma F, Li W, Liu C et al (2017) MiR-23a promotes TGF-beta1-induced EMT and tumor metastasis in breast cancer cells by directly targeting CDH1 and activating Wnt/beta-catenin signaling. Oncotarget 8(41):69538–69550

Majidinia M, Yousefi B (2016) DNA damage response regulation by microRNAs as a therapeutic target in cancer. DNA Repair (Amst) 47:1–11

Mao S, Li J, Huang J et al (2025) Therapeutic potential of microRNA-506 in cancer treatment: mechanisms and therapeutic implications. Front Oncol 15:1524763

Meng F, Henson R, Wehbe-Janek H et al (2007) MicroRNA-21 regulates expression of the PTEN tumor suppressor gene in human hepatocellular cancer. Gastroenterology 133(2):647–658

Moskwa P, Buffa FM, Pan Y et al (2011) miR-182-mediated downregulation of BRCA1 impacts DNA repair and sensitivity to PARP inhibitors. Mol Cell 41(2):210–220

Moyret-Lalle C, Prodhomme MK, Burlet D et al (2022) Role of EMT in the DNA damage response, double-strand break repair pathway choice, and its implications in cancer treatment. Cancer Sci 113(7):2214–2223

Nakashima H, Hashimoto N, Aoyama D et al (2012) Involvement of the transcription factor twist in phenotype alteration through epithelial-mesenchymal transition in lung cancer cells. Mol Carcinog 51(5):400–410

Natarajan V (2016) Regulation of DNA repair by non-coding miRNAs. Noncoding RNA Res 1(1): 64–68

Nguyen VHL, Yue C, Du KY et al (2020) The role of microRNAs in epithelial ovarian cancer metastasis. Int J Mol Sci 21(19):7093

Nieto MA, Huang RY, Jackson RA et al (2016) EMT: 2016. Cell 166(1):21–45

Nussinov R, Yavuz BR, Jang H (2025) Drug resistance and tumor heterogeneity: cells and ensembles. Biophys Rev 17(3):759–779

O'Brien SJ, Fiechter C, Burton J et al (2021) Long non-coding RNA ZFAS1 is a major regulator of epithelial-mesenchymal transition through miR-200/ZEB1/E-cadherin, vimentin signaling in colon adenocarcinoma. Cell Death Discov 7(1):61

Ors-Kumoglu G, Gulce-Iz S, Biray-Avci C (2019) Therapeutic microRNAs in human cancer. Cytotechnology 71(1):411–425

Pecina-Slaus N, Kafka A, Salamon I et al (2020) Mismatch repair pathway, genome stability and cancer. Front Mol Biosci 7:122

Pei YF, Lei Y, Liu XQ (2016) MiR-29a promotes cell proliferation and EMT in breast cancer by targeting ten eleven translocation 1. Biochim Biophys Acta 1862(11):2177–2185

Peraza-Vega RI, Valverde M, Rojas E (2022) Interactions between miRNAs and double-Strand breaks DNA repair genes, pursuing a fine-tuning of repair. Int J Mol Sci 23(6):3231

Prahm KP, Hogdall CK, Karlsen MA et al (2021) MicroRNA characteristics in epithelial ovarian cancer. PLoS One 16(6):e0252401

Prakash R, Zhang Y, Feng W et al (2015) Homologous recombination and human health: the roles of BRCA1, BRCA2, and associated proteins. Cold Spring Harb Perspect Biol 7(4):a016600

Reda El Sayed S, Cristante J, Guyon L et al (2021) MicroRNA therapeutics in cancer: current advances and challenges. Cancer 13(11):2680

Romano G, Acunzo M, Nana-Sinkam P (2021) microRNAs as novel therapeutics in cancer. Cancer 13(7):1526

Roszkowska M (2024) Multilevel mechanisms of cancer drug resistance. Int J Mol Sci 25(22): 12402

Sadeghi Z, Malekzadeh M, Sharifi M et al (2025) The role of miR-16 and miR-34a family in the regulation of cancers: a review. Heliyon 11(4):e42733
Setton J, Zinda M, Riaz N et al (2021) Synthetic lethality in cancer therapeutics: the next generation. Cancer Discov 11(7):1626–1635
Shang Y, Chen H, Ye J et al (2017) HIF-1alpha/Ascl2/miR-200b regulatory feedback circuit modulated the epithelial-mesenchymal transition (EMT) in colorectal cancer cells. Exp Cell Res 360(2):243–256
Shibue T, Weinberg RA (2017) EMT, CSCs, and drug resistance: the mechanistic link and clinical implications. Nat Rev Clin Oncol 14(10):611–629
Siemens H, Jackstadt R, Hünten S et al (2014) miR-34 and SNAIL form a double-negative feedback loop to regulate epithelial-mesenchymal transitions. Cell Cycle 10(24):4256–4271
Slabáková E, Culig Z, Remšík J et al (2017) Alternative mechanisms of miR-34a regulation in cancer. Cell Death Dis 8(10):e3100-e
Slaby O, Svoboda M, Fabian P et al (2007) Altered expression of miR-21, miR-31, miR-143 and miR-145 is related to clinicopathologic features of colorectal cancer. Oncology 72(5–6): 397–402
Song SZ, Lin S, Liu JN et al (2019) Targeting of SPP1 by microRNA-340 inhibits gastric cancer cell epithelial-mesenchymal transition through inhibition of the PI3K/AKT signaling pathway. J Cell Physiol 234(10):18587–18601
Spies J, Polasek-Sedlackova H, Lukas J et al (2021) Homologous recombination as a fundamental genome surveillance mechanism during DNA replication. Genes (Basel) 12(12):1960
Sreekumar R, Al-Saihati H, Emaduddin M et al (2021) The ZEB2-dependent EMT transcriptional programme drives therapy resistance by activating nucleotide excision repair genes ERCC1 and ERCC4 in colorectal cancer. Mol Oncol 15(8):2065–2083
Stinson BM, Loparo JJ (2021) Repair of DNA double-Strand breaks by the nonhomologous end joining pathway. Annu Rev Biochem 90:137–164
Sudhanva MS, Hariharasudhan G, Jun S et al (2022) MicroRNA-145 impairs classical non-homologous end-joining in response to ionizing radiation-induced DNA double-Strand breaks via targeting DNA-PKcs. Cells 11(9):1509
Szczepanek J, Skorupa M, Tretyn A (2022) MicroRNA as a potential therapeutic molecule in cancer. Cells 11(6):1008
Takahashi R, Prieto-Vila M, Kohama I et al (2019) Development of miRNA-based therapeutic approaches for cancer patients. Cancer Sci 110(4):1140–1147
Tan J, Sun X, Zhao H et al (2023) Double-strand DNA break repair: molecular mechanisms and therapeutic targets. MedComm (2020). 4(5):e388
Tang J, Li Y, Wang J et al (2016) Molecular mechanisms of microRNAs in regulating epithelial–mesenchymal transitions in human cancers. Cancer Lett 371(2):301–313
Tessitore A, Cicciarelli G, Del Vecchio F et al (2014) MicroRNAs in the DNA damage/repair network and cancer. Int J Genomics 2014:820248
Valastyan S, Chang A, Benaich N et al (2011) Activation of miR-31 function in already-established metastases elicits metastatic regression. Genes Dev 25(6):646–659
Valenti F, Sacconi A, Ganci F et al (2019) The miR-205-5p/BRCA1/RAD17 Axis promotes genomic instability in head and neck squamous cell carcinomas. Cancers (Basel) 11(9):1347
Valeri N, Gasparini P, Braconi C et al (2010a) MicroRNA-21 induces resistance to 5-fluorouracil by down-regulating human DNA MutS homolog 2 (hMSH2). Proc Natl Acad Sci USA 107(49): 21098–21103
Valeri N, Gasparini P, Fabbri M et al (2010b) Modulation of mismatch repair and genomic stability by miR-155. Proc Natl Acad Sci USA 107(15):6982–6987
Wallace SS, Murphy DL, Sweasy JB (2012) Base excision repair and cancer. Cancer Lett 327(1–2): 73–89
Wang Z, Wang X (2020) miR-122-5p promotes aggression and epithelial-mesenchymal transition in triple-negative breast cancer by suppressing charged multivesicular body protein 3 through mitogen-activated protein kinase signaling. J Cell Physiol 235(3):2825–2835

Wang L, Mezencev R, Svajdler M et al (2014) Ectopic over-expression of miR-429 induces mesenchymal-to-epithelial transition (MET) and increased drug sensitivity in metastasizing ovarian cancer cells. Gynecol Oncol 134(1):96–103

Wang Y, Feng J, Zang W et al (2015) MiR-499 enhances the cisplatin sensitivity of esophageal carcinoma cell lines by targeting DNA polymerase beta. Cell Physiol Biochem 36(4): 1587–1596

Wang H, Wu Q, Zhang Y et al (2017) TGF-beta1-induced epithelial-mesenchymal transition in lung cancer cells involves upregulation of miR-9 and downregulation of its target, E-cadherin. Cell Mol Biol Lett 22:22

Wang Y, Wu Z, Hu L (2018) The regulatory effects of metformin on the [SNAIL/miR-34]: [ZEB/miR-200] system in the epithelial-mesenchymal transition(EMT) for colorectal cancer (CRC). Eur J Pharmacol 834:45–53

Williams RS, Williams JS, Tainer JA (2007) Mre11-Rad50-Nbs1 is a keystone complex connecting DNA repair machinery, double-strand break signaling, and the chromatin template. Biochem Cell Biol 85(4):509–520

Wu J, Li WZ, Huang ML et al (2017) Regulation of cancerous progression and epithelial-mesenchymal transition by miR-34c-3p via modulation of MAP3K2 signaling in triple-negative breast cancer cells. Biochem Biophys Res Commun 483(1):10–16

Wu Y, Zhang J, Zheng Y et al (2018) miR-216a-3p inhibits the proliferation, migration, and invasion of human gastric cancer cells via targeting RUNX1 and activating the NF-kappaB signaling pathway. Oncol Res 26(1):157–171

Wu K-L, Tsai Y-M, Lien C-T et al (2019) The roles of MicroRNA in lung cancer. Int J Mol Sci 20(7):1611

Xiao M, Cai J, Cai L et al (2017) Let-7e sensitizes epithelial ovarian cancer to cisplatin through repressing DNA double strand break repair. J Ovarian Res 10(1):24

Xiao M, Guo J, Xie L et al (2020) Let-7e suppresses DNA damage repair and sensitizes ovarian cancer to cisplatin through targeting PARP1. Mol Cancer Res 18(3):436–447

Xu M, Qin S, Cao F et al (2017) MicroRNA-379 inhibits metastasis and epithelial-mesenchymal transition via targeting FAK/AKT signaling in gastric cancer. Int J Oncol 51(3):867–876

Yang Q, Wang Y, Lu X et al (2015) MiR-125b regulates epithelial-mesenchymal transition via targeting Sema4C in paclitaxel-resistant breast cancer cells. Oncotarget 6(5):3268–3279

Yang H, Liu Y, Chen L et al (2023) MiRNA-based therapies for lung cancer: opportunities and challenges? Biomolecules 13(6):877

You J, Li M, Tan Y et al (2017) Snail1-expressing cancer-associated fibroblasts induce lung cancer cell epithelial-mesenchymal transition through miR-33b. Oncotarget 8(70):114769–114786

Yu T, Wang LN, Li W et al (2018) Downregulation of miR-491-5p promotes gastric cancer metastasis by regulating SNAIL and FGFR4. Cancer Sci 109(5):1393–1403

Zhang P, Wei Y, Wang L et al (2014) ATM-mediated stabilization of ZEB1 promotes DNA damage response and radioresistance through CHK1. Nat Cell Biol 16(9):864–875

Zhang N, Liu Y, Wang Y et al (2017) Decitabine reverses TGF-beta1-induced epithelial-mesenchymal transition in non-small-cell lung cancer by regulating miR-200/ZEB axis. Drug Des Devel Ther 11:969–983

Zhang H, Zhang Y, Chen C et al (2018a) A double-negative feedback loop between DEAD-box protein DDX21 and Snail regulates epithelial-mesenchymal transition and metastasis in breast cancer. Cancer Lett 437:67–78

Zhang X, Zhang Z, Zhang Q et al (2018b) ZEB1 confers chemotherapeutic resistance to breast cancer by activating ATM. Cell Death Dis 9(2):57

Zhang L, Liao Y, Tang L (2019) MicroRNA-34 family: a potential tumor suppressor and therapeutic candidate in cancer. J Exp Clin Cancer Res 38(1):53

Zhang H, Steed A, Co M et al (2021) Cancer stem cells, epithelial-mesenchymal transition, ATP and their roles in drug resistance in cancer. Cancer Drug Resist 4(3):684–709

Zhang A, Miao K, Sun H et al (2022) Tumor heterogeneity reshapes the tumor microenvironment to influence drug resistance. Int J Biol Sci 18(7):3019–3033

Zhao YF, Han ML, Xiong YJ et al (2018) A miRNA-200c/cathepsin L feedback loop determines paclitaxel resistance in human lung cancer A549 cells in vitro through regulating epithelial-mesenchymal transition. Acta Pharmacol Sin 39(6):1034–1047

Zhou T, Yi F, Wang Z et al (2019) The functions of DNA damage factor RNF8 in the pathogenesis and progression of cancer. Int J Biol Sci 15(5):909–918

Zhou W, Xu Y, Zhang J et al (2022) MiRNA-363-3p/DUSP10/JNK axis mediates chemoresistance by enhancing DNA damage repair in diffuse large B-cell lymphoma. Leukemia 36(7): 1861–1869

# miRNA Modulation of Cancer Therapeutic Resistance Due to Failed Autophagy and Cancer Stemness

7

Kazi Tasnuva Alam, Zarin Tasnim Tisha, and Tanvir Ahmed

**Abstract**

Therapeutic resistance in cancer remains a substantial barrier to effective and long-term treatment outcomes. According to recent research, cancer stem cell (CSC) populations and dysregulated autophagy play a key role in the development of resistance, which in turn causes tumor recurrence and metastasis. Since they affect vital pathways linked to autophagy and the maintenance of cancer stem cells (CSCs), microRNAs (miRNAs) are acknowledged as significant regulators of gene expression. By developing feedback loops and constitutively activating carcinogenic signaling pathways, miRNA dysregulation promotes cell stemness and aids in tumor metastasis and therapeutic resistance. More significantly, certain miRNAs might be viable targets for cancer treatment, diagnosis, and prognosis. The intricate relationships between miRNAs, failed autophagy, and cancer stem cell-driven resistance mechanisms in cancer treatment are discussed in this chapter. We discuss how certain miRNAs can either promote or inhibit autophagy, hence influencing cellular survival in response to targeted and chemotherapeutic therapies. Furthermore, we highlight how miRNAs control the properties of CSCs, such as self-renewal, plasticity, and the epithelial-to-mesenchymal transition (EMT), all of which enhance tumor resistance. In addition to discover a potential strategy for overcoming therapeutic resistance, research is currently being conducted on miRNA-based therapeutic approaches, such as miRNA mimics and inhibitors. Through innovative combination

K. T. Alam · T. Ahmed (✉)
Department of Pharmaceutical Sciences, School of Health and Life Sciences, North South University, Dhaka, Bangladesh
e-mail: tanvir.ahmed111@northsouth.edu

Z. T. Tisha
School of Chemistry and Chemical Sciences, University of Surrey, Guildford, UK



S. Tiwari et al. (eds.), *microRNA Modulation: A Promising Combination Therapy for Cancer*, https://doi.org/10.1007/978-981-95-5989-3_7

treatment approaches, integrating miRNA regulation with existing cancer therapies may improve therapeutic efficacy and reduce relapse rates. This chapter explores the translational effects of miRNA-targeted therapies and offers novel concepts for addressing cancer resistance through the control of cancer stem cells and autophagy.

**Keywords**

miRNA · Cancer Stem Cell · Therapeutic Resistance · Autophagy · Cancer

## Abbreviations

| | |
|---|---|
| ABC | ATP-Binding Cassette |
| ABCB1 | ATP-Binding Cassette Subfamily B Member 1 |
| ABCG2 | ATP-Binding Cassette Subfamily G Member 2 |
| AKT | Protein Kinase B |
| ALDH1 | Aldehyde Dehydrogenase 1 |
| AMPK | AMP-Activated Protein Kinase |
| ATG | Autophagy-Related Gene |
| ATG5 | Autophagy-Related Protein 5 |
| aPKC | Atypical Protein Kinase C |
| BCL-2 | B-Cell Lymphoma 2 |
| BECN1 | Beclin-1 |
| BH3 | BCL-2 Homology 3 |
| bHLH | Basic Helix-Loop-Helix |
| BRCA1/2 | Breast Cancer Gene 1/2 |
| CAFs | Cancer-Associated Fibroblasts |
| CD44 | Cluster of Differentiation 44 |
| CD133 | Cluster of Differentiation 133 |
| CMA | Chaperone-Mediated Autophagy |
| COX-2 | Cyclooxygenase-2 |
| CSC | Cancer Stem Cell |
| CTLA-4 | Cytotoxic T-Lymphocyte-Associated Protein 4 |
| CXCR4 | C-X-C Chemokine Receptor Type 4 |
| DNMT | DNA Methyltransferase |
| EMT | Epithelial-to-Mesenchymal Transition |
| EMT-TFs | EMT-Transcription Factors |
| EZH2 | Enhancer of Zeste Homolog 2 |
| FDA | Food and Drug Administration |
| GLI | Glioma-Associated Oncogene Homolog |
| GLUT1 | Glucose Transporter 1 |
| GLOBOCAN | Global Cancer Observatory |
| HDAC | Histone Deacetylase |
| HES/HEY | Hairy and Enhancer of Split / Hairy/Enhancer-of-Split Related with YRPW Motif |

| | |
|---|---|
| HIF | Hypoxia-Inducible Factor |
| HIF-1α | Hypoxia-Inducible Factor 1-Alpha |
| Hh | Hedgehog |
| HR | Homologous Recombination |
| ICSC | Induced Cancer Stem Cell |
| IL-6 | Interleukin-6 |
| JAK/STAT | Janus Kinase/Signal Transducer and Activator of Transcription |
| KLF4 | Krüppel-Like Factor 4 |
| LC3 | Microtubule-Associated Protein 1A/1B-Light Chain 3 |
| LDHA | Lactate Dehydrogenase A |
| MAPK | Mitogen-Activated Protein Kinase |
| MDR1 | Multidrug Resistance Protein 1 |
| miRNA | microRNA |
| mTOR | Mechanistic Target of Rapamycin |
| mTORC1 | mTOR Complex 1 |
| MYC | MYC Proto-Oncogene |
| NF-κB | Nuclear Factor Kappa-Light-Chain-Enhancer of Activated B Cells |
| NICD | Notch Intracellular Domain |
| NPs | Nanoparticles |
| OCT4 | Octamer-Binding Transcription Factor 4 |
| OXPHOS | Oxidative Phosphorylation |
| PAR3/6 | Partitioning Defective 3/6 |
| PARP1 | Poly(ADP-Ribose) Polymerase 1 |
| PD-L1 | Programmed Death-Ligand 1 |
| PI3K | Phosphoinositide 3-Kinase |
| PORCN | Porcupine Homolog |
| PRC2 | Polycomb Repressive Complex 2 |
| RAD51 | RAD51 Recombinase |
| ROS | Reactive Oxygen Species |
| SMO | Smoothened |
| SOX2 | SRY-Box Transcription Factor 2 |
| SQSTM1/p62 | Sequestosome 1 |
| STAT3 | Signal Transducer and Activator of Transcription 3 |
| TEAD | TEA Domain Transcription Factor |
| TGF-β | Transforming Growth Factor Beta |
| TNF-α | Tumor Necrosis Factor Alpha |
| TRAIL | TNF-Related Apoptosis-Inducing Ligand |
| TWIST1/2 | Twist Family BHLH Transcription Factor 1/2 |
| UPR | Unfolded Protein Response |
| XIAP | X-Linked Inhibitor of Apoptosis Protein |
| YAP | Yes-Associated Protein |
| ZEB1/2 | Zinc Finger E-Box Binding Homeobox 1/2 |

**Learning Objectives**

- *Explain* the role of miRNAs in regulating autophagy pathways and how their dysregulation contributes to therapeutic resistance in cancer.
- *Analyze* how miRNAs influence cancer stem cell properties, including self-renewal, differentiation, and survival under therapy-induced stress.
- *Evaluate* the interplay between failed autophagy, cancer stemness, and miRNA modulation in driving resistance to chemotherapy, radiotherapy, and targeted therapies.
- *Assess* therapeutic opportunities and challenges in targeting miRNAs associated with autophagy failure and stemness to overcome resistance and improve cancer treatment outcomes.

## 7.1 Introduction

Cancer poses a significant global health problem, identified as the leading or secondary cause of death before age 70 in over 60% of countries worldwide. GLOBOCAN estimates that 20 million new cancer cases were recorded worldwide in 2022. By 2050, this figure is predicted to exceed 35 million, representing a 77% rise, highlighting the critical necessity for enhanced preventive, diagnostic, and treatment approaches (2- Bray et al. 2021; Bray et al. 2024). Decades of scientific research and experimentation have been dedicated to successfully treating this potentially lethal disease with various types of treatment strategies, namely, chemotherapy, surgery, radiotherapy, immunotherapy, and targeted therapy. Despite the development and implementation of several treatment options, no therapeutic approach has proven to be completely effective in curing cancer or ensuring the absence of its recurrence. A frequent cause of treatment failure in current treatment is resistance to anticancer therapies. Cancer therapeutic resistance denotes the capacity of cancer cells to evade the effects of anticancer therapies, resulting in treatment failure and disease progression. This resistance is driven by multifaceted mechanisms, categorized into intrinsic and acquired factors. Intrinsic resistance comprises of preexisting cellular characteristics, including drug inactivation via metabolic pathways, inadequate intracellular drug levels, modifications in drug targets, improved DNA repair mechanisms, and the activation of autophagic pathways, which allows cancer cells to endure therapy-induced stress. Cancer stem cells (CSCs) are a distinct subpopulation characterized by their capacity for self-renewal and differentiation, significantly contributing to intrinsic resistance by evading conventional therapies and facilitating tumor relapse. Conversely, acquired resistance arises from therapeutic pressure. It involves activating alternate signaling pathways, tumor plasticity, and changes in the tumor microenvironment, including extracellular matrix remodeling, exosome release, and angiogenesis. The intricate processes that enable cancer cells to adapt, survive, and multiply shows how challenging it is to treat cancer therapy resistance. This chapter specifically explores the function of autophagy and cancer stem cells in the development of treatment resistance in cancer (Lei et al. 2023).

Autophagy is known as a conserved process mediated by lysosomes that degrades proteins, aiding in the metabolism of organelles and long-lived proteins, which is crucial for sustaining homeostasis under conditions of stress or nutrient deficiency. The relationship between autophagy and cancer is well defined; however, the dual role of autophagy as both a tumor suppressor and promoter has sparked significant debate regarding its implications in cancer pathology. During the initial stages of tumorigenesis, autophagy contributes to tumor prevention by eliminating defective cells and sustaining energy balance in response to stress conditions. In established cancers, it facilitates survival, progression, and drug resistance through nutrient recycling. Numerous anticancer therapies induce autophagy as a stress response, potentially compromising their effectiveness. The paradoxical nature of autophagy presents challenges in cancer treatment, highlighting the necessity of understanding the molecular mechanisms that drive therapeutic resistance associated with autophagy, thus making it as a potential target for effective anticancer therapy (Jalali et al. 2025).

A subpopulation of cells termed "stem cells" comprises rare, immortal cells within a tumor that exhibit self-renewal and differentiation, leading to the generation of various cell types constituting the tumor. CSCs shows intrinsic resistance to radiotherapy, chemotherapy, and immunotherapy, unlike bulk tumor cells. This resistance is mediated by mechanisms including enhanced DNA repair, overexpression of ATP-binding cassette transporters, and effective scavenging of reactive oxygen species (ROS). Recent findings suggest that traditional therapies might unintentionally enhance cancer stem cell populations through the process of dedifferentiation. The tumor microenvironment, which includes cancer-associated fibroblasts, inflammatory mediators, and mesenchymal stromal cells, facilitates the survival of cancer stem cells and contributes to drug resistance. Additionally, factors including slow-cycling behavior, epithelial–mesenchymal transition, and microRNA regulation contribute to their persistence, underscoring cancer stem cells as essential therapeutic targets (Makena et al. 2020).

Among the numerous driving factors of cancer development and treatment resistance, microRNAs serve as essential regulators of gene expression, influencing numerous cellular processes, including autophagy and the maintenance of CSCs, which are significant contributors to cancer progression and therapeutic resistance. MicroRNAs are short noncoding RNA molecules that regulate gene expression posttranscriptionally by binding to target messenger RNAs, thereby restricting their translation or promoting their degradation. Specific miRNAs modulate autophagy-related genes, such as Beclin-1, ATG5, and mTOR; their inhibition might facilitate cancer cell survival under stress, whereas excessive autophagy may trigger dormancy and resistance. Dysregulated miRNAs concurrently influence CSC maintenance by affecting essential stemness pathways such as Notch, Wnt/β-catenin, and PI3K/Akt. Moreover, tumor suppressor miRNAs like miR-34a, let-7 hinder stemness, whereas oncogenic miRNAs, including miR-138, miR-181, and miR-155, enhance it. This dual regulatory axis emphasizes miRNAs as potential therapeutic targets for addressing cancer resistance. Since the discovery of the first miRNA, cancer biology has discovered its vital roles in cancer. Multiple cancer types show miRNA dysregulation, which can act as either tumor suppressors or oncogenes, suggesting the therapeutic potential of miRNAs in cancer treatment. This chapter aims to explain the role of miRNAs in cancer treatment resistance, emphasizing disrupted mechanisms in autophagy and cancer stem cell dynamics, as well as their potential as therapeutic targets for overcoming treatment resistance (MicroRNA-mediated autophagy and drug resistance in cancer.pdf n.d.; Takahashi et al. 2013). Figure 7.1 illustrates the use of miRNA in cancer research.

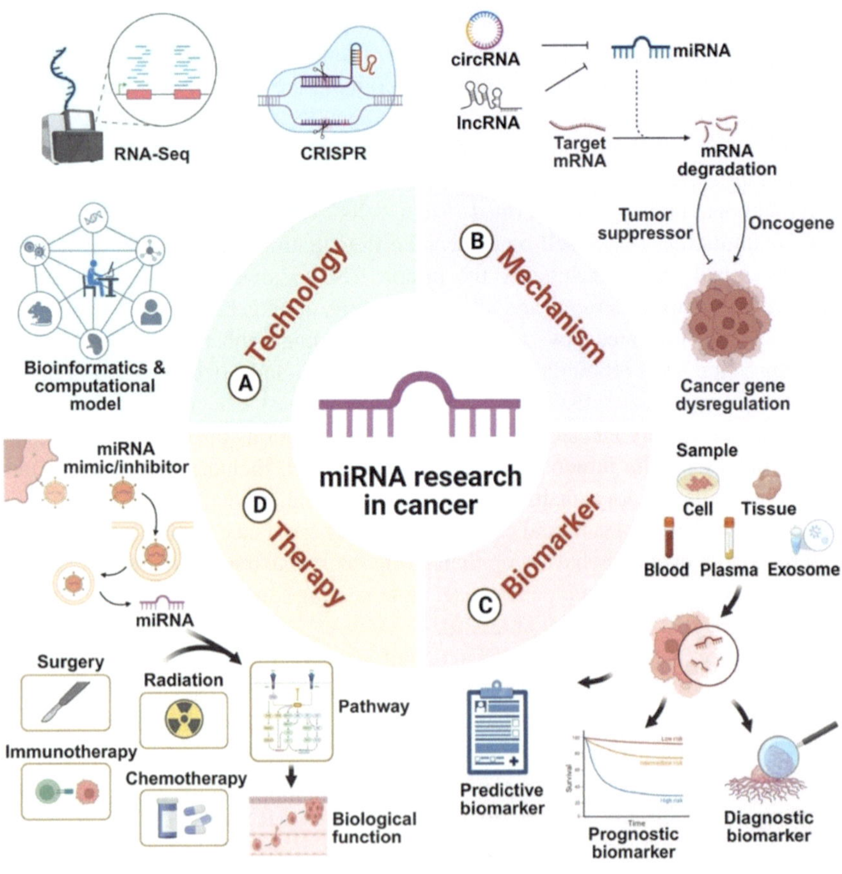


**Fig. 7.1** We have improved our knowledge of MicroRNA (miRNA) interactions by combining RNA sequencing, CRISPR, bioinformatics, and computational models. The development of cancer is closely linked with miRNAs that are regulated bycompeting endogenous RNAs (ceRNAs). They function as biomarkers for cancer diagnosis, prognosis, and prediction, and can be found in a variety of samples. miRNA-based therapies have the potential to improve the results of cancer treatment by combining mimics or inhibitors with conventional therapies. (Reprinted from Wang et al. (2024), Copyright © 2024 The Author(s). MedComm—Oncology is published by John Wiley & Sons Australia, Ltd on behalf of Sichuan International Medical Exchange & Promotion Association (SCIMEA). This is an open-access article distributed under the terms of the Creative Commons Attribution License (CC BY))

## 7.2 The Mechanistic Role of Autophagy in Cancer Therapy

### 7.2.1 Overview of Autophagy and Its Dual Role in Cancer

Autophagy is defined as a catabolic mechanism responsible for the breakdown and recycling of damaged cellular components. It is essential for sustaining cellular homeostasis, especially during stress conditions like nutrient deprivation, hypoxia, and exposure to chemotherapeutic agents. Autophagy in cancer shows a dual role, acting as both a tumor suppressor and a tumor promoter, depending upon the precise setting and stage of carcinogenesis. Autophagy functions as a tumor suppressor by eliminating damaged organelles and proteins, including through mechanisms such as mitophagy and chaperone-mediated autophagy (CMA), thus preserving genomic stability and preventing tumor initiation. It also diminishes EMT, metastasis, and populations of CSCs. Moreover, the inhibition of autophagy-related processes may increase chemosensitivity by resensitizing cancer cells to anticancer agents. Conversely, autophagy may facilitate tumor development in advanced malignancies by enhancing survival under metabolic and therapeutic stress. Autophagy often promotes metastasis by increasing invasion under stress and preserving cancer stem cell characteristics. Additionally, it contributes to chemoresistance, as autophagy and related pathways play a key role in the resistance mechanisms to diverse anticancer drugs. Therefore, it is essential to understand this double-edged, sword-like nature of autophagy for the development of effective treatment techniques that may either block or promote autophagy to mitigate tumor growth and surmount drug resistance (Yun et al. 2021).

### 7.2.2 Autophagy and Cell Survival Under Therapeutic Stress

Autophagy serves a vital cytoprotective function in cancer cells facing therapeutic stress, such as chemotherapy or radiation, by facilitating metabolic adaptation and enhancing stress tolerance. In response to oxidative damage, nutrient deprivation, or organelle dysfunction, tumor cells activate autophagy to eliminate damaged components, thereby maintaining homeostasis and enhancing survival. This process supports cancer cells by repurposing cellular debris into metabolic substrates, reducing therapy-induced cytotoxicity, and inhibiting DNA damage responses. Autophagy influences key stress response pathways, such as mTOR, Unfolded Protein Response (UPR), AMPK, and HIF, thereby enhancing stress resistance. The UPR reduces endoplasmic reticulum (ER) stress, mTOR is regulated by nutrient and energy stress, AMPK is activated under low ATP conditions, and HIF responds to hypoxia. Autophagy enables tumor cells to avoid apoptosis, thereby contributing to resistance against therapy. Inhibition of autophagy has been demonstrated to increase the sensitivity of cancer cells to treatment, suggesting that targeting this process may improve therapeutic effectiveness. Consequently, while autophagy functions as a pro-survival mechanism in stressed cancer cells, its limitation may provide a viable approach to diminish resistance and improve clinical outcomes (Yang 2017; White et al. 2021).

### 7.2.3 Autophagy Failure and Its Consequences in Cancer Progression

Autophagy failure denotes the dysfunction or suppression of the cellular autophagy mechanism, which may result from genetic abnormalities in essential autophagy-related genes, including ATG genes and Beclin-1, or from environmental perturbations. This failure leads to the buildup of damaged proteins and structures, which exacerbates cellular stress. In cancer, autophagy failure leads to genomic instability by allowing reactive oxygen species (ROS) generated by dysfunctional mitochondria to cause DNA damage and mutations, hence facilitating tumor initiation and spread. The accumulation of cellular debris induces chronic inflammation, fostering a tumor-promoting milieu that facilitates cancer growth and metastasis. Autophagy generally acts as a tumor suppressor by removing oncogenic proteins and dysfunctional organelles; however, its dysfunction abolishes this protective process, facilitating unrestrained cancer cell proliferation. Moreover, autophagy insufficiency precipitates metabolic stress, prompting certain cancer cells to engage alternate survival pathways, frequently culminating in more aggressive tumors. Autophagy malfunction significantly contributes to resistance against chemotherapy and radiation by altering drug metabolism and survival signaling pathways. Disruption of autophagy undermines cellular quality control and creates an environment favorable to cancer growth, progression, aggressiveness, and resistance to treatment (Lee and Schmitt 2019; Mukherjee et al. 2020; White et al. 2021).

### 7.2.4 Interplay Between Autophagy, Apoptosis, and Therapy Resistance

The association among autophagy, apoptosis, and therapy resistance in cancer cells is intricate and highly regulated. Inhibition of autophagy in cells possessing functional apoptotic pathways frequently results in enhanced apoptosis. Many tumors demonstrate impaired apoptosis, compelling cells to depend on alternative cell death pathways, such as autophagy. Autophagy functions as a form of survival system; however, in apoptosis-resistant cells, it may also promote cell death, although the precise conditions under which this transpires remain inadequately defined. There is interaction between the two processes, involving common regulators and mechanisms. The autophagy adaptor protein p62/SQSTM1 enhances apoptosis by binding and activating caspase-8, particularly in response to TRAIL, a cytokine under clinical investigation for its tumor-selective apoptotic activity. Furthermore, antiapoptotic Bcl-2 family proteins, often overexpressed in cancer, impede apoptosis and autophagy. They inhibit autophagy through interaction with Beclin 1, a critical autophagy protein characterized by a BH3 domain. Small-molecule BH3 mimetics, such as ABT-737 and obatoclax, can interfere with this interaction, thereby reinstating autophagy and enhancing cellular sensitivity to apoptosis. These compounds demonstrate potential in promoting cancer cell death, especially when used in conjunction with other agents. Celecoxib, a COX-2 inhibitor, promotes apoptosis in colon cancer cells, an effect that is augmented by BH3 mimetics and the suppression of autophagy. Investigating the relationship between autophagy and apoptosis presents a viable strategy for addressing therapeutic resistance in cancer (Maiuri et al. 2007; Levine et al. 2008; Jin et al. 2009; Huang and Sinicrope 2010). Figure 7.2 illustrates the mechanisms of miRNA in the development of cancer.

## 7.3 Cancer Stem Cells (CSCs) and Therapy Resistance

### 7.3.1 CSCs and Their Characteristics

Cancer stem cells (CSCs) are a tumorigenic subpopulation characterized by stem-like properties, such as self-renewal via asymmetric division and differentiation capability, contributing to tumor heterogeneity and development. The cells can be classified into three primary categories: (1) tissue-specific CSCs originating from progenitor cells, (2) dedifferentiated CSCs that develop from mature cancer cells that reacquire stem-like properties, and (3) hybrid cancer stem cells demonstrate dynamic phenotypic transitions between stem and non-stem states in response to microenvironmental stimuli. CSCs display inherent resistance to therapy due to improved DNA repair mechanisms, drug efflux via ABC transporters, and quiescent states. Their tumor-initiating capacity is evidenced by their ability to form tumors at low cellular densities in xenograft models. The malignant characteristics are sustained through the constitutive activation of essential developmental pathways, including Wnt/β-catenin, Hedgehog, PI3K/AKT/mTOR, and Notch, as well as through phenotypic plasticity via epithelial–mesenchymal transition (EMT), which enables metastatic spread. Moreover, CSCs show significant metabolic flexibility, marked by transitions between glycolytic and oxidative phosphorylation pathways, which facilitate survival in hypoxic and nutrient-limited environments. Their ability to evade the immune system, through mechanisms like PD-L1 upregulation and the recruitment of immunosuppressive cells, further contributes to their persistence in tumors.

Due to their critical involvement in drug resistance and tumor recurrence, current therapeutic approaches concentrate on targeting CSC-specific surface markers (CD44, CD133), interrupting stemness-related signaling pathways, and altering the tumor microenvironment. Eliminating this robust cell population constitutes a promising strategy for enhancing long-term therapeutic results by preventing relapse and metastasis (Lim et al. 2021; Chu et al. 2024).

### 7.3.2 Self-Renewal, Plasticity, and Tumor Heterogeneity

Cancer stem cells (CSCs) preserve their tumorigenic capacity via evolutionarily conserved self-renewal mechanisms that are disrupted in cancer. The fundamental regulatory networks includes signaling pathways such as Wnt/β-catenin, Hedgehog, and Notch that initiate stemness-related transcriptional programs; epigenetic

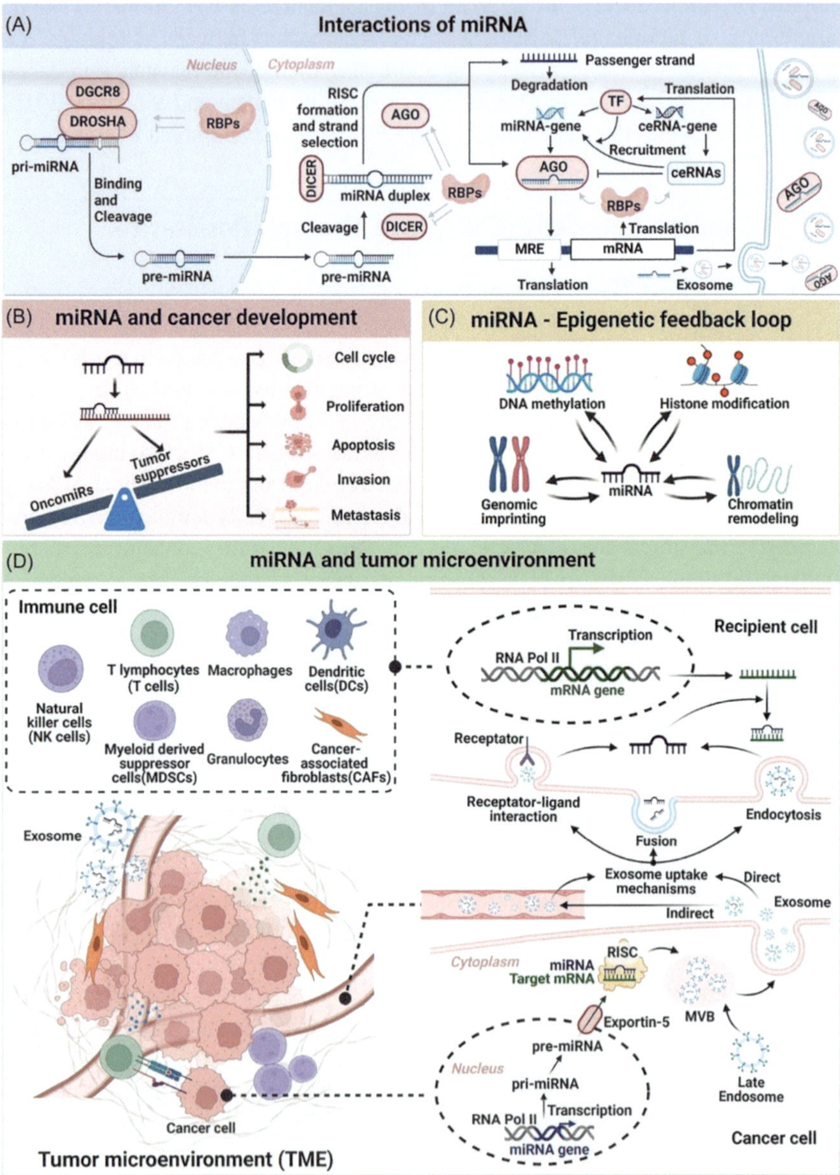


**Fig. 7.2** miRNA mechanisms in the development of cancer. (**a**) A number of important processes are involved in the biogenesis of miRNAs: pri-miRNA is cleaved in the nucleus by Drosha and DGCR8; it is transported to the cytoplasm by Exportin5 and RanGTP; it is generated in the cytoplasm by Dicer and TRBP; and it is bound to pre-miRNA by Argonautes (AGO) to form the minimal miRNA-induced silencing complex (miRISC) to control the expression of target mRNA. By affecting the ceRNA–miRNA–mRNA axis, TFs can produce a feedback loop. Furthermore, vesicles can carry miRNAs to other cells so they can have an impact. (**b**) Through pathways such as cell cycle and proliferation, metastasis and invasion, apoptosis, EMT, immunological modulation,

mechanisms involving PRC2-mediated gene silencing and microRNA networks that regulate differentiation; asymmetric cell division governed by polarity complexes (PAR3/6-aPKC) and cell fate determinants (NUMB); and metabolic adaptations including HIF-1α-driven glycolysis and autophagy-mediated nutrient recycling. This requires the formulation of therapeutic strategies that concentrate on pathway inhibitors (e.g., PORCN for Wnt), epigenetic modulators, and differentiation therapies to address these redundant self-renewal pathways (Niwa 2014).

Cancer stem cell (CSC) plasticity represents the dynamic and reversible capacity of cancer cells to shift between stem-like and non-stem-like conditions. This adaptability is essential in tumor heterogeneity, therapy resistance, and metastasis. CSCs demonstrate two plasticity mechanisms: intrinsic (cell autonomous) and extrinsic (microenvironment-driven), facilitating dynamic and reversible transitions between stem-like and differentiated states. Intrinsic plasticity surrounds epigenetic reprogramming, involving alterations to DNA and histones, alongside the function of noncoding RNAs. It also surrounds metabolic transitions between glycolysis and oxidative phosphorylation, alongside the activation of essential stemness pathways such as Notch, Wnt, Hippo, and NF–κB. In contrast, extrinsic plasticity is affected by hypoxic niches, inflammatory cytokines such as TNF-α and IL-6, and the stiffness of the extracellular matrix, collectively promoting epithelial–mesenchymal transition (EMT) and therapeutic resistance. This adaptive plasticity enables CSCs to circumvent treatment, withstand immune elimination, and promote recurrence, especially when conventional therapies prompt non-CSCs to regain stem-like characteristics. Plasticity is highly essential in the development of drug-tolerant persister states during therapy and in tumor regeneration following treatment. To tackle these challenges, therapeutic strategies should concurrently target both types of plasticity by using epigenetic modulators (DNMT/HDAC inhibitors), niche disruptors (hypoxia-targeting agents), and differentiation therapies, in conjunction with conventional treatments aimed at preventing dedifferentiation and eradicating resilient CSC populations to enhance clinical outcomes (Warrier et al. 2023).

CSCs serve as fundamental architects of tumor heterogeneity, attributable to their inherent plasticity, ability for differentiation and self-renewal into different cell

**Fig. 7.2** (continued) angiogenesis, and drug resistance, miRNA can function as both tumor suppressors and oncomiRs, influencing cancer. (**c**) A miRNA-epigenetic feedback loop is formed when miRNA and epigenetic effects interact; its dysregulation is strongly linked to the development of cancer. (**d**) Immune cells like natural killer cells (NK cells), T lymphocytes (T cells), macrophages, dendritic cells (DCs), myeloid-derived suppressor cells (MDSCs), granulocytes, and cancer-associated fibroblasts (CAFs) are all impacted by miRNA dysregulation, which creates a complex tumorigenic symbiosis with tumor cells and the tumor microenvironment. Through blood arteries, exosomal miRNA can influence the development of cancer by acting on distant immune cells. RISC stands for RNA-induced silencing complex; MVB stands for multivesicular bodies. (Reprinted from Wang et al. (2024), Copyright © 2024 The Author(s). MedComm—Oncology published by John Wiley & Sons Australia, Ltd on behalf of Sichuan International Medical Exchange & Promotion Association (SCIMEA). This is an open-access article distributed under the terms of the Creative Commons Attribution License (CC BY))

types, and hierarchical organization. Cellular diversity is generated by CSCs through asymmetric division that yields phenotypically distinct progeny, epigenetic reprogramming that results in transcriptional variability, differentiation responsive to the microenvironment (such as hypoxia and cytokines), and genomic instability that facilitates clonal evolution. This heterogeneity is evident in distinct CSC subpopulations exhibiting varying metastatic potentials, with mesenchymal-like CSCs promoting dissemination and epithelial-like CSCs sustaining primary tumors, alongside differing drug sensitivities. The CSC niche significantly influences heterogeneity via paracrine signaling (TGF-β, IL-6) and extracellular matrix remodeling, whereas therapy exerts selective pressures that dynamically modify CSC clonal composition. The functional heterogeneity observed explains metastatic organotropism and therapeutic failures, highlighting the necessity for multi-targeted strategies aimed at plastic cancer stem cell populations and their supportive microenvironments (Lee and Hall 2016).

### 7.3.3 Role of CSCs in Chemoresistance and Radioresistance

CSCs are crucial in facilitating chemoresistance and radioresistance, significant factors in cancer relapse and treatment failure. Chemotherapy or radiation therapy applied to a bulk tumor results in the elimination of the majority of tumor cells, including both differentiated and progenitor cells. CSCs with a nucleus-like center persist owing to their intrinsic resistance mechanisms. Factors contributing to this phenomenon include slow proliferation rates, improved DNA repair capabilities, expression of drug-efflux pumps, and resistance to apoptosis. This selective survival results in a transient decrease in tumor size. Posttreatment, cancer stem cells (CSCs) experience asymmetric division, leading to two outcomes: self-renewal, which preserves the CSC population, and differentiation into progenitor and differentiated cells, thereby replenishing the heterogeneous tumor mass. This capability supports tumor recurrence, as the surviving cancer stem cells repopulate the tumor, reinstating its original cellular complexity and size, leading to chemoresistance and radioresistance. Incomplete eradication of cancer stem cells (CSCs) during therapy may result in their persistence in a quiescent or minimally active state, which can subsequently lead to tumor relapse. Recurrent tumors frequently demonstrate increased aggressiveness and heightened resistance to standard therapies (Das et al. 2017).

### 7.3.4 CSC-Associated Signaling Pathways in Therapy Resistance

CSCs significantly contribute to therapeutic resistance by maintaining the activation of interconnected signaling pathways that govern stemness, survival, and plasticity. The Wnt/β-catenin pathway enhances chemoresistance by stabilizing β-catenin, which leads to the upregulation of MYC and Cyclin D1, as well as the enhancement of DNA repair mechanisms. Conversely, Notch signaling preserves undifferentiated states through NICD-mediated activation of HES/HEY and the upregulation of drug efflux pumps like ABCG2. The Hedgehog (Hh) pathway, through GLI effectors, promotes EMT and essential stemness factors (OCT4, SOX2, NANOG). The PI3K/AKT/mTOR signaling pathway enhances cancer stem cell survival through the inhibition of apoptosis and the promotion of metabolic adaptation to stress. Concurrently, the JAK/STAT3 and NF–κB pathways create pro-survival microenvironments through inflammatory cytokines and antiapoptotic proteins, whereas the Hippo/YAP-TAZ pathway increases plasticity and metastasis through TEAD-dependent transcription. Epigenetic regulators such as EZH2 and DNMTs contribute to the stabilization of resistant phenotypes through the silencing of tumor suppressor and differentiation genes. Targeted inhibitors for these pathways, like γ-secretase inhibitors for Notch, SMO antagonists for Hh, and AKT inhibitors, demonstrate preclinical promise; however, their efficacy is constrained by compensatory crosstalk, niche-mediated protection, and cancer stem cell plasticity. Combinatorial approaches that integrate pathway inhibitors, epigenetic modulators including HDAC and DNMT inhibitors, and microenvironment disruptors (including hypoxia-targeting agents and CXCR4 antagonists) are crucial for eradicating therapy-resistant cancer stem cell subpopulations and preventing tumor recurrence (Manni and Min 2022; Talukdar et al. 2023).

This Table 7.1 outlines the critical roles of autophagy and cancer stem cells in influencing cancer therapy and the development of treatment resistance.

## 7.4 miRNA Regulation of Autophagy in Cancer Therapy Resistance

### 7.4.1 miRNAs as Autophagy Modulators in Cancer

miRNA participates in different steps of the autophagy process, working as a modulator. It regulates upstream pathways, the developmental and degradation stages of autophagy, thus influencing the different proteins and genes related to autophagy (Shan et al. 2021). As autophagy can suppress the expression of cancer, it has both physiological and pathological implications. Any abnormalities in miRNA can play a role in tumor progression as they can influence the tumor microenvironment and apoptosis. The stress caused by chemotherapy can produce cytotoxic chemicals, which can promote autophagy in favor of a rapid increase in cancer cells (Wei et al. 2024). For instance, both miR-378 and miRNA-124 can inhibit the progression of colorectal cancer. Moreover, miR-378 can help in cervical cancer progression by regulating autophagy. Both can work as cancer biomarkers in tumor diagnosis (Kadhim et al. 2023).

**Table 7.1** Mechanistic insights into autophagy and cancer stem cells in cancer therapy and therapeutic resistance

| SL No. | Biological process | Regulatory molecules | Key pathway | Therapeutic implications | References |
|---|---|---|---|---|---|
| 1. | Autophagy activation | Beclin-1, LC3, ATG5, p62/SQSTM1 | PI3K/AKT/mTOR, AMPK, ULK1 | Boosts cancer cell survival under stress; inhibition could sensitize tumors to therapy | Maiuri et al. (2007, Yun et al. (2021) |
| 2. | Autophagy inhibition | mTOR, Bcl-2, Rubicon | mTORC1, NF-κB | Overactivation may lead to tumor progression | Mukherjee et al. (2020), White et al. (2021) |
| 3. | CSC maintenance | Oct4, Sox2, Nanog, CD44, ALDH1 | Wnt/β-catenin, Hedgehog, Notch | Targeting cancer stem cells diminishes tumor recurrence and chemoresistance | Lim et al. (2021), Talukdar et al. (2023) |
| 4. | Autophagy in CSC survival | HIF-1α, BNIP3, PARK2 | Hypoxia-induced autophagy | Inhibiting the miRNA-hypoxia-autophagy loop surmounts resistance | Yun et al. (2021), Chu et al. (2024) |
| 5. | EMT and metastasis | Zeb1, N-cadherin, Snail | TGF-β, MAPK | Autophagy facilitates CSC plasticity; dual-targeting may inhibit metastasis | Dong et al. (2021, Warrier et al. (2023) |
| 6. | Chemoresistance in CSCs | ABC transporters, Bmi-1 | NF-κB, STAT3 | Combining autophagy inhibition with chemotherapy may reduce resistance conferred by cancer stem cells | Manni and Min (2022) |
| 7. | Immune evasion by CSCs | PD-L1, CD47, CTLA-4 | JAK/STAT, immune checkpoint pathways | The integration of autophagy modulators with immunotherapy could improve anti-tumor immunity | White et al. (2021), Talukdar et al. (2023) |
| 8. | Metabolic reprogramming | HK2, LDHA, GLUT1 | Glycolysis, OXPHOS | Targeting autophagy impairs CSC metabolism, enhancing therapeutic response | Yadav et al. (2020), Chu et al. (2024) |
| 9. | miRNA regulation | miR-21, miR-7, miR-125b, miR-214, miR-302 family | PI3K/AKT, NF-κB, Wnt/β-catenin, EMT | miRNA-based treatments can mitigate resistance and enhance tumor sensitivity | Frankel and Lund (2012), Gozuacik et al. (2017) |
| 10. | Therapeutic targeting | Metformin, Salinomycin | Combined pathway inhibition | Utilizing autophagy inhibitors in conjunction with CSC-targeting drugs enhances therapeutic efficacy | Makena et al. (2020), Mukherjee et al. (2020) |

### 7.4.2 Pro-autophagic miRNAs and Their Therapeutic Implications

Cancer cells have rapid proliferation, thus having higher metabolic rates and increased levels of glycolytic metabolism. So, anti-autophagy treatments can significantly contribute in inhibiting cancer cells (Koustas et al. 2022). But during radiotherapy in ovarian cancer, miRAN-214 becomes upregulated, making tumor cells resistant to radiotherapy (Ding et al. 2024). Several miRNAs have shown their influence in breast cancer by influencing both autophagy and apoptosis. For example, increased miR-125 b-5p makes cancer cells sensitive to docetaxel. So monitoring the level of this miRNA can give information about the tumor progression and thus tailor the therapeutics (Biomed et al. 2021).

### 7.4.3 Anti-autophagic miRNAs Contributing to Resistance

Different types of mRNA have been proven to influence autophagy, such as miR96, miR128, miR100, miR124, etc. (Akkoc and Gozuacik 2020). Recently, various studies have suggested that irregularities in autophagy influence multidrug resistance. Significant changes in miRNA expression have been noted in drug-resistant cancer cells. Many studies have found that ABCB1 expression can be regulated by miR, and overexpression of p-glycoprotein can be due to activation of the ABCB1 gene. And this can result in downregulation of miR-138, causing resistance to Adriamycin in breast cancer treatment (An et al. 2017). Additionally, miR-495-3p may be targeted to address multidrug resistance in gastric cancer (Chen et al. 2018).

### 7.4.4 miRNA-Mediated Crosstalk Between Autophagy and Apoptosis

Several miRNAs regulate programmed cell death caused by autophagy and apoptosis. The miR-21 influences the regulation of the extrinsic pathway of apoptosis (Su et al. 2015). Both autophagy and apoptosis significantly contribute in cellular homeostasis, cellular functions, and oncogenesis. The miRNA can inhibit common regulators shared by both autophagy and apoptosis pathways. When cellular stress occurs, BECN1 and BCL2, which are antiapoptotic, become dissociated and promote autophagy and inhibit apoptosis. The miR-30A can inhibit the expression of both genes, thus influencing both processes. The miR 204 can inhibit autophagy by blocking the LC3-II protein during hypoxia, and this miR 304 can also facilitate apoptosis related to chemotherapeutic drugs by regulating BCL2 (Xu et al. 2012).

The overexpression or inhibition of various miRNAs modulates the autophagic pathway, influencing the development or combat of chemoresistance in several anticancer therapies.

Table 7.2 represents the key miRNAs that play vital roles in regulating autophagy in cancer. Figure 7.3 illustrates the function of miRNAs regulating different stages in autophagy.

**Table 7.2** Key miRNAs regulating autophagy in cancer therapy resistance

| SL No. | miRNA | Cancer type | Autophagy target(s) | Mechanism of action | Therapeutic effect | References |
|---|---|---|---|---|---|---|
| 1. | miR-30a | Lung, breast, renal cell carcinoma | Beclin-1, ATG5 | Inhibition | Resistant to Sorafenib, Cis-DDP, Taxol | Zou et al. (2012), Zheng et al. (2015) |
| 2. | miR-101 | Prostate, liver, osteosarcoma | STMN1, ATG4D, RAB5A, | Inhibition | Resistant to Cisplatin, Doxorubicin | Chang et al. (2014) |
| 3. | miR-17 | Lung | Beclin-1 | Inhibition | Resistant to Paclitaxel | Chatterjee et al. (2015) |
| 4. | miR-495-3p | Gastric | GRP78/mTOR | Inhibition | Inhibits Multidrug resistance | Chen et al. (2018) |
| 5. | miR-34A | Retinoblastoma | HMGB1 | Inhibition | Resistant to Carboplatin, Vincristine, Etoposide | Liu et al. (2014) |
| 6. | miR-125b | Hepatocellular carcinoma | EVA1A | Inhibition | Resistant to Oxaliplatin | Ren et al. (2018) |
| 7. | miR-204 | Ovarian cancer | IL-6R/STAT3 | Inhibition | Resistant to Cisplatin | Zhu et al. (2017) |
| 8. | miR-15a/16 | Cervical carcinoma | Rictor | Promotion | Enhances the efficacy of Camptothecin | Huang et al. (2015) |
| 9. | miR-181 | NSCLC | PTEN/PI3K/AKT | Promotion | Resistant to Cisplatin | Liu et al. (2018) |
| 10. | miR-193b | Esophageal cancer | Stathmin 1 | Promotion | Enhances the chemosensitivity of 5-Fluorouracil | Nyhan et al. (2016) |
| 11. | miR-138 | Glioblastoma | BIM | Promotion | Resistant to Temozolomide | Stojcheva et al. (2016) |

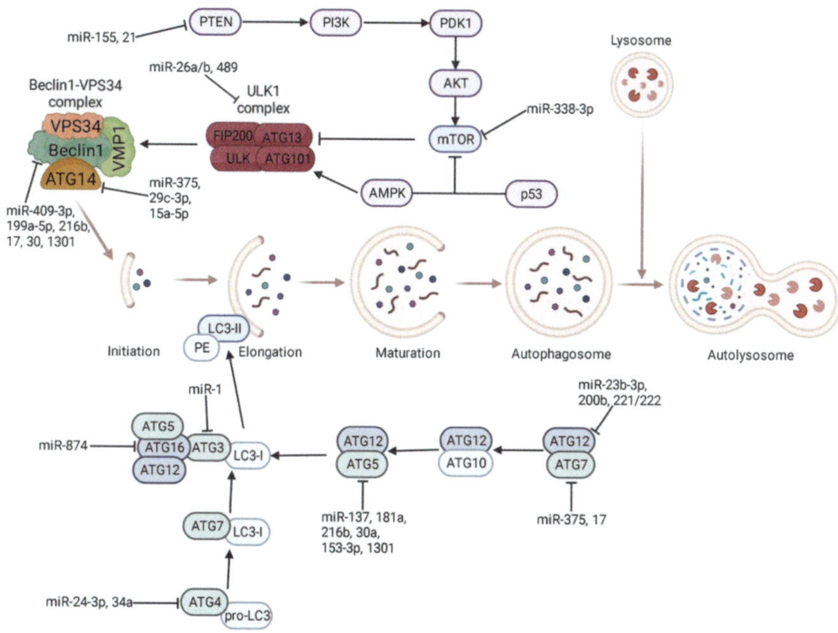


**Fig. 7.3** The function of miRNAs in regulating autophagy at every step. Every stage of autophagy, including phagophore initiation and elongation, autophagosome maturation, and lysosomal fusion, is associated with core proteins and signaling networks. To take part in all phases of autophagy, certain important miRNAs target autophagy-related genes at the post-transcriptional level. An inhibitory impact is denoted by ⊥, and a promoting effect by →. (Reprinted from Lei et al. (2022), Copyright © 2022 Lei, Chen, Liu, Zhong and Deng. This is an open-access article distributed under the terms of the Creative Commons Attribution License (CC BY))

## 7.5 miRNA Modulation of Cancer Stemness and Tumor Plasticity

### 7.5.1 Role of miRNAs in CSC Maintenance and Differentiation

Through genetic modifications and epigenetic alterations, cancer cells can achieve the cell renewal properties and differentiation characteristics. The miRNAs are also a crucial part of cell renewal and differentiation. Irregularities in miRNA can cause abnormal cell division, drug resistance, and benefit tumorigenesis (Wang et al. 2013; Drug and Rev 2016). There are chromosomal regions that can code miRNAs, which are directly involved in downregulation of tumor suppressor genes, thereby promoting the proliferation of cancer cells. This type of amplification can increase the number of oncogenic miRNAs that silence the tumor suppressor genes. Also, miRNA can accumulate at the disease site, and their dysregulation can increase the metastasis, tissue invasion, and multidrug resistance (Takahashi et al. 2014). So, some miRNAs work as tumor suppressors, and some benefit tumor cells. Let-7 is the most common miRNA that works as a tumor suppressor. The miRNA and CSC have direct involvement in some common pathways, for instance, JAK/STAT, NF–kB, and Wnt/β-catenin, which are prominent pathways in cell renewal. So, the miRNAs can be targeted to inhibit the cell renewal and differentiation of cancer cells (Yoshida et al. 2021).

### 7.5.2 Key miRNAs Regulating Self-Renewal Pathways

The miRNAs participate in the cell renewal process by modulating the cell reprogramming cascade. Increased amount of miR 302 or miR290 family can increase the cell renewal process and reprogram it. Other miRNAs, including miR-372, the miR-106b-25 cluster, and miR-17-92, have been demonstrated to enhance cell renewal pathways. For instance, miR-302 has the capability to convert human cancer cells into pluripotent cells, thereby enhancing cell proliferation (Li et al. 2017). Other miRNAs are involved in several cell renewal pathways, such as PI3K/Akt (He et al. 2021), Wnt/β-catenin (Liu et al. 2022) and NF–kB (Manuscript 2015). The miRNA is considered an important pathway in cancer stem cell production (Xia and Xu 2015). In these pathways, miRNAs can work to inhibit cancer cells. So, putative genes related to miRNA can work for targeted treatment. Downregulation of miR-7 has shown that it can inhibit the production of stem cells in prostate cancer by adversely affecting the KLF4/PI3K/AKT/p21 pathway (Akbarzadeh et al. 2021). The Wnt/β-catenin pathway is a significant mechanism that can be pivotal in cancer. The miRNAs can inhibit or activate these pathways by modifying some components. The miR-135b has been found to increase this signaling pathway by enhancing the expression of β-catenin in osteocarcinoma (Song et al. 2024). The NF–kB can orchestrate different pathological conditions in our body. The diminished expression of miR-9 has been demonstrated to augment the activities of the NF–kB pathway, hence facilitating ovarian cancer. So, these pathways can work as therapeutic targets in cancer treatment (Zheng 2018).

### 7.5.3 miRNA Influence on Epithelial-to-Mesenchymal Transition (EMT) and Metastasis

The EMT process is associated with both wound healing and metastasis. Epithelial–mesenchymal transition (EMT) is a biological process in which epithelial cells progressively lose their characteristics and acquire properties associated with a mesenchymal phenotype, including enhanced invasiveness. MiRNAs are believed to participate in regulating genes involved in the EMT process. For example, miR-220 b has several binding sites where 60 genes of EMT can bind. The miR-200 family primarily functions to inhibit the epithelial–mesenchymal transition

(EMT) and metastasis and their overexpression may reduce the invasiveness of cancer cells. The EMT can activate various transcription factors such as bHLH (E47, E2-2, and TWIST1/TWIST2), (EMT–TFs)—SNAIL1/SNAIL2 and ZEB1/ZEB2, which can work as repressors of E-cadherin. As a result, downregulation of E-cadherin decreases cell adhesion, thus increasing metastasis (Díaz-López et al. 2014; Dong et al. 2021). The transition of cancer cells from epithelial to mesenchymal phenotypes results in chemoresistance, indicating the significant role of epithelial-mesenchymal transition (EMT) in cancer treatment (Dong et al. 2021).

### 7.5.4 miRNAs Targeting CSC Metabolic Adaptations for Therapy Resistance

CSCs are essential elements that play a significant role in the development and progression of various cancer types, including pancreatic, colon, prostate, brain, and breast cancers. They can make tumors more resistant to chemotherapy and radiotherapy and help in metastasis, leading to greater tumor invasion. Both normal and somatic stem cells can turn into CSCs when they undergo abnormal genetic or epigenetic changes (Niu et al. 2021). Tumor microenvironment helps to increase CSC production by modulating different factors like Sox-2 and Klf4, and also customizes the CSC properties to make cancer cells chemoresistant. The cytokines and chemokines can help in the production of iCSC, which can lead to weak prognosis, thus limiting the efficacy of cancer treatments and increasing cellular invasion (Richard et al. 2021). Multiple research studies have discovered that overexpression of some miRNA (miR-1296, miR-38, miR-365a) can make the malignancy more chemoresistant by enhancing cell renewal, proliferation, and differentiation capacity (Bhagtaney et al. 2024). Besides, CSC can switch between glycolysis and oxidative phosphorylation during stressed environments to survive, and miRNA can be involved to help the tumor cells switch their metabolic pathways to survive (Yadav et al. 2020).

The following Table 7.3 outlines a few key miRNAs that act as suppressors and promoters of CSCs in different types of cancer, indicating their role in therapeutic resistance.

**Table 7.3** Key miRNAs regulating cancer stemness in cancer therapy resistance

| SL No. | miRNA | Cancer type | Cancer stem cell target(s) | Therapeutic impact | References |
|---|---|---|---|---|---|
| 1. | miR-122 | Lung cancer | Suppressor at H460, HCC827, A549 cell lines | More susceptible to chemotherapy, inducing apoptosis, | Jiao et al. (2020) |
| 2. | miR-128 | Lung cancer | Suppressor at A549/paclitaxel cell line | Reduction of self-renewal, more susceptible to paclitaxel | Jiao et al. (2020) |

(continued)

**Table 7.3** (continued)

| SL No. | miRNA | Cancer type | Cancer stem cell target(s) | Therapeutic impact | References |
|---|---|---|---|---|---|
| 3. | miR-34a | Breast cancer | Suppressor in the MCF-7 cell line | Inhibits tumor sphere formation | Jiao et al. (2020) |
| 4. | miR-590-5p | Breast cancer | Suppressor at MCF-7, ZR75-1 cell line | Suppresses SOX2 | Jiao et al. (2020) |
| 5. | miR-19 | Lung cancer | Promoter at A549, H1299 cell lines | Inhibition of glycogen synthase kinase | Jiao et al. (2020) |
| 6. | miR-148a | Colorectal cancer | Suppressor at SW480 cell lines | Reduces cell renewal and drug resistance | Jiao et al. (2020) |
| 7. | miR-449a | Prostate cancer | Suppressor at PC3, LNCaP cell lines | Enhancement of chemosensitivity | Jiao et al. (2020) |
| 8. | miR-448 | Liver cancer | Suppressor at Hep3B | Reduction of stemness and tumorsphere | Jiao et al. (2020) |
| 9. | miR-17-5p | Esophagus cancer | Suppressor at OE33 P, OE33 R cell lines | Increased radiosensitivity and inhibition of the expression of stem-related genes | Jiao et al. (2020) |
| 10. | miR-137 | Pancreatic cancer | Suppressor at AsPC-1, PANC-1 cell lines | Inhibition of stem markers | Jiao et al. (2020) |

## 7.6 miRNA-Based Therapeutic Strategies for Overcoming Cancer Resistance

### 7.6.1 miRNA Mimics and Inhibitors as Therapeutic Agents

In hematological melanoma, dysregulation of miRNA can cause tumor progression and drug resistance. The miR-21 is a notably common miRNA that is a common oncogenic factor in hematological malignancies. So, miRNA has therapeutic potential to target. Although overexpression of miRNA is correlated with cancer progression, no miRNA genes are related to cancer. Some miRNAs can play a dual role by being both oncogenes and tumor suppressor genes. So, recognizing the functions of specific miRNAs can facilitate the development of targeted treatment strategies (Dong et al. 2021). Therapeutic based on miRNA can be utilized as a replacement or an inhibitor. Some miRNAs are being developed to mimic the natural miRNA and are designed to replace the miRNAs that can work as tumor suppressors. Synthetic miRNAs are injected into tumor cells to restore tumor suppressive miRNAs such as let-7, miR-29b, and miR-34a (Bailey et al. 2013).

On the other hand, other miRNAs are developed that work as inhibitors or anti-miRNAs. Their main function is to block the miRNAs that help in the upregulation of cancer cells. These miRNAs inhibit oncogenic miRNAs such as miR-221. So, these miRNA-based therapeutics are showing potential in novel cancer treatment (Diener et al. 2022). Currently, several microRNAs have been patented for their roles in diagnosis, management, and treatment of various cancer types, suggesting commercial interest in transforming miRNA biology into viable cancer therapies (Table 7.4). Additionally, multiple clinical trials have been completed and many are in the pipeline involving miRNA-based therapeutic strategies for treating a number of cancers (Table 7.5).

### 7.6.2 Delivery Systems for miRNA Therapeutics

miRNAs, which are small transcripts, are related to different cellular functions, including metabolic pathways, apoptosis, and autophagy, which are related to the progression of cancer. Due to their dual functionality, they pose a prominent therapeutic target in novel cancer treatment. For safe and effective delivery of these small transcripts, nanoparticles (NPs) are being investigated to protect them from the external environment (Gareev et al. 2023; Ghafouri-Fard et al. 2023). The miR-7 has been proven as effective against solid tumors, and cationic liposomes are being studied for the delivery of miR-7 to the target cells for ovarian cancers (Cui et al. 2021). Also, liposomes have been approved to deliver drugs due to fewer side effects and targeted delivery, and that makes them a perfect candidate for miRNA delivery (Tenchov et al. 2021). Nanovesicles, like exosomes, are also potential candidates for miRNA delivery. Exosomes loaded with miRNA can ensure cell specificity and safe delivery (Zhang et al. 2016). Loading miRNA in exosomes can be done by transfecting from donor cells, electroporation, or nanoporation. Exosomes can be a suitable vector for delivering miRNA due to their safety and efficacy (Volpini et al. 2023).

### 7.6.3 Combination Therapies Integrating miRNA Modulation with Conventional Treatments

The sustainability of cancer cells against various treatment modalities represents a significant obstacle to the success of cancer therapies. To overcome this problem, the utilization of miRNA in combination with other cancer therapies, such as chemotherapy, radiotherapy, photodynamic therapy, etc., has garnered significant interest (van Beijnum et al. 2017; Seo et al. 2019). Developing miRNAs that will

**Table 7.4** Patents associated with miRNA therapeutics in different cancers

| SL No. | Patent No. | Application No. | Inventor | Country | Status | References |
|---|---|---|---|---|---|---|
| 1. | US20090253780A1 | 12/412,087 | Fumitaka Takeshita, David Brown, Takahiro Ochiya | United States | Abandoned | Data and Brown (2009) |
| 2. | US8507195B2 | 13/390,460 | Jennifer Richer, Dawn Cochrane, Steven M. Anderson | United States | Active | Cochrane and Us (2013) |
| 3. | US20070299030A1 | 11/730,570 | Ethan Dmitrovsky Xi Liu, Sarah Freemantle, Lorenzo Sempere, Charles Cole, Sakari Kauppinen, Mads Bak | United States | Active | Data et al. (2007) |
| 4. | US11236337B2 | US16/912,932 | Jingfang Ju, Andrew Fesler, Younghwa Song, Jun Chung | United States | Active | Yager (1988) |
| 5. | US10851372B2 | US15/760,874 | Derrick Gibbings, James Andrew Taylor | United States | Active | Taylor and Ca (2020) |
| 6. | US11147828B2 | US16/212,416 | Judy Lieberman, Erwei Song, Fengyan Yu, Xiaoqu Hu | United States | Active | Lieberman et al. (2018) |

**Table 7.5** Clinical trials of miRNA-based therapeutics for the treatment and management of cancer

| SL No. | Clinical trial identifier | Study title | Status | Sponsor | Condition | Enrolment | Intervention |
|---|---|---|---|---|---|---|---|
| 1. | NCT01829971 | Phase I Study of MRX34 (miR-34a Mimic) in Patients with Advanced Solid Tumors | Terminated | Mirna Therapeutics, Inc. | Liver cancer, NSCLC, Renal cell carcinoma, Melanoma | 155 | miR-34a mimic (MRX34) delivered via liposomal nanoparticles |
| 2. | NCT02369198 | A Phase I Study of TargomiRs as 2nd or 3rd Line Treatment for Patients with Recurrent MPM and NSCLC | Completed | Asbestos Diseases Research Foundation | Malignant Pleural Mesothelioma, NSCLC | 27 | miR-16 mimic in EGFR-targeted bacterial minicells (TargomiRs) |
| 3. | NCT02580552 | Safety, Tolerability and Pharmacokinetics of MRG-106 in Patients with Mycosis Fungoides (MF), CLL, DLBCL or ATLL | Completed | miRagen Therapeutics, Inc. | Chronic Lymphocytic Leukemia, Mycosis Fungoides, Diffuse Large B-Cell Lymphoma | 66 | Cobomarsen (anti-miR-155) |
| 4. | NCT02705196 | LOAd703 Oncolytic Virus Therapy for Pancreatic Cancer | Active, not recruiting | Lokon Pharma AB | Pancreatic cancer | 51 | Genetic: delolimogene mupadenorepvec<br>Drug: gemcitabine<br>Drug: nab-paclitaxel<br>Biological: atezolizumab |
| 5. | NCT04435756 | A Study of MiRNA 371 in Patients with Germ Cell Tumors | Active, not recruiting | SWOG Cancer Research Network | Germ cell tumor, Testicular cancer, Seminoma | 956 | Biomarker analysis of miR-371 |

(continued)

**Table 7.5** (continued)

| SL No. | Clinical trial identifier | Study title | Status | Sponsor | Condition | Enrolment | Intervention |
|---|---|---|---|---|---|---|---|
| 6. | NCT05854030 | Serum Exosomal miRNA Predicting the Therapeutic Efficiency in Lung Squamous Carcinoma | Unknown | Tianjin Medical University Cancer Institute and Hospital | Lung neoplasm, Squamous cell carcinoma | 60 | Diagnostic test: collect plasma samples and clinical features |
| 7. | NCT04675996 | First-in-Human Study of INT-1B3 in Patients with Advanced Solid Tumors | Terminated | InteRNA | Solid tumor | 25 | miR-193a-3p mimic (Drug: INT-1B3) |
| 08 | NCT01849952 | Evaluating the Expression Levels of MicroRNA-10b in Patients with Gliomas | Recruiting | Dartmouth-Hitchcock Medical Center | Oligodendroglioma, Glioblastoma, Anaplastic Astrocytoma, Brain tumor | 200 | Observational study on miR-10b expression |
| 09 | NCT03601052 | Efficacy, Safety, and Tolerability of Remlarsen (MRG-201) Following Intradermal Injection in Subjects with a History of Keloids | Completed | miRagen Therapeutics, Inc. | Keloid scars | 14 | miR-29 mimic (MRG-201)<br>Drug: Remlarsen<br>Drug: Placebo |
| 10 | NCT04811898 | A Dose Escalation Study of LNA-i-Mir-221 for Cancer Treatment (LNA-i-miR221) | Completed | Azienda Ospedaliera Universitaria Mater Domini, Catanzaro | Hepatocarcinoma, Multiple Myeloma, Solid tumor | 17 | Biological: LNA-i-miR-221 |

From: https://clinicaltrials.gov

either inhibit oncogenes or mimic tumor suppressive miRNAs can enhance the efficacy of standard treatments (Miroshnichenko and Patutina 2019). For example, miRNA that inhibits cancer cell progression can help chemotherapeutic agents by sensitizing the cancer cells to chemotherapy and limiting the side effects (Shah et al. 2016).

### 7.6.4 Challenges and Future Directions in miRNA-Based Cancer Therapy

In the advancement of miRNA-based therapeutics, various challenges must be addressed. The primary challenge involves the targeted delivery of miRNA to diseased sites while minimizing off-target effects and enhancing cellular uptake of miRNA. Also, it is crucial to ensure that these will not be affected by biological fluids or tissues, or enzymes (Bailey et al. 2013). Despite having advantages, it is challenging to convert the clinical development into clinical trials. Some pharmaceutical companies are attempting to conduct clinical trials; however, these efforts have been halted by the Food and Drug Administration (FDA) due to significant adverse effects such as hyperbilirubinemia. Also, the selection of the delivery system and route of administration is quite challenging. Although the utilization of NPs is being investigated, these studies have yet to be translated to clinical trials (El Sayed et al. 2021). Despite having many limitations, miRNAs still offer unique advantages in cancer treatment as they can single-handedly suppress multiple oncogenic pathways. One of the most unique characteristics of miRNAs as therapeutics is that they can be utilized against multidrug resistant. The miRNA also has the potential to boost our immune system, thus helping in immunotherapy. So, to utilize such a potential component, more research needs to be done on its impact on various cancers. By modifying the strategies, in the future, it will be possible to utilize multi-targeted miRNA to develop more targeted cancer treatment (Richard 2011; Micha 2017).

This Table 7.6 summarizes different approaches employed to develop miRNA-based therapies that combat treatment resistance in cancer treatment.

**Table 7.6** Emerging miRNA-based therapeutic approaches to counter cancer therapy resistance

| SL No. | Strategy | miRNA | Delivery vehicle | Combination therapy | Challenges | References |
|---|---|---|---|---|---|---|
| 1. | miRNA Mimics | miR-34a | Lipid nanoparticles | PD-1 inhibitors | Immune-related toxicity, off-target side effects | Sharma et al. (2020) |
| 2. | Anti-miRs | miR-155 | Polymeric nanoparticles | Chemotherapy | Poor tissue penetration, rapid degradation | Li et al. (2021) |
| 3. | CRISPR-miR | miR-21 | Viral vector | PARP inhibitors | Immunogenicity, off-target side effects | Vaghari-Tabari et al. (2022) |
| 4. | Dual miRNA Therapy | miR-200c + miR-34a | Gold nanoparticles | Paclitaxel | High cost | Liu et al. (2016) |
| 5. | Exosome Delivery | miR-let-7a | Tumor-derived exosomes | Radiotherapy | Scalability issues | Sousa and Conde (2022) |
| 6. | Epigenetic miRNA Activation | miR-1292 | CRISPR-dCas9/VPR system | 5-Flurouracil | Off-target side-effects | Khan and Torchilin (2022) |
| 7. | Tissue-Specific Targeting | miR-145 | Hyaluronic acid-coated liposomes | EGFR inhibitors | Hepatotoxicity | Ortíz et al. (2021) |

**Acknowledgments** Not applicable.

**Author Contributions CRediT Author Statement:** Conceptualization: Tanvir Ahmed; Writing—Original draft: Kazi Tasnuva Alam, Zarin Tansim Tisha, Tanvir Ahmed; Writing—Review and editing: Zarin Tansim Tisha, Tanvir Ahmed; Visualization: Tanvir Ahmed.

**Funding Statement** This research did not receive any specific grant from funding agencies in the public, commercial, or not-for-profit centers.

**Data Availability Statement** Data sharing is not applicable to this chapter as no datasets were generated or analysed during the current study period.

**Ethics Statement** Not applicable.

**Conflict of Interest** The authors declare no conflict of interest.

## References

2- Bray, F, Laversanne M, Weiderpass E, Soerjomataram I (2021) The ever-increasing importance of cancer as a leading cause of premature death worldwide. Cancer 127(16):3029–3030

Akbarzadeh M et al (2021) Crosstalk between miRNA and PI3K / AKT / mTOR signaling pathway in cancer. Life Sci 285(September):119984. https://doi.org/10.1016/j.lfs.2021.119984

Akkoc Y, Gozuacik D (2020) Molecular cell research MicroRNAs as major regulators of the autophagy pathway. BBA Mol Cell Res 1867(5):118662. https://doi.org/10.1016/j.bbamcr.2020.118662

An X et al (2017) Regulation of multidrug resistance by microRNAs in anti-cancer therapy. Acta Pharm Sin B 7(1):38–51. https://doi.org/10.1016/j.apsb.2016.09.002

Bailey J, Oliveri A, Levin E (2013) Targeting MicroRNAs in cancer: rationale, strategies and challenges. Bone 23(1):1–7. https://doi.org/10.1038/nrd3179.Targeting

Bhagtaney L, Dharmarajan A, Warrier S (2024) miRNA on the battlefield of cancer: significance in cancer stem cells, WNT pathway, and treatment. Cancer 16(5):1–17. https://doi.org/10.3390/cancers16050957

Biomed J et al (2021) Regulation of autophagy by microRNAs in human breast cancer. J Biomed Sci:1–18. https://doi.org/10.1186/s12929-021-00715-9

Bray F et al (2024) Global cancer statistics 2022: GLOBOCAN estimates of incidence and mortality worldwide for 36 cancers in 185 countries. CA Cancer J Clin 74(3):229–263. https://doi.org/10.3322/caac.21834

Chang Z et al (2014) Blocked autophagy by mir-101 enhances osteosarcoma cell chemosensitivity in vitro. Sci World J 2014. https://doi.org/10.1155/2014/794756

Chatterjee A, Chattopadhyay D, Chakrabarti G (2015) MiR-16 targets Bcl-2 in paclitaxel-resistant lung cancer cells and overexpression of miR-16 along with miR-17 causes unprecedented sensitivity by simultaneously modulating autophagy and apoptosis. Cell Signal 27(2):189–203. https://doi.org/10.1016/j.cellsig.2014.11.023

Chen S et al (2018) MicroRNA-495-3p inhibits multidrug resistance by modulating autophagy through GRP78 / mTOR axis in gastric cancer. Cell Death Dis. https://doi.org/10.1038/s41419-018-0950-x

Chu X et al (2024) Cancer stem cells: advances in knowledge and implications for cancer therapy, signal transduction and targeted therapy. Springer US. https://doi.org/10.1038/s41392-024-01851-y

Cochrane D, Us CO (2013) '(12) United States Patent (10) Patent No. 2(12)

Cui X et al (2021) Liposomal delivery of MicroRNA-7 targeting EGFR to inhibit the growth, invasion, and migration of ovarian cancer. ACS Omega 6(17):11669–11678. https://doi.org/10.1021/acsomega.1c00992

Das A et al (2017) Cancer stem cells, their origin and niche: a search for the therapeutic target. J Stem Cell Res Med 2(1). https://doi.org/10.15761/jscrm.1000118

Data RUSA, Brown D (2009) '(12) Patent Application Publication (10) Pub. No.: US 2009/0253780 A1', 1(19)

Data RUSA et al (2007) '(12) Patent Application Publication (10) Pub. No.: US 2007/0299030 A1', 1(19)

Díaz-López A, Moreno-Bueno G, Cano A (2014) Role of microRNA in epithelial to mesenchymal transition and metastasis and clinical perspectives. Cancer Manag Res 6(1):205–216. https://doi.org/10.2147/CMAR.S38156

Diener C, Keller A, Meese E (2022) Emerging concepts of miRNA therapeutics: from cells to clinic. Trends Genet 38(6):613–626. https://doi.org/10.1016/j.tig.2022.02.006

Ding Y et al (2024) The emerging roles of miRNA-mediated autophagy in ovarian cancer:1–15. https://doi.org/10.1038/s41419-024-06677-8

Dong B et al (2021) MiRNA-mediated EMT and CSCs in cancer chemoresistance. Exp Hematol Oncol 10(1):1–12. https://doi.org/10.1186/s40164-021-00206-5

Drug A, Rev D (2016) Role of microRNAs in maintaining cancer stem cells. 44(0):53–61. https://doi.org/10.1016/j.addr.2014.11.014.Role

El Sayed SR et al (2021) Microrna therapeutics in cancer: current advances and challenges. Cancer 13(11):1–28. https://doi.org/10.3390/cancers13112680

Frankel LB, Lund AH (2012) MicroRNA regulation of autophagy. Carcinogenesis 33(11): 2018–2025. https://doi.org/10.1093/carcin/bgs266

Gareev I et al (2023) Methods of miRNA delivery and possibilities of their application in neuro-oncology. Non-coding RNA Res 8(4):661–674. https://doi.org/10.1016/j.ncrna.2023.10.002

Ghafouri-Fard S et al (2023) Nanoparticle-mediated delivery of microRNAs-based therapies for treatment of disorders. Pathol Res Pract 248:154667. https://doi.org/10.1016/j.prp.2023.154667

Gozuacik D et al (2017) Autophagy-regulating microRNAs and cancer. Front Oncol 7(APR):1–22. https://doi.org/10.3389/fonc.2017.00065

He Y et al (2021) Targeting PI3K/Akt signal transduction for cancer therapy. https://doi.org/10.1038/s41392-021-00828-5

Huang S, Sinicrope FA (2010) Celecoxib-induced apoptosis is enhanced by ABT-737 and by inhibition of autophagy in human colorectal cancer cells. Autophagy 6(2):256–269. https://doi.org/10.4161/auto.6.2.11124

Huang N et al (2015) MiR-15a and miR-16 induce autophagy and enhance chemosensitivity of Camptothecin. Cancer Biol Ther 16(6):941–948. https://doi.org/10.1080/15384047.2015.1040963

Jalali P et al (2025) The role of autophagy in cancer: from molecular mechanism to therapeutic window. Front Immunol 16(April). https://doi.org/10.3389/fimmu.2025.1528230

Jiao X et al (2020) MicroRNA: the impact on cancer stemness and therapeutic resistance. Cells 9(1). https://doi.org/10.3390/cells9010008

Jin Z et al (2009) Cullin3-based Polyubiquitination and p62-dependent aggregation of Caspase-8 mediate extrinsic apoptosis signaling. Cell 137(4):721–735. https://doi.org/10.1016/j.cell.2009.03.015

Kadhim MM et al (2023) Autophagy as a self-digestion signal in human cancers: regulation by microRNAs in affecting carcinogenesis and therapy response. 189(February). https://doi.org/10.1016/j.phrs.2023.106695

Khan MM, Torchilin VP (2022) Recent trends in nanomedicine-based strategies to overcome multidrug resistance in tumors. Cancer 14(17). https://doi.org/10.3390/cancers14174123

Koustas E et al (2022) The emerging role of MicroRNAs and autophagy mechanism in pancreatic cancer progression: future therapeutic approaches. Genes 13:1868

Lee G, Hall R (2016) Cancer stem cells: cellular plasticity, niche, and its clinical relevance. J Stem Cell Res Ther 06(10). https://doi.org/10.4172/2157-7633.1000363
Lee S, Schmitt CA (2019) The dynamic nature of senescence in cancer. Nat Cell Biol 21(1):94–101. https://doi.org/10.1038/s41556-018-0249-2
Lei Y et al (2022) The MicroRNA-based strategies to combat cancer Chemoresistance via regulating autophagy. Front Oncol 12(February):1–16. https://doi.org/10.3389/fonc.2022.841625
Lei ZN et al (2023) Understanding and targeting resistance mechanisms in cancer. MedComm 4(3): 1–38. https://doi.org/10.1002/mco2.265
Levine B, Sinha S, Kroemer G (2008) BCL-2 family members. Autophagy 4(5):600–606
Li N et al (2017) microRNAs: important regulators of stem cells. Stem Cell Res Ther:1–7. https://doi.org/10.1186/s13287-017-0551-0
Li Y et al (2021) MicroRNA-155-5p promotes tumor progression and contributes to paclitaxel resistance via TP53INP1 in human breast cancer. Pathol Res Pract 220. https://doi.org/10.1016/j.prp.2021.153405
Lieberman J, et al (2018) Let-7 Microrna and Mimetics there of as therapeutics for cancer Au, 2
Lim JR et al (2021) Cancer stem cell characteristics and their potential as therapeutic targets. Med Oncol 38(7):1–12. https://doi.org/10.1007/s12032-021-01524-8
Liu K et al (2014) MIR34A regulates autophagy and apoptosis by targeting HMGB1 in the retinoblastoma cell. Autophagy 10(3):442–452. https://doi.org/10.4161/auto.27418
Liu J et al (2016) MicroRNA-200c delivered by solid lipid nanoparticles enhances the effect of paclitaxel on breast cancer stem cell. Int J Nanomedicine 11:6713–6725. https://doi.org/10.2147/IJN.S111647
Liu J, Xing Y, Rong L (2018) MiR-181 regulates cisplatin-resistant non-small cell lung cancer via downregulation of autophagy through the PTEN/PI3K/AKT pathway. Oncol Rep 39(4): 1631–1639. https://doi.org/10.3892/or.2018.6268
Liu J et al (2022) Wnt/β-catenin signalling: function, biological mechanisms, and therapeutic opportunities. Signal Transduct Target Ther. https://doi.org/10.1038/s41392-021-00762-6
Maiuri MC et al (2007) Self-eating and self-killing: crosstalk between autophagy and apoptosis. Nat Rev Mol Cell Biol 8(9):741–752. https://doi.org/10.1038/nrm2239
Makena MR et al (2020) Cancer stem cells: road to therapeutic resistance and strategies to overcome resistance. Biochim Biophys Acta Mol basis Dis 1866(4):165339. https://doi.org/10.1016/j.bbadis.2018.11.015
Manni W, Min W (2022) Signaling pathways in the regulation of cancer stem cells and associated targeted therapy. MedComm 3(4):1–34. https://doi.org/10.1002/mco2.176
Manuscript A (2015) NF-κB, an active player in human cancers. 2(9):823–830. https://doi.org/10.1158/2326-6066.CIR-14-0112.NF-
Micha R (2017) Challenges identifying efficacious miRNA therapeutics for cancer. Physiol Behav 176(1):100–106. https://doi.org/10.1177/0022146515594631.Marriage
MicroRNA-mediated autophagy and drug resistance in cancer.pdf (n.d.)
Miroshnichenko S, Patutina O (2019) Enhanced inhibition of tumorigenesis using combinations of miRNA-targeted therapeutics. Front Pharmacol 10(MAY):1–17. https://doi.org/10.3389/fphar.2019.00488
Mukherjee DD, Choudhury SD, Chakrabarti G (2020) Targeting autophagy in cancer: therapeutic implications. Autophagy Tumor Tumor Microenviron 17(9):249–264. https://doi.org/10.1007/978-981-15-6930-2_12
Niu T, Zhang W, Xiao W (2021) MicroRNA regulation of cancer stem cells in the pathogenesis of breast cancer. Cancer Cell Int 21(1):1–14. https://doi.org/10.1186/s12935-020-01716-8
Niwa H (2014) Mechanisms of stem cell self-renewal. In: Essentials of stem cell biology: third edition, 3rd edn. Elsevier Inc. https://doi.org/10.1016/B978-0-12-409503-8.00007-X
Nyhan MJ et al (2016) MiR-193b promotes autophagy and non-apoptotic cell death in oesophageal cancer cells. BMC Cancer 16(1):1–13. https://doi.org/10.1186/s12885-016-2123-6

Ortíz R et al (2021) Nanomedicine to overcome multidrug resistance mechanisms in colon and pancreatic cancer: recent progress. Cancer 13(9):1–26. https://doi.org/10.3390/cancers13092058

Ren WW et al (2018) MicroRNA-125b reverses oxaliplatin resistance in hepatocellular carcinoma by negatively regulating EVA1A mediated autophagy article. Cell Death Dis 9(5). https://doi.org/10.1038/s41419-018-0592-z

Richard J (2011) Challenges and opportunities in the delivery of cancer therapeutics. Ther Deliv 2(1):107–121. https://doi.org/10.4155/tde.10.89

Richard V, Kumar TRS, Pillai RM (2021) Transitional dynamics of cancer stem cells in invasion and metastasis. Transl Oncol 14(1):100909. https://doi.org/10.1016/j.tranon.2020.100909

Seo HA et al (2019) MicroRNA-based combinatorial cancer therapy: effects of MicroRNAs on the efficacy of anti-cancer therapies. Cells 9(1):29. https://doi.org/10.3390/cells9010029

Shah MY et al (2016) microRNA therapeutics in cancer — An emerging concept. EBioMedicine 12:34–42. https://doi.org/10.1016/j.ebiom.2016.09.017

Shan C et al (2021) The emerging roles of autophagy-related MicroRNAs in cancer. 17. https://doi.org/10.7150/ijbs.50773

Sharma P et al (2020) Nanomaterials for autophagy-related miRNA-34a delivery in cancer treatment. Front Pharmacol 11(July). https://doi.org/10.3389/fphar.2020.01141

Song P et al (2024) Wnt/β-catenin signaling pathway in carcinogenesis and cancer therapy. J Hematol Oncol 4:1–30

Sousa DP, Conde J (2022) Gold Nanoconjugates for miRNA modulation in cancer therapy: from miRNA silencing to miRNA mimics. ACS Mater Au 2(6):626–640. https://doi.org/10.1021/acsmaterialsau.2c00042

Stojcheva N et al (2016) MicroRNA-138 promotes acquired alkylator resistance in glioblastoma by targeting the Bcl-2-interacting mediator BIM. Oncotarget 7(11):12937–12950. https://doi.org/10.18632/oncotarget.7346

Su Z et al (2015) MicroRNAs in apoptosis, autophagy and necroptosis. Oncotarget 6(11):8474

Takahashi RU, Miyazaki H, Ochiya T (2013) The role of microRNAs in the regulation of cancer stem cells. Front Genet 4(JAN):1–11. https://doi.org/10.3389/fgene.2013.00295

Takahashi R et al (2014) The role of microRNAs in the regulation of cancer stem cells. 4 (January):1–11. https://doi.org/10.3389/fgene.2013.00295

Talukdar J et al (2023) Cancer stem cells: Signaling pathways and therapeutic targeting. MedComm Oncol. https://doi.org/10.1002/mog2.62

Taylor JA, Ca G (2020) (12) United States Patent, 2

Tenchov R et al (2021) Lipid nanoparticles from liposomes to mRNA vaccine delivery, a landscape of research diversity and advancement. ACS Nano 15(11):16982–17015. https://doi.org/10.1021/acsnano.1c04996

Vaghari-Tabari M et al (2022) CRISPR/Cas9 gene editing: a new approach for overcoming drug resistance in cancer. Cell Mol Biol Lett 27(1). https://doi.org/10.1186/s11658-022-00348-2

van Beijnum JR et al (2017) miRNAs: micro-managers of anticancer combination therapies. Angiogenesis 20(2):269–285. https://doi.org/10.1007/s10456-017-9545-x

Volpini L et al (2023) Advances in RNA cancer therapeutics: new insight into exosomes as miRNA delivery. Aspects Mol Med 1(January):100005. https://doi.org/10.1016/j.amolm.2023.100005

Wang Z et al (2013) MicroRNAs are involved in the self-renewal and differentiation of cancer stem cells. Nature Publishing Group, pp 1374–1380. https://doi.org/10.1038/aps.2013.134

Wang Z et al (2024) miRNA interplay: mechanisms and therapeutic interventions in cancer. MedComm Oncol 3(4). https://doi.org/10.1002/mog2.93

Warrier NM et al (2023) Understanding cancer stem cells and plasticity: towards better therapeutics. Eur J Cell Biol 102(2):151321. https://doi.org/10.1016/j.ejcb.2023.151321

Wei J et al (2024) MicroRNA - mediated autophagy and drug resistance in cancer: mechanisms and therapeutic strategies. Discov Oncol 2880. https://doi.org/10.1007/s12672-024-01525-9

White E, Lattime EC, Guo JY (2021) Autophagy regulates stress responses, metabolism, and anticancer immunity. Trends Cancer 7(8):778–789. https://doi.org/10.1016/j.trecan.2021.05.003

Xia P, Xu X (2015) PI3K/Akt/mTOR signaling pathway in cancer stem cells: from basic research to clinical application. Am J Cancer Res 5(5):1602–1609

Xu J et al (2012) MicroRNAs in autophagy and their emerging roles in crosstalk with apoptosis. Autophagy 8:873–882

Yadav UP et al (2020) Metabolic adaptations in cancer stem cells. Front Oncol 10(June). https://doi.org/10.3389/fonc.2020.01010

Yager TA (1988) United States patent: 5861366 United States patent : 5861366, New York

Yang (2017) Infantile hemangiomas: epidemiology, pathogenesis, clinical features, and complica … Seite 1 von 31. 10(9):1–8. https://doi.org/10.1158/1535-7163.MCT-11-0047.The

Yoshida K, Yamamoto Y, Ochiya T (2021) miRNA signaling networks in cancer stem cells. Regen Ther 17:1–7. https://doi.org/10.1016/j.reth.2021.01.004

Yun CW et al (2021) The dual role of autophagy in cancer development and a therapeutic strategy for cancer by targeting autophagy. Int J Mol Sci 22(1):1–22. https://doi.org/10.3390/ijms22010179

Zhang D et al (2016) Enrichment of selective miRNAs in exosomes and delivery of exosomal miRNAs in vitro and in vivo. Am J Physiol Lung Cell Mol Physiol 312(1):L110–L121. https://doi.org/10.1152/ajplung.00423.2016

Zheng Y (2018) MicroRNA roles in the nuclear factor Kappa B signaling pathway in cancer. 9 (March). https://doi.org/10.3389/fimmu.2018.00546

Zheng B et al (2015) MiRNA-30a-mediated autophagy inhibition sensitizes renal cell carcinoma cells to sorafenib. Biochem Biophys Res Commun 459(2):234–239. https://doi.org/10.1016/j.bbrc.2015.02.084

Zhu X et al (2017) IL-6R/STAT3/miR-204 feedback loop contributes to cisplatin resistance of epithelial ovarian cancer cells. Oncotarget 8(24):39154–39166. https://doi.org/10.18632/oncotarget.16610

Zou Z et al (2012) MicroRNA-30a sensitizes tumor cells to cis-platinum via suppressing beclin 1-mediated autophagy. J Biol Chem 287(6):4148–4156. https://doi.org/10.1074/jbc.M111.307405

# Development of Resistance in Tumor for the Available Conventional Therapies

8

Payal Sharma, Jahnavi Sharma, Bharti Nayak, Shikha Tiwari, and Ashutosh Kumar

**Abstract**

Resistance to conventional cancer therapies such as chemotherapy, radiotherapy, immunotherapy, and targeted therapy remains a major obstacle in effective treatment. Tumor cells employ diverse mechanisms to evade therapeutic effects, including drug efflux via ATP-binding cassette transporters, alterations in drug targets, increased DNA damage repair, inhibition of apoptosis, and activation of pro-survival signaling pathways. Additionally, tumor heterogeneity, epithelial to mesenchymal transition (EMT), cancer stem cells, and interactions within the tumor microenvironment contribute to both intrinsic and acquired resistance. These mechanisms not only reduce treatment efficacy but also promote relapse, metastasis, and poor patient outcomes. Understanding the molecular, and cellular basis of resistance is critical for designing novel therapeutic interventions, and combinatorial strategies. This chapter explains the pathways underlying resistance to conventional therapies, highlights the role of tumor plasticity, and discusses emerging approaches aimed at overcoming resistance to improve long-term cancer management.

P. Sharma
Centre of biotechnology, Nana ji Deshmukh Veterinary Science University, Jabalpur, India

J. Sharma
Department of Biochemistry, All India Institute of Medical Sciences, Bhopal, India

B. Nayak · S. Tiwari (✉)
School of Sciences, Sanjeev Agrawal Global Educational University, Bhopal, India

A. Kumar (✉)
Department of Electrical Engineering, University of Notre Dame, Notre Dame, IN, USA





S. Tiwari et al. (eds.), *microRNA Modulation: A Promising Combination Therapy for Cancer*, https://doi.org/10.1007/978-981-95-5989-3_8

**Keywords**

Tumor resistance · Conventional cancer therapies · miRNA-based combination therapy · Overcoming tumor resistance

**Learning Objectives**

1. The molecular and cellular mechanisms underlying tumor resistance to conventional cancer therapies
2. The role of tumor heterogeneity, EMT, and cancer stem cells in cancer therapeutic resistance
3. The contribution of tumor microenvironment to treatment failure and relapse
4. Emerging strategies designed to overcome conventional therapy resistance in cancer

## 8.1 Introduction

One of the biggest causes of death globally is still cancer, a complicated and multifaceted illness. Conventional therapies, such as chemotherapy, radiotherapy, and hormonal therapy, have been the mainstay of cancer treatment for the past few decades. Even while many patients' survival rates have increased dramatically as a result of these strategies, a serious problem still exists: the emergence of therapeutic resistance, which frequently results in treatment failure, disease recurrence, and a poor prognosis.

There are two types of resistance to conventional therapies: acquired resistance, which arises over time as a result of ongoing therapeutic exposure, and intrinsic resistance, in which cancer cells are non-responsive from the start (Holohan et al. 2013). This resistance is caused by a variety of biological processes, including metabolic reprogramming, increased drug efflux through membrane transporters, enhanced DNA repair mechanisms, genetic alterations in drug targets, and apoptosis (programmed cell death) (Al-Lazikani et al. 2012). Additionally, the tumor microenvironment (TME) can create a protective niche, aiding in protecting tumor cells from harm and enabling them to endure treatment-induced stress. This protection can be achieved by a number of methods, including reducing immunological attack, sending out survival signals, or changing the way medications are delivered and work (Meads et al. 2009). One of the major reasons is tumor heterogeneity. This signal tumor is not made up of identical cells but rather contains many different types of cancer cells, each with their own genetic mutations and physical or behavioral traits (this is called phenotypic diversity). This diversity can arise during tumor growth as cells mutate and evolve in response to their environment and treatment (Marusyk et al. 2012). The existence of cancer stem cells (CSCs) in tumors, a unique subset of cancer cells that exhibits characteristics of normal stem cells in the body, including the capacity to self-renew (create copies of themselves) and differentiate into different types of tumor cells. CSCs are believed to play a key role in initiating

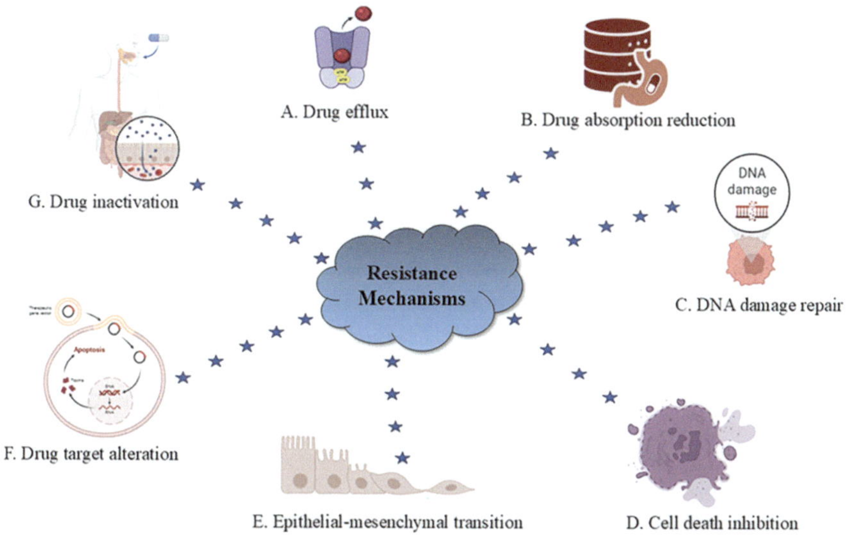


**Fig. 8.1** Molecular mechanism of tumor resistance- (**a**) Drugs are pumped out of the cell due to abnormal ABC family expression, which results in intracellular drug concentrations that are too low for drug sensitivity. (**b**) A reduction in drug absorption rate and chemoresistance results from a modification in the expression of transport proteins involved in drug absorption. (**c**) The rapid repair of DNA damaged by radiation and chemotherapy is intimately linked to the development of radioresistance and chemoresistance. (**d**) The inhibition of cell death suggests that the key gene families, including p53 and Bcl, are affecting the equilibrium between apoptosis and cell proliferation. (**e**) The development of cellular EMT characteristics results in resistance to radiation and chemotherapy. (**f**) Tumor cells can change the molecular target of medications. (**g**) Drug activation occurs when certain detoxification-related proteins render medicines inactive within cells. This is followed by the acquisition of chemoresistance

and sustaining the tumor, but they are also inherently more resistant to standard treatments like radiation and chemotherapy (Dean et al. 2005).

Resistance still exists despite the fresh hope that has been brought about by developments in targeted medicines and molecular oncology. To create more successful tactics, it is crucial to comprehend the common and unique resistance mechanisms in traditional treatments. Finding predictive biomarkers, creating combination treatments, and focusing on the adaptive processes behind resistance are the main goals of current research initiatives. To overcome treatment resistance, a multimodal strategy that incorporates basic research, translational studies, and clinical innovation is needed.

This chapter aims to explore the *mechanisms responsible for resistance to conventional cancer therapies and* assess their impact on clinical outcomes.

Molecular mechanism of tumour resistance is shown in Fig. 8.1.

## 8.2 Overview of Conventional Therapy

### 8.2.1 Types of Conventional Therapy

(a) *Chemotherapy*—Cytotoxic medications are used in chemotherapy to destroy cancer cells that divide quickly. These medications cause apoptosis (programmed cell death) or stop cell division by interfering with the DNA replication or mitotic processes of cells.
*Administration Methods*: Intravenous is the most common, but oral and localized techniques (such as intrathecal and intraperitoneal) are also employed.
*Applications:* Induction therapy, adjuvant, neoadjuvant, palliative, and curative.
*Limitations:* Causes adverse consequences (e.g., hair loss, myelosuppression, gastrointestinal toxicity) by affecting quickly dividing cells, both malignant and healthy. Drug resistance may emerge over time (DeVita and Chu 2008).

(b) *Radiotherapy*—Cytotoxic medications are used in chemotherapy to destroy cancer cells that divide quickly. These medications cause apoptosis (programmed cell death) or stop cell division by interfering with the DNA replication or mitotic processes of cells.
High-energy ionizing radiation, like X-rays or gamma rays, is used in radiotherapy to harm cancer cells' DNA, which results in their death or loss of ability to divide.
Delivery options include internal (brachytherapy) or external beam radiation therapy (EBRT).
*Precision:* More accurate targeting is possible with more recent methods such as stereotactic radiosurgery and intensity-modulated radiation therapy (IMRT).
*Use:* Good for palliative care to stop bleeding or pain, as well as for localized tumors.
*Limitations:* Over time, radiation exposure may result in secondary malignancies and harm nearby healthy tissues (Baskar et al. 2012).

(c) *Hormonal therapy*—Breast, prostate, and endometrial cancers are among the hormone-sensitive malignancies that are treated with hormonal therapy. It functions by preventing the synthesis of hormones or by interfering with the hormone receptors on cancerous cells (Osborne and Schiff 2011).
*Mechanism:* Drugs that block hormone-driven tumor growth include androgen deprivation therapy (ADT) for prostate cancer and tamoxifen (anti-estrogen).
*Timing:* As a maintenance, neoadjuvant, or adjuvant treatment.
*Limitations:* Only works on hormone-dependent tumors. Over time, resistance frequently develops, and adverse effects could include metabolic abnormalities, heat flushes, or bone loss (Harris et al. 2009).

### 8.2.2 Initial Success Rates and Long-Term Limitations

Conventional cancer treatments—such as chemotherapy, radiotherapy, and hormonal therapy—have played a crucial role in cancer management for decades. When initially used, several of these treatments have encouraging initial outcomes, particularly in early-stage malignancies or quickly dividing tumors. In many cases, patients report symptom improvement, tumor shrinking, and even brief remission. Conventional therapy has demonstrated good cure rates for testicular cancer, Hodgkin lymphoma, and juvenile leukemia.

Long-term results, however, are usually unsatisfactory, particularly in cases of advanced or metastatic malignancies, despite these early achievements. Many tumors recur over time, frequently in a more aggressive form that is resistant to treatment. The primary cause of this is the emergence of resistance, which can be acquired (formed following therapy exposure) or intrinsic (existed before treatment). Furthermore, the dose and duration of treatment are frequently limited due to nonspecific toxicity to healthy tissues, which results in serious side effects and reduces the quality of life.

Additionally, while not all cancer cells divide at the same speed, conventional therapies target those that divide quickly. Cancer stem cells and dormant tumor cells, for example, may withstand early therapy, stay undetected, and subsequently cause recurrence. Likewise, the tumor microenvironment and tumor heterogeneity give cancer cells additional defense, which permits some of them to escape and proliferate.

Therefore, even if traditional treatments are still vital and have the potential to save lives, their long-term drawbacks underscore how critical it is to comprehend and get past resistance mechanisms in order to increase patient survival and treatment benefit.

## 8.3 Types of Therapeutic Resistance

Therapeutic resistance in cancer refers to the ability of tumor cells to evade or withstand the effects of anticancer treatments, including chemotherapy, targeted therapy, and other modalities. This resistance arises through multiple mechanisms, such as drug delivery barriers, genetic mutations, cancer stem cells, tumor heterogeneity, epithelial–mesenchymal transition (EMT), and interactions with the tumor microenvironment (TME), ultimately leading to reduced treatment effectiveness and disease progression (Wei et al. 2025). Cancer cells may have the innate capacity to endure clinically relevant drug doses due to intrinsic resistance, or after initially reacting to a therapeutic treatment (such as chemotherapy, targeted therapy, or immunotherapy), cancer cells may eventually acquire the capacity to avoid or tolerate the effects of that treatment with acquired resistance.

Genetic mutations, epigenetic modifications, changes in cellular signaling pathways, or interactions with the tumor microenvironment can all cause it (Soragni et al. 2025). One of the most common causes of resistance to targeted cancer therapy is when mutations occur in the target molecule itself, such as a receptor or enzyme that the drug is meant to block. These mutations change the structure of the target, making it harder for the drug to bind and do its job. This kind of mutation-driven resistance also affects therapies targeting EGFR, FLT3, FGFR, SRC, and MEK1. In some cases, resistance also arises due to gene amplification (an increase in the number of copies of the target gene), which leads to overproduction of the target protein, making the drug less effective (Ramos and Bentires-Alj 2015). Another prominent mechanism by which tumors develop resistance to targeted therapy—beyond mutations in the drug target itself—is the reactivation of the oncogenic signalling pathway either upstream or downstream of the inhibited target. This leads to continued pathway activation and tumor survival, rendering the therapy ineffective (Ramos and Bentires-Alj 2015). A well-characterized example of this is observed in HER2-amplified breast cancer, where patients initially respond well to targeted therapies such as trastuzumab or lapatinib (Nahta et al. 2006).

## 8.4 Molecular Mechanisms of Cancer Chemotherapy Resistance

### 8.4.1 Drug Efflux Transporters and Multidrug Resistance

The upregulation of drug efflux transporters, which aggressively pump chemotherapeutic drugs out of cancer cells, lowering intracellular drug accumulation and cytotoxicity, is one of the most well-studied mechanisms of chemotherapy resistance. Most of these transporters belong to the ATP-binding cassette (ABC) transporter family, which uses the energy from ATP hydrolysis to move a variety of substrates across the plasma membrane, including chemotherapeutic drugs. ABC transporters like P-glycoprotein (P-gp, encoded by ABCB1/MDR1), Multidrug Resistance-Associated Proteins (MRPs, e.g., ABCC1), and Breast Cancer Resistance Protein (BCRP, encoded by ABCG2) are often implicated in the phenomenon of multidrug resistance (MDR), in which tumor cells develop resistance to structurally and functionally unrelated drugs (Gottesman 2002).

#### 8.4.1.1 P-glycoprotein and Broad-Spectrum Drug Resistance

P-glycoprotein (P-gp) is the most well-characterized of the efflux transporters. The 170-kDa membrane glycoprotein P-gp provides resistance to a variety of chemotherapeutic medications, such as taxanes (paclitaxel, docetaxel), anthracyclines (doxorubicin, daunorubicin), vinca alkaloids (vincristine, vinblastine), and epipodophyllotoxins (etoposide). Numerous cancer types, including leukemia, breast, ovarian, lung, and gastrointestinal malignancies, have been shown to overexpress P-gp. P-gp lowers drug retention in tumor cells below cytotoxic levels by binding to lipophilic medicines within the plasma membrane and actively extruding them into the extracellular environment. This poses a serious clinical problem since cancers with elevated P-gp frequently respond poorly to chemotherapy and have higher recurrence rates (Gottesman 2002; Luqmani 2005).

#### 8.4.1.2 Multidrug Resistance-Associated Proteins (MRPs)

Apart from P-gp, chemotherapy resistance is largely influenced by the MRP family. Because MRPs preferentially transport medicines that are glutathione-, glucuronate-, or sulfate-conjugated, their substrate specificity varies significantly. For example, by exporting medications in conjugated forms, MRP1 (ABCC1) provides resistance to medications like doxorubicin and etoposide. The transport of physiological substrates, including leukotrienes, prostaglandins, and cyclic nucleotides, is also linked to MRPs, which is intriguing since it suggests that MRPs' involvement in cancer goes beyond drug resistance to include altering tumor biology and microenvironmental signalling (Luqmani 2005). MRP overexpression has been documented in glioblastoma, acute myeloid leukemia, and non-small cell lung cancer, where it dramatically diminishes the effectiveness of conventional chemotherapies (Elfadadny et al. 2021).

#### 8.4.1.3 Breast Cancer Resistance Protein (BCRP/ABCG2)

Another important ABC transporter, BCRP, has been extensively linked to malignancies' resistance to tyrosine kinase inhibitors, mitoxantrone, topotecan, and irinotecan. In contrast to P-gp and MRPs, BCRP is frequently linked to resistance in populations of cancer cells that resemble stem cells, where it helps to preserve the phenotype of cancer stem cells. Poor clinical outcomes have been linked to the presence of BCRP in gastrointestinal cancers, breast cancer, hepatocellular carcinoma, and hematological malignancies. Crucially, BCRP is expressed in the blood–brain barrier, which prevents many anticancer medications from entering the central nervous system and prevents brain cancers and metastases from being effectively treated (Chen et al. 2024). Studies showing that individuals whose malignancies overexpress P-gp or BCRP frequently have lower progression-free survival and overall survival after chemotherapy underscore the therapeutic importance of efflux transporters (Pilotto Heming et al. 2022).

#### 8.4.1.4 Therapeutic Strategies to Overcome Efflux-Mediated Resistance

Efflux transporters are appealing candidates for therapeutic intervention because of their important function. Clinical trials have examined a number of P-gp inhibitors to restore chemosensitivity, including tariquidar, verapamil, and cyclosporin A. Newer generations of inhibitors have demonstrated better selectivity and tolerance, whereas first-generation inhibitors, such as verapamil, were constrained by toxicity at effective concentrations. Likewise, preclinical efficacy in overcoming resistance to topoisomerase inhibitors has been shown for BCRP inhibitors like Ko143. It has been difficult to implement these inhibitors in standard clinical practice, nevertheless, because many medications that target efflux pumps also block transporters in healthy tissues, which can result in altered pharmacokinetics and systemic toxicities. Alternative tactics to direct inhibition include drug delivery systems based on nanoparticles, which can get around efflux transporters by promoting the intracellular accumulation of chemotherapeutics. For instance, by avoiding P-gp-mediated drug extrusion, liposomal doxorubicin (Doxil) and other nanocarrier systems have demonstrated effectiveness in preclinical and clinical contexts (Xue and Liang

2012). Additionally, efflux pump inhibitors are now being investigated in combination with immune checkpoint blockade or targeted therapy as a way to reduce resistance and increase antitumor activity (Robey et al. 2018).

## 8.4.2 Epigenetic Modifications in Chemotherapy Resistance

One of the main regulatory layers in cancer biology is epigenetic modifications, which are heritable shifts in gene expression that take place without changes to the DNA sequence. Epigenetic modifications are reversible and therefore responsive to therapeutic intervention, in contrast to genetic mutations that permanently affect DNA. DNA methylation, histone modifications, and noncoding RNA-mediated control are the three main types of epigenetic regulation. Collectively, these processes have a significant impact on transcriptional programs, modify chromatin structure, and control the activity of pathways related to drug metabolism, DNA repair, apoptosis, and cell survival. Aberrant epigenetic modifications in chemotherapy resistance can alter the cellular response to cytotoxic medicines by either activating oncogenic signaling pathways or silencing tumor suppressor genes. Each of these pathways is thoroughly examined in this section, with an emphasis on how they affect drug resistance and potential treatments (Song et al. 2025).

### 8.4.2.1 DNA Methylation and Chemotherapy Resistance

One of the most researched epigenetic changes is DNA methylation. DNA methyltransferases (DNMTs) catalyze the addition of methyl groups to cytosine residues in CpG dinucleotides. DNA methylation patterns in healthy cells control tissue-specific gene expression and preserve genomic stability. However, a dual environment that promotes carcinogenesis is created in cancer by the coexistence of global hypomethylation and gene-specific hypermethylation. Chemotherapy-resistant cancers frequently exhibit hypermethylation of tumor suppressor genes. For instance, there is a correlation between decreased susceptibility to platinum-based medicines and BRCA1 hypermethylation in ovarian cancer cells. Interestingly, BRCA1 silencing through methylation paradoxically enhances platinum sensitivity, but BRCA1 deletion by mutation confers resistance to platinum therapy (Stefansson et al. 2012). This suggests that the connection between methylation and drug responsiveness is context-dependent. Likewise, hypermethylation of TGF-β receptor I (TGF-βRI) in gastric cancer promotes resistance by decreasing responsiveness to TGF-β signaling (Kang et al. 1999).

On the other hand, global hypomethylation can activate genes linked to drug metabolism and oncogenes, making treatment outcomes even more challenging. Because DNA methylation is reversible, attempts have been made to restore chemosensitivity in resistant tumors by using DNA demethylating adjuvants, such as azacitidine and decitabine (Nervi et al. 2015). Clinical studies in myelodysplastic syndromes, acute myeloid leukemia, and some solid malignancies have demonstrated promising outcomes when demethylating drugs are used in conjunction with traditional chemotherapy (Linnekamp et al. 2017).

#### 8.4.2.2 Histone Modifications and Chromatin Remodeling

Histone alterations have a major influence on gene expression in addition to DNA methylation. Acetylation, methylation, phosphorylation, ubiquitination, and sumoylation are examples of posttranslational changes that occur in histone proteins, which serve as the structural foundation of nucleosomes. These chromatin conformational changes result in either a compacted, repressive state (heterochromatin) or an open, transcriptionally active state (euchromatin). Acetylation is one of the most significant histone changes in chemotherapy resistance. Histone acetyltransferases (HATs) mediate this process, which relaxes chromatin and encourages transcriptional activation. Histone deacetylases (HDACs), on the other hand, remove acetyl groups, which suppresses transcription. Drug resistance has been directly associated with the dysregulation of HATs and HDACs. For example, because p300-driven acetylation increases the transcription of genes involved in survival signaling, p300 overexpression has been linked to gemcitabine resistance in pancreatic cancer. Similarly, HDAC overexpression has been shown in melanoma and ovarian cancer, where it increases chemoresistance and inhibits tumor suppressor pathways (Ono et al. 2016).

Histone deacetylase inhibitors (HDACis), including belinostat, vorinostat, and panobinostat, have demonstrated the ability to resensitize tumors to chemotherapy, underscoring the therapeutic potential of targeting histone alterations. HDACis can reawaken tumor suppressor genes that have been silenced and make resistant cancer cells more susceptible to cytotoxic treatments by restoring a transcriptionally permissive chromatin state (Ono et al. 2016).

Another important regulator is histone methylation. Cisplatin-resistant ovarian cancer cells overexpress the histone methyltransferase enzyme Enhancer of Zeste Homolog 2 (EZH2), which catalyzes H3K27 trimethylation. By suppressing DNA repair inhibitors, EZH2 activity increases DNA repair capacity and decreases the effectiveness of cisplatin. It has been demonstrated that chemosensitivity can be restored by inhibiting EZH2 with small-molecule inhibitors or genetic knockdown. Histone methyltransferases have been found to control cancer stemness, migration, and the epithelial–mesenchymal transition (EMT) in lung cancer, all of which lead to resistance (Bandyopadhyay et al. 2004; Kim et al. 2012).

#### 8.4.2.3 Therapeutic Potential of Epigenetic Interventions

Because epigenetic modifications are reversible, they are desirable targets for therapy. DNMT inhibitors, HDAC inhibitors, EZH2 inhibitors, and ncRNA-based therapies are examples of contemporary strategies. For instance, by restoring hypermethylation-induced resistance, the combination of decitabine and platinum drugs has demonstrated encouraging outcomes in preclinical ovarian cancer models. Similarly, HDAC inhibitors in combination with cytotoxic medications are being tested in clinical trials for a number of cancers, such as gastrointestinal tumors, lung cancer, and lymphomas (Fang et al. 2010).

### 8.4.3 Noncoding RNAs and Chemoresistance

Noncoding RNAs (ncRNAs), in particular long noncoding RNAs (lncRNAs) and microRNAs (miRNAs), have become important epigenetic regulators of drug resistance in recent years. Small, ~22 nucleotide RNAs known as microRNAs (miRNAs) control the expression of genes after transcription by attaching to complementary regions in target mRNAs and causing translational repression or mRNA destruction. Dysregulated miRNA expression modifies important pathways such as apoptosis, drug metabolism, DNA repair, and efflux transport in the context of chemotherapy resistance. For instance, miRNAs control drug resistance genes in gallbladder cancer; some miRNAs suppress apoptotic mediators and increase survival when chemotherapeutics are present (Pavlíková et al. 2022; Rana et al. 2021).

Extending beyond 200 nucleotides, long noncoding RNAs (lncRNAs) exercise regulatory control via a variety of mechanisms, including miRNA sponging, transcriptional regulation, and chromatin remodeling. By encouraging autophagy, which shields cells from drug-induced apoptosis, the lncRNA GBCDRlnc1 has been linked to doxorubicin resistance in gallbladder cancer. The intricacy of RNA-mediated networks in chemoresistance is further demonstrated by the way the lncRNA HOTTIP influences chemotherapy sensitivity in cholangiocarcinoma via the HOTTIP/miR-637/LASP1 axis (Cai et al. 2019; Gao et al. 2021).

Although ncRNA-targeting therapies are still in their infancy, methods to modify dysregulated miRNA expression are being explored, including antagomirs (anti-miRNA oligonucleotides) and miRNA mimics. Another frontier is represented by LncRNAs, whose suppression may help drug-resistant tumors become more sensitive to chemotherapy, according to new research (Rana et al. 2021).

### 8.4.4 Altered Drug Metabolism and Detoxification Pathways

The capacity of cancer cells to metabolize and detoxify cytotoxic chemicals is a critical factor in determining the effectiveness of chemotherapy. Like the majority of xenobiotics, chemotherapy medications are enzymatically modified in cells, mostly via phase I and phase II metabolic pathways, and then eliminated through phase III transport processes. Although these pathways typically serve to shield healthy tissues from toxic substances, cancer cells have the ability to take advantage of them to neutralize or inactivate anticancer medications, which can result in treatment failure. Chemotherapy resistance in a variety of tumor types has been closely associated with changes in drug-metabolizing enzymes, such as cytochrome P450s (CYPs), glutathione S-transferases (GSTs), uridine 5′-diphospho-glucuronosyltransferases (UGTs), and aldehyde dehydrogenases (ALDHs) (Verma et al. 2019).

#### 8.4.4.1 Phase I Enzymes: Cytochrome P450-mediated Resistance

Drugs undergo oxidative, reductive, or hydrolytic modification during phase I metabolism, with the cytochrome P450 (CYP) family being a key player. CYP enzymes change the solubility and action of drugs by adding polar or reactive groups to them. Overexpression of CYP enzymes in cancer can improve medication clearance or change chemotherapeutics into less harmful byproducts. For example, paclitaxel, docetaxel, and vincristine are among the many chemotherapeutics that are metabolized by CYP3A4, one of the most prevalent CYP enzymes. In some malignancies, CYP3A4 overexpression lowers drug accumulation to levels that are not fatal, allowing tumor cells to withstand treatment. Similar to this, CYP1B1 has been linked to resistance to cisplatin and docetaxel through its effects on apoptotic and survival pathways, as well as drug metabolism (Hofman et al. 2021).

Most importantly, CYP1B1 overexpression has been found in a variety of solid tumors, including lung, breast, ovarian, and prostate malignancies, suggesting that it may be a biomarker of resistance. By blocking medication metabolism and restoring cytotoxic activity, CYP inhibitors including fluvoxamine, ritonavir, and ketoconazole have been used to try to target CYP-mediated resistance. However, systemic inhibition of CYP enzymes can complicate clinical translation by interfering with normal xenobiotic metabolism and causing drug–drug interactions (Martinez et al. 2008; Zhu et al. 2015).

#### 8.4.4.2 Phase II Enzymes: Detoxification by Conjugation

Conjugation processes that improve drug solubility and ease excretion are a part of phase II metabolism. Glutathione S-transferases (GSTs) are one of the most significant phase II enzymes in drug resistance. These enzymes counteract the cytotoxic effects of reduced glutathione (GSH) by conjugating it to electrophilic drug intermediates.

Resistance to platinum medications (cisplatin, carboplatin), anthracyclines, and alkylating medicines (cyclophosphamide, melphalan) has been closely linked to increased GST activity. For instance, because GST-mediated conjugation makes it easier to detoxify and remove platinum–DNA adducts, increased GSTP1 expression is associated with cisplatin resistance in ovarian and lung malignancies. Furthermore, by altering the intracellular redox state, GST activity can indirectly control apoptosis, thereby improving tumor survival (Sawers et al. 2014).

Uridine diphosphate glucuronosyltransferases (UGTs) and other phase II enzymes conjugate medications with glucuronic acid to improve their solubility and excretion. Resistance to medications like irinotecan (via SN-38 glucuronidation) and epirubicin is facilitated by UGT-mediated metabolism. In colorectal and liver tumors, elevated UGT expression has been shown to decrease effective drug exposure (Oguri et al. 2004).

#### 8.4.4.3 Aldehyde Dehydrogenases (ALDHs) and Cancer Stem Cell-Associated Resistance

The dual function of aldehyde dehydrogenases (ALDHs) in drug metabolism and the control of cancer stem cell (CSC) phenotypes makes them stand out among detoxifying enzymes. ALDHs detoxify reactive intermediates by oxidizing intracellular aldehydes to carboxylic acids. CSC populations have been shown to exhibit elevated ALDH activity, which imparts resistance to chemotherapeutics such as doxorubicin and cyclophosphamide.

For instance, ALDH1A1 expression identifies chemoresistant CSCs in lung, ovarian, and breast malignancies, where it shields cells from apoptosis and oxidative damage brought on by drugs. Additionally, poor clinical outcomes, increased metastasis, and recurrence have been linked to high ALDH activity, highlighting its dual significance as a prognostic biomarker and a functional modulator of resistance.

One possible strategy for reducing metabolic resistance is to target ALDH activity, either by disrupting CSC habitats or by using pharmacologic inhibitors (like disulfiram) (Landen et al. 2010; Varisli et al. 2025).

#### 8.4.4.4 Glutathione (GSH) Homeostasis and Redox-Mediated Resistance

The function of glutathione (GSH), a tripeptide that preserves cellular redox balance, is intimately related to GST-mediated detoxification. Drug-resistant cancer cells are characterized by elevated intracellular GSH levels. GSH shields tumor cells from apoptosis by neutralizing reactive oxygen species (ROS) produced during treatment and offering reducing equivalents for GST conjugation. For example, resistance to doxorubicin and cisplatin has been linked to elevated GSH levels in lung cancer.

Preclinical results have demonstrated that GSH depletion with drugs such as buthionine sulfoximine (BSO) can restore chemosensitivity; however, systemic toxicity is still a problem (Meijer et al. 1990).

### 8.4.5 DNA Damage Response and Repair Mechanisms in Chemotherapy Resistance

DNA damage is how many frontline chemotherapeutic medicines, such as antimetabolites, topoisomerase inhibitors, alkylating agents, and platinum medications, cause their cytotoxic effects. Single-strand breaks (SSBs), double-strand breaks (DSBs), interstrand crosslinks (ICLs), and bulky DNA adducts are among the DNA lesions produced; these affect transcription and replication, ultimately leading to apoptosis. However, by triggering strong DNA damage response (DDR) and repair mechanisms, cancer cells frequently avoid this fatal consequence and become resistant to chemotherapy treatments that damage DNA (Ya et al. 2021).

Damage detection, signal transduction, and the hiring of repair equipment are all part of the highly coordinated DDR network. DDR abnormalities contribute to genomic instability and tumor growth in addition to shielding cancer cells from drug-induced cytotoxicity. Nucleotide excision repair (NER), mismatch repair (MMR), homologous recombination repair (HRR), non-homologous end joining (NHEJ), and translesion DNA synthesis (TLS) are important processes implicated in chemoresistance (Li et al. 2023) .

### 8.4.6 Evasion of Apoptosis and Cell Death Pathways in Chemotherapy Resistance

Programmed cell death, also known as apoptosis, is a strictly controlled biological process that gets rid of injured or aberrant cells and preserves tissue homeostasis. The majority of chemotherapeutic drugs ultimately cause cytotoxicity by activating apoptotic signaling cascades, whether through causing oxidative stress, DNA damage, or disruption of mitosis. However, one of the most important reasons for chemotherapy resistance is the frequent acquisition of mutations or adaptive alterations by cancer cells that enable them to avoid apoptosis.

The intrinsic (mitochondrial) and extrinsic (death receptor-mediated) pathways are the two main mechanisms that regulate the apoptotic machinery (Tian et al. 2024).

### 8.4.7 Tumor Microenvironment-Mediated Chemotherapy Resistance

It is becoming more well acknowledged that the tumor microenvironment (TME) plays a crucial role in determining the course, metastasis, and responsiveness to treatment of cancer. Instead of serving as a passive background, the TME is a dynamic ecosystem that actively shapes tumor activity. It is made up of fibroblasts, endothelial cells, stromal cells, immune cells, extracellular matrix (ECM), and soluble signaling chemicals. Crucially, interactions between cancer cells and the TME either directly shield tumor cells from cytotoxic damage or indirectly promote adaptive survival mechanisms, which both greatly increase chemotherapy resistance.

In addition to killing tumor cells, chemotherapy also significantly changes the TME, sometimes in ways that ironically increase resistance. Tumor survival under therapeutic stress is made possible by a protective niche that is created by a number of processes, including angiogenesis, hypoxia, soluble factor secretion, cell–cell interactions, and extracellular matrix remodeling (Senthebane et al. 2017) (Fig. 8.2).

#### 8.4.7.1 Role of Cancer-Associated Fibroblasts (CAFs)

One of the most prevalent stromal cell types in the TME, cancer-associated fibroblasts (CAFs), are crucial in fostering chemoresistance. CAFs influence cancer cell survival and treatment responsiveness by secreting growth factors, cytokines, and extracellular vesicles. For instance, CAF-derived hepatocyte growth factor (HGF) causes tumor cells to become resistant to EGFR inhibitors and chemicals that damage DNA by activating the MET signaling pathway. Similarly, the STAT3 and PI3K/AKT pathways are activated by CAF-secreted IL-6 and CXCL12, preventing tumor cells from undergoing apoptosis in response to doxorubicin and cisplatin. In addition to paracrine signaling, CAFs produce collagen and fibronectin, which stiffen tissue and prevent drugs from penetrating tumors. One known obstacle to gemcitabine administration in pancreatic ductal adenocarcinoma (PDAC) is the thick fibrotic stroma created by CAFs. Although completely removing CAFs may

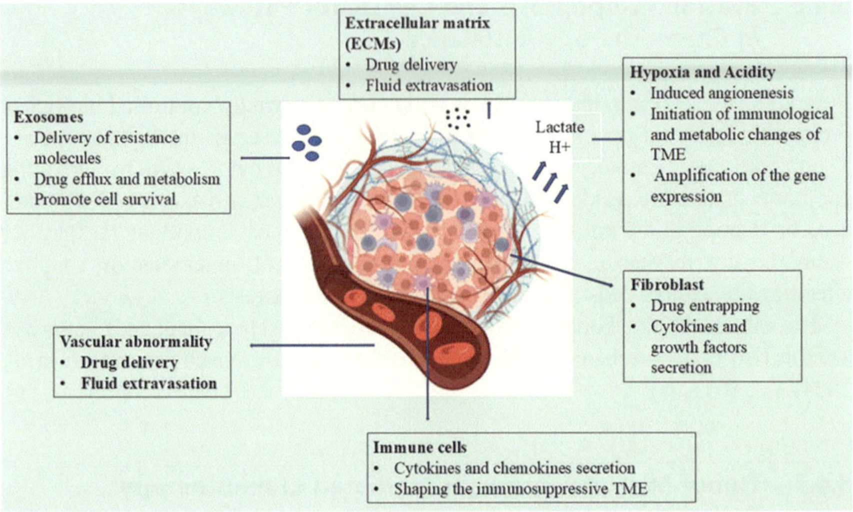


**Fig. 8.2** CAFs, immune cells, the vascular system, extracellular matrix, exosomes, hypoxia, and acidity are the primary adaptive mechanisms for therapeutic resistance supported by the TME (tumor microenvironment)

have unanticipated tumor-promoting effects, targeting them or their signaling pathways has been suggested as a technique to improve chemotherapy efficacy (Kittirat et al. 2022).

#### 8.4.7.2 Hypoxia and the Hypoxia-Inducible Factor (HIF) Pathway

Chemotherapy resistance is significantly impacted by hypoxia, a typical characteristic of solid tumors caused by disordered vasculature. Hypoxia-inducible factors (HIF-1α and HIF-2α) are stabilized by oxygen deprivation and subsequently control the transcription of genes related to angiogenesis, glycolysis, drug efflux, and survival signaling. P-glycoprotein (MDR1/ABCB1) and other ABC transporters are upregulated in response to hypoxia, which improves drug efflux and lowers the intracellular concentration of cytotoxic drugs like doxorubicin and paclitaxel. Furthermore, hypoxic circumstances protect tumor cells from apoptosis by activating pro-survival pathways such as NF–κB and PI3K/AKT/mTOR. Crucially, hypoxia also affects the ability to repair DNA. For example, it has been demonstrated that HIF-1α increases RAD51, a crucial mediator of homologous recombination, providing resistance to substances that damage DNA, such as cisplatin (141). HIF inhibitors and hypoxia-activated prodrugs, such as tirapazamine, are being investigated therapeutically to tackle this resistance mechanism (Altwerger et al. 2024; Liu et al. 2020).

#### 8.4.7.3 Extracellular Matrix (ECM) and Physical Barriers

When it comes to drug distribution, the extracellular matrix (ECM) creates a biochemical and physical barrier. Increased interstitial fluid pressure due to high quantities of collagen, hyaluronan, and proteoglycans inhibits medication penetration into the tumor core. Furthermore, ECM proteins can activate pathways like FAK, SRC, and PI3K/AKT, which provide survival benefits under chemotherapy-induced stress, by engaging integrin-mediated signaling in tumor cells. Integrin β1 has been linked to doxorubicin and paclitaxel resistance in breast cancer. Therefore, one intriguing tactic to improve chemotherapeutic sensitivity is to interfere with ECM–integrin interactions (Pang et al. 2023; Liu et al. 2023).

### 8.4.8 Epithelial–Mesenchymal Transition (EMT) and Stemness-Associated Chemotherapy Resistance

A dynamic biological mechanism known as the epithelial–mesenchymal transition (EMT) causes epithelial cells to lose their polarity and adhesion traits while gaining mesenchymal traits like motility, invasiveness, and increased survival. In addition to being crucial for tumor invasion and metastasis, EMT is closely related to treatment resistance. The development of characteristics similar to those of cancer stem cells (CSCs), such as self-renewal, flexibility, and treatment resistance, is a closely linked phenomenon. Following chemotherapy, EMT and stemness work together to increase the risk of tumor recurrence and minimum residual disease (Mitra et al. 2015).

Mesenchymal markers including vimentin, fibronectin, and N-cadherin are elevated during EMT, but epithelial markers like E-cadherin are downregulated. Transcription factors including as SNAIL, SLUG, TWIST, and ZEB1/2 are responsible for this phenotypic change. The inhibition of apoptosis is one way that EMT imparts drug resistance. It has been demonstrated that EMT-inducing transcription factors upregulate survival factors like BCL-2 and MCL-1 while downregulating pro-apoptotic proteins like BAX and BIM. Overexpression of TWIST is linked to doxorubicin and paclitaxel resistance in breast cancer. Similarly, in pancreatic cancer, ZEB1-driven EMT activates survival signaling via the PI3K/AKT pathway, promoting resistance to gemcitabine. Improved DNA damage repair is another mechanism. The expression of homologous recombination proteins, like RAD51, which lessen chemotherapy-induced DNA damage, is frequently elevated in EMT cells. Because of this, mesenchymal-like cells are more able to withstand genotoxic substances like platinum medicines than their epithelial counterparts. Additionally, EMT triggers the development of drug efflux pumps, such as the MDR1/ABCB1 and ABCG2 transporters, which actively eliminate harmful substances from cells (Jiang et al. 2025). This characteristic, which is also present in CSCs, emphasizes how EMT and stemness-associated resistance overlap (Tanabe et al. 2020).

### 8.4.9 Genetic Responses and Resistance to Chemotherapy

Genetic changes are another way for cancer cells to resist chemotherapy, which allows them to survive drug exposure. For example, drugs like methotrexate work by blocking enzymes involved in cell growth and division. However, if cancer cells produce more of the enzyme than usual—through a process called gene amplification—the drug becomes less effective because it cannot block all of the excess enzyme. Gene amplification happens when a specific DNA region is copied multiple times, leading to overproduction of the target protein and reducing the drug's ability to inhibit cell growth.

In addition to these mechanisms, defects in the apoptotic (cell death) pathway also contribute to resistance. The p53 protein plays a central role in detecting DNA damage and triggering cell cycle arrest or programmed cell death (apoptosis). Many anticancer drugs work by damaging DNA, which would normally activate p53 to kill the cell. However, if p53 is mutated or lost, which is common in various cancers, damaged cells can continue to divide, making them resistant to DNA-damaging therapies. Studies have shown that the loss or inactivation of p53 is associated with increased drug resistance, reduced apoptosis, and enhanced genomic instability—all of which allow cancer cells to survive and evolve.

Other genes involved in cell survival and death, such as h-ras, bcl-2, and bax, also play roles in drug resistance. When these genes are altered, they can reduce the ability of drugs to trigger apoptosis. This type of resistance can make cancer cells less responsive to a wide range of anticancer drugs, not just one type. Over time, this promotes the survival of mutant, drug-resistant cells, leading to greater tumor heterogeneity and making treatment even more difficult (Luqmani 2005).

## 8.5 Molecular Mechanisms of Cancer Radioresistance

### 8.5.1 Radiotherapy, DNA Damage, and Cancer Cell Resistance

Genotoxic cancer therapy, such as radiotherapy (RT), works by causing severe DNA damage in cancer cells, leading to their death. RT is one of the most common treatments in oncology. It harms DNA in two ways: directly, by radiation energy damaging the DNA itself, and indirectly, by creating harmful free radicals from surrounding molecules that then damage the DNA (on behalf of the NCRI CTRad Academia-Pharma Joint Working Group et al. 2016; Azzam et al. 2012). This process leads to various types of DNA injuries, such as single- and double-strand breaks (DSBs), base damage, and interstrand crosslinks, with DSBs being the most dangerous. While such damage is often lethal to cancer cells, they can survive by activating complex DNA repair mechanisms, including base excision repair, nucleotide excision repair, and especially DSB repair through two main pathways: non-homologous end joining (NHEJ) and homologous recombination (HR). These pathways are tightly regulated by DNA damage checkpoints, which delay cell division to give cells time to repair the damage (Li et al. 2019). However, cancer cells can become resistant to radiotherapy by enhancing their DNA repair abilities. Specialized proteins and complexes such as ATRIP, Rad24p, γH2AX, NBS1, BRCA1/2, Ku70/80, and RNA polymerase detect DNA damage and activate important enzymes like ATM, ATR, and DNA-PK. These, in turn, start a chain of reactions that repair the damaged DNA (Khanna and Jackson 2001; Fojta et al. 2016). For example, in rectal cancer, the Mre11-Rad50-Nbs1 (MRN) complex is often overexpressed, making these cancer cells better at repairing damage and thus more resistant to radiation. Similarly, increased levels of Ku70/80, which are part of the NHEJ repair system, are linked to poor outcomes and radioresistance (Zhang et al. 2016).

### 8.5.2 Cell Cycle Arrest

The cell cycle is the process by which cells grow and divide, and it includes four main phases: G1 (growth before DNA replication), S (DNA synthesis), G2 (preparation for division), and M (mitosis) (Fu et al. 2022). Two important checkpoints, at the G1/S and G2/M phases, act like security checks that can pause the cycle if the cell is damaged (Luserna et al. 2017). When cancer cells are exposed to ionizing radiation, these checkpoints help the cells pause their cycle to repair DNA damage, a keyway cancer cells become resistant to radiotherapy. In normal cells, the G1/S checkpoint is regulated by a protein called p53, which stops the cell from entering the DNA replication phase if there's damage. This gives the cell time to activate DNA repair mechanisms. However, many tumor cells have defective or missing p53, so they skip this checkpoint and continue to multiply, even when their DNA is damaged. Interestingly, when radiation causes DNA damage in cancer cells, it can also trigger the production of an enzyme called Caspase-activated DNase (CAD). CAD contributes to G2 phase arrest, meaning the cells pause in the phase just before mitosis (Larsen et al. 2022). This delay gives cancer cells more time to repair their DNA and survive the effects of radiotherapy.

### 8.5.3 Epithelial–Mesenchymal Transition

Epithelial–mesenchymal transition (EMT) is a process where cancer cells lose their original epithelial characteristics, such as tight cell-to-cell connections, and gain mesenchymal traits that make them more mobile and invasive (Ribatti et al. 2020). This change helps tumors grow, spread (metastases), and also makes them more resistant to radiotherapy. One key feature of EMT is the loss of E-cadherin, a protein that keeps cells attached to each other. The proteins Snail and Slug, which are regulated by specific genes, help reduce E-cadherin levels and promote EMT (Lamouille et al. 2014). Studies have shown that radiation activates signaling pathways like Wnt/β-catenin, which trigger EMT in cancers such as esophageal squamous cell carcinoma (ESCC) and colorectal cancer (CRC). In radiation-resistant ESCC cells (e.g., KYSE-150RR), radiation was found to reduce levels of PTEN, a tumor suppressor gene, and activate the Akt/Snail pathway, further pushing cells into EMT. Normally, PTEN helps stop tumors from growing and spreading by controlling another pathway called PI3K, and when PTEN is lost, cells become more resistant to radiation (He et al. 2015). In nasopharyngeal carcinoma (NPC), cancer cells that survived radiation not only became more resistant to further radiation but also to chemotherapy drugs like paclitaxel and cisplatin. These cells also showed lower levels of E-cadherin and higher levels of vimentin, another marker of EMT. Similar patterns have been observed in other cancers. For example, non-small-cell lung cancer (NSCLC) and prostate cancer cells that survive radiation often develop EMT traits, which makes treatment more difficult (Stark et al. 2017). In head and neck squamous cell carcinoma (HNSCC), EMT is linked to cells with high CD44 and low EGFR levels—markers that reduce the effectiveness of radiation therapy.

### 8.5.4 Autophagy

Autophagy is a natural process in cells that helps remove damaged components and recycle proteins to maintain cellular health (Nam et al. 2013). Under stress conditions, such as exposure to radiation, autophagy becomes more active and helps cancer cells survive by providing them with energy and protecting them from damage. This survival mechanism can limit the effectiveness of radiotherapy (Chaachouay et al. 2011). For example, studies have shown that radiation-induced autophagy can protect breast cancer cells, making them more resistant to radiation. Also, low oxygen conditions (hypoxia), which are common in tumors, can stimulate autophagy and further increase radiation resistance. In lung cancer, hypoxia triggers autophagy through a protein called c-Jun, which increases Beclin1, a key autophagy regulator. Beclin1 helps start the autophagy process by blocking another protein called osteopontin. In nasopharyngeal carcinoma (NPC), a protein called LAPTM4B interacts with EGFR and Beclin1 to promote autophagy. Reducing LAPTM4B levels has been shown to decrease autophagy and make cancer cells more sensitive to radiation. In esophageal squamous cell carcinoma (ESCC), using an autophagy inhibitor called 3-methyladenine (3-MA) prevented the formation of autophagosomes (the structures that carry out autophagy), increased cancer cell death, and reduced tumor growth, both in lab settings and in animals (Chen et al. 2015).

### 8.5.5 Cancer Stem Cells

Cancer stem cells (CSCs) are a special type of cancer cell that are undifferentiated, meaning they haven't fully matured, and they can self-renew, grow into different types of cancer cells, and form new tumors. These cells are often more resistant to radiotherapy than regular cancer cells. As a result, even after most of the tumor is destroyed by radiation, CSCs can survive, begin to grow again, and lead to tumor recurrence or metastasis (spread to other parts of the body). Radiation treatment itself might trigger the formation of new CSCs, which adds another layer of resistance. Normally, radiation causes DNA damage in cancer cells, leading to cell death. However, in CSCs, this damage is often less effective because they have stronger

DNA repair systems, high expression of stem cell-related genes, and altered control of the cell cycle. These features allow CSCs to resist the effects of treatment and continue driving cancer growth (Liu et al. 2021).

## 8.6 Hormone Therapy Resistance in Cancer

Hormonal therapy has been highly successful in treating cancers that depend on hormones, like breast and prostate cancer. However, over time, many patients develop resistance to these treatments. This resistance can be present from the start (intrinsic) or can develop during treatment (acquired), and it happens because cancer cells find ways to adapt and survive despite therapy. One major cause is mutations in hormone receptors. For example, in metastatic breast cancer, changes in the ESR1 gene can cause the estrogen receptor to stay active even without estrogen, making drugs like aromatase inhibitors ineffective. In prostate cancer, shortened versions of the androgen receptor (AR-V7) keep the receptor active without needing testosterone, leading to resistance against AR-targeted therapies.

Another way cancer cells escape hormone therapy is by losing the receptors entirely. Some tumors stop producing estrogen or androgen receptors, becoming hormone-independent. This change is more common in metastatic tumors and is linked to more aggressive cancer behavior. Cancer cells can also bypass hormone pathways by activating other survival routes, like the PI3K/AKT/mTOR pathway or increasing HER2 signaling, which help the cells keep growing even when hormone signals are blocked.

In advanced prostate cancer, cancer cells can even make their own androgens from cholesterol, using enzymes like CYP17A1. This process allows the cancer to grow even when the body's testosterone levels are very low, as seen in castration-resistant prostate cancer (CRPC). These resistance mechanisms are usually rare in early-stage tumors but become common in metastatic disease as cancer cells evolve under treatment pressure (https://oncodaily.com/oncolibrary/hormonal-therapy).

## 8.7 Immunotherapy Resistance in Cancer

### 8.7.1 Reductions in Antigen Expression

Tumour-specific antigens play a key role in helping the immune system detect and attack cancer cells. Tumors with high levels of these antigens, such as microsatellite-instable (MSI) colon cancers, are more immunogenic and respond better to immunotherapy. However, some tumors either naturally express low antigen levels or reduce antigen expression over time to escape immune detection, leading to resistance. Additionally, tumors may decrease MHC class I receptor expression, which impairs antigen presentation to T cells. This mechanism is seen in cancers like melanoma and colorectal cancer. Despite this, natural killer (NK) cells can still attack tumors lacking MHC I expression, so the loss of MHC I alone doesn't fully explain why some tumors resist immunotherapy.

### 8.7.2 Alterations in Cell Receptor Expression

Another way tumors resist immunotherapy is by changing the expression of certain cell surface receptors. Both normal and cancer cells have proteins on their surfaces that either help or block immune cell attacks. For example, cancer cells can express PD-L1, which turns off immune cells like T cells and NK cells. They may also reduce molecules like Fas and TRAIL, which are needed for immune cells to kill them. Drugs like PD-1 and CTLA-4 inhibitors are used in immunotherapy to block these suppressive signals and boost the immune response.

Tumors can also reduce levels of MICA and MICB, which are stress signals recognized by NK cells through a receptor called NKG2D. These markers help NK cells kill cancer cells. Low levels of MICA or MICB make it harder for NK cells to do this. Some studies show that higher levels of MICA in cancer patients are linked to better outcomes with certain treatments, like cytokine-induced killer (CIK) cells and chemotherapy. For example, gastric and lung cancer patients with high MICA had better survival rates. However, in colorectal cancer, low MICA levels were linked to worse outcomes, possibly because MICA gets shed off the cell surface, preventing NK cells from recognizing the tumor.

To improve treatment, researchers are exploring drugs that increase MICA expression. One such drug is all-trans retinoic acid (ATRA), which, when combined with CIK cell therapy, has shown promise in boosting immune cell activity against lung and liver tumors. Other substances like sodium butyrate and matrix metalloproteinase inhibitors have also been found to raise MICA levels in some cancer cells. More research is needed to fully understand how these treatments work and whether they can be effective in other types of cancer.

### 8.7.3 Alterations in CD28 Expression

CD28 is an important receptor on T cells that helps activate the immune response, especially during cancer immunotherapy. It works in contrast to CTLA-4, even though both bind to the same molecules (CD80 or CD86). While CTLA-4 turns off immune responses, CD28 turns them on. CD28 triggers several pathways that help T cells grow, produce IL-2, and kill cancer cells. Without CD28, T cells become exhausted—they stop multiplying and lose their ability to fight tumors.

CD28 also plays a role in shaping the cell's structure, which further boosts T cell activity. Studies in mice have shown that when CD28 is blocked, tumors like colon cancer grow faster, even when treated with checkpoint inhibitors like PD-L1 blockers. This shows that CD28 is essential for effective immunotherapy. In the future, boosting CD28 expression may help revive exhausted T cells and strengthen the body's ability to fight cancer.

### 8.7.4 Alterations in Cellular Enzymes and Metabolic Pathways

Immunotherapy resistance is directly caused by alterations in receptor expression and antigen synthesis. However, a variety of other tumour cell modifications including metabolism and enzymatic activity can further modify the tumor microenvironment and prevent an immunotherapy response. Four such pathways have been suggested: mutations in the interferon gamma (IFN-) route, loss of phosphatase and tensin homologue (PTEN) expression, activation of indoleamine 2,3-dioxygenase (IDO), and uncontrolled expression of the Wnt—catenin pathway (Rieth and Subramanian 2018).

## 8.8 Summary

Cancer remains one of the leading causes of death globally, and while conventional therapies—such as chemotherapy, radiotherapy, and hormonal therapy—have significantly improved patient outcomes, their effectiveness is frequently compromised by the development of therapeutic resistance. This chapter provides an in-depth exploration of the mechanisms underlying resistance to conventional cancer treatments, their clinical implications, and potential strategies to overcome these challenges.

Two main types of resistance are discussed: intrinsic resistance, where cancer cells are unresponsive to treatment from the outset, and acquired resistance, which develops over time due to therapy-induced selective pressure. The chapter highlights the molecular and cellular mechanisms that contribute to resistance, including altered drug transport (e.g., overexpression of ABC transporters such as P-gp), genetic mutations in drug targets, enhanced DNA repair, defects in apoptosis pathways, and tumor heterogeneity.

*Cancer stem cells (CSCs)* are identified as key contributors to both chemo- and radioresistance due to their abilities to self-renew, evade cell death, and regenerate tumors after treatment. Additionally, *the tumor microenvironment (TME)*—including immune cells, fibroblasts, blood vessels, and extracellular matrix—plays a critical role in shielding tumor cells from therapy through cell adhesion, cytokine signaling, and modulation of drug availability.

The chapter also details the phenomenon of *tumor heterogeneity*, which exists both spatially (differences between cells in different regions of the tumor) and temporally (evolution of genetic changes over time), further complicating treatment. Such diversity within tumors promotes the survival of resistant subclones, making single-biopsy diagnostics and uniform treatments insufficient.

Moreover, *specific resistance mechanisms* related to each treatment modality are addressed (Table 8.1):

- For *chemotherapy*, resistance is often due to drug efflux, gene amplification, DNA repair, and anti-apoptotic signaling.

**Table 8.1** Mechanism of chemoresistance, radio resistance, and immune evasion for tumor survival

| S. No. | Mechanism | Examples of associated cancers | References |
|---|---|---|---|
| 1 | Reduced neoantigen generation | Colorectal, kidney, clear cell | Rieth and Subramanian (2018) |
| 2 | Reduced Fas expression | Melanoma | Rieth and Subramanian (2018) |
| 3 | Reduced TRAIL expression | Melanoma | Rieth and Subramanian (2018) |
| 5 | Induction of IDO | AML, colorectal, endometrial, SCLC, melanoma, ovarian | Rieth and Subramanian (2018) |
| 6 | Loss of PTEN expression | Melanoma | Rieth and Subramanian (2018) |
| 7 | Deregulation of Wnt–β-catenin pathway | Melanoma | Rieth and Subramanian (2018) |
| 8 | Cisplatin resistance | Lung, ovarian, head & neck, bladder | Ouyang et al. (2018), Safaei et al. (2009) |
| 9 | 5-FU resistance | Colorectal, gastric, pancreatic | Zhang et al. (2019), Moore Joshi et al. (2005) |
| 10 | Taxol resistance | Breast, ovarian, lung | Fu et al. (2007), Zunino et al. (1999) |
| 11 | Oxaliplatin resistance | Colorectal, gastric | Wang et al. (2015), Jarhad et al. (2018) |
| 12 | Multidrug resistance | Breast, ovarian, leukemia, colon | Eid et al. (1998), Chen and Tiwari (2011) |
| 13 | DNA damage repair | Breast, ovarian, prostate, lung | Crucitta et al. (2022) |
| 14 | Cell cycle redistribution | Leukemia, breast, colorectal | Crucitta et al. (2022) |
| 15 | Apoptosis escape | Melanoma, lung, colorectal, pancreatic | Crucitta et al. (2022) |
| 16 | Tumor microenvironment | Breast, pancreatic, glioblastoma | Crucitta et al. (2022) |
| 17 | CSCs | Breast, colorectal, glioblastoma | Crucitta et al. (2022) |
| 18 | Metabolic reprogramming | Lung, breast, pancreatic | Crucitta et al. (2022) |
| 19 | Exosomes | Melanoma, colorectal, breast | Crucitta et al. (2022) |
| 20 | Ferroptosis resistance | Hepatocellular carcinoma, pancreatic, lung | Crucitta et al. (2022) |

- For *radiotherapy*, resistance arises from enhanced DNA repair, altered cell cycle regulation, EMT (epithelial–mesenchymal transition), autophagy, and the persistence of CSCs.
- For *hormonal therapy*, resistance is commonly driven by receptor mutations, receptor loss, bypass signaling, and intracrine hormone production.

  The clinical implications of resistance are profound, often resulting in treatment failure, disease relapse, and poor prognosis. One of the key challenges is the early detection of resistance, which is hindered by intratumoral heterogeneity and limitations in traditional diagnostic tools. To address these issues, the chapter emphasizes the critical role of molecular profiling and liquid biopsies in enabling personalized therapy. These tools help identify actionable mutations and monitor tumor evolution, allowing clinicians to tailor treatment strategies and adjust them in response to emerging resistance.

## References


Al-Lazikani B, Banerji U, Workman P (2012) Combinatorial drug therapy for cancer in the post-genomic era. Nat Biotechnol 30(7):679–692

Altwerger G, Ghazarian M, Glazer PM (2024) Harnessing the effects of hypoxia-like inhibition on homology-directed DNA repair. Semin Cancer Biol 98:11–18

Azzam EI, Jay-Gerin JP, Pain D (2012) Ionizing radiation-induced metabolic oxidative stress and prolonged cell injury. Cancer Lett 327(1–2):48–60

Bandyopadhyay D, Mishra A, Medrano EE (2004) Overexpression of histone deacetylase 1 confers resistance to sodium butyrate–mediated apoptosis in melanoma cells through a p53-mediated pathway. Cancer Res 64(21):7706–7710

Baskar R, Lee KA, Yeo R, Yeoh KW (2012) Cancer and radiation therapy: current advances and future directions. Int J Med Sci 9(3):193–199

Cai Q, Wang S, Jin L, Weng M, Zhou D, Wang J et al (2019) Long non-coding RNA GBCDRlnc1 induces chemoresistance of gallbladder cancer cells by activating autophagy. Mol Cancer 18(1): 82

Chaachouay H, Ohneseit P, Toulany M, Kehlbach R, Multhoff G, Rodemann HP (2011) Autophagy contributes to resistance of tumor cells to ionizing radiation. Radiother Oncol J Eur Soc Ther Radiol Oncol 99(3):287–292

Chen ZS, Tiwari AK (2011) Multidrug resistance proteins (MRPs/ABCCs) in cancer chemotherapy and genetic diseases. FEBS J 278(18):3226–3245

Chen Y, Li X, Guo L, Wu X, He C, Zhang S et al (2015) Combining radiation with autophagy inhibition enhances suppression of tumor growth and angiogenesis in esophageal cancer. Mol Med Rep 12(2):1645–1652

Chen R, Yu Y, Liu R, Chen Q (2024) Targeting breast cancer resistance protein (BCRP/ABCG2) in cancer. Transl Cancer Res 13(11):6550–6564

Crucitta S, Cucchiara F, Mathijssen R, Mateo J, Jager A, Joosse A et al (2022) Treatment-driven tumour heterogeneity and drug resistance: lessons from solid tumours. Cancer Treat Rev 104: 102340

Dean M, Fojo T, Bates S (2005) Tumour stem cells and drug resistance. Nat Rev Cancer 5(4): 275–284

DeVita VT, Chu E (2008) A history of cancer chemotherapy. Cancer Res 68(21):8643–8653

Eid H, Géczi L, Mágori A, Bodrogi I, Institoris E, Bak M (1998) Drug resistance and sensitivity of germ cell testicular tumors: evaluation of clinical relevance of MDR1/Pgp, p53, and metallothionein (MT) proteins. Anticancer Res 18(4C):3059–3064

Elfadadny A, El-Husseiny HM, Abugomaa A, Ragab RF, Mady EA, Aboubakr M et al (2021) Role of multidrug resistance-associated proteins in cancer therapeutics: past, present, and future perspectives. Environ Sci Pollut Res 28(36):49447–49466

Fang F, Balch C, Schilder J, Breen T, Zhang S, Shen C et al (2010) A phase 1 and pharmacodynamic study of decitabine in combination with carboplatin in patients with recurrent, platinum-resistant, epithelial ovarian cancer. Cancer 116(17):4043–4053

Fojta M, Daňhel A, Havran L, Vyskočil V (2016) Recent progress in electrochemical sensors and assays for DNA damage and repair. TrAC Trends Anal Chem 79:160–167

Fu Y, Ye D, Chen H, Lu W, Ye F, Xie X (2007) Weakened spindle checkpoint with reduced BubR1 expression in paclitaxel-resistant ovarian carcinoma cell line SKOV3-TR30. Gynecol Oncol 105(1):66–73

Fu H, Wu ZX, Lei ZN, Teng QX, Yang Y, Ashby CR et al (2022) The resistance of cancer cells to Palbociclib, a cyclin-dependent Kinase 4/6 inhibitor, is mediated by the ABCB1 transporter. Front Pharmacol 13:861642

Gao K, Chen S, Yang X (2021) HOTTIP enhances gemcitabine and cisplatin resistance through sponging miR-637 in Cholangiocarcinoma. Front Oncol 11:664916

Gottesman MM (2002) Mechanisms of cancer drug resistance. Annu Rev Med 53(1):615–627

Harris WP, Mostaghel EA, Nelson PS, Montgomery B (2009) Androgen deprivation therapy: progress in understanding mechanisms of resistance and optimizing androgen depletion. Nat Clin Pract Urol 6(2):76–85

He E, Pan F, Li G, Li J (2015) Fractionated ionizing radiation promotes epithelial-mesenchymal transition in human Esophageal cancer cells through PTEN deficiency-mediated Akt activation. PLoS One 10(5):e0126149

Hofman J, Vagiannis D, Chen S, Guo L (2021) Roles of CYP3A4, CYP3A5 and CYP2C8 drug-metabolizing enzymes in cellular cytostatic resistance. Chem Biol Interact 340:109448

Holohan C, Van Schaeybroeck S, Longley DB, Johnston PG (2013) Cancer drug resistance: an evolving paradigm. Nat Rev Cancer 13(10):714–726

Jarhad DB, Mashelkar KK, Kim HR, Noh M, Jeong LS (2018) Dual-Specificity Tyrosine Phosphorylation-Regulated Kinase 1A (DYRK1A) inhibitors as potential therapeutics. J Med Chem 61(22):9791–9810

Jiang M, Wang J, Li Y, Zhang K, Wang T, Bo Z et al (2025) EMT and cancer stem cells: drivers of therapy resistance and promising therapeutic targets. Drug Resist Updat 83:101276

Kang SH, Bang YJ, Im YH, Yang HK, Lee DA, Lee HY et al (1999) Transcriptional repression of the transforming growth factor-β type I receptor gene by DNA methylation results in the development of TGF-β resistance in human gastric cancer. Oncogene 18(51):7280–7286

Khanna KK, Jackson SP (2001) DNA double-strand breaks: signaling, repair and the cancer connection. Nat Genet 27(3):247–254

Kim MG, Pak JH, Choi WH, Park JY, Nam JH, Kim JH (2012) The relationship between cisplatin resistance and histone deacetylase isoform overexpression in epithelial ovarian cancer cell lines. J Gynecol Oncol 23(3):182

Kittirat Y, Suksawat M, Thongchot S, Padthaisong S, Phetcharaburanin J, Wangwiwatsin A et al (2022) Interleukin-6-derived cancer-associated fibroblasts activate STAT3 pathway contributing to gemcitabine resistance in cholangiocarcinoma. Front Pharmacol 13:897368

Lamouille S, Xu J, Derynck R (2014) Molecular mechanisms of epithelial-mesenchymal transition. Nat Rev Mol Cell Biol 15(3):178–196

Landen CN, Goodman B, Katre AA, Steg AD, Nick AM, Stone RL et al (2010) Targeting aldehyde dehydrogenase cancer stem cells in ovarian cancer. Mol Cancer Ther 9(12):3186–3199

Larsen BD, Benada J, Yung PYK, Bell RAV, Pappas G, Urban V et al (2022) Cancer cells use self-inflicted DNA breaks to evade growth limits imposed by genotoxic stress. Science 376(6592): 476–483

Li J, Sun H, Huang Y, Wang Y, Liu Y, Chen X (2019) Pathways and assays for DNA double-strand break repair by homologous recombination. Acta Biochim Biophys Sin 51(9):879–889

Li Q, Qian W, Zhang Y, Hu L, Chen S, Xia Y (2023) A new wave of innovations within the DNA damage response. Signal Transduct Target Ther 8(1):338

Linnekamp JF, Butter R, Spijker R, Medema JP, Van Laarhoven HWM (2017) Clinical and biological effects of demethylating agents on solid tumours—a systematic review. Cancer Treat Rev 54:10–23

Liu R, Chen Y, Liu G, Li C, Song Y, Cao Z et al (2020) PI3K/AKT pathway as a key link modulates the multidrug resistance of cancers. Cell Death Dis 11(9):797

Liu YP, Zheng CC, Huang YN, He ML, Xu WW, Li B (2021) Molecular mechanisms of chemo- and radiotherapy resistance and the potential implications for cancer treatment. MedComm 2(3): 315–340

Liu F, Wu Q, Dong Z, Liu K (2023) Integrins in cancer: emerging mechanisms and therapeutic opportunities. Pharmacol Ther 247:108458

Luqmani YA (2005) Mechanisms of drug resistance in cancer chemotherapy. Med Princ Pract 14 (Suppl. 1):35–48

Luserna G, di Rora A, Iacobucci I, Martinelli G (2017) The cell cycle checkpoint inhibitors in the treatment of leukemias. J Hematol Oncol 10(1):77

Martinez VG, O'Connor R, Liang Y, Clynes M (2008) CYP1B1 expression is induced by docetaxel: effect on cell viability and drug resistance. Br J Cancer 98(3):564–570

Marusyk A, Almendro V, Polyak K (2012) Intra-tumour heterogeneity: a looking glass for cancer? Nat Rev Cancer 12(5):323–334

Meads MB, Gatenby RA, Dalton WS (2009) Environment-mediated drug resistance: a major contributor to minimal residual disease. Nat Rev Cancer 9(9):665–674

Meijer C, Mulder N, Hospers G, Uges D, De Vries E (1990) The role of glutathione in resistance to cisplatin in a human small cell lung cancer cell line. Br J Cancer 62(1):72–77

Mitra A, Mishra L, Li S (2015) EMT, CTCs and CSCs in tumor relapse and drug-resistance. Oncotarget 6(13):10697–10711

Moore Joshi MB, Shirota Y, Danenberg KD, Conlon DH, Salonga DS, Herndon JE et al (2005) High gene expression of *TS1*, *GSTP1*, and *ERCC1* are risk factors for survival in patients treated with trimodality therapy for Esophageal cancer. Clin Cancer Res 11(6):2215–2221

Nahta R, Yu D, Hung MC, Hortobagyi GN, Esteva FJ (2006) Mechanisms of disease: understanding resistance to HER2-targeted therapy in human breast cancer. Nat Clin Pract Oncol 3(5): 269–280

Nam HY, Han MW, Chang HW, Kim SY, Kim SW (2013) Prolonged autophagy by MTOR inhibitor leads radioresistant cancer cells into senescence. Autophagy 9(10):1631–1632

Nervi C, De Marinis E, Codacci-Pisanelli G (2015) Epigenetic treatment of solid tumours: a review of clinical trials. Clin Epigenetics 7(1):127

Oguri T, Takahashi T, Miyazaki M, Isobe T, Kohno N, Mackenzie PI et al (2004) UGT1A10 is responsible for SN-38 glucuronidation and its expression in human lung cancers. Anticancer Res 24(5A):2893–2896

on behalf of the NCRI CTRad Academia-Pharma Joint Working Group, Sharma RA, Plummer R, Stock JK, Greenhalgh TA, Ataman O et al (2016) Clinical development of new drug–radiotherapy combinations. Nat Rev Clin Oncol 13(10):627–642

Ono H, Basson MD, Ito H (2016) P300 inhibition enhances gemcitabine-induced apoptosis of pancreatic cancer. Oncotarget 7(32):51301–51310

Osborne CK, Schiff R (2011) Mechanisms of endocrine resistance in breast cancer. Annu Rev Med 62(1):233–247

Ouyang Q, Liu Y, Liu Y (2018) Research progress of copper transporter 1 in platinum-based chemotherapy. Zhong Nan Da Xue Xue Bao Yi Xue Ban 43(12):1376–1379

Pang X, He X, Qiu Z, Zhang H, Xie R, Liu Z et al (2023) Targeting integrin pathways: mechanisms and advances in therapy. Signal Transduct Target Ther 8(1):1

Pavlíková L, Šereš M, Breier A, Sulová Z (2022) The roles of microRNAs in cancer multidrug resistance. Cancer 14(4):1090

Pilotto Heming C, Muriithi W, Wanjiku Macharia L, Niemeyer Filho P, Moura-Neto V, Aran V (2022) P-glycoprotein and cancer: what do we currently know? Heliyon 8(10):e11171

Ramos P, Bentires-Alj M (2015) Mechanism-based cancer therapy: resistance to therapy, therapy for resistance. Oncogene 34(28):3617–3626

Rana V, Parama D, Khatoon E, Girisa S, Sethi G, Kunnumakkara AB (2021) Reiterating the emergence of noncoding RNAs as regulators of the critical hallmarks of gall bladder cancer. Biomolecules 11(12):1847

Ribatti D, Tamma R, Annese T (2020) Epithelial-mesenchymal transition in cancer: a historical overview. Transl Oncol 13(6):100773

Rieth J, Subramanian S (2018) Mechanisms of intrinsic tumor resistance to immunotherapy. Int J Mol Sci 19(5):1340

Robey RW, Pluchino KM, Hall MD, Fojo AT, Bates SE, Gottesman MM (2018) Revisiting the role of ABC transporters in multidrug-resistant cancer. Nat Rev Cancer 18(7):452–464

Safaei R, Maktabi MH, Blair BG, Larson CA, Howell SB (2009) Effects of the loss of Atox1 on the cellular pharmacology of cisplatin. J Inorg Biochem 103(3):333–341

Sawers L, Ferguson MJ, Ihrig BR, Young HC, Chakravarty P, Wolf CR et al (2014) Glutathione S-transferase P1 (GSTP1) directly influences platinum drug chemosensitivity in ovarian tumour cell lines. Br J Cancer 111(6):1150–1158

Senthebane DA, Rowe A, Thomford NE, Shipanga H, Munro D, Mazeedi MAMA et al (2017) The role of tumor microenvironment in chemoresistance: to survive, keep your enemies closer. Int J Mol Sci 18(7):1586

Song J, Yang P, Chen C, Ding W, Tillement O, Bai H et al (2025) Targeting epigenetic regulators as a promising avenue to overcome cancer therapy resistance. Signal Transduct Target Ther 10(1): 219

Soragni A, Knudsen ES, O'Connor TN, Tognon CE, Tyner JW, Gini B et al (2025) Acquired resistance in cancer: towards targeted therapeutic strategies. Nat Rev Cancer [Internet]. [cited 2025 Jul 14]; Available from: https://www.nature.com/articles/s41568-025-00824-9

Stark TW, Hensley PJ, Spear A, Pu H, Strup SS, Kyprianou N (2017) Predictive value of epithelial-mesenchymal-transition (EMT) signature and PARP-1 in prostate cancer radioresistance. Prostate 77(16):1583–1591

Stefansson OA, Villanueva A, Vidal A, Martí L, Esteller M (2012) *BRCA1* epigenetic inactivation predicts sensitivity to platinum-based chemotherapy in breast and ovarian cancer. Epigenetics 7(11):1225–1229

Tanabe S, Quader S, Cabral H, Ono R (2020) Interplay of EMT and CSC in cancer and the potential therapeutic strategies. Front Pharmacol 11:904

Tian X, Srinivasan PR, Tajiknia V, Sanchez Sevilla Uruchurtu AF, Seyhan AA, Carneiro BA et al (2024) Targeting apoptotic pathways for cancer therapy. J Clin Invest 134(14):e179570

Varisli L, Zoumpourlis P, Spandidos D, Zoumpourlis V, Vlahopoulos S (2025) ALDH1A1 in breast cancer: a prospective target to overcome therapy resistance (review). Oncol Lett 29(5):1–17

Verma H, Singh Bahia M, Choudhary S, Kumar Singh P, Silakari O (2019) Drug metabolizing enzymes-associated chemo resistance and strategies to overcome it. Drug Metab Rev 51(2): 196–223

Wang X, Jiang P, Wang P, Yang CS, Wang X, Feng Q (2015) EGCG enhances cisplatin sensitivity by regulating expression of the copper and cisplatin influx transporter CTR1 in ovary cancer. PLoS One 10(4):e0125402

Wei JR, Lu MY, Wei TH, Fleishman JS, Yu H, Chen XL et al (2025) Overcoming cancer therapy resistance: from drug innovation to therapeutics. Drug Resist Updat 81:101229

Xue X, Liang XJ (2012) Overcoming drug efflux-based multidrug resistance in cancer with nanotechnology. Chin J Cancer 31(2):100–109

Ya LL, Di GY, Sha CX, Ming YJ, Cheng Y (2021) DNA repair pathways in cancer therapy and resistance. Front Pharmacol 11:629266

Zhang Q, Karnak D, Tan M, Lawrence TS, Morgan MA, Sun Y (2016) FBXW7 facilitates nonhomologous end-joining via K63-linked Polyubiquitylation of XRCC4. Mol Cell 61(3): 419–433

Zhang Q, Liu RX, Chan KW, Hu J, Zhang J, Wei L, et al (2019) Exosomal transfer of p-STAT3 promotes acquired 5-FU resistance in colorectal cancer cells. J Exp Clin Cancer Res [Internet] [cited 2025 Jul 14];38(1). Available from: https://jeccr.biomedcentral.com/articles/10.1186/s13046-019-1314-9

Zhu Z, Mu Y, Qi C, Wang J, Xi G, Guo J et al (2015) CYP1B1 enhances the resistance of epithelial ovarian cancer cells to paclitaxel in vivo and in vitro. Int J Mol Med 35(2):340–348

Zunino F, Cassinelli G, Polizzi D, Perego P (1999) Molecular mechanisms of resistance to taxanes and therapeutic implications. Drug Resist Updat 2(6):351–357

# miRNA Delivery Systems, Improvements, and Challenges

9

Deepika Tuteja, Amit K. Yadav, Dhiraj Bhatia, and Abhishek Kumar

**Abstract**

MicroRNAs (miRNAs) are critical post-transcriptional regulators of gene expression involved in a wide array of physiological and pathological processes. Since their discovery, miRNAs have been recognized for their capacity to modulate diverse gene networks, impacting cell proliferation, differentiation, apoptosis, and metabolism. Their dysregulation is strongly implicated in cancer, cardiovascular, and neurological diseases, making them attractive biomarkers and therapeutic targets. However, miRNA therapeutics face substantial challenges in clinical translation, including instability, degradation, delivery inefficiency, rapid systemic clearance, and off-target effects. To address these, both viral and non-viral delivery systems have been engineered, ranging from lipid nanoparticles and polymers to conjugates and exosomes, to enhance miRNA stability, specificity, and cellular uptake. Advances such as chemical modifications, ligand-mediated targeting, and stimuli-responsive carriers have significantly improved delivery efficacy. Despite notable progress, key obstacles, including tissue penetration, especially across the blood–brain barrier, immunogenicity, and safety, remain. Continued innovation in targeted, responsive delivery technologies is essential to realize the clinical potential of miRNA therapeutics fully.

D. Tuteja
Institute of Nuclear Medicine and Allied Sciences, DRDO, New Delhi, India

A. K. Yadav · D. Bhatia
Department of Biological Sciences and Engineering, Indian Institute of Technology Gandhinagar, Gandhinagar, Gujarat, India

A. Kumar (✉)
Albert Einstein College of Medicine, Bronx, NY, USA



S. Tiwari et al. (eds.), *microRNA Modulation: A Promising Combination Therapy for Cancer*, https://doi.org/10.1007/978-981-95-5989-3_9

**Keywords**

miRNAs · Delivery Systems · Endosomal Escape · Blood Brain Barrier · Therapeutics

**Learning Objectives**

- *Describe* the key delivery systems developed for therapeutic miRNAs, including viral vectors, lipid nanoparticles, polymer-based carriers, and exosome-mediated delivery.
- *Explain* the major challenges in miRNA delivery, such as stability, off-target effects, immune activation, and tissue-specific targeting.
- *Evaluate* recent technological improvements in delivery strategies that enhance efficiency, biocompatibility, and therapeutic potential.
- *Assess* the translational prospects of miRNA delivery systems, considering pre-clinical progress, clinical trials, and regulatory barriers.

## 9.1 Introduction

MicroRNAs (miRNAs) are some of the essential regulators of gene expression, influencing several physiological and pathological processes. Identified in the early 1990s, miRNAs are brief, noncoding RNA molecules, which are generally 18 to 24 nucleotides long. miRNAs regulate gene expression post-transcriptionally by binding to complementary sequences on target messenger RNAs (mRNAs) (Ying et al. 2008; Vyas et al. 2025). This binding results in translational repression or mRNA destruction, hence regulating the amounts of numerous proteins across different cellular settings. miRNA biogenesis is a tightly regulated, multistep process that begins in the nucleus, where RNA polymerase II generates primary transcripts (pri-miRNAs) (Garzon et al. 2010). The Drosha processes these pri-miRNAs–DGCR8 complex into precursor molecules (pre-miRNAs), which are exported to the cytoplasm via Exportin-5. In the cytoplasm, the RNase III enzyme Dicer further cleaves pre-miRNAs to produce mature miRNA duplexes. One strand of this duplex is then incorporated into the RNA-induced silencing complex (RISC), which directs sequence-specific binding to target mRNAs, leading to either their degradation or translational repression, ultimately influencing protein expression (Melo and Esteller 2011). Through this mechanism, miRNAs regulate diverse cellular processes, including proliferation, apoptosis, and differentiation. Aberrations in their expression are frequently associated with tumorigenesis (Yadav et al. 2023, 2025a). Depending on their function, miRNAs can act as oncomiRs, which promote cancer progression by suppressing tumor suppressor genes, or as tumor-suppressive miRNAs, which inhibit oncogenes to restrain malignant growth. These dual roles position miRNAs as attractive therapeutic candidates. Strategies under investigation include the use of antisense oligonucleotides (ASOs) to inhibit oncogenic miRNAs and the application of synthetic miRNA mimics to restore the activity of tumor-suppressive miRNAs, thereby offering novel approaches for cancer treatment

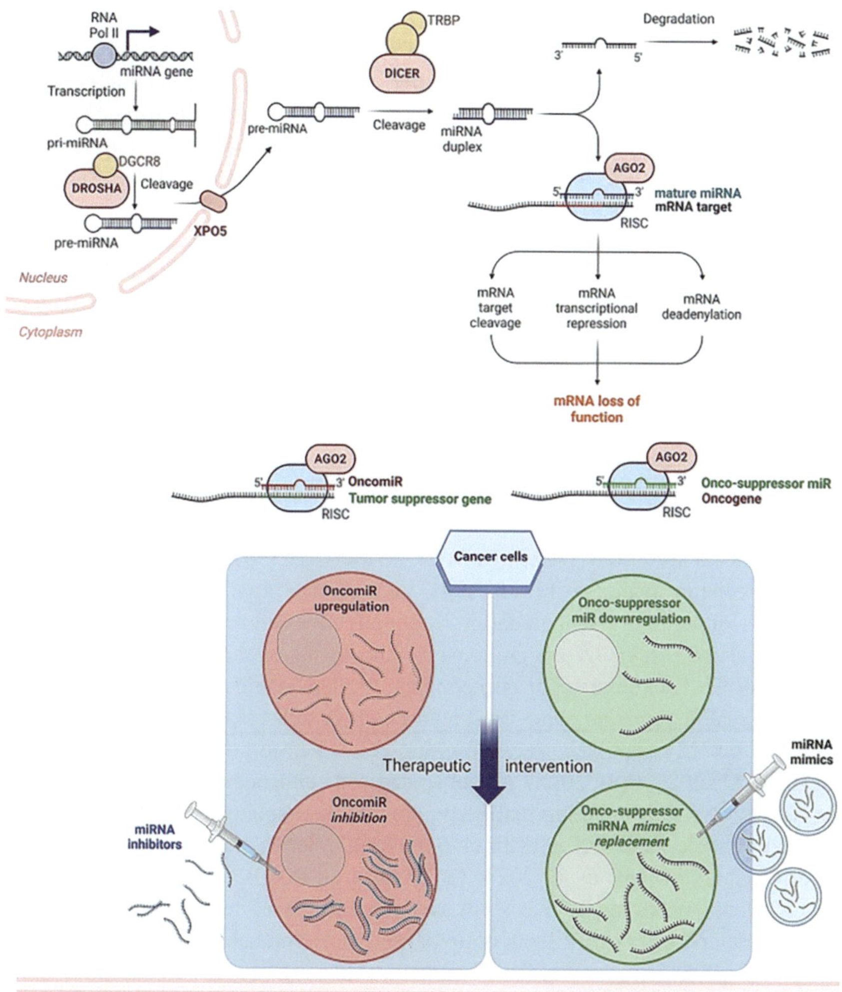


**Fig. 9.1** Overview of miRNA biogenesis and its relevance as a therapeutic target in cancer. The figure illustrates the key steps in miRNA maturation and highlights how dysregulation of this process contributes to tumor development, offering opportunities for therapeutic intervention. (Reproduced with permission from MartinoDi et al. (2025)), Copyright @ 2025 SpringerNature)

(O'Brien et al. 2018). Figure 9.1 depicts the miRNA biogenesis pathway, highlights the consequences of their dysregulation in cancer cells, and underscores their promise as targets for therapeutic intervention.

The miRNAs function as essential molecular switches within cells. By selectively targeting specific mRNAs, they assimilate information from the environment and developmental indicators, facilitating tissue and organ development, immunological responses, and the preservation of cellular homeostasis. Genetic alterations or dysregulation of miRNA expression are often associated with the initiation and advancement of several illnesses (Wang et al. 2016; Huang 2017). Significant evidence has associated abnormal miRNA expression with a broad range of human diseases. Altered miRNA profiles are indicative of numerous malignancies, facilitating carcinogenesis, metastasis, and treatment resistance (Sharma and Gupta 2020). Similarly, miRNAs regulate pathways critical to cardiovascular diseases, such as atherosclerosis and myocardial infarction (Lu et al. 2018; Zhou et al. 2018), and are integral to the etiology of neurological disorders, including Alzheimer's and Parkinson's diseases (Awuson-David et al. 2023; Karnati et al. 2015). The distinctive expression patterns and regulatory roles of miRNAs render them appealing biomarkers and therapeutic targets for disease diagnosis, prognosis, and intervention.

Therapies that utilize or regulate miRNA function offer substantial potential. In contrast to conventional pharmaceuticals that typically focus on individual proteins, miRNAs affect comprehensive molecular networks, resulting in more integrated and powerful therapeutic outcomes. The synthetic manipulation of miRNAs, via mimics to replenish deficient miRNAs or antagomirs to suppress overexpressed ones, provides the opportunity for precision regulation of disease-related gene circuits with significant specificity. The miRNA-based therapies present considerable potential for precision medicine; nevertheless, their clinical application is hindered by instability, vulnerability to degradation, and ineffective cellular transport (Chen et al. 2015; Segal and Slack 2020). Significant advancements, such as chemical alterations of the miRNA backbone sugar components, implementation of delivery vehicles, ligand-mediated targeting, and the creation of stimuli-responsive systems, have improved miRNA stability, specificity, and cellular absorption (Chen et al. 2015; Segal and Slack 2020; Ju 2025). Novel strategies to enhance endosomal escape and targeting specificity, including antibody, peptide, and aptamer conjugation, alongside nanoparticle-mediated and stimulus-triggered release, are essential for effective therapy (Ghafouri-Fard et al. 2023; Orellana et al. 2019). Notwithstanding these advancements, obstacles such as swift systemic clearance, restricted tissue penetration, particularly across the blood–brain barrier, and apprehensions over immunogenicity and off-target gene regulation remain. Although specific miRNA therapies have advanced to clinical trials for a variety of disorders, careful attention to efficacy, safety, and overall ethical problems, such as informed consent and transparency, is critical as the field moves closer to clinical implementation (Daby et al. 2025).

The miRNAs are an emerging class of regulators that have considerable potential for use in diagnostics and treating diseases. However, their successful use in medicine depends upon finding ways to get beyond all the challenges that come with their potent delivery. This is why the development of advanced delivery systems is such an essential part of the growing field of miRNA therapies. Further advancement of sophisticated miRNA delivery systems is a crucial aspect of the expanding domain of miRNA treatments (Yadav et al. 2025b).

## 9.2 Overview of miRNA Delivery Systems

To overcome the intrinsic instability and rapid degradation of miRNAs, a wide range of delivery platforms has been developed to enhance their therapeutic applicability. Broadly, these delivery strategies are categorized into viral and non-viral systems, each offering unique advantages and limitations (Fig. 9.2). Viral vectors such as adeno-associated viruses (AAVs), adenoviruses, and lentiviruses provide efficient cellular uptake, high transduction rates, and long-term expression. However, their clinical use is often restricted by immunogenicity, safety concerns, production challenges, and size limitations. In contrast, non-viral delivery systems, including liposomes, nanoparticles, extracellular vesicles, and synthetic polymers, are generally associated with lower immunogenicity and greater versatility. These carriers can be engineered with tunable physicochemical properties to achieve targeted delivery, cost-effective production, and the ability to co-deliver multiple therapeutics. Nonetheless, issues such as limited delivery efficiency, non-specific binding, bioaccumulation, and variable clearance remain persistent obstacles. Overall, the

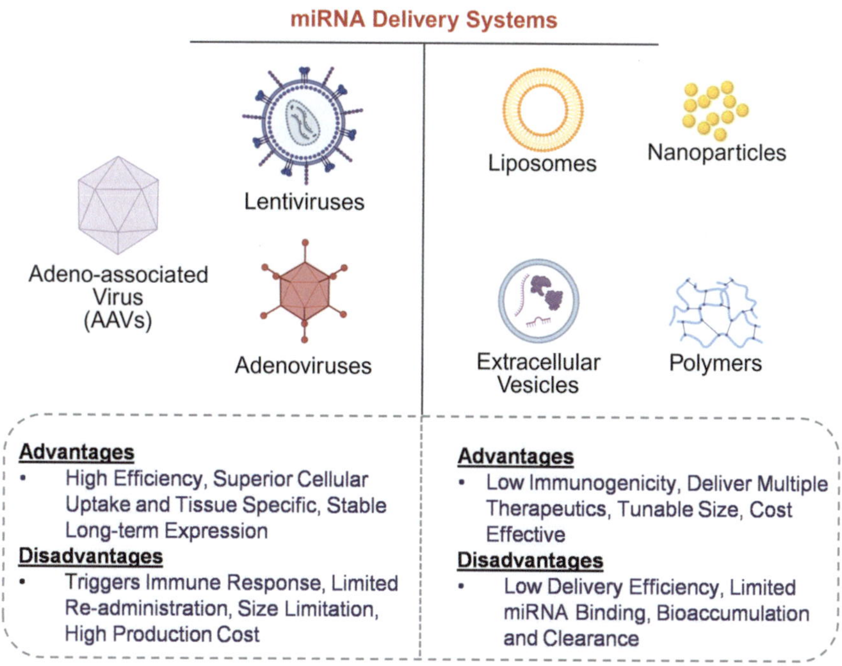


**Fig. 9.2** Viral and non-viral delivery strategies for miRNA therapeutics, along with their respective advantages and disadvantages. Viral vectors (AAVs, adenoviruses, lentiviruses) exhibit strong cellular uptake and long-term expression but are hindered by immunogenicity, size constraints, and high production costs. Non-viral systems (liposomes, nanoparticles, extracellular vesicles, and polymers) offer lower immunogenicity, tunable size, and cost-effectiveness, yet often face challenges with delivery efficiency, off-target binding, and clearance. (Illustration was made using biorender.com)

choice of delivery system must balance the therapeutic efficacy, safety, and manufacturability, underscoring the importance of continued innovation in this area to advance miRNA-based therapeutics. Understanding these delivery methods allows researchers to tailor approaches based on specific treatment goals, patient needs, and the characteristics of the target disease (Kumar et al. 2025).

### 9.2.1 Viral Vector-Based Systems

Viral vector-based systems are powerful tools that exploit the natural ability of viruses to deliver genetic material into host cells. By modifying viruses to be replication-deficient and safe, researchers use them as efficient carriers for therapeutic genes, vaccines, and research applications (Pan et al. 2012). These systems are particularly valued for their high transduction efficiency, stable gene expression, and capacity to target specific cell types, making them central to modern gene therapy and biotechnology (Geisler and Fechner 2016).

(i) *Adenoviral (Ad) vectors,* particularly oncolytic adenoviruses, have demonstrated high transfection efficiency and the ability to provide long-term miRNA expression. Studies have shown the successful delivery of miR-34a and anti-HBV pri-miRNA mimics for hepatocellular carcinoma and HBV infection treatment (Ren et al. 2018). However, these systems face challenges, including strong immunogenicity and potential liver toxicity at high doses.
(ii) *Adeno-associated virus (AAV) vectors* have emerged as particularly promising due to their ability to cross the blood-brain barrier and provide tissue-specific delivery. Research has demonstrated the successful delivery of miR-196a to motor neurons and miR-181a/b sponge constructs for retinal diseases (Brillante et al. 2024). The development of rAAV5-mediated delivery systems has progressed to clinical trials, representing a significant milestone in viral vector-based miRNA therapy (Borel et al. 2014).
(iii) *Lentiviral and retroviral vectors* offer the advantage of stable genome integration and long-term expression, with studies showing remarkable increases in miRNA expression levels exceeding 1000-fold (Milone and O'Doherty 2018). However, concerns about the potential for integration mutagenesis and oncogenesis limit their clinical application.

### 9.2.2 Lipid-Based Delivery Systems

Lipid-based delivery systems are innovative platforms that harness the properties of lipids to facilitate the transport of therapeutic compounds, including genetic material, drugs, and vaccines, into target cells. These systems, such as liposomes, solid lipid nanoparticles (SLNs), and lipid nanoparticles (LNPs), are valued for their ability to protect cargo from degradation, improve cellular uptake, and promote controlled release within the body (Falsini et al. 2014). Recent advances have enabled the successful use of lipid nanoparticles in genetic therapies and vaccines, notably highlighted by their role in delivering mRNA for COVID-19 vaccines. Their biocompatibility, customizable formulations, and versatility make them essential in the development of next-generation gene therapies, although challenges like stability, targeted delivery, and immunogenicity remain active areas of research (Jogdeo et al. 2023).

(i) *LNPs* have gained significant attention due to their established clinical track record and ability to protect miRNA from degradation while facilitating cellular uptake. Recent studies have demonstrated the successful delivery of miR-200b for intestinal regeneration (Wei et al. 2023a), miR-146a for immune modulation (Gilyazova et al. 2023), and miR-124 for neuroinflammation treatment (Zhou et al. 2022). The technology benefits from well-established manufacturing processes and regulatory pathways, though cytotoxicity concerns and serum protein interactions remain challenges.

(ii) SLNs are an important class of lipid-based carriers with growing application in the field of miRNA delivery. Characterized by a solid lipid core at body temperature, SLNs offer enhanced stability, biocompatibility, and capacity for encapsulating both hydrophilic and lipophilic molecules (Shi et al. 2012). In the context of miRNA delivery, SLNs have been engineered often with cationic lipids to facilitate the efficient binding and encapsulation of negatively charged miRNAs, protecting them from enzymatic degradation during systemic circulation (Guney Eskiler et al. 2018; Shi et al. 2014). These nanoparticles enable improved cellular uptake and controlled intracellular release of miRNA cargo, overcoming challenges faced by other carriers like premature degradation or limited cell entry. For example, SLNs have been successfully used to deliver miR-200c and paclitaxel together with breast cancer stem cells, resulting in superior miRNA expression and therapeutic effect when compared to conventional agents. Importantly, SLNs can efficiently co-deliver miRNAs with chemotherapeutic drugs, harnessing synergistic anti-cancer effects while minimizing toxicity (Küçüktürkmen and Bozkır 2018).

    Despite these advantages, challenges remain, such as optimizing encapsulation efficiency for anionic miRNAs, achieving precise tissue targeting, and minimizing potential cytotoxicity associated with cationic lipid components. Continuous innovations, such as PEGylation, functionalization with targeting ligands, and hybrid lipid-polymer designs, are being explored to address these issues and maximize the therapeutic potential of SLNs in miRNA delivery (German-Cortés et al. 2023).

(iii) *Cationic liposomes* provide biocompatible alternatives with controllable size characteristics typically ranging from 80 to 150 nm for optimal cellular uptake. Studies have shown the successful delivery of miR-140 for chondrogenic differentiation and miR-122 for hepatocellular carcinoma treatment (Rajagopal et al. 2021; Hsu et al. 2013). However, these systems face limitations, including short half-life and potential immune responses.

### 9.2.3 Polymer-Based Systems

Polymer-based delivery systems offer exceptional versatility and tunability for miRNA applications.

(i) *Poly(lactic-co-glycolic acid) (PLGA)* nanoparticles represent the gold standard in this category, since they are biodegradable, FDA-approved, and capable of sustained release, with 60–85% of their payload being released over 6–14 days. Research has demonstrated the successful delivery of multiple miRNAs, including miR-150, for pancreatic cancer (Arora et al. 2014) miR-34a for multiple myeloma (Cosco et al. 2015), and miR-99a for hepatic carcinoma (Cai et al. 2017).
(ii) *Polyethylenimine (PEI)* systems provide high transfection efficiency through the proton sponge effect for endosomal escape. While traditional PEI faces cytotoxicity concerns, innovative modifications such as disulfide cross-linked PEI (PEI–SS) have shown improved safety profiles for breast cancer gene therapy applications (Dai and Zhang 2019).
(iii) *Chitosan nanoparticles* offer excellent biocompatibility and biodegradability with high entrapment efficiency reaching 92.9%. Studies have demonstrated successful applications in cholesterol metabolism modulation with miR-33 (Nguyen et al. 2019), and wound healing applications.

### 9.2.4 Exosome-Based Systems

Exosomes represent natural biological carriers that offer unique advantages for miRNA delivery, including low immunogenicity and natural cell-to-cell communication capabilities.

(i) *Natural exosomes* derived from mesenchymal stem cells and other cell types have shown success in delivering miR-15a for immune modulation and miR-26a for hepatocellular carcinoma treatment (Zhang et al. 2018; Mahati et al. 2021).
(ii) *Engineered exosomes* enhance targeting specificity through surface modifications such as scFv (single-chain variable fragment) targeting moieties. These systems maintain the natural advantages of exosomes while providing enhanced loading efficiency and improved therapeutic targeting (Mahati et al. 2021).

### 9.2.5 Inorganic Nanoparticle Systems

Inorganic nanoparticles offer unique properties for miRNA delivery applications. These properties enhance the stability and cellular uptake of miRNAs, potentially improving therapeutic outcomes in various diseases.

(i) *Gold nanoparticles (AuNPs)* provide excellent stability and easy surface modification capabilities. Studies have demonstrated the successful delivery of miR-214 for calcific aortic valve disease (Xu et al. 2025) miR-206 for breast cancer (Chaudhari et al. 2022), and various miRNAs through self-assembled monolayer (SAM) coated systems (Hoang et al. 2023).
(ii) *Mesoporous silica nanoparticles (MSNs)* offer high loading capacity and controlled release characteristics. Research has shown successful immunomodulation with miR-26a delivery, dual miRNA therapy for triple-negative breast cancer using miR-34a-mimic/antisense-miR-10b combinations, and tumor-targeted delivery of miR-200c-3p.
(iii) *Quantum dots (QDs)* provide unique dual functionality combining miRNA delivery with real-time imaging capabilities. PEI-coated CdSe/ZnS quantum dots have demonstrated successful miR-26a delivery for hepatocellular carcinoma treatment with simultaneous fluorescent tracking (Liang et al. 2016).

### 9.2.6 Hydrogel-Based Systems

Hydrogel systems offer sustained release capabilities and excellent biocompatibility for local miRNA delivery applications.

(i) *Natural hydrogels*, including hyaluronic acid, chitosan, and gelatin, provide biodegradable matrices for controlled miRNA release. Studies have demonstrated successful corneal regeneration with miR-24-3p delivery (Sun et al. 2023), wound healing with miR-126-3p, and various inflammatory disease treatments (Hu et al. 2024).
(ii) *Composite hydrogels* combine multiple materials to enhance mechanical properties and delivery capabilities. Examples include graphene oxide-PEG-PEI chitosan systems for miR-29b delivery and gold nanoparticle-polymer hydrogels for has-miR-214 delivery (Qin et al. 2022; Van der Ven et al. 2021).
(iii) *Stimuli-responsive hydrogels* provide precise spatiotemporal control through external triggers such as UV light, pH changes, or temperature variations (Luo et al. 2025). These systems enable on-demand miRNA release for applications including controlled bone regeneration and deep tissue therapy.

### 9.2.7 Peptide-Based Systems

Peptide-based delivery systems leverage the natural properties of amino acid sequences for miRNA transport. These systems can enhance the stability and bioavailability of miRNAs, allowing for more effective therapeutic applications. By optimizing the peptide sequence, researchers can improve targeting and uptake in specific tissues, potentially providing targeted treatments for various diseases (Chauhan et al. 2025).

(i) *Cell-penetrating peptides (CPPs)* offer direct cytoplasmic delivery capabilities, avoiding endosomal entrapment. The MPG system and other CPPs have shown success in clinical trials for various therapeutic applications (Kim et al. 2022).
(ii) *Self-assembling peptides (SAPs)* provide unique advantages, including self-organization into ordered nanofibers and stem cell homing capabilities. The SKPPGTSS-functionalized SAP system has demonstrated successful agomir-29b-5p delivery for joint inflammation treatment with minimal invasive injection protocols (Zhu et al. 2022).

Upon thorough investigations and clinical evaluations, it has been determined that, in terms of delivery efficiency, viral vectors such as adenoviral, lentiviral, and AAV systems are among the most effective. Their evolved cellular entry mechanisms allow them to achieve 90–100% transfection rates and provide stable, long-term expression, which is ideal for chronic disease treatment (Yang 2015). These vectors may target dividing and non-dividing cells, carry large genetic payloads (up to 38 KB for adenoviruses), and defend against nuclease degradation because of their capsid shapes. Due to preexisting immunity from natural viral infections, insertional mutagenesis, dose-dependent systemic inflammatory responses, and acquired immunity, 30–70% of patients are excluded from repeated administration, making it a limitation.

Whereas, non-viral delivery techniques often trade efficiency for biocompatibility, however, they address numerous safety concerns (Fig. 9.2). Lipid-based vectors, including commercially available lipofection reagents, have low immunogenicity, excellent safety, and tunable properties for co-delivery applications, but transfection efficiencies (5–60%) are low, transient expression requires repeated dosing, and endosomal entrapment reduces bioavailability (Yang 2015). Chitosan and PAMAM dendrimers have regulated release kinetics and protective encapsulation, but therapeutic cytotoxicity and complicated optimization requirements make them challenging to use (Wang et al. 2022). Inorganic nanoparticles, especially gold and silica-based systems, are stable and theranostic for combined therapy and imaging, although transfection efficiency is low and bioaccumulation is a concern. Extracellular vesicles offer natural biocompatibility, immune evasion through CD47 expression, and tissue-targeting capabilities that allow crossing biological barriers like the blood–brain barrier (Zeng et al. 2024), but production scalability, heterogeneous composition, and limited experimental data hinder their clinical development.

## 9.3 Enhancing miRNA Delivery Potency and Stability

### 9.3.1 Chemical Modifications of miRNAs

The miRNA is an oligonucleotide mimic, which has specific properties that make it less stable within a biological system. These oligonucleotide mimics are prone to degradation by nucleases present within the cell. Furthermore, their uptake within the cell is poor, and they have unpredictable pharmacokinetic profiles (Shen and

Corey 2018). Therefore, to overcome these limitations, certain chemical modifications are required, which can improve the potency of these therapeutics. Chemical modifications can reduce the off-target activity of miRNA by improving antisense strand selectivity for RISC loading. These modifications can also enhance cellular uptake and facilitate their escape from cells. miRNA can be made more stable by introducing changes in its phosphodiester bond or ribose backbone (Setten et al. 2019).

(i) *Ribose Sugar Modification:* Modification within the ribose sugar of miRNA can improve the binding affinity of miRNA with its target, reduce the immunological activations, and make it resistant to degradation from nuclease attacks. The most common modification within the ribose sugar includes replacement of 2-hydroxyl with more stable groups- 2′-O-methyl, 2′-fluoro (2′-F), and 2′-O-(2-methoxyethyl) (2′-MOE) (Manoharan 1999).

(ii) *Phosphorothioate Modifications (PS):* It is a type of modification in which oxygen within the phosphodiester bond is replaced with the sulfur atom. This modification increases the resistance toward nucleases like RNAse H and increases the half-life of miRNA within the biological system. (PS) Linkages introduce chiral centres within the antisense oligonucleotides (ASO), which improve their efficacy and pharmacokinetic properties (Dowdy 2017; Iwamoto et al. 2017). Other modifications that can be done within the phosphodiester bond of miRNA are methyl phosphonate-containing oligonucleotides and boranophosphate-containing oligonucleotides, which contain a methyl or borano group instead of oxygen (Fu et al. 2019)

(iii) *Locked Nucleic Acid (LNA) Oligonucleotides:* These are RNAs that have a 2′,4′ methylene bridge within the ribose sugar. This C3′-*endo* conformation within LNA enhances its affinity to target sequences (Fu et al. 2019; Stenvang et al. 2008).

### 9.3.2 Enhancing Cellular Uptake and Endosomal Escape

The cellular uptake of miRNA within the biological system is relatively poor. This condition occurs due to the low permeability of the cell membrane towards oligonucleotides. Strategies that can be used to improve miRNA delivery within targeted cells include the use of carriers such as liposomes, polymers, extracellular vesicles, and nanocomplex-based formulations. These carriers can be designed to respond to specific stimulus changes, enabling targeted release. Other methods include chemical modification of miRNA and ligand-mediated targeting of miRNA. Apart from this, direct delivery of miRNA can be done by intra-tumoral or topical injections of miRNA (Diener et al. 2022). Overcoming endosomal entrapment is essential for the delivery of miRNA inside cells. Usually, these miRNAs enter cells via endocytosis and get trapped within the endosomes and lysosomes. The acidic pH of these organelles degrades the miRNA. Thus, to improve their endosomal escape, several strategies have been exploited (Pei and Buyanova 2018).

1. *Proton sponges* are the polymers that absorb the protons released upon entry of polymer-linked miRNA within the endosome. As the polymer neutralizes these protons, more protons are pumped into the endosome to lower the pH. This results in the build-up of a positive charge within the endosome. To balance the protons, chloride ions also enter, creating an osmotic pressure that causes swelling of the endosome and rupture. The rupture of the endosome releases the miRNA cargo within the cytoplasm (Funhoff et al. 2004).
2. *Membrane Fusion:* This is another strategy to overcome endosomal entrapment. It involves the merging of the endosomal membrane with a carrier, resulting in the release of miRNA from the endosome to the cytoplasm of the targeted cell (Zelphati and Szoka Jr 1996; Wittrup et al. 2015)
3. Photochemical Internalization: In this approach, light is used as a trigger to burst endosomes, which causes the subsequent release of miRNA into the cytoplasm (Miyoshi et al. 2020).
4. *Ionophore-Mediated Targeted Release:* This method exploits the ion gradient differences between the endosome and the cytoplasm. Nigericin is an ionophore that exchanges potassium ions with hydrogen ions in the endosomal membrane. This causes osmotic imbalance, leading to endosomal swelling and rupture (Orellana et al. 2019).

### 9.3.3 Ligand-Mediated Targeting (Antibodies, Peptides, Aptamers)

The major challenge in miRNA-directed therapy is its systemic distribution, cellular uptake, penetration within the tissue, and endosomal escape. To improve target specificity, miRNA is often conjugated with ligands like aptamers, peptides, and antibodies. The ligands should be designed to interact specifically with the receptor expressed only on the diseased or target cell. These ligands are usually linked to the miRNA via linker molecules (Abdelaal and Kasinski 2021).

(i) Antibodies conjugated to RNAi have been studied in cancer therapeutics. ExomiR-Tracker is one of the studied antibody-conjugated miRNAs. ExomiT-Tracker contains anti-exosome antibodies (against CD63), which deliver the anti-miR-21 into exosome recipient cells. miRNA (MiTNA-101-5p) conjugated with HER2-targeting monoclonal antibodies such as trastuzumab and lapatinib has been shown to reduce breast cancer cell viability. By high-throughput screening, eight miRNAs have been identified that can reduce the viability of cancer cells when conjugated with trastuzumab and lapatinib. MiTNA-101-5p alone can reduce the cell viability by 50%, and in combination with lapatinib, its efficacy is increased by 75% (Normann et al. 2021).

(ii) Aptamers are short oligonucleotides that are designed to interact with their target with high specificity. Properties such as low immunogenicity and smaller size make them a suitable ligand for miRNA-directed therapy (Abdelaal and Kasinski 2021). Aptamer (GL21.T)-conjugated miRNA (let-7g) has been studied in targeting cancer cells overexpressing oncogenic receptor tyrosine kinase

Axl. This allows specific entry of *let*-7g miRNA into the oncogenic cells expressing Axl (Esposito et al. 2014). The miR-26a is a miRNA that targets breast cancer cells and hematopoietic stem/progenitor cells (HSPC) by targeting proapoptotic protein (Bak1). A gene that is significantly expressed in human cancers is c-KIT. Therefore, miR-26a conjugated with aptamers, which recognize c-KIT, has been designed for the specific entry of miRNA in cancer cells (Tanno et al. 2017).

### 9.3.4 Stimuli-Responsive Delivery (pH, Temperature, Light, Enzymes)

Several nanoparticles have been designed that recognize internal and external stimuli such as changes in pH, temperature, enzymes, hypoxia, redox conditions, and light. These nanoparticles, when conjugated with target miRNA, offload the miRNA to the site of action (Zhou et al. 2023).

(i) Glutathione tripeptide (GSH) is an antioxidant that is expressed in higher amounts in cancer cells (1–10 mM) as compared to normal cells (2–20 μm). A glutathione-activated DNA nanoprobe has been designed that uses disulfide bonds in DNA hairpin probes tethered on a metal-organic framework. The cleavage of the disulfide bond by endogenous GSH present in cancer cells releases the targeted delivery of miRNA (Wei et al. 2023b).
(ii) The ROS level of tumor cells is generally high, which can be used for the selective entry of miRNA. ROS-responsive micelles have been designed that sense the higher ROS within the cancer cell and release volasertib—a PLK1 inhibitor—and miR-34 within the targeted cell (Xin et al. 2019).
(iii) Metalloproteinase-2 (MMP2) is an enzyme highly expressed in tumor cells and tissue. A PEGylated cationic vector has been designed that contains an MMP2-sensitive peptide substrate. Upon reaching the tumor cell, MMP2 cleaves the peptide substrate, thus releasing the therapeutic miRNA-34a within the cell (Zeng et al. 2017).

## 9.4 Challenges in miRNA Delivery

The clinical translation of miRNA-based therapeutics remains highly challenging due to intertwined regulatory, biological, and manufacturing barriers. Despite their therapeutic promise, fewer than 20 miRNA drugs have entered clinical trials, with none advancing to phase III, in contrast to the more mature field of small interfering RNA (siRNA) therapeutics, where over sixty candidates are under investigation, and a few have already secured regulatory approval (Zhang et al. 2021). A higher percentage of terminated or suspended studies for miRNA drugs compared to siRNA (50% vs. 35%) highlights the unique difficulties that miRNA therapeutics face. While siRNAs are designed with strict sequence complementarity to a single target gene, miRNAs naturally regulate multiple transcripts, raising the risk of off-target interactions that could affect healthy tissues and contribute to systemic toxicity (Zhang et al. 2021).

In addition to biological complexity, regulatory approval presents another major obstacle. Agencies such as the FDA and EMA require robust demonstrations of safety, efficacy, and specificity, which are especially demanding for molecules with pleiotropic functions (Chakraborty et al. 2021). Delivery systems further complicate this landscape. Platforms including lipid nanoparticles, viral vectors, and chemically conjugated oligonucleotides are employed to enhance miRNA stability and targeting; however, they are scrutinized for potential immunogenicity, biodistribution, and long-term accumulation in nontarget tissues. Comprehensive pharmacokinetic (PK), pharmacodynamic (PD), and toxicological assessments under Good Laboratory Practice (GLP) conditions are mandatory before advancing to human trials (Bader et al. 2010).

Manufacturing also poses significant hurdles. Unlike small-molecule drugs, miRNAs are prone to rapid degradation and require chemical modifications (e.g., 2'-O-methylation or phosphorothioate linkages) to improve stability and reduce immune activation (van Rooij and Kauppinen 2014). These modifications, while beneficial, must be carefully validated to ensure they do not compromise therapeutic efficacy or introduce unforeseen toxicities. Batch-to-batch reproducibility is another critical issue, as even subtle variations in synthesis or formulation can affect drug performance. Regulatory authorities demand strict adherence to Good Manufacturing Practice (GMP), mandating detailed characterization of purity, potency, and stability, which substantially increases production costs and technical challenges (Mitragotri et al. 2014). Together, these factors explain the slower progress of miRNA therapeutics relative to siRNA counterparts. The combination of biological complexity, regulatory rigor, and manufacturing limitations underscores the need for innovative delivery strategies, advanced analytical tools, and harmonized global regulatory frameworks to unlock the therapeutic potential of miRNAs.

### 9.4.1 Systemic Barriers

Although systemic delivery of miRNA has been significantly improved, there are still challenges for its effective use. Some of the challenges include:

(i) *Rapid Clearance by Reticuloendothelial System (RES)*: RES is a mononuclear phagocyte system that plays a central role in clearing the nanoparticles present in the liver and spleen. Unprotected miRNA is generally cleared up via renal excretion, resulting in a shorter life span in the bloodstream (Chen et al. 2015; Holjencin and Jakymiw 2022).

(ii) *Blood–Brain Barrier (BBB) Penetration*: miRNA targeting brain tumors faces a significant drawback in its penetration. The blood–brain barrier is an interconnected network of endothelial cells. Pericytes and astrocytes are also present within this barrier. All these cells are tightly bonded, making the blood–brain barrier impermeable to many molecules, such as miRNA (Kim et al. 2016; Gareev et al. 2023).

### 9.4.2 Immunogenicity and Toxicity Concerns

(i) *Innate Immune Responses to Nucleic Acids and Vectors*: miRNA delivery within the biological system can activate innate immune responses by specifically acting as a ligand for toll-like receptors (TLR). miR-21 and miR-29a can bind with TLR7 and TLR8 via the GU-rich motif. This binding activates NF–kB pathways, leading to the secretion of cytokines, IL-6, and TNF-α. This recognition causes local cancer growth and metastasis (Chen et al. 2015).
(ii) *Off-Target Effects of miRNAs*: Since miRNAs bind via imperfect base pairing, they can regulate multiple genes of mRNA. This multiple binding of miRNA can result in unintentional gene regulation. Resulting in adverse symptoms (Segal and Slack 2020).
(iii) *Efficacy in* In Vivo *Models* vs. *Human Trials*: Due to the ability of miRNA to change gene expression in human disease, the research on miRNA has increased, leading to its entry in preclinical and clinical studies. The following are some of the miRNAs that are under clinical trials (Table 9.1):

### 9.4.3 Ethical Consideration

Therapeutics based on RNAi interference are developing at a faster rate. Clinical trials on potentially significant miRNAs has already begun. One of the essential criteria for any potential drug to reach the market is that it must pass through ethical evaluation. Some of the key ethical considerations are as follows:

(i) It is crucial to identify off-target effects due to the ability of miRNA to interact with multiple genes (Cullen 2006).
(ii) Patients participating in clinical trials should be well informed about the experimental nature and risks associated with miRNA interventions. Throughout the procedure, the consent of the patient should be considered paramount.
(iii) There should be proper transparency between the patients and the physicians about the risks associated with the therapeutics and potential side effects (Beauchamp and Childress 1994).
(iv) Promising but unapproved miRNA should be provided to patients on compassionate grounds, to patients with no alternatives (Ebbesen et al. 2008)

The successful development of miRNA-based therapeutics relies on several critical steps: identifying disease-associated miRNAs with strong biological relevance, designing modulators that can efficiently reach the intended cells or tissues, ensuring precise regulation of target gene expression, and minimizing unintended

**Table 9.1** miRNA-based therapeutics currently in clinical development, including their molecular design, developers, therapeutic targets, and clinical trial status

| S. No. | miRNA drug | Structure | Company | Target | Status | References |
|---|---|---|---|---|---|---|
| 1. | Miravirsen | 15-mer LNA-PS-modified ASO antagomir targeting miR-122 | Santaris Pharma and Roche Pharmaceuticals | hepatitis C virus (HCV) | Phase IIa | Gebert et al. (2014), Huang et al. (2021), Iacomino (2023), Janssen et al. (2013) |
| 2. | RG-012 | anti-miR-21 | Regulus Therapeutics | Alport syndrome | Phase II | Seyhan (2024) |
| 3. | MRX34 | miR-34a encapsulated into a liposome-formulated nanoparticle (NOV40) | miRNA Therapeutics Inc. | melanoma, NSCLC, hepatocellular carcinoma, and renal carcinoma. | Phase I | Beg et al. (2017) |
| 4. | Cobomarsen | LNA-based antagomir of miR-155 | Miragen Therapeutics | cutaneous T-cell lymphoma (CTCL), chronic lymphocytic leukemia, diffuse large B-cell lymphoma, and mycosis fungoides | Phase II | Seyhan (2024), Seto et al. (2018) |
| 5. | Remlarsen | LNA RNA mimic of miR-29 | MiRagen Therapeutics | fibrosis | Phase II | Seyhan (2024), Gallant-Behm et al. (2019) |
| 6. | TargomiRs | Mimic of miR-16 | EnGeneIC Ltd | leptomeningeal metastasis | Phase I | Brillante et al. (2024), Van Zandwijk et al. (2017) |
| 7. | ABX464/ Obefazimod | Enhancer of mir-124 | Abivax | HIV Infections, Rheumatoid Arthritis, COVID-19, Ulcerative Colitis, Ulcerative Colitis | Phase III | Brillante et al. (2024), Vermeire et al. (2021) |
| 8. | CDR132L | PS-modified LNA/DNA mixmer | Cardior Pharmaceuticals | Heart Failure, Myocardial Infarction/Heart Failure | Phase II | Vermeire et al. (2021), Batkai et al. (2021), Foinquinos et al. (2020) |
| 9. | AMT-130/ rAAV5-miHTT | Pri-miR-451 backbone-AAV5-delivered | uniQure | Huntington Disease | Phase I/II | Brillante et al. (2024), Caron et al. (2020) |

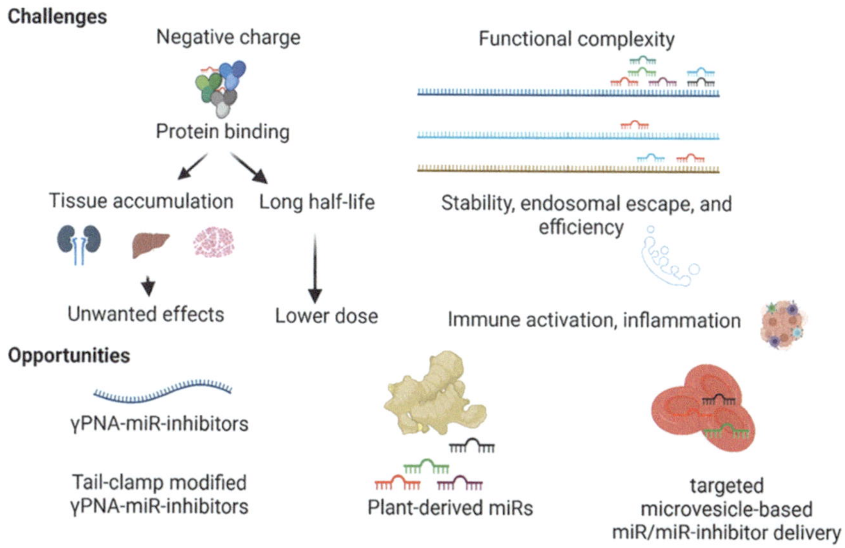


**Fig. 9.3** Key barriers and potential avenues in advancing miRNA therapeutics, highlighting delivery, safety, regulatory, and translational aspects. (Reproduced with permission from Momin et al. (2021)), Copyright @ 2021, MDPI)

off-target effects. However, the inherent biological complexity of miRNAs, along with their distinct physicochemical properties, presents significant obstacles in therapeutic design and delivery (Mitragotri et al. 2014). These challenges are summarized in Fig. 9.3. Negatively charged miRNA inhibitors often interact non-specifically with serum proteins, which can prolong their circulation time but also lead to unintended tissue accumulation and off-target effects. The complexity of miRNA–mRNA interactions further complicates therapeutic design, as a single miRNA can regulate numerous genes, while multiple miRNAs may converge on the same transcript, making precise target selection highly challenging (Fig. 9.3) (Momin et al. 2021; Bhattacharya et al. 2024). Additional barriers, including poor stability, limited endosomal escape, lack of specificity, and immune activation, further restrict clinical translation. To address these limitations, innovative strategies such as γ-modified peptide nucleic acid (PNA)-based inhibitors, tail–clamp structural modifications, plant-derived miRNAs, and targeted microvesicle-mediated delivery systems are being explored, offering promising avenues to enhance the safety and efficacy of miRNA therapeutics.

## 9.5 Summary and Conclusion

miRNA delivery systems have emerged as a central focus in advancing gene therapy, diagnostics, and novel therapeutics. This chapter has explored a range of delivery platforms, including viral vectors, lipid-based carriers like SNPs, polymer-based vehicles, inorganic nanoparticles, and hybrid systems, each offering unique advantages and limitations. Significant improvements in formulation chemistry, targeting strategies, and nanoparticle engineering have led to enhanced miRNA stability, protection against nuclease degradation, and efficient cellular uptake. Moreover, tailored surface modifications and ligand attachment have substantially increased the specificity of delivery, reducing off-target effects and improving therapeutic outcomes. Despite these advances, persistent challenges remain. Issues with immunogenicity, cytotoxicity, efficient release, endosomal escape, and precise tissue or cellular targeting continue to limit the full clinical potential of miRNA-based therapies. Balancing safety and efficacy, minimizing systemic toxicity, and scaling up production for widespread clinical use are active areas of research. Ongoing innovations such as stimuli-responsive releases, co-delivery with other therapeutic agents, and adaptive biomaterial designs promise to further enhance the translation of miRNA delivery platforms from research to real-world medical applications. The clinical translation of miRNA-based therapeutics depends on addressing ongoing issues related to stability, delivery, specificity, and safety. Despite considerable progress in areas like advanced nanocarrier design, chemical alterations, precision targeting, and stimuli-responsive systems, challenges persist, including quick clearance, inadequate tissue penetration, and off-target effects. Ongoing research and interdisciplinary collaboration are essential to enhance delivery strategies and mitigate toxicity and immunogenicity issues. In conclusion, while remarkable progress has been achieved in the field of miRNA delivery, overcoming the remaining obstacles will require interdisciplinary efforts and continued collaboration between material scientists, clinicians, and molecular biologists. The growing understanding of miRNA biology, coupled with next-generation nanotechnology, holds immense promise to realize the clinical potential of miRNA-based therapies, paving the way for highly targeted and personalized medical interventions. As these challenges are mitigated, miRNA therapies possess significant potential for precision medicine, disease diagnostics, and innovative interventions, likely revolutionizing the future of personalized healthcare.

**Acknowledgments** D.T. would like to thank Department of Biotechnology (DBT) for providing financial support through DBT-JRF fellowship. A.K.Y. sincerely thanks IIT Gandhinagar for the financial support provided through the Early Career Fellowship (ECF) award. D.B. acknowledges research funding support from the Science and Engineering Research Board (SERB) under the Core Research Grant (CRG), Gujarat Council on Science and Technology (GUJCOST) through the GSBTM program, and the STARS–MoES initiative.

**Conflict of Interest** The authors declare no conflict of interest.

## References

Abdelaal AM, Kasinski AL (2021) Ligand-mediated delivery of RNAi-based therapeutics for the treatment of oncological diseases. NAR. Cancer 3:zcab030

Arora S, Swaminathan SK, Kirtane A, Srivastava SK, Bhardwaj A, Singh S et al (2014) Synthesis, characterization, and evaluation of poly (D, L-lactide-co-glycolide)-based nanoformulation of miRNA-150: potential implications for pancreatic cancer therapy. Int J Nanomedicine 9:2933–2942

Awuson-David B, Williams AC, Wright B, Hill LJ, Di Pietro V (2023) Common microRNA regulated pathways in Alzheimer's and Parkinson's disease. Front Neurosci 17:1228927

Bader AG, Brown D, Winkler M (2010) The promise of microRNA replacement therapy. Cancer Res 70:7027–7030. https://doi.org/10.1158/0008-5472.CAN-10-2010/656532/P/THE-PROMISE-OF-MICRORNA-REPLACEMENT

Batkai S, Genschel C, Viereck J, Rump S, Bär C, Borchert T et al (2021) CDR132L improves systolic and diastolic function in a large animal model of chronic heart failure. Eur Heart J 42: 192–201

Beauchamp TL, Childress JF (1994) Principles of biomedical ethics. Edicoes Loyola

Beg MS, Brenner AJ, Sachdev J, Borad M, Kang Y-K, Stoudemire J et al (2017) Phase I study of MRX34, a liposomal miR-34a mimic, administered twice weekly in patients with advanced solid tumors. Investig New Drugs 35:180–188

Bhattacharya T, Tiwari S, Nandi A, Lin WY, Arumugam S (2024) Development of microRNA as prognostic markers in head and neck cancer. In: Diagnostic, prognostic, and therapeutic role of MicroRNAs in head and neck cancer. Academic Press, pp 193–205

Borel F, Kay MA, Mueller C (2014) Recombinant AAV as a platform for translating the therapeutic potential of RNA interference. Mol Ther 22:692–701

Brillante S, Volpe M, Indrieri A (2024) Advances in MicroRNA therapeutics: from preclinical to clinical studies. Hum Gene Ther 35:628–648

Cai C, Xie Y, Wu L, Chen X, Liu H, Zhou Y et al (2017) PLGA-based dual targeted nanoparticles enhance miRNA transfection efficiency in hepatic carcinoma. Sci Rep 7:46250

Caron NS, Southwell AL, Brouwers CC, Cengio LD, Xie Y, Black HF et al (2020) Potent and sustained huntingtin lowering via AAV5 encoding miRNA preserves striatal volume and cognitive function in a humanized mouse model of Huntington disease. Nucleic Acids Res 48:36–54

Chakraborty C, Sharma AR, Sharma G, Lee SS (2021) Therapeutic advances of miRNAs: a preclinical and clinical update. J Adv Res 28:127–138. https://doi.org/10.1016/J.JARE.2020.08.012

Chaudhari R, Nasra S, Meghani N, Kumar A (2022) MiR-206 conjugated gold nanoparticle based targeted therapy in breast cancer cells. Sci Rep 12:4713

Chauhan D, Yadav AK, Bhatia D, Solanki PR (2025) Global impact of vitamin D deficiency and innovative biosensing technologies. Chem Eng J 506:159790. https://doi.org/10.1016/J.CEJ.2025.159790

Chen Y, Gao D-Y, Huang L (2015) In vivo delivery of miRNAs for cancer therapy: challenges and strategies. Adv Drug Deliv Rev 81:128–141

Cosco D, Cilurzo F, Maiuolo J, Federico C, Di Martino MT, Cristiano MC et al (2015) Delivery of miR-34a by chitosan/PLGA nanoplexes for the anticancer treatment of multiple myeloma. Sci Rep 5:17579

Cullen BR (2006) Enhancing and confirming the specificity of RNAi experiments. Nat Methods 3: 677–681

Daby TPM, Modi U, Yadav AK, Bhatia D, Solanki R (2025) Bioimaging and therapeutic applications of multifunctional carbon quantum dots: recent progress and challenges. Next Nanotechnol 8:100158. https://doi.org/10.1016/J.NXNANO.2025.100158

Dai Y, Zhang X (2019) MicroRNA delivery with bioreducible polyethylenimine as a non-viral vector for breast cancer gene therapy. Macromol Biosci 19:1800445

Diener C, Keller A, Meese E (2022) Emerging concepts of miRNA therapeutics: from cells to clinic. Trends Genet 38:613–626

Dowdy SF (2017) Overcoming cellular barriers for RNA therapeutics. Nat Biotechnol 35:222–229

Ebbesen M, Jensen TG, Andersen S, Pedersen FS (2008) Ethical perspectives on RNA interference therapeutics. Int J Med Sci 5:159

Esposito CL, Cerchia L, Catuogno S, De Vita G, Dassie JP, Santamaria G et al (2014) Multifunctional aptamer-miRNA conjugates for targeted cancer therapy. Mol Ther 22:1151–1163

Falsini S, Ciani L, Ristori S, Fortunato A, Arcangeli A (2014) Advances in lipid-based platforms for RNAi therapeutics. J Med Chem 57:1138–1146. https://doi.org/10.1021/JM400791Q/ASSET/IMAGES/LARGE/JM-2013-00791Q_0005.JPEG
Foinquinos A, Batkai S, Genschel C, Viereck J, Rump S, Gyöngyösi M et al (2020) Preclinical development of a miR-132 inhibitor for heart failure treatment. Nat Commun 11:633
Fu Y, Chen J, Huang Z (2019) Recent progress in microRNA-based delivery systems for the treatment of human disease. ExRNA 1:1–14
Funhoff AM, van Nostrum CF, Koning GA, Schuurmans-Nieuwenbroek NME, Crommelin DJA, Hennink WE (2004) Endosomal escape of polymeric gene delivery complexes is not always enhanced by polymers buffering at low pH. Biomacromolecules 5:32–39
Gallant-Behm CL, Piper J, Lynch JM, Seto AG, Hong SJ, Mustoe TA et al (2019) A microRNA-29 mimic (Remlarsen) represses extracellular matrix expression and fibroplasia in the skin. J Invest Dermatol 139:1073–1081
Gareev I, Beylerli O, Tamrazov R, Ilyasova T, Shumadalova A, Du W et al (2023) Methods of miRNA delivery and possibilities of their application in neuro-oncology. Noncoding RNA Res 8:661–674
Garzon R, Marcucci G, Croce CM (2010) Targeting microRNAs in cancer: rationale, strategies and challenges. Nat Rev Drug Discov 9:10:9:775–9:789. https://doi.org/10.1038/nrd3179
Gebert LFR, Rebhan MAE, Crivelli SEM, Denzler R, Stoffel M, Hall J (2014) Miravirsen (SPC3649) can inhibit the biogenesis of miR-122. Nucleic Acids Res 42:609–621
Geisler A, Fechner H (2016) MicroRNA-regulated viral vectors for gene therapy. World J Exp Med 6:37. https://doi.org/10.5493/WJEM.V6.I2.37
German-Cortés J, Vilar-Hernández M, Rafael D, Abasolo I, Andrade F (2023) Solid lipid nanoparticles: multitasking nano-carriers for cancer treatment. Pharmaceutics 15:831. https://doi.org/10.3390/PHARMACEUTICS15030831
Ghafouri-Fard S, Shoorei H, Noferesti L, Hussen BM, Moghadam MHB, Taheri M et al (2023) Nanoparticle-mediated delivery of microRNAs-based therapies for treatment of disorders. Pathol Res Pract 248:154667
Gilyazova I, Asadullina D, Kagirova E, Sikka R, Mustafin A, Ivanova E et al (2023) MiRNA-146a—a key player in immunity and diseases. Int J Mol Sci 24:12767
Guney Eskiler G, Cecener G, Dikmen G, Egeli U, Tunca B (2018) Solid lipid nanoparticles: reversal of tamoxifen resistance in breast cancer. Eur J Pharm Sci 120:73–88. https://doi.org/10.1016/J.EJPS.2018.04.040
Hoang J, Patil SL, Srinoi P, Liu T, Marquez MD, Khantamat O et al (2023) Transfection of unmodified MicroRNA using monolayer-coated Au nanoparticles as gene-delivery vehicles. ACS Appl Bio Mater 7:230–237
Holjencin C, Jakymiw A (2022) MicroRNAs and their big therapeutic impacts: delivery strategies for cancer intervention. Cells 11:2332
Hsu S, Yu B, Wang X, Lu Y, Schmidt CR, Lee RJ et al (2013) Cationic lipid nanoparticles for therapeutic delivery of siRNA and miRNA to murine liver tumor. Nanomedicine 9:1169–1180
Hu S, Liang Y, Chen J, Gao X, Zheng Y, Wang L et al (2024) Mechanisms of hydrogel-based microRNA delivery systems and its application strategies in targeting inflammatory diseases. J Tissue Eng 15:20417314241265896
Huang W (2017) MicroRNAs: biomarkers, diagnostics, and therapeutics. In: Bioinformatics in MicroRNA research, pp 57–67
Huang P-S, Liao C-J, Huang Y-H, Yeh C-T, Chen C-Y, Tang H-C et al (2021) Functional and clinical significance of dysregulated microRNAs in liver cancer. Cancers (Basel) 13:5361
Iacomino G (2023) miRNAs: the road from bench to bedside. Genes (Basel) 14:314
Iwamoto N, Butler DCD, Svrzikapa N, Mohapatra S, Zlatev I, Sah DWY et al (2017) Control of phosphorothioate stereochemistry substantially increases the efficacy of antisense oligonucleotides. Nat Biotechnol 35:845–851
Janssen HLA, Reesink HW, Lawitz EJ, Zeuzem S, Rodriguez-Torres M, Patel K et al (2013) Treatment of HCV infection by targeting microRNA. N Engl J Med 368:1685–1694

Jogdeo CM, Panja S, Kanvinde S, Kapoor E, Siddhanta K, Oupický D (2023) Advances in lipid-based codelivery systems for cancer and inflammatory diseases. Adv Healthc Mater 12: 2202400. https://doi.org/10.1002/ADHM.202202400;REQUESTEDJOURNAL:JOURNAL:21922659

Ju J (2025) Challenges and opportunities in microRNA-based cancer therapeutics. Cell Rep Med 6: 102057

Karnati HK, Panigrahi MK, Gutti RK, Greig NH, Tamargo IA (2015) miRNAs: key players in neurodegenerative disorders and epilepsy. J Alzheimer's Dis 48:563–580

Kim DG, Kim KH, Seo YJ, Yang H, Marcusson EG, Son E et al (2016) Anti-miR delivery strategies to bypass the blood-brain barrier in glioblastoma therapy. Oncotarget 7:29400

Kim Y, Kim H, Kim EH, Jang H, Jang Y, Chi S-G et al (2022) The potential of cell-penetrating peptides for mRNA delivery to cancer cells. Pharmaceutics 14:1271

Küçüktürkmen B, Bozkır A (2018) Development and characterization of cationic solid lipid nanoparticles for co-delivery of pemetrexed and miR-21 antisense oligonucleotide to glioblastoma cells. Drug Dev Ind Pharm 44:306–315. https://doi.org/10.1080/03639045.2017.1391835;WGROUP:STRING:PUBLICATION

Kumar A, Patel DD, Banerjee J, Sharma M, Bhatia D, Banerjee S et al (2025) Next-generation cancer theragnostic: applications of carbon quantum dots. ChemNanoMat 11:e202500061. https://doi.org/10.1002/CNMA.202500061;WGROUP:STRING:PUBLICATION

Liang G, Li Y, Feng W, Wang X, Jing A, Li J et al (2016) Polyethyleneimine-coated quantum dots for miRNA delivery and its enhanced suppression in HepG2 cells. Int J Nanomedicine 11:6079–6088

Lu Y, Thavarajah T, Gu W, Cai J, Xu Q (2018) Impact of miRNA in atherosclerosis. Arterioscler Thromb Vasc Biol 38:e159–e170

Luo W, Cheng W, Ni S, Zhu X (2025) Wu M. Int J Biol Macromol, A review of stimuli-responsive hydrogels for RNA delivery: From material innovations to clinical barriers, p 144862

Mahati S, Fu X, Ma X, Zhang H, Xiao L (2021) Delivery of miR-26a using an exosomes-based nanosystem inhibited proliferation of hepatocellular carcinoma. Front Mol Biosci 8:738219

Manoharan M (1999) 2′-Carbohydrate modifications in antisense oligonucleotide therapy: importance of conformation, configuration and conjugation. Biochim Biophys Acta 1489:117–130

MartinoDi MT, Tagliaferri P, Tassone P (2025) MicroRNA in cancer therapy: breakthroughs and challenges in early clinical applications. J Exp Clin Cancer Res 44:1–14. https://doi.org/10.1186/S13046-025-03391-X

Melo SA, Esteller M (2011) Dysregulation of microRNAs in cancer: playing with fire. FEBS Lett 585:2087–2099. https://doi.org/10.1016/J.FEBSLET.2010.08.009;SUBPAGE:STRING:FULL

Milone MC, O'Doherty U (2018) Clinical use of lentiviral vectors. Leukemia 32:1529–1541

Mitragotri S, Burke PA, Langer R (2014) Overcoming the challenges in administering biopharmaceuticals: formulation and delivery strategies. Nat Rev Drug Discov 13:655–672. https://doi.org/10.1038/NRD4363

Miyoshi Y, Kadono M, Okazaki S, Nishimura A, Kitamatsu M, Watanabe K et al (2020) Endosomal escape of peptide-photosensitizer conjugates is affected by amino acid sequences near the photosensitizer. Bioconjug Chem 31:916–922

Momin MY, Gaddam RR, Kravitz M, Gupta A, Vikram A (2021) The challenges and opportunities in the development of MicroRNA therapeutics: a multidisciplinary viewpoint. Cells 10:3097. https://doi.org/10.3390/CELLS10113097

Nguyen M-A, Wyatt H, Susser L, Geoffrion M, Rasheed A, Duchez A-C et al (2019) Delivery of microRNAs by chitosan nanoparticles to functionally alter macrophage cholesterol efflux in vitro and in vivo. ACS Nano 13:6491–6505

Normann LS, Aure MR, Leivonen S-K, Haugen MH, Hongisto V, Kristensen VN et al (2021) MicroRNA in combination with HER2-targeting drugs reduces breast cancer cell viability in vitro. Sci Rep 11:10893

O'Brien J, Hayder H, Zayed Y, Peng C (2018) Overview of microRNA biogenesis, mechanisms of actions, and circulation. Front Endocrinol (Lausanne) 9:402

Orellana EA, Abdelaal AM, Rangasamy L, Tenneti S, Myoung S, Low PS et al (2019) Enhancing microRNA activity through increased endosomal release mediated by nigericin. Mol Ther Nucleic Acids 16:505–518

Pan Y, Zhang Y, Jia T, Zhang K, Li J, Wang L (2012) Development of a microRNA delivery system based on bacteriophage MS2 virus-like particles. FEBS J 279:1198–1208. https://doi.org/10.1111/J.1742-4658.2012.08512.X

Pei D, Buyanova M (2018) Overcoming endosomal entrapment in drug delivery. Bioconjug Chem 30:273–283

Qin H, Ji Y, Li G, Xu X, Zhang C, Zhong W et al (2022) MicroRNA-29b/graphene oxide–polyethyleneglycol–polyethylenimine complex incorporated within chitosan hydrogel promotes osteogenesis. Front Chem 10:958561

Rajagopal K, Arjunan P, Marepally S, Madhuri V (2021) Controlled differentiation of mesenchymal stem cells into hyaline cartilage in miR-140-activated collagen hydrogel. Cartilage 13: 571S–581S

Ren F-H, Yang H, He R, Lu J, Lin X, Liang H-W et al (2018) Analysis of microarrays of miR-34a and its identification of prospective target gene signature in hepatocellular carcinoma. BMC Cancer 18:12

Segal M, Slack FJ (2020) Challenges identifying efficacious miRNA therapeutics for cancer. Expert Opin Drug Discov 15:987–991

Seto AG, Beatty X, Lynch JM, Hermreck M, Tetzlaff M, Duvic M et al (2018) Cobomarsen, an oligonucleotide inhibitor of miR-155, co-ordinately regulates multiple survival pathways to reduce cellular proliferation and survival in cutaneous T-cell lymphoma. Br J Haematol 183: 428–444

Setten RL, Rossi JJ, Han S (2019) The current state and future directions of RNAi-based therapeutics. Nat Rev Drug Discov 18:421–446

Seyhan AA (2024) Trials and tribulations of MicroRNA therapeutics. Int J Mol Sci 25:1469

Sharma PC, Gupta A (2020) MicroRNAs: potential biomarkers for diagnosis and prognosis of different cancers. Transl Cancer Res 9:5798

Shen X, Corey DR (2018) Chemistry, mechanism and clinical status of antisense oligonucleotides and duplex RNAs. Nucleic Acids Res 46:1584–1600

Shi SJ, Zhong ZR, Liu J, Zhang ZR, Sun X, Gong T (2012) Solid lipid nanoparticles loaded with anti-microRNA oligonucleotides (AMOs) for suppression of microRNA-21 functions in human lung cancer cells. Pharm Res 29:97–109. https://doi.org/10.1007/S11095-011-0514-6/FIGURES/10

Shi S, Han L, Deng L, Zhang Y, Shen H, Gong T et al (2014) Dual drugs (microRNA-34a and paclitaxel)-loaded functional solid lipid nanoparticles for synergistic cancer cell suppression. J Control Release 194:228–237. https://doi.org/10.1016/J.JCONREL.2014.09.005

Stenvang J, Silahtaroglu AN, Lindow M, Elmen J, Kauppinen S (2008) The utility of LNA in microRNA-based cancer diagnostics and therapeutics. Semin Cancer Biol 18, Elsevier:89–102

Sun X, Song W, Teng L, Huang Y, Liu J, Peng Y et al (2023) MiRNA 24-3p-rich exosomes functionalized DEGMA-modified hyaluronic acid hydrogels for corneal epithelial healing. Bioact Mater 25:640–656

Tanno T, Zhang P, Lazarski CA, Liu Y, Zheng P (2017) An aptamer-based targeted delivery of miR-26a protects mice against chemotherapy toxicity while suppressing tumor growth. Blood Adv 1:1107–1119

Van der Ven CFT, Tibbitt MW, Conde J, van Mil A, Hjortnaes J, Doevendans PA et al (2021) Controlled delivery of gold nanoparticle-coupled miRNA therapeutics via an injectable self-healing hydrogel. Nanoscale 13:20451–20461

van Rooij E, Kauppinen S (2014) Development of micro RNA therapeutics is coming of age. EMBO Mol Med 6:851–864. https://doi.org/10.15252/EMMM.201100899/ASSET/E5588EDD-54CB-4613-B844-67AD162D4C0D/ASSETS/GRAPHIC/EMMM201100899-FIG-0002-M.PNG

Van Zandwijk N, Pavlakis N, Kao SC, Linton A, Boyer MJ, Clarke S et al (2017) Safety and activity of microRNA-loaded minicells in patients with recurrent malignant pleural mesothelioma: a first-in-man, phase 1, open-label, dose-escalation study. Lancet Oncol 18:1386–1396

Vermeire S, Hébuterne X, Tilg H, De Hertogh G, Gineste P, Steens J-M et al (2021) Induction and long-term follow-up with ABX464 for moderate-to-severe ulcerative colitis: results of phase IIa trial. Gastroenterology 160:2595–2598

Vyas A, Kumar K, Sharma A, Verma D, Bhatia D, Wahi N et al (2025) Advancing the frontier of artificial intelligence on emerging technologies to redefine cancer diagnosis and care. Comput Biol Med 191:110178. https://doi.org/10.1016/J.COMPBIOMED.2025.110178

Wang J, Chen J, Sen S (2016) MicroRNA as biomarkers and diagnostics. J Cell Physiol 231:25–30

Wang J, Li B, Qiu L, Qiao X, Yang H (2022) Dendrimer-based drug delivery systems: history, challenges, and latest developments. J Biol Eng 16:18

Wei X, Yu S, Zhang T, Liu L, Wang X, Wang X et al (2023a) MicroRNA-200 loaded lipid nanoparticles promote intestinal epithelium regeneration in canonical microRNA-deficient mice. ACS Nano 17:22901–22915

Wei R, Wang K, Liu X, Shi M, Pan W, Li N et al (2023b) Stimuli-responsive probes for amplification-based imaging of miRNAs in living cells. Biosens Bioelectron 239:115584

Wittrup A, Ai A, Liu X, Hamar P, Trifonova R, Charisse K et al (2015) Visualizing lipid-formulated siRNA release from endosomes and target gene knockdown. Nat Biotechnol 33:870–876

Xin X, Lin F, Wang Q, Yin L, Mahato RI (2019) ROS-responsive polymeric micelles for triggered simultaneous delivery of PLK1 inhibitor/miR-34a and effective synergistic therapy in pancreatic cancer. ACS Appl Mater Interfaces 11:14647–14659

Xu R, Sun Y, Tian F, Zhao M (2025) LncRNA NEAT1 sponges miR-214-3p to promote osteoblast differentiation through regulating the PI3K/AKT/mTOR pathway in aortic valve calcification. Sci Rep 15:13665

Yadav AK, Verma D, Tabassum S, Solanki PR (2023) Functionalized graphitic carbon nitride based aptasensing platform for electrochemical detection of Helicobacter Pylori. IEEE Sens Lett 7:1–4. https://doi.org/10.1109/LSENS.2023.3327589

Yadav SK, Yadav AK, Kaushik A, Solanki PR (2025a) Functionalized graphitic carbon nitride as an efficient electro-analytical platform for the label-free electrochemical sensing of interleukin-8 in saliva samples. Nanoscale 17:7926–7944. https://doi.org/10.1039/D4NR02039A

Yadav KA, Damini Verma R, Solanki P (2025b) Electrochemical biosensing of sperm protein 17 in clinical samples using a hydrothermally synthesized MoS 2 @rGO@MWCNTs ternary system without prior pre-treatment. J Mater Chem B 13:10187–10205. https://doi.org/10.1039/D5TB00741K

Yang N (2015) An overview of viral and nonviral delivery systems for microRNA. Int J Pharm Investig 5:179

Ying S-Y, Chang DC, Lin S-L (2008) The microRNA (miRNA): overview of the RNA genes that modulate gene function. Mol Biotechnol 38:257–268

Zelphati O, Szoka FC Jr (1996) Mechanism of oligonucleotide release from cationic liposomes. Proc Natl Acad Sci 93:11493–11498

Zeng Y, Zhou Z, Fan M, Gong T, Zhang Z, Sun X (2017) PEGylated cationic vectors containing a protease-sensitive peptide as a miRNA delivery system for treating breast cancer. Mol Pharm 14:81–92

Zeng B, Li Y, Xia J, Xiao Y, Khan N, Jiang B et al (2024) Micro Trojan horses: engineering extracellular vesicles crossing biological barriers for drug delivery. Bioeng Transl Med 9: e10623. https://doi.org/10.1002/btm2.10623

Zhang D, Lee H, Wang X, Rai A, Groot M, Jin Y (2018) Exosome-mediated small RNA delivery: a novel therapeutic approach for inflammatory lung responses. Mol Ther 26:2119–2130

Zhang S, Cheng Z, Wang Y, Han T (2021) The risks of mirna therapeutics: in a drug target perspective. Drug Des Devel Ther 15:721–733. https://doi.org/10.2147/DDDT.S288859; PAGE:STRING:ARTICLE/CHAPTER

Zhou S, Jin J, Wang J, Zhang Z, Freedman JH, Zheng Y et al (2018) miRNAS in cardiovascular diseases: potential biomarkers, therapeutic targets and challenges. Acta Pharmacol Sin 39:1073–1084

Zhou S, Li J, Zhang X, Xiong W (2022) MicroRNA-124 modulates neuroinflammation in acute methanol poisoning rats via targeting Krüppel-like factor-6. Bioengineered 13:13507–13519

Zhou H, Chen DS, Hu CJ, Hong X, Shi J, Xiao Y (2023) Stimuli-responsive nanotechnology for RNA delivery. Adv Sci 10:2303597

Zhu J, Yang S, Qi Y, Gong Z, Zhang H, Liang K et al (2022) Stem cell–homing hydrogel-based miR-29b-5p delivery promotes cartilage regeneration by suppressing senescence in an osteoarthritis rat model. Sci Adv 8:eabk0011

# miRNA-Based Combinatorial Approach for Therapeutic Efficacy Improvement of Chemotherapy in Relapse and Multidrug Resistance Tumor

10

Foram Lomesh Contractor, Karan Panchal, Raghu Solanki, Dhiraj Bhatia, and Unnati Modi

**Abstract**

MicroRNAs (miRNAs/miRs) are endogenous noncoding RNAs that regulate gene expression at the post-transcriptional level. In cancer biology, microRNAs (miRNAs) have emerged as key regulators of carcinogenesis, progression, and treatment resistance. Due to their ability to modulate multiple oncogenic and tumor-suppressive pathways, they are considered promising diagnostic biomarkers and therapeutic agents. However, despite of significant progress, the clinical translation of miRNA-based biomarkers and therapeutics remains limited due to challenges in stability, delivery, immune activation, and patient-specific responses. Additionally, naked miRNA therapeutics are particularly constrained by rapid degradation and poor bioavailability, underscoring the need for advanced therapeutic strategies including delivery systems. To overcome these limitations, combination therapies involving chemotherapeutic drugs,

Foram Lomesh Contractor and Karan Panchal contributed equally to this work.

F. L. Contractor
Department of Biotechnology, Parul Institute of Technology, Parul University, Vadodara, Gujarat, India

K. Panchal
Sarvajanik University of Engineering and Technology, Athwa, Surat, Gujarat, India

R. Solanki · D. Bhatia (✉)
Department of Biological Sciences and Engineering, Indian Institute of Technology, Gandhinagar, Palaj, Gujarat, India
e-mail: dhiraj.bhatia@iitgn.ac.in

U. Modi (✉)
School of Biotechnology & Bioengineering, Institute of Advanced Research, Gandhinagar, Gujarat, India
e-mail: unnati.modi@iar.ac.in





S. Tiwari et al. (eds.), *microRNA Modulation: A Promising Combination Therapy for Cancer*, https://doi.org/10.1007/978-981-95-5989-3_10

radiotherapy, immunotherapy, small molecules, antibodies, peptides, and nanotechnology-based formulations have demonstrated enhanced efficacy, reduced drug resistance, and improved therapeutic stability. Preclinical and clinical evidence supports the potential of such combination approaches in amplifying the anti-tumor effects of miRNAs. This chapter highlights the role of miRNAs in cancer initiation and progression, the limitations of miRNA therapies, and the emerging promise of miRNA-based combination therapies. We discuss how miRNA-based combination approaches offer an innovative therapeutic paradigm for effective, patient-friendly, and personalized cancer treatment; however, the need for further in-depth studies are required to optimize their clinical applications.

**Keywords**

microRNA · Cancer therapy · Combination therapy

**Learning Objectives**

- *Explain* the molecular mechanisms by which dysregulated miRNAs contribute to relapse and develop multi-drug resistance in tumors.
- *Analyze* how combinatorial approaches using miRNAs with chemotherapeutic agents enhance drug sensitivity and reduce resistance.
- *Evaluate* preclinical and clinical evidences supporting miRNA-based adjuvant therapies to improve chemotherapy outcomes.
- *Assess* the opportunities and challenges in translating miRNA–chemotherapy combinations into clinical practice, including delivery, specificity, and safety concerns.

## 10.1 Introduction

Cancer remains the second leading cause of death globally, after cardiovascular diseases (CVDs). Despite many advancements in cancer research, recent trends and statistics suggest that cancer may outperform CVDs to become the leading cause of mortality by the end of this century (Bray et al. 2021). The recent advances, particularly in the development of chemotherapy and targeted cancer therapy, have significantly enhanced the survival rates of patients. However, these cancer therapies are also limited by multidrug resistance (MDR) and cancer recurrence (Maimaitijiang et al. 2024). Slow-proliferative cancer stem cells (CSCs) often limits the efficacy of chemotherapeutic drugs like 5-flurouracil and gemcitabine, necessitating the development of alternative approaches. Multiple evidences indicates that micro-RNAs (miRNAs) play crucial roles in most cancer types. Recent cancer research has identified a wide range of miRNA that are dysregulated in specific cancers (Kim and Croce 2023; Sempere et al. 2021). These findings highlight the potential of miRNA as a biomarker for cancer diagnosis and prognosis.

Moreover, many miRNAs exhibit tumor-suppressive or oncogenic functions, making them attractive candidates for therapeutic targeting in clinical applications (Kim and Croce 2023).

miRNAs are small noncoding RNAs, typically 20–25 nucleotides in length (Ambros 2004). Since the discovery of the first miRNA, which was lin-4 in *Caenorhabditis elegans* in 1993 (Kay 2024), thousands of miRNAs have been identified across species, and their progressing pathways and functional mechanisms have been elucidated. Beyond regulating gene expression at the post-transcriptional level, studies suggest that miRNAs can also directly interact with proteins to modulate their activity (Castell-Auví et al. 2013). In oncology, miRNAs exhibit a dual role; functioning as tumor suppressors (e.g., miRNA-200c-3p, which inhibits epithelial–mesenchymal transition) (Nie et al. 2024) or as oncogenes (e.g., miR-21, which promotes cell proliferation and metastasis) (da Silvia Lima et al. 2024). Furthermore, miRNAs can directly suppress ATP-binding cassette (ABC) transporters, which mediate MDR, thereby sensitizing cancer cells to therapy (Mesci̇ 2025). Collectively, these features highlight the potential of miRNAs as powerful tools for overcoming MDR and advancing cancer treatment.

Nowadays, miRNA-based treatment strategies are gaining increasing attention. miRNA-based therapies can be employed not only as monotherapies but also in combination with immunotherapy and conventional chemotherapy to enhance their therapeutic efficacy (Chitkara et al. 2016). For example, in vivo studies demonstrated that vesicular co-delivery of miR159 and doxorubicin effectively silenced the T cell-factor 7 (TCF-7), resulting in enhanced anticancer efficacy without added toxicity (Gong et al. 2019). This synergistic approach highlights the potential of miRNA–drug combinations to inhibit key oncogenic pathways and improve therapeutic outcomes in cancer. Furthermore, combinational immunotherapies incorporating miRNAs hold significant promise in enhancing the effectiveness of immune checkpoint inhibitors. Additionally, miRNAs targeting CSCs can reduce tumor recurrence and enhance drug sensitivity, thereby amplifying the effect of chemotherapy.

Despite these advancements, the clinical application of miRNA-based strategies remains a major concern due to the instability of miRNAs in circulation, which limits their efficient delivery to target cancer cells. To address this issue, nanotechnology-based approaches have been explored to develop effective delivery systems such as lipid nanoparticles, exosomes, and polymeric carriers that encapsulate therapeutic agents, protect miRNAs from degradation, and enhance their stability for targeted delivery (Fawzy et al. 2024).

This chapter provides an overview of miRNA-based combination approaches aimed at enhancing the efficacy of chemotherapy in MDR and relapsed tumors. First, we describe miRNAs and their dual role as tumor suppressors and tumor promoters. Then, we discuss miRNAs' therapeutic potential and the development of combination approaches with miRNAs. The chapter also highlights future prospects and presents a promising outlook on miRNA-based therapies.

## 10.2 Link of miRNA and Cancer

Cancer develops when cells grow uncontrollably and lose the ability to undergo apoptosis. Several studies have demonstrated that miRNAs regulate cell proliferation and apoptosis (Zhang et al. 2007; Shenouda and Alahari 2009). The first study linking miRNAs to cancer was conducted in patients with chronic lymphocytic leukaemia, where miR-15a and miR-16-1 were found to be either downregulated or deleted in most cases (Calin et al. 2002). Another study showed that many human miRNA genes are located within genomic regions and fragile sites associated with cancer, suggesting that miRNAs may contribute to tumorigenesis (Calin et al. 2004). miRNAs can be involved in oncogenesis and can also act as onco-suppressors. Whether a particular miRNA is beneficial, detrimental, or disease-associated depends on its target genes and the cellular context. For example, miR-15 and miR-16 promote apoptosis by targeting the anti-apoptotic gene B-cell lymphoma 2 (BCL2) mRNA (Cimmino et al. 2005). BCL2 is a key factor in the pathogenesis of several human malignancies, including leukaemia, lymphomas, and carcinoma (Esquela-Kerscher and Slack 2006). Figure 10.1 illustrates the association of specific miRNAs with different types of cancers. Each miRNA is linked to a distinct malignancy, which highlights their potential as diagnostic biomarkers and therapeutic targets (Table 10.1).

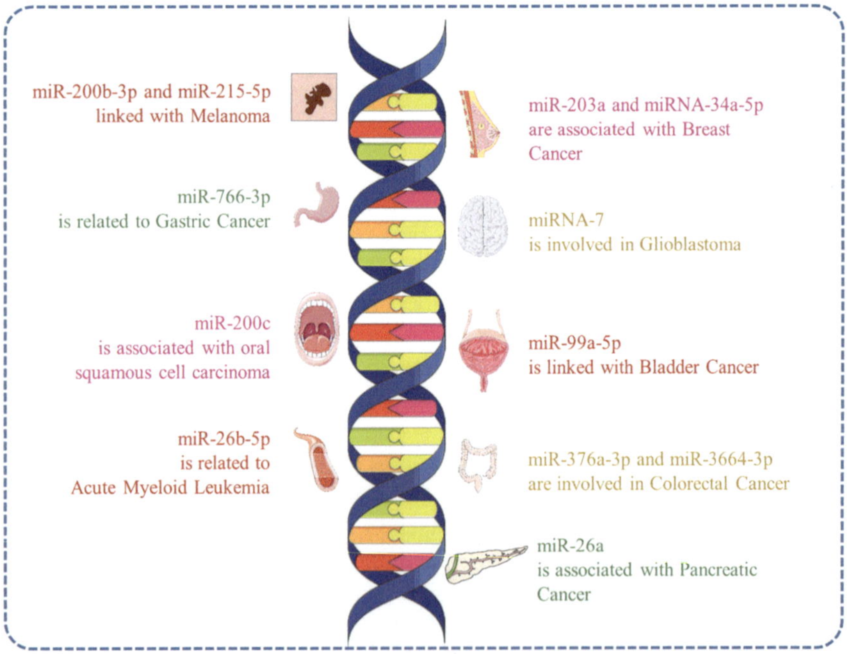


**Fig. 10.1** Representation of cancer-associated miRNAs and their correlation with different cancer types, including melanoma, breast, gastric, glioblastoma, bladder, and colorectal cancers

**Table 10.1** Examples of therapeutic miRNA, its role, molecular targets, and downstream effects in cancer

| Sr. No. | miRNA | Cancer involved | Role | miRNA targets | Downstream effects | References |
|---|---|---|---|---|---|---|
| 1 | miRNA-34a-5p | Breast cancer | Tumor suppressor | 3′ UTR of ABCC1 mRNA | Increased sensitivity of breast cancer cells to doxorubicin | Yahya et al. (2024) |
| 2 | miR-99a-5p | Bladder cancer | Tumor suppressor | FZD8 receptor of bladder cancer cells | Controlled Wnt/β-catenin pathway; reduced cell proliferation and invasion | Liu et al. (2024) |
| 3 | miR-26b-5p | Acute myeloid leukemia | Cancer progression | USP48 enzyme | Dysregulation of the Wnt/β-catenin pathway | Xie et al. (2024) |
| 4 | miR-203a | Breast cancer | Tumor suppressor | 3′ UTR of PIK3CA gene | Inhibits PI3K/Akt pathway; suppresses cell progression and migration | Entezari et al. (2024) |
| 5 | miR-766-3p | Gastric cancer | Tumor suppressor | COL1A1 gene | Regulates PI3K/Akt pathway; controls tumor growth and EMT | Ding et al. (2024a) |
| 6 | miR-200b-3p, miR-215-5p | Melanoma | Tumor suppressor | PI3K, AKT and ZEB1 | Suppresses the PI3K/AKT/ZEB1 axis; lowers cancer cell viability | Bhartiya et al. (2025) |
| 7 | miR-376a-3p | Colorectal cancer | Tumor suppressor | IGF1R gene | Increased sensitivity of cancer cells to CPT-11 | Jassi et al. (2024) |
| 8 | MiR-3664-3p | Colorectal cancer | Tumor suppressor | CYP3A4, ABCG2, MCL1, and MLH1 genes | Increased sensitivity of cancer cells to Irinotecan | Farrokhnazar et al. (2025) |
| 9 | miR-26a | Pancreatic cancer | Dual function depending on the radiation exposure | HMGA2 gene | Can lead to cancer progression or apoptosis of cancer cells, depending on the exposure | Jie et al. (2024) |
| 10 | miRNA-7 | Glioblastoma | Tumor suppressor | mTOR PI3K/Akt pathway | Inhibition of cell proliferation, cell cycle progression, invasion, migration, and PCNA expression while inducing apoptosis of cancer cells and tumor growth | Kutwin et al. (2024) |

(continued)

**Table 10.1** (continued)

| Sr. No. | miRNA | Cancer involved | Role | miRNA targets | Downstream effects | References |
|---|---|---|---|---|---|---|
| 11 | miR-200c | Oral squamous cell carcinoma | Tumor suppressor | Notch1, FAT1, CDKN2A, Zeb1, PIK3CA, p53, FADD, and CCND genes | Drastic reduction in cancer cell migration | Ding et al. (2024b) |

### 10.2.1 miRNA as Oncogenes and Suppressor Genes

Oncogenic miRNAs or oncomirs are defined as miRNAs that are significantly overexpressed in malignancies. Although only a few have been extensively characterized, they consistently function as oncogenes by negatively regulating tumor suppressor genes and genes involved in cell differentiation or apoptosis, thereby promoting tumor growth. Evidence from multiple studies and clinical research suggests that miRNAs can act as a novel class of oncogenes or tumor suppressor genes (Manikandan et al. 2008). miRNA-21 has been shown to play a crucial role in several types of cancer, including breast, colorectal, pancreatic, prostate, gastric, oral cancers, glioblastoma, and hepatocellular carcinoma (Yu et al. 2012). miRNA-21 located on chromosome band 17q23.2 targets genes such as SMAD7, PDCD4, tropomyosin (TPM1), and PTEN, thereby functioning as an oncogene and contributing to pathological phenotypes such as enhanced cell proliferation and inhibition of apoptosis. It has been demonstrated that miRNA-21 promotes carcinogenesis and is associated with histopathological features, including advanced tumor stage, lymph node metastasis, and reduced patient survival. Furthermore, miRNA-21 has been reported as a reliable indicator for breast cancer (Diansyah et al. 2021). It regulates LZTFL1 such that decreased level of miRNA-21 leads to increased LZTFL1 expression, which in turn inhibits breast cancer growth and metastasis (Wang et al. 2019).

In lung adenocarcinoma, reduced LRIG1 protein is linked to increased expression of hsa-miR-21(*Homo sapiens-miRNA-21*), which in turn promotes EGFR tyrosine kinase signalling (**MIR21** 2025). For treatment, the level of LRIG1 can be increased, which would decrease the level of hsa-miR-21 and hence inhibit the EGFR tyrosine kinase signalling pathway (Liu et al. 2025). In colorectal cancer (CRC), miRNAs such as miR-19b, miR-20a, and miR-21 are upregulated, where miR-20a alone, or in combination with miR-21 and miR-542, serves as a valuable diagnostic marker. Oncogenic miRNAs regulate multiple genes critical for tumor progression, migration, and drug resistance, including NTRK2, TrkB, CDC42, and EGFR (Schetter et al. 2012). In CRC, miR-21 promotes cancer growth by interacting with AGO2, a key mediator of RNA silencing, while other miRNAs modulate genes such as PRKCA, HSP90AA1, TLR4, IGF1, ESR1, SMAD2, SMAD4, and NEDD4L, all of which are involved in oncogenic signalling, cell survival and metastasis (Schetter et al. 2012; Cooke et al. 2022; Mezher et al. 2023; Catellani et al. 2021; Zuehlke et al. 2015). The SMAD2 and SMAD4 family located on chromosome band 18q21, are commonly deleted tumor suppressor gene in cancer (Fang et al. 2021). Similarly, NEDD4L9 (Neural precursor cell expressed, developmentally downregulated 4 Like) plays an important role in maintaining cell stability, and its downregulation leads to genomic instability and cancer progression. Collectively, these 12 genes are among the most frequent targets of oncogenic miRNAs, particularly through the tyrosine kinase and MAPK signalling pathways (Jamialahmadi et al. 2024).

Glioblastoma is a very aggressive brain tumor in which miRNAs play an important role in disease progression and treatment. In glioblastoma, elevated levels of certain miRNAs are associated with STAT3, a protein that promotes tumor growth and angiogenesis. Additionally, by modulating p53 activity, miRNAs can interfere with apoptosis. Measuring circulating miRNA levels in body fluids like blood may therefore aid in the early detection of glioblastoma (MicroRNAs 2025). In hepatocellular carcinoma (HCC), disruption in the balance between oncomirs (e.g., miRNA-21) and tumor suppressors (e.g., miRNA-122) contributes to cancer progression. Cancer-associated fibroblasts (CAFs) further support tumor growth by promoting immune evasion angiogenesis and metastasis. CAFs derived exosomes, enriched with proteins, lipids, and genetic material, travel through body fluids and influence distant cells, enhancing tumor development and metastasis (Hsu et al. 2025).

Sengupta, P., revealed that the metastatic long noncoding RNA (lncRNA) LINC00273 regulates cancer progression through its miRNA sponging mechanism (Sengupta et al. 2024). It functions as a master regulator of oncogenes involved in metastasis, stemness, and chemoresistance. They identified TGF-β1, SIX-1, and MAPK12 as its direct downstream targets. The strong association between TNBC aggressiveness and aberrant LINC00273 expression highlights its potential as a target for the development of lncRNA-based antimetastatic therapies.

## 10.3 miRNAs as Therapeutic Targets: Progress and Challenges

miRNAs have emerged as promising therapeutic targets due to their ability to regulate gene expression across multiple oncogenic and tumor suppressive pathways (Chakrabortty et al. 2023; Li and Rana 2014). By modulating key signaling networks, they can influence cell proliferation, apoptosis, metastasis, and treatment resistance. Both mimics and antagonists have been developed as miRNA-based therapeutic strategies for cancer management. According to recent clinical trial data, anti-miR compounds and specific inhibitors of miRNA hold significant potential as a novel class of cancer therapeutic (McLeod et al. 2011; Chakraborty et al. 2021; Chakraborty et al. 2017). For example, MRX34, a synthetic mimic of miR-34, was evaluated in clinical trials for liver cancer (Beg et al. 2017). Tuschl et al. obtained a patent for nucleic acids (18–25 nucleotides) that showed at least 90% identity to the complementary mature sequence of miR-122 (Tuschl et al. 2018). This innovation, referred to as the "Tuschl III" series, is considered one of the first patents in the field of miRNA therapeutics.

Despite their potential, several challenges limit the clinical translation of miRNA-based therapies. These include off-target effects, immune activation, and delivery inefficiencies (Di Martino et al. 2025). Single miRNA therapy often lacks sufficient efficacy due to the complexity of tumor biology, redundancy in signaling pathways, and tumor heterogeneity. Moreover, the instability of naked miRNAs due to susceptibility to nuclease-mediated degradation in biological systems, poor penetration across cell membranes, entrapment within endosomes, limited binding affinity to complementary sequence, and inefficient transport to target tissues (Segal and Slack 2020) are some of the key challenges hindering clinical translation of miRNA-based therapies. Tumors frequently adapt by activating compensatory mechanisms,

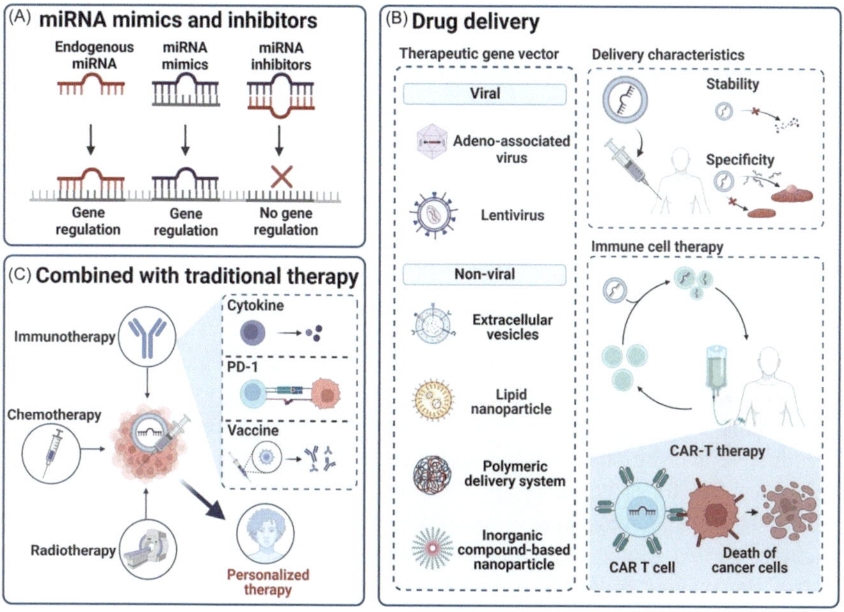


**Fig. 10.2** Overview of miRNA-based therapies, including mimics and inhibitors (**a**), delivery approaches through viral, nonviral, and immune cell-based systems (**b**), and their integration with conventional treatments for personalized cancer therapy (**c**) (Figure reprinted from Wang et al. (2024a), Copyright © 2024, John Wiley & Sons)

limiting the effectiveness of naked miRNA-based therapies and raising the risk of relapse. To overcome these limitations, combination therapies involving miRNAs have been explored. The rationale lies in the multi-targeting nature of miRNAs, which allows them to act synergistically with other miRNAs or conventional therapies to inhibit multiple pathways simultaneously. Combining miRNA therapeutics with chemotherapy, immunotherapy, or targeted agents enhances therapeutic efficacy, reduces drug resistance, and lowers the chances of tumor recurrence (Fig. 10.2) (Segal and Slack 2020; Yang et al. 2023; Wang et al. 2024a).

## 10.4 Combination Therapies with miRNAs

Combination therapies employing miRNAs have shown promising outcomes in preclinical and clinical studies. Co-delivery of miRNA mimics or inhibitors with chemotherapeutic drugs can suppress oncogenic pathways while sensitizing cancer cells to treatment (Yang et al. 2023). For instance, co-administration of miRNAs with doxorubicin has been shown to synergistically inhibit tumor growth and overcome multidrug resistance (Torki et al. 2021; Jamialahmadi et al. 2021). Similarly, combining miRNA-based therapies with nanotechnology enhances antitumor effects (Ghafouri-Fard et al. 2023; Bravo-Vázquez et al. 2023). For example,

MRX34, the first-in-class liposomal nanoparticles loaded with miR-34 mimic, which entered clinical trials as a cancer therapy (Beg et al. 2017). Such approaches highlight the feasibility of harnessing miRNAs as novel therapeutic agents in oncology. These strategies underscore the potential of miRNA-based combination therapies as a next-generation approach for more effective and durable cancer treatment.

### 10.4.1 miRNA with Chemotherapy, Radiotherapy, and Immunotherapy

**Chemotherapy** Chemotherapy remains a primary treatment for cancer, particularly in patients with metastases, improving both survival and quality of life. However, its use is limited by severe side effects due to damage to bone marrow, hair follicles, and reproductive and digestive systems cells (Solanki et al. 2022). To overcome these limitations, the focus is increasing toward a combination of chemotherapy with miRNA-based therapies (Seo et al. 2019). miRNAs influence cancer treatment outcomes by regulating genes involved in DNA repair, apoptosis, and cell-cycle control. For example, miR-155 and miR-21 disrupt DNA repair pathways, leading to 5-fluorouracil resistance in colorectal cancer (Valeri et al. 2010). On the other hand, certain miRNAs enhance drug sensitivity. For example, Li. et al. demonstrated that miR-22, a positive regulator of the tumor suppressor PTEN, increases sensitivity to paclitaxel in colon tumors with p53 mutations (Li et al. 2011).

In chemotherapy, miRNA can be used in combinations with chemotherapeutic drugs to precisely target molecules or signaling pathways, minimizing harm to healthy cells and promoting the death of malignant cells (van Beijnum et al. 2017). For example, in an in vitro study by Passadouro et al. the combination of anti-miR-21 and sunitinib was found to be more effective in inhibiting the growth of pancreatic ductal adenocarcinoma (PDAC) cells than anti-miR-21 with the standard chemotherapeutic agent gemcitabine (Passadouro et al. 2014). Inhibition of miR-21 has also been shown to enhance the efficacy of chemotherapeutic agents and trastuzumab in reducing the viability of erb-b2 receptor tyrosine kinase (ERBB2; HER2)-positive breast cancer cells (De Mattos-Arruda et al. 2015). This finding supports the use of miR-21 expression as a biomarker to identify patients with HER2-positive breast cancer who are resistant to trastuzumab chemotherapy and who may benefit from immunomodulatory or PI3K inhibitor-based therapies.

**Radiotherapy** Radiation response is determined by multiple processes such as autophagy, apoptosis, DNA damage response (DDR) activation, cell cycle checkpoint regulation, engagement of survival pathways, mutations in oncogenes or tumor suppressors, and hypoxia within the tumor microenvironment (Konoshenko et al. 2021). Most of these mechanisms are regulated by miRNAs, whose altered expression following ionizing radiation (IR) contributes to the development of either radio resistance or radiosensitivity. Radiation resistance is considered one of the major causes of local failure after radiotherapy and tumor recurrence (Mueller et al. 2016).

However, precise mechanisms underlying tumor radio resistance remain poorly understood. Recent studies have highlighted miRNAs as promising therapeutic and diagnostic agents in cancer, with growing evidence of their involvement in modulating the response to radiation (Korpela et al. 2015).

Metheetrairut and Slack reported correlations between altered miRNA expression levels and cancer radio resistance (Metheetrairut and Slack 2013). Supporting this, knock down of key miRNA biogenesis proteins such as Drosha and Dicer was shown to reduce DNA damage response (DDR) foci following IR, although effects on cell survival beyond disrupted cell-cycle checkpoints and proliferation were not assessed (Francia et al. 2012). Further evidence has been obtained in glioblastoma, where miR-181a was found to be downregulated. Whereas its transient overexpression significantly increased the radiosensitivity of malignant glioma cells by suppressing Bcl-2 (Chen et al. 2010). Such findings underscore the potential of miRNA-based approaches to improve therapeutic outcomes by enhancing sensitivity to irradiation or reducing the risk of tumor recurrence when applied alongside surgery or radiotherapy. For example, miR-34a (MRX34) has been shown to exert a radio-sensitizing effect in non-small cell lung cancer (NSCLC) by targeting the DNA damage response gene RAD51 (Curtin 2012).

**Immunotherapy** The roles of miRNAs in immunotherapy are diverse. They can modulate the immunosuppressive tumor microenvironment, regulate dendritic cell (DC) antigen representation and control immune checkpoint molecules, and enhance immunotherapy response (van Beijnum et al. 2017). Investigations into novel targets for miRNA mimic-based therapies are ongoing in combination with immunotherapies (Fig. 10.3) (Cortez et al. 2019; Zabeti Touchaei and Vahidi 2024).

For example, miR-124 mimics have been shown to inhibit STAT3, which is a key regulator of immunosuppression in the tumor microenvironment (TME) (Wei et al. 2013). Administration of miR-124 mimics enhances the production of immunogenic cytokines such as IFN-$\gamma$, TNF-$\alpha$, and IL-2 in the glioma microenvironment, resulting in potent anti-glioma effects. In addition to miR-124 and miR-15a/16 have also been implicated as immune regulator in cancers and inflammatory diseases (Xing et al. 2021). Yang et al. revealed the unique role of the tumor-associated miR-15a/16 cluster in modulating $CD8^+$ T cell (CTL)-mediated anti-tumor activity and tumor progression (Yang et al. 2017). In a glioma model using GL261 cells, CTLs displayed elevated levels of miR-15a/16. Conversely, CTLs deficient in miR-15a/16 showed enhanced cytokine release, cell proliferation, and more aggressive phenotypes within gliomas.

These findings highlight the critical role of miRNAs as modulators of the tumor-immune interface and underscore their potential as powerful adjuncts to chemotherapy, radiotherapy, and immunotherapy to improve cancer treatment outcomes. Combining miRNA-targeting strategies with these traditional cancer therapies may help overcome existing limitations. However, to fully realize this potential, further fundamental research and translational studies are required to better understand the roles of miRNAs in cancer therapy.

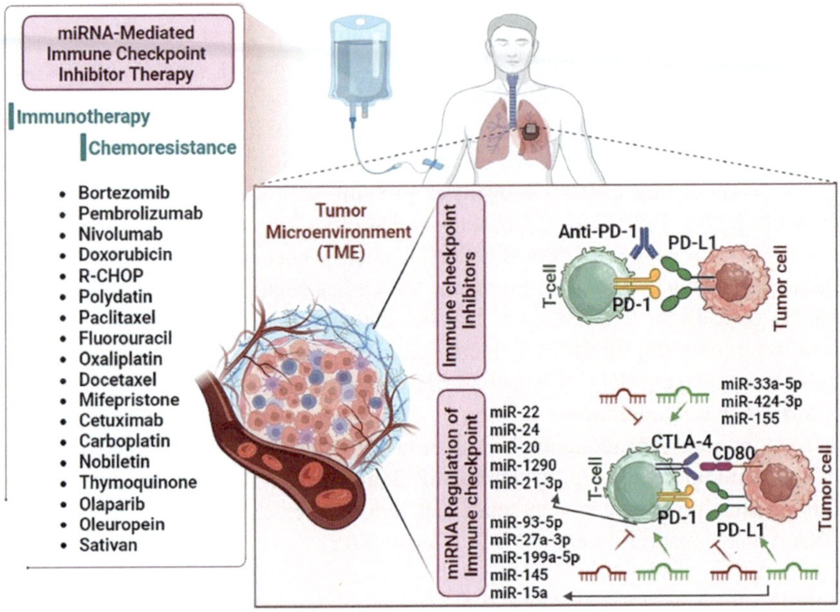


**Fig. 10.3** Schematic representation showing how miRNAs regulate immune checkpoint pathways and modulate tumor microenvironment responses (Figure reprinted from Zabeti Touchaei and Vahidi (2024), Copyright© 2024, Cancer Cell International, Springer Nature)

## 10.4.2 miRNA with Small Molecules, Antibodies, and Peptides

**Small Molecules** Small molecules are organic compounds having low molecular weight that perform diverse biological functions (Chen et al. 2020; Kumar et al. 2024). Studies have shown that small molecules can influence disease outcomes positively or negatively by blocking specific functions of multifunctional proteins or interfering with protein–protein interactions. The secondary structures of miRNAs make them "druggable," as the bulges and stem-loop structures in pre-miRNAs provide accessible binding sites for small molecules (Zhang et al. 2010). Alexander miRNAs must undergo three key processing stages; transcriptional, pre-transcriptional, and posttranscriptional—to fulfil their biological functions, and small molecules can inhibit miRNA maturation at each of these stages (Deiters 2010). For more detailed information on how small molecule inhibitors target miRNA at different stages, see the published review articles (Fig. 10.4) (Deiters 2010; Wen et al. 2015; Monroig et al. 2015).

Melo and Calin et al. first introduced the term small molecule inhibitors of miRNAs "SMIR" (small molecule that inhibits specific miRNA) to describe the interaction between small molecules and miRNAs (Melo et al. 2011). This idea foreshadowed a new direction of more targeted cancer therapy development. Gumireddy et al. identified the first small molecule inhibitor of miRNA function

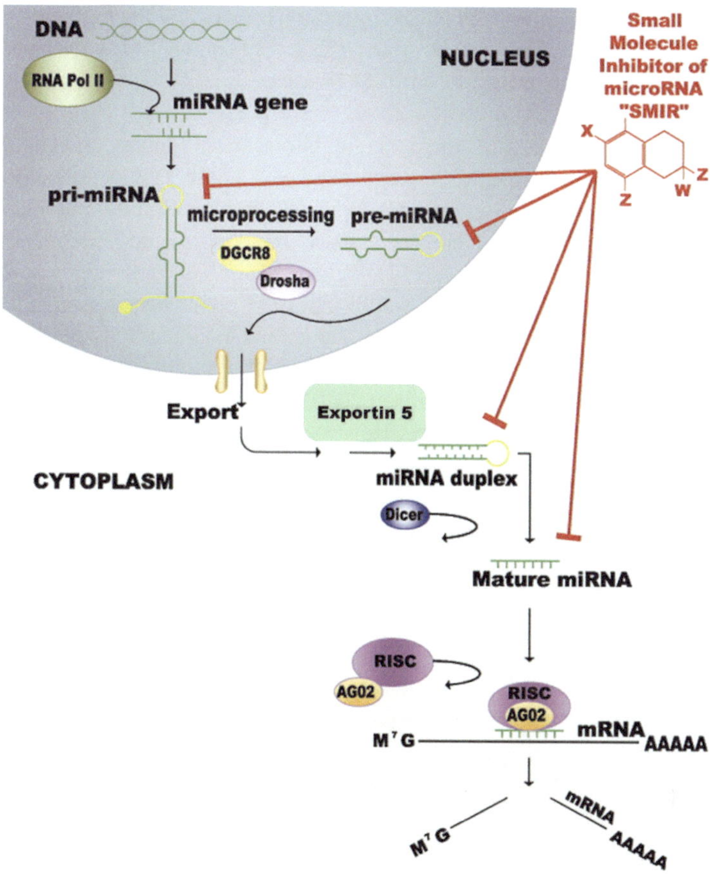


**Fig. 10.4** Schematic representation of miRNA biogenesis and the mechanism of small molecule inhibitors of miRNAs (SMIRs) targeting different stages to reduce mature miRNA levels (Figure reprinted from Monroig et al. (2015) with permission, Copyright© 2015, Elsevier B.V)

and developed a cellular screening assay for miRNA-pathway inhibitors (Gumireddy et al. 2008). Their study reported the creation of an assay capable of detecting inhibited miRNAs upon treatment with small molecules that disrupt and target their biogenesis pathway. Scott et al. reported that 22 miRNAs were downregulated following treatment with the histone deacetylase inhibitor LAQ824 (Scott et al. 2006). Lünse et al. identified an aptamer targeting the apical-loop domain of pri-miRNAs, which altered the maturation processing of miR-17, miR-18a, and miR-19 (Lunse et al. 2010). Using a fluorescence polarization-based high-throughput screening approach (HTS), Tan et al. discovered aurintricarboxylic acid as an inhibitor of RNA-Induced Silencing Complex (RISC) loading (Tan et al. 2012). Shi et al. used in silico HTS approach to design a more selective SMIR against miR-21 (Shi et al. 2013). By screening 1990 compounds with the help of MC-Fold/MC-Sym and Autodock programs, they analyzed the three-dimensional structure of the Dicer binding site on pre-miR-21. This strategy led to the identification of AC1MMYR2, an SMIR that effectively prevented Dicer-mediated processing of the precursor form of miR-21 in silico.

**Antibodies** In addition to small molecule inhibitors, several monoclonal antibody-based therapies are available for cancer treatment. These include unconjugated monoclonal antibodies, antibody-drug conjugates, immunotoxins, and bispecific single-chain antibodies that redirect cytotoxic lymphocytes to specific surface antigens on cancer cells (Kantarjian et al. 2012; Mihăilă 2020; Mehta et al. 2025). Immune checkpoint therapy is a well-established treatment option that has transformed the field of oncology. In 2011, the monoclonal antibody anti-CTLA-4 (Ipilimumab) was approved as the first immune checkpoint inhibitor (ICI) for metastatic melanoma (Sharma and Allison 2020). Since then, ICIs have transformed the cancer immunotherapies landscape. However, their benefits are limited as many patients fail to respond due to variations in the immune microenvironment and reduced tumor immunogenicity. To address this challenge, researchers are exploring the role of miRNAs in improving antibody-based therapies (Di Martino et al. 2021). Several miRNAs have been linked to the response and resistance of ICIs, making them valuable as predictive biomarkers.

Since tumor immune evasion is largely driven by the PD-1/PD-L1 signaling pathway, modulation of PD-L1 expression offers opportunities to reshape the tumor microenvironment (Zabeti Touchaei and Vahidi 2024). For example, Jiang and Zou reported that miR-20b-5p enhances tumor cell radiosensitivity in the presence of pembrolizumab, as it directly targets and negatively regulates PD-L1 (Jiang and Zou 2022). Nelson et al. reported the development of 2A8, a novel anti-Argonaute (AGO) monoclonal antibody that effectively immunoprecipitates miRNA and recognizes both human and mouse AGO proteins (Nelson et al. 2007). They further described its characterization and application in cloning miRNAs from human brain and polymorphonuclear leukocyte (neutrophil) preparations, leading to the identification of a common miRNA with unique characteristics.

Liu et al. developed nanobubbles (NBs) loaded with the miR-424 gene and programmed death-ligand 1 (PD-L1) antibody to assess their combined anti-tumor effect (Liu et al. 2022). Through target gene verification and bioinformatics analysis, miR-424 was identified as an anti-tumor gene capable of targeting both PD-L1 and Bcl-2. Using an H22 hepatoma-transplanted tumor model in mice, they evaluated the synergistic immunotherapeutic effect and underlying mechanism of the combined anti-PD-L1 antibody and miR-424 therapy. The study demonstrated that the targeted NBs facilitated the PD-L1 antibody in blocking the PD-1/PD-L1 signaling pathway, while the transfected miR-424 gene downregulated PD-L1 expression in tumor cells, thereby simultaneously enhancing the T cell-mediated antitumor immune response. Thus, combining miRNA therapeutics with antibody-based immunotherapy offers a promising strategy to overcome resistance and achieve better outcomes for cancer patients.

**Peptides** Peptide biology has gained increasing attention in recent years (Ormancey et al. 2024; Vishwanath et al. 2025). The discovery of complementary peptides (cPEPs) and microRNA-encoded peptides (miPEPs) has revealed novel peptide functions (Lauressergues et al. 2015; Ormancey et al. 2023). miPEPs are natural peptides encoded by primary microRNA transcripts (pri-miRNAs) that enhance the transcription of their corresponding microRNAs by directly interacting with their nascent mRNAs. Lauressergues et al. discovered and re-annotated several Arabidopsis thaliana pri-miRNAs to identify the ORFs encoding miPEPs (Lauressergues et al. 2022). Their findings demonstrated that miPEP activity depends on the presence of its own miORF, which explains both the lack of strong selection pressure on miPEP sequences and the ability of non-conserved peptides to trigger the expression of their corresponding miRNAs.

In another study, Lee et al. utilized tumor-targeting T7 peptides to develop AMO21 co-micelles, which were administered intranasally (Lee et al. 2023). Compared to naked AMO21c and AMO21c/scrambled T7c, these co-micelles showed superior distribution efficiency into the brain in intracranial GBM models, suggesting that T7c/AMO21c co-micelles may represent an effective intranasal delivery strategy for brain tumors. Furthermore, Kim et al. evaluated the combined biological effects of miRNA and peptides using a peptide/miRNA nanocomplex (Kim et al. 2019). They employed the TAT-Bim peptide, a cell-penetrating peptide with intrinsic apoptotic activity, as an RNA carrier. To investigate the synergistic effects, TAT-Bim peptide was complexed with the pro-apoptotic miRNA miR-34a, thereby enhancing the therapeutic potential of the nanocomplex. These findings highlight the emerging potential of peptide biology in enhancing miRNA-based therapies, where miPEPs, tumor-targeting peptides, and peptide/miRNA nanocomplexes collectively provide novel avenues for effective and targeted cancer treatment.

### 10.4.3 miRNA with Nanotechnology Approaches

miRNA therapies have shown promise in cancer management; however, the direct administration of exogenous miRNA mimics or anti-miRNAs faces major challenges, including rapid degradation by RNases, non-specific tissue absorption (e.g., hepatic uptake), clearance via the reticuloendothelial systems' (RES) effects, and potential toxicity (Fig. 10.5) (Wang et al. 2024b; Lee et al. 2019). Nanotechnology offers advanced drug delivery platforms that can protect miRNAs, ensure targeted delivery, and enhance their therapeutic efficacy, thereby accelerating their clinical translation (Wang et al. 2024b; Sharma et al. 2025). The properties of nanoparticles, such as small size, biocompatibility, biodegradability, and the ability to encapsulate multiple drugs, can enhance the circulation time, stability, and controlled release. Various nanocarriers have been engineered for miRNA delivery, including lipid-based (e.g., liposomes), inorganic nanoparticles (e.g., gold nanoparticles), and polymeric nanoparticles (e.g., chitosan and polyethyleneimine (PEI)) (Ganju et al. 2017).

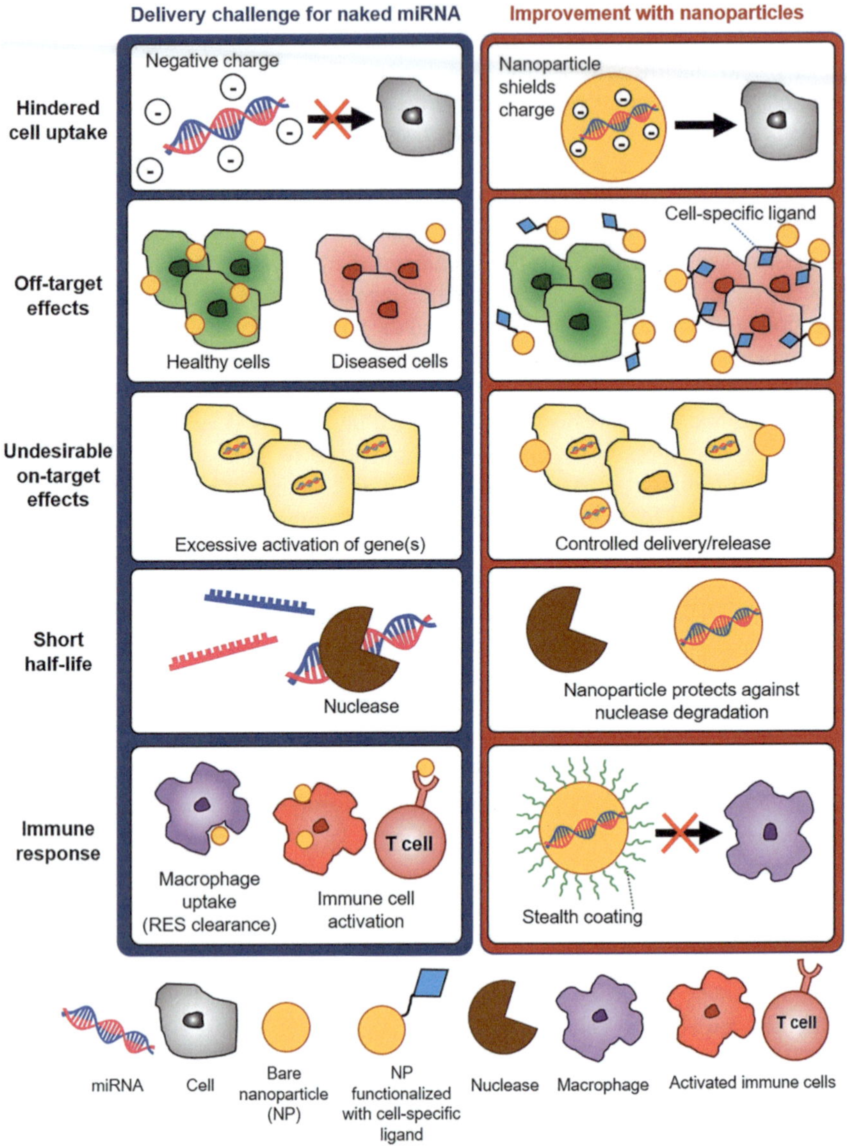


**Fig. 10.5** Schematic illustration representing challenges of naked miRNA delivery and nanoparticle-based solutions for efficient delivery and activity of miRNA. Naked miRNAs show poor uptake, off-target effects, rapid degradation, and immune clearance in vivo. Nanotechnology-based approaches overcome these limitations by shielding charge, enabling targeted delivery, protecting against nuclease degradation, and evading immune systems (Figure reprinted with permission from Lee et al. (2019), Copyrights© 2019, Elsevier B.V)

Among these, cationic polymers are widely used because their positive charge enables electrostatic conjugation with negatively charged nucleic acids. Moreover, compared to other polymer-based gene delivery systems, they exhibit relatively lower toxicity and reduced immunogenicity. For example, Conte et al. synthesized biodegradable polyethyleneimine (PEI)-functionalized polyhydroxybutyrate nanoparticles (PHB–PEI NPs) via aminolysis and employed them as cationic nonviral vectors for miR-124 delivery in PC3 (human prostate adenocarcinoma) cells. The PHB–PEI NP/miR-124 complex effectively protected miR-124 from RNase degradation, achieving a 30% higher delivery efficiency compared to the commercial transfection agent Lipofectamine RNAiMAX. Moreover, NP-delivered miR-124 inhibited key hallmarks of tumorigenicity, including cell motility, colony formation, and proliferation, through modulation of Carnitine Palmitoyltransferase 1A (CPT1A) (Conte et al. 2020).

This evidence suggests that polymeric nanocarriers are in demand for the delivery of miRNA. However, in physiological environments, nanoparticles can interact with biomolecules such as proteins and lipids, which can compromise their performance. PEGylation has emerged as an effective strategy to overcome these limitations by improving the nanoparticles' stability, prolonging circulation, and providing a stealth effect against immune recognition (Makharadze et al. 2025; Otsuka et al. 2003). For example, Devulapally et al. developed PEGylated–PLGA nanoparticles co-encapsulating gemcitabine (GEM) and antisense-miRNA-21 (Devulapally et al. 2016). Compared to nanoparticles loaded with either GEM or antisense-miRNA-21 alone at equal concentrations, the co-encapsulated system showed superior therapeutic efficacy in hepatocellular carcinoma (HCC) cells. These findings suggest that downregulating endogenous miRNA-21 enhances the sensitivity of HCC cells to GEM, thereby reducing cell viability and proliferation. Similarly, Zeng et al. designed a tailored delivery system by attaching cationic β-cyclodextrin–PEI conjugates to enzyme-cleavable polyethylene glycol (PEG) derivatives, which allowed condensation of the therapeutic payload while reducing the inherent toxicity of PEI (Zeng et al. 2017). Their study investigated the therapeutic potential of the tumor suppressor miRNA-34a (known to inhibit the growth and metastasis of various cancers) for breast cancer treatment. The PEG coating effectively minimized nonspecific interactions between serum proteins and cationic particles, facilitating desired accumulation of nanoparticles at the target site for appropriate delivery of the drugs or miRNAs.

## 10.5 Conclusion

miRNA-based approaches are considered a promising alternative to conventional cancer therapies. However, several challenges limit clinical translation of miRNA-based therapies, including off-target effects, immune system activation, and delivery inefficiencies. To address these limitations, diverse combination strategies have been explored, integrating miRNAs with chemotherapy, radiotherapy, immunotherapy, small molecules, peptides, and several nanotechnology-based platforms. Such strategies hold the potential to enhance the efficacyof both the conventional as well as next-generation cancer treatments. However, further fundamental research and translational studies are essential to fully understand the roles of miRNAs in cancer therapy and realize their promising clinical impact.

**Acknowledgments** The authors sincerely thank IIT Gandhinagar and IAR Gandhinagar for providing the necessary facilities. R.S. acknowledges the Anusandhan National Research Foundation (ANRF), Government of India, for financial support through the National Post-Doctoral Fellowship (NPDF), file no. PDF/2023/000365. D.B. thanks SERB-DST, GUJCOST, GSBTM, and STARS–MoES for research grants.

**Authors Contributions** F.C., K. P., and R.S. wrote the main manuscript text and prepared the figures. Figures are prepared using Biorender.com. U.M. and D.B. reviewed and edited the manuscript. All authors reviewed the final version of the manuscript.

**Competing Interests** The authors declare no competing interests.

## References

Ambros V (2004) The functions of animal microRNAs. Nature 431:350–355

Beg MS, Brenner AJ, Sachdev J, Borad M, Kang Y-K, Stoudemire J, Smith S, Bader AG, Kim S, Hong DS (2017) Phase I study of MRX34, a liposomal miR-34a mimic, administered twice weekly in patients with advanced solid tumors. Investig New Drugs 35:180–188

Bhartiya P, Jaiswal A, Negi M, Kaushik N, Ha Choi E, Kumar Kaushik N (2025) Unlocking melanoma suppression: insights from plasma-induced potent miRNAs through PI3K-AKT-ZEB1 axis. J Adv Res 68:147–161

Bravo-Vázquez LA, Méndez-García A, Rodríguez AL, Sahare P, Pathak S, Banerjee A, Duttaroy AK, Paul S (2023) Applications of nanotechnologies for miRNA-based cancer therapeutics: current advances and future perspectives. Front Bioeng Biotechnol 11:1208547

Bray F, Laversanne M, Weiderpass E, Soerjomataram I (2021) The ever-increasing importance of cancer as a leading cause of premature death worldwide. Cancer 127:3029–3030

Calin GA, Dumitru CD, Shimizu M, Bichi R, Zupo S, Noch E, Aldler H, Rattan S, Keating M, Rai K (2002) Frequent deletions and down-regulation of micro-RNA genes miR15 and miR16 at 13q14 in chronic lymphocytic leukemia. Proc Natl Acad Sci 99:15524–15529

Calin GA, Sevignani C, Dumitru CD, Hyslop T, Noch E, Yendamuri S, Shimizu M, Rattan S, Bullrich F, Negrini M (2004) Human microRNA genes are frequently located at fragile sites and genomic regions involved in cancers. Proc Natl Acad Sci 101:2999–3004

Castell-Auví A, Cedó L, Movassat J, Portha B, Sánchez-Cabo F, Pallarès V, Blay M, Pinent M, Ardévol A (2013) Procyanidins modulate microRNA expression in pancreatic islets. J Agric Food Chem 61:355–363

Catellani C, Ravegnini G, Sartori C, Angelini S, Street ME (2021) GH and IGF system: the regulatory role of miRNAs and lncRNAs in cancer. Front Endocrinol (Lausanne) 12:701246

Chakrabortty A, Patton DJ, Smith BF, Agarwal P (2023) miRNAs: potential as biomarkers and therapeutic targets for cancer. Genes (Basel) 14:1375

Chakraborty C, Sharma AR, Sharma G, Doss CGP, Lee S-S (2017) Therapeutic miRNA and siRNA: moving from bench to clinic as next generation medicine. Mol Ther Nucleic Acids 8: 132–143

Chakraborty C, Sharma AR, Sharma G, Lee S-S (2021) Therapeutic advances of miRNAs: a preclinical and clinical update. J Adv Res 28:127–138

Chen G, Zhu W, Shi D, Lv L, Zhang C, Liu P, Hu W (2010) MicroRNA-181a sensitizes human malignant glioma U87MG cells to radiation by targeting Bcl-2. Oncol Rep 23:997–1003
Chen X, Guan N-N, Sun Y-Z, Li J-Q, Qu J (2020) MicroRNA-small molecule association identification: from experimental results to computational models. Brief Bioinform 21:47–61
Chitkara D, Singh S, Mittal A (2016) Nanocarrier-based co-delivery of small molecules and siRNA/miRNA for treatment of cancer. Ther Deliv 7:245–255
Cimmino A, Calin GA, Fabbri M, Iorio MV, Ferracin M, Shimizu M, Wojcik SE, Aqeilan RI, Zupo S, Dono M (2005) miR-15 and miR-16 induce apoptosis by targeting BCL2. Proc Natl Acad Sci 102:13944–13949
Conte R, Valentino A, Di Cristo F, Peluso G, Cerruti P, Di Salle A, Calarco A (2020) Cationic polymer nanoparticles-mediated delivery of miR-124 impairs tumorigenicity of prostate cancer cells. Int J Mol Sci 21:869
Cooke M, Zhang X, Zhang S, Eruslanov E, Lal P, Daniel RE, Feldman MD, Abba MC, Kazanietz MG (2022) Protein kinase C alpha is a central node for check for updates tumorigenic transcriptional networks in human prostate cancer. Cancer Res Commun 2:1372–1387
Cortez MA, Anfossi S, Ramapriyan R, Menon H, Atalar SC, Aliru M, Welsh J, Calin GA (2019) Role of miRNAs in immune responses and immunotherapy in cancer. Genes Chromosomes Cancer 58:244–253
Curtin NJ (2012) DNA repair dysregulation from cancer driver to therapeutic target. Nat Rev Cancer 12:801–817
da Silvia Lima RQ, Vasconcelos CFM, Gomes JPA, Bezerra de Menezes E da S, de Oliveira Silva B, Montenegro C, Paiva Júnior S de SL, Pereira MC (2024) miRNA-21, an oncomiR that regulates cell proliferation, migration, invasion and therapy response in lung cancer. Pathol Res Pract 263:155601
De Mattos-Arruda L, Bottai G, Nuciforo PG, Di Tommaso L, Giovannetti E, Peg V, Losurdo A, Pérez-Garcia J, Masci G, Corsi F (2015) MicroRNA-21 links epithelial-to-mesenchymal transition and inflammatory signals to confer resistance to neoadjuvant trastuzumab and chemotherapy in HER2-positive breast cancer patients. Oncotarget 6:37269
Deiters A (2010) Small molecule modifiers of the microRNA and RNA interference pathway. AAPS J 12:51–60
Devulapally R, Foygel K, Sekar TV, Willmann JK, Paulmurugan R (2016) Gemcitabine and antisense-microRNA co-encapsulated PLGA–PEG polymer nanoparticles for hepatocellular carcinoma therapy. ACS Appl Mater Interfaces 8:33412–33422
Di Martino MT, Riillo C, Scionti F, Grillone K, Polerà N, Caracciolo D, Arbitrio M, Tagliaferri P, Tassone P (2021) miRNAs and lncRNAs as novel therapeutic targets to improve cancer immunotherapy. Cancers (Basel) 13:1587
Di Martino MT, Tagliaferri P, Tassone P (2025) MicroRNA in cancer therapy: breakthroughs and challenges in early clinical applications. J Exp Clin Cancer Res 44:126
Diansyah MN, Prayogo AA, Sedana MP, Savitri M, Romadhon PZ, Amrita PNA, Wijaya AY, Hendrata WM, Bintoro UY (2021) Early detection breast cancer: role of circulating plasma miRNA-21 expression as a potential screening biomarker. Turk J Med Sci 51:562
Ding Y, Zhang M, Hu S et al (2024a) MiRNA-766-3p inhibits gastric cancer via targeting COL1A1 and regulating PI3K/AKT signaling pathway. J Cancer 15:990–998
Ding QJ, Remy MT, Upara C, Hu J, Mora Mata AV, Haes AJ, Lanzel E, Sun H, Buchakjian MR, Hong L (2024b) CaCO3 nanoparticles delivering microRNA-200c suppress oral squamous cell carcinoma. J Dent Res 103:147–155
Entezari M, Soltani BM, Sadeghizadeh M (2024) MicroRNA-203a inhibits breast cancer progression through the PI3K/Akt and Wnt pathways. Sci Rep 2024 14:1 14:1–10
Esquela-Kerscher A, Slack FJ (2006) Oncomirs-microRNAs with a role in cancer. Nat Rev Cancer. https://doi.org/10.1038/nrc1840
Fang T, Liang T, Wang Y, Wu H, Liu S, Xie L, Liang J, Wang C, Tan Y (2021) Prognostic role and clinicopathological features of SMAD4 gene mutation in colorectal cancer: a systematic review and meta-analysis. BMC Gastroenterol 21:297

Farrokhnazar E, Moghbelinejad S, Najafipour R, Teimoori-Toolabi L (2025) MiR-3664-3p through suppressing ABCG2, CYP3A4, MCL1, and MLH1 increases the sensitivity of colorectal cancer cells to irinotecan. Heliyon 11:41933

Fawzy MP, Hassan HAFM, Sedky NK, Nafie MS, Youness RA, Fahmy SA (2024) Revolutionizing cancer therapy: nanoformulation of miRNA-34—enhancing delivery and efficacy for various cancer immunotherapies: a review. Nanoscale Adv 6:5220–5257

Francia S, Michelini F, Saxena A, Tang D, De Hoon M, Anelli V, Mione M, Carninci P, D'adda Di Fagagna F (2012) Site-specific DICER and DROSHA RNA products control the DNA-damage response. Nature 488:231–235

Ganju A, Khan S, Hafeez BB, Behrman SW, Yallapu MM, Chauhan SC, Jaggi M (2017) miRNA nanotherapeutics for cancer. Drug Discov Today 22:424–432

Ghafouri-Fard S, Shoorei H, Noferesti L, Hussen BM, Moghadam MHB, Taheri M, Rashnoo F (2023) Nanoparticle-mediated delivery of microRNAs-based therapies for treatment of disorders. Pathol Res Practice 248:154667

Gong C, Tian J, Wang Z, Gao Y, Wu X, Ding X, Qiang L, Li G, Han Z, Yuan Y (2019) Functional exosome-mediated co-delivery of doxorubicin and hydrophobically modified microRNA 159 for triple-negative breast cancer therapy. J Nanobiotechnol 17:93

Gumireddy K, Young DD, Xiong X, Hogenesch JB, Huang Q, Deiters A (2008) Small molecule inhibitors of microRNA miR-21 function. Angew Chem Int Ed Engl 47:7482

Hsu CY, Ahmed AT, Rab SO, Uthirapathy S, Ballal S, Kalia R, Arya R, Nathiya D, Kariem M, Kadhim AJ (2025) Cancer-associated fibroblast-derived exosomes in cancer progression: a focus on hepatocellular carcinoma. Exp Cell Res 445:114424

Jamialahmadi K, Zahedipour F, Karimi G (2021) The role of microRNAs on doxorubicin drug resistance in breast cancer. J Pharm Pharmacol 73:997–1006

Jamialahmadi H, Asadnia A, Khalili-Tanha G et al (2024) Identification of miR-20a as a diagnostic and prognostic biomarker in colorectal cancer: microRNA sequencing and machine learning analysis. MicroRNA 14:73–91

Jassi C, Kuo WW, Chang YC, Wang TF, Ho TJ, Hsieh DJY, Kuo CH, Chen MC, Li CC, Huang CY (2024) MicroRNA-376a-3p sensitizes CPT-11-resistant colorectal cancer by enhancing apoptosis and reversing the epithelial-to-mesenchymal transition (EMT) through the IGF1R/PI3K/AKT pathway. Transl Oncol 50:102125

Jiang K, Zou H (2022) microRNA-20b-5p overexpression combing Pembrolizumab potentiates cancer cells to radiation therapy via repressing programmed death-ligand 1. Bioengineered 13:917–929

Jie JM, Jing LC, Rao LF, Mei Z, Na GD, Tian L (2024) Pancreatic cancer cells hijack tumor suppressive microRNA-26a to promote radioresistance and potentiate tumor repopulation. Heliyon 10:31346

Kantarjian H, Thomas D, Wayne AS, O'Brien S (2012) Monoclonal antibody-based therapies: a new dawn in the treatment of acute lymphoblastic leukemia. J Clin Oncol 30:3876

Kay MA (2024) Ruvkun and Ambros recognized for miRNAs. Mol Ther Nucleic Acids 35:102379

Kim T, Croce CM (2023) MicroRNA: trends in clinical trials of cancer diagnosis and therapy strategies. Exp Mol Med 55:1314–1321

Kim H, Kitamatsu M, Ohtsuki T (2019) Combined apoptotic effects of peptide and miRNA in a peptide/miRNA nanocomplex. J Biosci Bioeng 128:110–116

Konoshenko MY, Bryzgunova OE, Laktionov PP (2021) miRNAs and radiotherapy response in prostate cancer. Andrology 9:529–545

Korpela E, Vesprini D, Liu SK (2015) MicroRNA in radiotherapy: miRage or miRador? Br J Cancer 112:777–782

Kumar R, Goel H, Solanki R, Rawat L, Tabasum S, Tanwar P, Pal S, Sabarwal A (2024) Recent developments in receptor tyrosine kinase inhibitors: a promising mainstay in targeted cancer therapy. Med Drug Discov 23:100195

Kutwin M, Sosnowska-ławnicka M, Wierzbicki M, Jaworski S, Nasiłowska B, Lange A (2024) The delivery of mimic miRNA-7 into glioblastoma cells and tumour tissue by graphene oxide nanosystems. Nanotechnol Sci Appl 17:167–188

Lauressergues D, Couzigou J-M, Clemente HS, Martinez Y, Dunand C, Bécard G, Combier J-P (2015) Primary transcripts of microRNAs encode regulatory peptides. Nature 520:90–93

Lauressergues D, Ormancey M, Guillotin B, San Clemente H, Camborde L, Duboé C, Tourneur S, Charpentier P, Barozet A, Jauneau A (2022) Characterization of plant microRNA-encoded peptides (miPEPs) reveals molecular mechanisms from the translation to activity and specificity. Cell Rep 38

Lee SWL, Paoletti C, Campisi M, Osaki T, Adriani G, Kamm RD, Mattu C, Chiono V (2019) MicroRNA delivery through nanoparticles. J Control Release 313:80–95

Lee Y, Ha J, Kim M, Kang S, Kang M, Lee M (2023) Antisense-oligonucleotide co-micelles with tumor targeting peptides elicit therapeutic effects by inhibiting microRNA-21 in the glioblastoma animal models. J Adv Res 53:249–260

Li Z, Rana TM (2014) Therapeutic targeting of microRNAs: current status and future challenges. Nat Rev Drug Discov 13:622–638

Li J, Zhang Y, Zhao J, Kong F, Chen Y (2011) Overexpression of miR-22 reverses paclitaxel-induced chemoresistance through activation of PTEN signaling in p53-mutated colon cancer cells. Mol Cell Biochem 357:31–38

Liu Y, Xie Q, Ma Y, Lin C, Li J, Hu B, Liu C, Zhao Y (2022) Nanobubbles containing PD-L1 Ab and miR-424 mediated PD-L1 blockade, and its expression inhibition to enable and potentiate hepatocellular carcinoma immunotherapy in mice. Int J Pharm 629:122352

Liu Y, Zhang K, Yang X (2024) CircMCTP2 enhances the progression of bladder cancer by regulating the miR-99a-5p/FZD8 axis. J Egypt Natl Canc Inst 36:1–9

Liu L, Liu X, Gao C, Liu M, Peng M, Wang L (2025) Hsa-miR-21 promoted the progression of lung adenocarcinoma by regulating LRIG1 expression. BMC Pulm Med 25:1–14

Lunse CE, Michlewski G, Hopp CS, Rentmeister A, Cáceres JF, Famulok M, Mayer G (2010) An aptamer targeting the apical-loop domain modulates pri-miRNA processing. Angew Chem Int Ed Engl 49:4674–4677

Maimaitijiang A, He D, Li D, Li W, Su Z, Fan Z, Li J (2024) Progress in research of nanotherapeutics for overcoming multidrug resistance in cancer. Int J Mol Sci 25:9973

Makharadze D, Del Valle LJ, Katsarava R, Puiggalí J (2025) The art of PEGylation: from simple polymer to sophisticated drug delivery system. Int J Mol Sci 26:3102

Manikandan J, Aarthi JJ, Kumar SD, Pushparaj PN (2008) Oncomirs: the potential role of non-coding microRNAs in understanding cancer. Bioinformation 2:330

McLeod BW, Hayman ML, Purcell AL, Marcus JS, Veitenheimer E (2011) The'real world'utility of miRNA patents: lessons learned from expressed sequence tags. Nat Biotechnol 29:129–133

Mehta A, Kulkarni N, Khot SS (2025) Monoclonal antibodies-based approaches in cancer immunotherapy. In: Nanotechnology based strategies for cancer immunotherapy: concepts, design, and clinical applications. Springer, pp 199–247

Melo S, Villanueva A, Moutinho C, Davalos V, Spizzo R, Ivan C, Rossi S, Setien F, Casanovas O, Simo-Riudalbas L (2011) Small molecule enoxacin is a cancer-specific growth inhibitor that acts by enhancing TAR RNA-binding protein 2-mediated microRNA processing. Proc Natl Acad Sci 108:4394–4399

Mesči S (2025) In silico identification of multidrug resistance gene (MDR)-targeted transposon miRNAs in human cancer. Mutat Res-Fund Mol Mech Mutagen 830:111903

Metheetrairut C, Slack FJ (2013) MicroRNAs in the ionizing radiation response and in radiotherapy. Curr Opin Genet Dev 23:12–19

Mezher M, Abdallah S, Ashekyan O, Al Shoukari A, Choubassy H, Kurdi A, Temraz S, Nasr R (2023) Insights on the biomarker potential of exosomal non-coding RNAs in colorectal cancer: an in silico characterization of related exosomal lncRNA/circRNA–miRNA–target Axis. Cells 12:1081

MicroRNAs as biomarkers for Glioblastoma and Therapeutic Opportunities Aashvi Dalmia, 2025

Mihăilă RG (2020) Monoclonal antibodies, bispecific antibodies and antibody-drug conjugates in oncohematology. Recent Pat Anticancer Drug Discov 15:272–292

MIR21 (human) | Gene Target—PubChem. https://pubchem.ncbi.nlm.nih.gov/gene/MIR21/human. Accessed 13 Aug 2025

Monroig P d C, Chen L, Zhang S, Calin GA (2015) Small molecule compounds targeting miRNAs for cancer therapy. Adv Drug Deliv Rev 81:104–116

Mueller A-K, Lindner K, Hummel R, Haier J, Watson DI, Hussey DJ (2016) MicroRNAs and their impact on radiotherapy for cancer. Radiat Res 185:668–677

Nelson PT, De Planell-Saguer M, Lamprinaki S, Kiriakidou M, Zhang P, O'Doherty U, Mourelatos Z (2007) A novel monoclonal antibody against human Argonaute proteins reveals unexpected characteristics of miRNAs in human blood cells. RNA 13:1787–1792

Nie F, Zhang Q, Ma WN, Yan J (2024) miRNA-200c-3p deficiency promotes epithelial-mesenchymal transition in triple-negative breast cancer by activating CRKL expression. Discov Oncol 15:1–11

Ormancey M, Guillotin B, Merret R, Camborde L, Duboé C, Fabre B, Pouzet C, Impens F, Van Haver D, Carpentier M-C (2023) Complementary peptides represent a credible alternative to agrochemicals by activating translation of targeted proteins. Nat Commun 14:254

Ormancey M, Tourneur S, Pouzet C, Thuleau P, Plaza S, Combier J-P (2024) Tips and rules for easy design of active microRNA-encoded peptides and complementary peptides. Plant Physiol 196: 2283–2285

Otsuka H, Nagasaki Y, Kataoka K (2003) PEGylated nanoparticles for biological and pharmaceutical applications. Adv Drug Deliv Rev 55:403–419

Passadouro M, Pedroso de Lima MC, Faneca H (2014) MicroRNA modulation combined with sunitinib as a novel therapeutic strategy for pancreatic cancer. Int J Nanomedicine:3203–3217

Schetter AJ, Okayama H, Harris CC (2012) The role of microRNAs in colorectal cancer. Cancer J 18:244

Scott GK, Mattie MD, Berger CE, Benz SC, Benz CC (2006) Rapid alteration of microRNA levels by histone deacetylase inhibition. Cancer Res 66:1277–1281

Segal M, Slack FJ (2020) Challenges identifying efficacious miRNA therapeutics for cancer. Expert Opin Drug Discov 15:987–991

Sempere LF, Azmi AS, Moore A (2021) microRNA-based diagnostic and therapeutic applications in cancer medicine. Wiley Interdiscip Rev RNA 12:e1662

Sengupta P, Roy A, Roy L, Bose D, Halder S, Jana K, Mukherjee G, Chatterjee S (2024) Long non-coding intergenic RNA, LINC00273 induces cancer metastasis and stemness via miRNA sponging in triple negative breast cancer. Int J Biol Macromol 274:132730

Seo HA, Moeng S, Sim S, Kuh HJ, Choi SY, Park JK (2019) MicroRNA-based combinatorial cancer therapy: effects of MicroRNAs on the efficacy of anti-cancer therapies. Cells 9:29

Sharma P, Allison JP (2020) Dissecting the mechanisms of immune checkpoint therapy. Nat Rev Immunol 20:75–76

Sharma R, Modi U, Kumar R, Sharma C, Srivastav AK, Bhatia D, Solanki R (2025) Bio-inspired, programmable biomacromolecules based nanostructures driven cancer therapy. Biomaterials Advances 171:214235

Shenouda SK, Alahari SK (2009) MicroRNA function in cancer: oncogene or a tumor suppressor? Cancer Metastasis Rev 28:369–378

Shi Z, Zhang J, Qian X, Han L, Zhang K, Chen L, Liu J, Ren Y, Yang M, Zhang A (2013) AC1MMYR2, an inhibitor of dicer-mediated biogenesis of oncomir miR-21, reverses epithelial–mesenchymal transition and suppresses tumor growth and progression. Cancer Res 73:5519–5531

Solanki R, Jodha B, Prabina KE, Aggarwal N, Patel S (2022) Recent advances in phytochemical based nano-drug delivery systems to combat breast cancer: a review. J Drug Deliv Sci Technol 77:103832

Tan GS, Chiu C-H, Garchow BG, Metzler D, Diamond SL, Kiriakidou M (2012) Small molecule inhibition of RISC loading. ACS Chem Biol 7:403–410

Torki Z, Ghavi D, Hashemi S, Rahmati Y, Rahmanpour D, Pornour M, Alivand MR (2021) The related miRNAs involved in doxorubicin resistance or sensitivity of various cancers: an update. Cancer Chemother Pharmacol 88:771–793

Thomas Tuschl et al (2018) MicroRNA molecules

Valeri N, Gasparini P, Braconi C, Paone A, Lovat F, Fabbri M, Sumani KM, Alder H, Amadori D, Patel T (2010) MicroRNA-21 induces resistance to 5-fluorouracil by down-regulating human DNA MutS homolog 2 (hMSH2). Proc Natl Acad Sci 107:21098–21103

van Beijnum JR, Giovannetti E, Poel D, Nowak-Sliwinska P, Griffioen AW (2017) miRNAs: micro-managers of anticancer combination therapies. Angiogenesis 20:269–285

Vishwanath R, Biswas A, Modi U, Gupta S, Bhatia D, Solanki R (2025) Programmable short peptides for modulating stem cell fate in tissue engineering and regenerative medicine. J Mater Chem B 13:2573–2591

Wang H, Tan Z, Hu H et al (2019) microRNA-21 promotes breast cancer proliferation and metastasis by targeting LZTFL1. BMC Cancer 19:738

Wang Z, Wang H, Zhou S, Mao J, Zhan Z, Duan S (2024a) miRNA interplay: mechanisms and therapeutic interventions in cancer. MedComm–Oncol 3:e93

Wang C, Zhang Y, Kong W, Rong X, Zhong Z, Jiang L, Chen S, Li C, Zhang F, Jiang J (2024b) Delivery of miRNAs using nanoparticles for the treatment of osteosarcoma. Int J Nanomedicine 19:8641–8660

Wei J, Wang F, Kong L-Y, Xu S, Doucette T, Ferguson SD, Yang Y, McEnery K, Jethwa K, Gjyshi O (2013) MiR-124 inhibits STAT3 signaling to enhance T cell–mediated immune clearance of glioma. Cancer Res 73:3913–3926

Wen D, Danquah M, Chaudhary AK, Mahato RI (2015) Small molecules targeting microRNA for cancer therapy: promises and obstacles. J Control Release 219:237–247

Xie Y, Tan L, Wu K, Li D, Li C (2024) miR-26b-5p affects the progression of acute myeloid leukemia by regulating the USP48-mediated Wnt/beta-catenin pathway. Crit Rev Eukaryot Gene Expr 34:33–44

Xing Y, Wang Z, Lu Z, Xia J, Xie Z, Jiao M, Liu R, Chu Y (2021) MicroRNAs: immune modulators in cancer immunotherapy. Immunotherapy Advances 1:ltab006

Yahya SMM, Nabih HK, Elsayed GH, Mohamed SIA, Elfiky AM, Salem SM (2024) Restoring microRNA-34a overcomes acquired drug resistance and disease progression in human breast cancer cell lines via suppressing the ABCC1 gene. Breast Cancer Res Treat 204:133–149

Yang J, Liu R, Deng Y, Qian J, Lu Z, Wang Y, Zhang D, Luo F, Chu Y (2017) MiR-15a/16 deficiency enhances anti-tumor immunity of glioma-infiltrating CD8+ T cells through targeting mTOR. Int J Cancer 141:2082–2092

Yang H, Liu Y, Chen L, Zhao J, Guo M, Zhao X, Wen Z, He Z, Chen C, Xu L (2023) MiRNA-based therapies for lung cancer: opportunities and challenges? Biomolecules 13:877

Yu Y, Kanwar SS, Patel BB, Oh PS, Nautiyal J, Sarkar FH, Majumdar APN (2012) MicroRNA-21 induces stemness by downregulating transforming growth factor beta receptor 2 (TGFβR2) in colon cancer cells. Carcinogenesis 33:68–76

Zabeti Touchaei A, Vahidi S (2024) MicroRNAs as regulators of immune checkpoints in cancer immunotherapy: targeting PD-1/PD-L1 and CTLA-4 pathways. Cancer Cell Int 24:102

Zeng Y, Zhou Z, Fan M, Gong T, Zhang Z, Sun X (2017) PEGylated cationic vectors containing a protease-sensitive peptide as a miRNA delivery system for treating breast cancer. Mol Pharm 14:81–92

Zhang B, Pan X, Cobb GP, Anderson TA (2007) microRNAs as oncogenes and tumor suppressors. Dev Biol 302:1–12

Zhang S, Chen L, Jung EJ, Calin GA (2010) Targeting microRNAs with small molecules: from dream to reality. Clin Pharmacol Ther 87:754–758

Zuehlke AD, Beebe K, Neckers L, Prince T (2015) Regulation and function of the human HSP90AA1 gene. Gene 570:8–16

# Combination of miRNA and Radiotherapy 11

Deepika Chauhan, Puja Patel, Bazil Gulzar Wani, and Tinku Basu

**Abstract**

The integration of microRNAs (miRNAs) with radiotherapy represents a novel paradigm in oncological treatment, aiming to enhance therapeutic efficacy while mitigating radiation resistance. miRNAs are small, noncoding RNA molecules that regulate gene expression posttranscriptionally and play a crucial role in modulating tumor radio sensitivity. Certain miRNAs act as radiosensitizers, enhancing radiation-induced apoptosis by targeting key regulators of DNA repair and cell survival. Notable examples include miR-34a, which suppresses DNA repair pathways and induces apoptosis through p53 activation, and miR-16, which downregulates Bcl-2, promoting radiation-induced cell death. Conversely, miRNA that confers radio-resistance, such as miR-210, miR-221/222, and miR-155, facilitate tumor cell survival by enhancing DNA damage repair, modulating hypoxia-associated signaling, and regulating inflammatory responses. The clinical application of miRNA-based therapy involves the use of miRNA mimics to enhance radiosensitivity and antagomirs to inhibit radio-resistance mechanisms. Additionally, nanoparticle-based delivery systems are being optimized to enhance stability, bioavailability, and targeted delivery of miRNA therapeutics. Besides, miRNAs hold promise as biomarkers for predicting radiotherapy outcomes, enabling more precise and individualized treatment strategies. However, critical challenges, including efficient delivery,

Deepika Chauhan and Puja Patel contributed equally with all other contributors.

D. Chauhan (✉) · B. G. Wani · T. Basu
Amity Centre for Nanomedicine, Amity University, Noida, UP, India
e-mail: deepikac@amity.edu

P. Patel
Department of Animal Health and Management, Alagappa University, Karaikudi, TN, India



S. Tiwari et al. (eds.), *microRNA Modulation: A Promising Combination Therapy for Cancer*, https://doi.org/10.1007/978-981-95-5989-3_11

off-target effects, and clinical validation, must be addressed. Rigorous preclinical and clinical investigations are essential to optimize the therapeutic feasibility and translational potential of miRNA-based radio-sensitization strategies, paving the way for their clinical implementation.

**Keywords**

microRNA · Radiotherapy · Radio-sensitization · Cancer therapy · miRNA mimics · Antagomirs

**Learning Objectives**

1. *Explain* the regulatory role of miRNAs in cellular responses to ionizing radiation, including DNA damage repair, apoptosis, and cell cycle control.
2. *Analyze* how dysregulated miRNAs contribute to radioresistance and tumor recurrence.
3. *Evaluate* strategies that combine therapeutic miRNA modulation with radiotherapy to enhance radiosensitivity and treatment efficacy.
4. *Assess* the translational potential and challenges of integrating miRNA-based interventions with radiotherapy in clinical oncology.

## 11.1 Introduction

Globally, cancer is still the leading cause of morbidity and mortality, necessitating ongoing advancements in therapeutic strategies. Radiotherapy is a keystone of cancer treatment, used exclusively or in conjunction with chemotherapy and surgery. This modality employs ionizing radiation to inflict DNA damage within malignant cells, triggering cell death through mechanisms such as apoptosis, mitotic catastrophe, or senescence (Chandra et al. 2021; Mathan et al. 2022). The therapeutic efficacy of radiotherapy is contingent upon multiple factors, including tumor oxygenation, DNA repair capacity, and the intrinsic radiosensitivity of cancer cells (Gao et al. 2023). However, the emergence of radioresistance in certain malignancies significantly limits treatment success, underscoring the necessity for novel radiosensitizers and synergistic combination therapies (Alamilla-Presuel et al. 2022).

One promising avenue for enhancing radiotherapy efficacy involves microRNAs (miRNAs), a class of small noncoding RNAs (18–25 nucleotides) that serve as critical regulators of posttranslational gene expression by targeting messenger RNAs (mRNAs) for degradation or translational repression (Darvish et al. 2023; Taeb et al. 2024). These regulatory molecules are integral to key cellular processes, including proliferation, apoptosis, differentiation, and stress responses. Within the context of oncogenesis, miRNAs function as either oncogenes (oncomiRs) or tumor suppressors, influencing tumor progression, metastatic potential, and responsiveness to therapeutic interventions (Otmani and Lewalle 2021). Aberrant miRNA expression in cancer has been implicated in dysregulation of gene expression networks that contribute to radioresistance, rendering miRNAs attractive candidates for therapeutic modulation (Otmani et al. 2024). Radiation therapy elicits a complex biological response in cancer cells, including DNA damage and repair mechanisms, apoptosis induction, and alterations in the tumor microenvironment. The DNA damage response (DDR) is crucial in determining the efficacy of radiotherapy, with homologous recombination (HR) and non-homologous end joining (NHEJ) serving as primary pathways for repairing radiation-induced DNA double-strand breaks (Huang and Zhou 2020). Dysregulation of these repair mechanisms contributes to radioresistance. Recent studies have revealed that miRNAs play a crucial role in modulating DNA repair pathways by targeting key repair proteins. Moreover, miRNAs influence radiation-induced cell death pathways, including apoptosis and autophagy, by regulating key apoptotic proteins such as B-cell lymphoma 2 (Bcl-2) and p-53 (Majidinia and Yousefi 2016; Jiao et al. 2022).

Beyond their role in intrinsic radiosensitivity, miRNA also regulates the tumor microenvironment, influencing hypoxia, angiogenesis, and immune responses to radiation therapy. Hypoxia, a well-established factor in radioresistance, is modulated by specific miRNAs that target hypoxia-inducible factors (HIFs) and angiogenic regulators (Hill et al. 2015). Additionally, miRNAs contribute to immune modulation post-radiation, affecting immune checkpoint pathways and inflammatory responses, which can either enhance or suppress therapeutic outcomes (Skafi et al. 2020).

Given their ability to regulate multiple radiation response pathways, miRNAs are being explored as potential therapeutic agents in combination with radiotherapy. miRNA-based therapeutics, including miRNA mimics and inhibitors, offer promising avenues to enhance radiosensitivity and counteract resistance mechanisms (Lo et al. 2020; Podralska et al. 2020). However, effective delivery remains a challenge, requiring the development of nanoparticle-based or viral vector-mediated strategies for clinical translation. Current research focuses on identifying miRNA biomarkers to predict radiation response and incorporating miRNA profiling into personalized medicine (Seyhan 2024; Hashemi et al. 2023).

Hence, this chapter will explore the intricate mechanisms by which miRNAs modulate radiation response in cancer, including their roles in DNA damage repair, apoptosis, cell cycle regulation, and the tumor microenvironment. Additionally, we discuss the clinical applications and emerging challenges associated with miRNA-based radiosensitization strategies, highlighting their potential in optimizing cancer therapy. Overall, the integration of miRNA-based therapy with radiotherapy represents an innovative approach to enhance therapeutic efficacy and mitigate resistance mechanisms.

## 11.2 Mechanism of Radiation Response in Cancer Cells

Radiation therapy exerts its therapeutic effects primarily by inducing DNA damage in cancer cells, ultimately leading to cell death. The response of cancer cells to radiation is governed by several key biological processes, including DDR, apoptosis, cell cycle regulation, and interactions with the tumor microenvironment (TME), as shown in Fig. 11.1 (Zhang et al. 2023; Wang et al. 2018). Radiation induces direct damage that occurs when radiation ionizes DNA molecules, causing strand breaks, while indirect damage results from the generation of reactive oxygen species (ROS), leading to oxidative stress and further DNA impairment (Jiao et al. 2022). Cells respond to this damage through two primary repair pathways: NHEJ, a quick but error-prone repair mechanism, and HR, a more accurate but slower repair process requiring a homologous DNA template. Deficiencies in these repair pathways can enhance radiosensitivity, whereas overactive repair mechanisms contribute to radioresistance (Szatkowska and Krupa 2020; Kim et al. 2015). Radiation-induced DNA damage also activates apoptotic pathways through p53-dependent and independent mechanisms. Key apoptosis regulators include p53, a tumor suppressor that triggers cell cycle arrest or apoptosis in response to DNA damage, Bcl-2 family proteins, which balance pro-apoptotic (Bax, Bak) and anti-apoptotic signals (Bcl-2, Bcl-xL), and caspases, executioner proteases that mediate apoptosis (Al-Arafat et al. 2024; Abuetabh et al. 2022). Additionally, radiation induces other types of cell

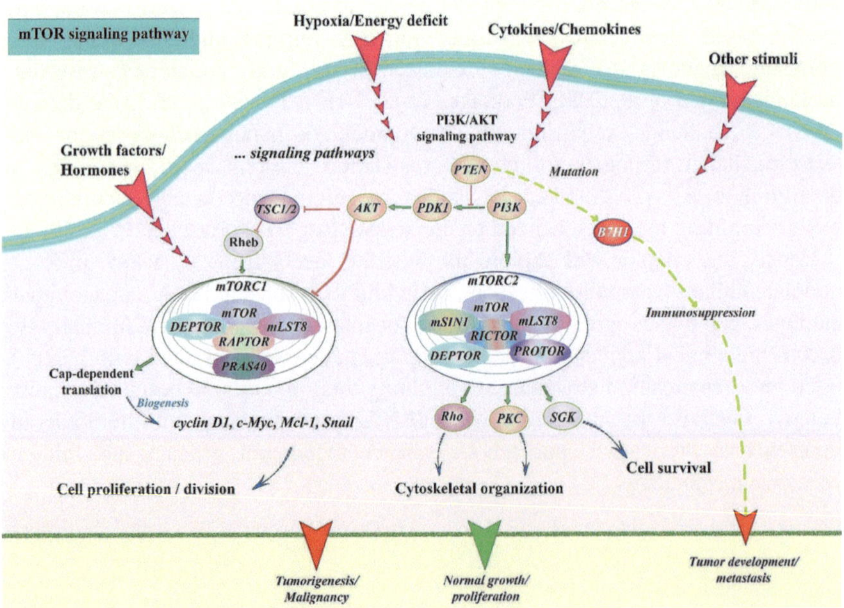


**Fig. 11.1** A comprehensive diagram of the signaling network for mTOR. (Adapted with permission from Taeb et al. (2024) under Creative Commons Attribution)

death, such as necrosis, autophagy, pyroptosis, and ferroptosis (Jiao et al. 2022). Another effective strategy enhancing tumor sensitivity involves targeting cell cycle checkpoint control, which regulates cell cycle progression and preserves genomic integrity. Inhibiting the kinases that govern these control mechanisms can disrupt cell cycle checkpoints and impair DNA damage repair, thereby increasing the susceptibility of tumor cells to radiation (Vlatkovic et al. 2022). The relative radiosensitivity is also influenced by cell cycle phase, as cells are most sensitive to radiation during G2-M phase (when the chromosomes are condensed and more vulnerable to damage), moderately sensitive during G1 phase, and least sensitive during the latter part of S phase (due to highly active DNA repair mechanisms and less condensed DNA) (Pawlik and Keyomarsi 2004). The TME further plays a crucial role in determining radiosensitivity and hypoxia, a key factor in TME, reduces the generation of ROS, leading to radioresistance. Radiation can also modulate the immune response by releasing tumor antigens, thereby stimulating immune activation, though tumors may evade immune detection through checkpoint pathways (Song et al. 2025; Derer et al. 2016; Yamazaki and Young 2022). Additionally, radiation influences tumor vasculature and angiogenesis, affecting oxygenation and nutrient supply. Thus, understanding these mechanisms is essential for developing strategies to enhance radiosensitivity and overcome resistance (Yamazaki and Young 2022). Emerging approaches, including miRNA-based therapies that target key molecular pathways involved in the radiation response, aiming to modulate DNA repair, apoptosis, cell cycle regulation, and interactions with TME, ultimately enhance the therapeutic efficacy of radiotherapy (He et al. 2020).

## 11.3 The Potential of miRNAs as Novel Markers of Tumor Radioresponse

miRNAs have become important modulators of cellular reactions to ionizing radiation in the context of radiotherapy. They play a crucial role in regulating radiation sensitivity or resistance in tumor and normal cells due to their capacity to affect gene networks. The exposure of both cancerous and healthy cells to ionizing radiation exhibits markedly different patterns of miRNA expression. Furthermore, new research suggests that miRNAs impact the radiosensitivity of cancer cells by modifying important cellular pathways that mediate the response to radiation, including cell cycle checkpoints, apoptosis, autophagy, epithelial–mesenchymal transition, and cancer stem cell survival. Hence, miRNAs can function as either radioresistance factors or radiosensitizers, depending on which genes they control. In addition to modulating irradiation response, miRNAs may be used as clinically meaningful indicators of response to radiation.

### 11.3.1 miRNAs as Indicators of Radiotherapy Response

If patients are unlikely to benefit from radiation therapy, the ability to anticipate response vs. nonresponse would enable treatment to be tailored, helping patients avoid needless risks of toxicity and side effects (Dittrick et al. 2012; Meredith et al. 2010). This is because patient responses to radiation treatment vary widely. To use miRNA profiling techniques to differentiate between responders and non-responders, each patient group's tumor miRNA expression patterns must differ significantly. Since most clinical and experimental studies looking at miRNAs and their relationship to treatment response include patients who have received combination chemotherapy and radiation, it is unfortunately difficult to identify miRNAs that specifically correlate with response to radiotherapy alone. However, it has been noted that miRNAs examined in the context of radiation therapy alone have demonstrated encouraging early findings in a small number of recent studies utilizing cancer cell lines or clinical samples (Korpela et al. 2015). These studies have shown that irradiation causes a significant change in the expression levels of numerous miRNAs, and that this change is repeatable across cell types and at varying dosages of X-ray or gamma-ray irradiation. Table 11.1 lists and cites these miRNAs, which indicates that miRNAs may be used as predictors of radiation response.

For instance, Wang and colleagues used samples from patients with non-small cell lung cancer to show that 7 miRNAs were downregulated in comparison to controls (miR-15b, 17-5p,19b, 21, 22, 106b, 130a) and 5 miRNAs were significantly upregulated in radiosensitive tumors (let-7a, miR-126, 128b, 451, 495) (Wang et al. 2011). In another investigation, tumors from patients who did not respond to radiation treatment had greater levels of miR-18a expression than tumors from responders ($P = 0.019$) (Shen et al. 2015). In the context of head and neck tumors, Qu et al. created a radiation-resistant nasopharyngeal carcinoma cell line and found that eight miRNAs were expressed differently from the parental cells (downregulation of miR-18a, 19b, 24, 93, 103; upregulation of let-7 g, miR-205, 224); the most notable expression changes between resistant and sensitive cells were observed for miR-205 (Qu et al. 2012). Another published study found that out of 41 differently expressed miRNAs in squamous cell carcinoma tumor samples from non-responders versus responders, miR-196a was the highest elevated miRNA (Suh et al. 2015). Similar findings have been shown in biopsies from squamous cervical carcinoma, where samples from radiation-resistant tumors taken six months after the end of radiation treatment showed downregulated levels of miR-16-2, 18a, 23a, 30, 221, and 378, while samples from radiation-resistant tumors showed upregulated levels of miR-21 and miR-181a. However, because patients received concomitant cisplatin treatment, this study's importance is limited (Ke et al. 2013). The research that has been published indicates that miRNAs may be used as predictors of radiation response and compiled in Table 11.1.

**Table 11.1** miRNAs as indicators of radiotherapy response

| S. No. | Type of miRNA | Up/down regulation vs. control | Radioresponse correlation | Type of cancer | Reference |
|---|---|---|---|---|---|
| 1. | let-7a | ↑ | ↑ Sensitivity | Lung cancer | Wang et al. (2011) |
| 2. | let-7 g | ↑ | ↑ Resistance | Nasopharyngeal carcinoma | Qu et al. (2012) |
| 3. | miR-15b | ↓ | ↑ Sensitivity | Lung cancer | Wang et al. (2011) |
| 4. | miR-17-5p | ↓ | ↑ Sensitivity | Lung cancer | Wang et al. (2011) |
| 5. | miR-18a | ↑ | ↑ Resistance | Lung cancer | Shen et al. (2015) |
| | | ↓ | ↑ Resistance | Nasopharyngeal carcinoma | Qu et al. (2012) |
| 6. | miR-19b | ↓ | ↑ Sensitivity | Lung cancer | Wang et al. (2011) |
| | | ↓ | ↑ Resistance | Nasopharyngeal carcinoma | Qu et al. (2012) |
| 7. | miR-21 | ↓ | ↑ Sensitivity | Lung cancer | Wang et al. (2011) |
| 8. | miR-22 | ↓ | ↑ Sensitivity | Lung cancer | Wang et al. (2011) |
| 9. | miR-24 | ↓ | ↑ Resistance | Nasopharyngeal carcinoma | Qu et al. (2012) |
| 10. | miR-106b | ↓ | ↑ Sensitivity | Lung cancer | Wang et al. (2011) |
| 11. | miR-93 | ↓ | ↑ Resistance | Nasopharyngeal carcinoma | Qu et al. (2012) |
| 12. | miR-126 | ↑ | ↑ Sensitivity | Lung cancer | Wang et al. (2011) |
| 13. | miR-128b | ↑ | ↑ Sensitivity | Lung cancer | Wang et al. (2011) |
| 14. | miR-130a | ↓ | ↑ Sensitivity | Lung cancer | Wang et al. (2011) |
| 15. | miR-181a | ↑ | ↑ Resistance | Cervical carcinoma | Ke et al. (2013) |

(continued)

**Table 11.1** (continued)

| S. No. | Type of miRNA | Up/down regulation vs. control | Radioresponse correlation | Type of cancer | Reference |
|---|---|---|---|---|---|
| 16. | miR-196a | ↑ | ↑ Resistance | Neck and head squamous cell carcinoma | Suh et al. (2015) |
| 17. | miR-451 | ↑ | ↑ Sensitivity | Lung cancer | Wang et al. (2011) |
| 18. | miR-205 | ↑ | ↑ Resistance | Nasopharyngeal carcinoma | Qu et al. (2012) |
| 19. | miR-495 | ↑ | ↑ Sensitivity | Lung cancer | Wang et al. (2011) |
| 20. | miR-224 | ↑ | ↑ Resistance | Nasopharyngeal carcinoma | Qu et al. (2012) |

### 11.3.2 miRNAs to Track Radiation Response

Since radiation therapy is typically administered over a period of weeks, early detection of patients who are not responding may offer a chance to prevent ineffective treatment or delays in switching to alternate forms of treatment. To use miRNA profiling techniques for "real-time" evaluation of radiation response, miRNA expression must alter in a distinctive way throughout a successful radiotherapy treatment. Recent research employing miRNA profiling methods, such as next-generation deep sequencing, real-time quantitative PCR, and microarrays, has shown that radiation exposure causes particular alterations in miRNA expression (Chaudhry 2014).

Broad alterations in miRNA expression during radiation therapy: Most published research has examined cancer tissue biopsies or (benign or malignant) cell lines to determine whether radiation exposure is linked to distinctive alterations in miRNA expression profiles (Metheetrairut and Slack 2013; Chaudhry et al. 2010a; Weidhaas et al. 2007; Johnson et al. 2005). The effects of ionizing radiation on miRNA expression in benign cells have been demonstrated to be substantial (Chaudhry et al. 2010a; Wagner-Ecker et al. 2010). For instance, in endothelial cells, radiation caused six miRNAs (miR-18a; 125a, 127, 148b, 189, 503) to be downregulated and five (let-7 g, miR-16; 20a, 21, 29c) to be upregulated (Wagner-Ecker et al. 2010).

Comparable outcomes have been shown in numerous cancer cell lines, Niemoeller et al. demonstrated that radiation treatment caused two- to three-fold alterations in the expression levels of several miRNAs, including let-7i, miR-24-1, miR-151-5p, and miR-1285 (Niemoeller et al. 2011). It's noteworthy that specific miRNAs have been found to regularly exhibit distinctive alterations after radiation treatment in a variety of distinct cancer types. For instance, radiation exposure was reported to impact the expression of multiple members of the let-7 family, a well-known tumor suppressor and regulator of the K-Ras oncogene (Metheetrairut and Slack 2013; Wagner-Ecker et al. 2010; Niemoeller et al. 2011; Chaudhry et al. 2010b, 2012; Saleh et al. 2011; Simone et al. 2009). The majority of let-7 family members aside from let-7 g have been shown to be downregulated in lung cancer cells (Weidhaas et al. 2007), colon cancer cells (Saleh et al. 2011) and irradiated fibroblasts (Simone et al. 2009). However, several let-7 family members are elevated in irradiation glioma and squamous cell cancer cell lines (Niemoeller et al. 2011; Chaudhry et al. 2010b). Following radiation exposure, two additional miRNAs, miR-21 and miR-34, are consistently increased in a variety of benign and malignant cell lines (Wagner-Ecker et al. 2010; Shi et al. 2012; Mert et al. 2012; Josson et al. 2008). The results of cell-line studies have only been used in clinical settings in two investigations, which also looked into miRNA responses in radiation patient tissues. The expression of 18 miRNAs was changed more than 1.5 times in irradiated gastric cancer samples (downregulation of let-7a, and miR-24-1, 141, 300, 377, 423-3p, 485-3P, 490-5p, 498, 615-3p, 625, 642, 657, 659, and 943; upregulation of miR-138-1 and miR-637) in comparison to nonirradiated samples (He et al. 2014). Following irradiation, qRT–PCR in nasopharyngeal cancer tissues demonstrated an increase of miR-205, which was in line with the authors' in vitro findings that miR-205 was upregulated in radioresistant cells (Qu et al. 2012).

### 11.3.3 miRNAs as Radiotherapy Response Modulators

Although miRNAs may be used as prognostic and diagnostic indicators in clinical settings, it is important to think about whether these chemicals present a new way to treat cancer. By manipulating miRNA levels, it is theoretically feasible to alter radiation sensitivity by altering the expression of genes crucial to radiation response. An increasing amount of research indicates that radioresistance in some human tumors is impacted by changes in miRNA levels that either increase or decrease the expression of specific miRNAs. For example, it has been shown that in vitro manipulation of levels of members of the let-7 family as well as miR-21, miR-181a, miR-221, and miR-222 affects the response to radiation therapy in head and neck squamous cell cancers, neuroblastoma, glioma, uveal melanoma, gastrointestinal tumors (esophageal, gastric, colorectal, and liver), tumors specific to gender (prostate, cervical, and breast cancer), renal cell cancer, lung cancer, and lymphoma. More details about how different miRNAs' up and downregulation affects radiosensitivity for in vitro research have been compiled by Muller et al. According to their preliminary in vivo research, radiosensitivity in melanoma, glioma, cervical cancer, esophageal cancer, renal cell cancer, and hepatoma is influenced by the let-7 family, miR-101, miR-145, miR-181a, miR-185, miR-210, and miR-38 (Mueller et al. 2016; Visser and Thomas 2021).

## 11.4 miRNA-Mediated DNA Damage Repair in Radiotherapy

Radiotherapy uses ionizing radiation (IR) to produce intermediate ions and free radicals, which harm tumor cells on several levels, particularly cellular DNA. It causes single-strand breaks or double-strand breaks (DSBs) in DNA, which triggers a variety of signaling networks to fix the damage (Li et al. 2022). The DNA repair system is thus triggered by many "damage sensors," such as ataxia-telangiectasia mutated (ATM), which activate cellular pathways and repair mechanisms like HR and NHEJ (Korpela et al. 2015; Long et al. 2019; Li et al. 2001). Radioresistance, which frequently obstructs effective treatment, can result from efficient repair. Many miRNAs appear to regulate and alter these DNA repair mechanisms. For instance, upregulating miR-18a in cervical cancer cell lines reduces the ATM level and reduces effective DSB repair following radiation therapy (Liu et al. 2015). Additionally, miR-101 and miR-421 are two other examples of how miRNAs interact with the DSB repair machinery through ATM regulation. These miRNAs have been shown to influence ATM expression, and in neuroblastoma cell lines, ectopic expression or overexpression of miR-421 enhances radiosensitivity. Other miRNAs, including miR-7, miR-101, and miR-182, have been demonstrated to radiosensitize cells in vitro and in xenograft models by targeting the transcripts of DSB repair genes, including BRCA1 (breast cancer 1) and DNA-dependent protein kinase catalytic subunit (DNA–PKcs), respectively (Yan et al. 2010; Hu et al. 2010).

As per the panel of human cancer cells, downregulation of epidermal growth factor receptor (EGFR) and Ak strain transforming (Akt) is caused by overexpression of miR-7 which was linked to delayed DNA repair and reduced phosphorylation of DNA–PKcs, resulted into radiation sensitivity (Lee et al. 2011). Other miRNAs, on the other hand, seem to primarily function through distinct DNA repair pathways for non-DSB damage or other DNA repair processes. It has been demonstrated that overexpression of miR-31 in radiation-resistant esophageal adenocarcinoma cells leads to altered expression of 13 DNA repair genes and re-sensitization to radiation (Lynam-Lennon et al. 2012) (Fig. 11.2).

## 11.5 miRNA and Radiation-Induced Cell Death Pathways

Apoptosis and the cell cycle are linked to the biological response to radiation, and many miRNAs are important in radiation-induced apoptosis. Many pro-apoptotic proteins are increased to carry out programmed cell death once a cell has committed to apoptosis (Yang et al. 2015). For example, miR-21 inhibits the pro-apoptotic gene caspase-3, which increases radioresistance. It also seems to be linked to the apoptotic pathway in lung cancer cells. miR-21 downregulation increases radiosensitivity, potentially by preventing cell cycle progression and proliferation (Xiao-chun et al. 2013). miR-125b, another miRNA that affects radioresistance through apoptotic

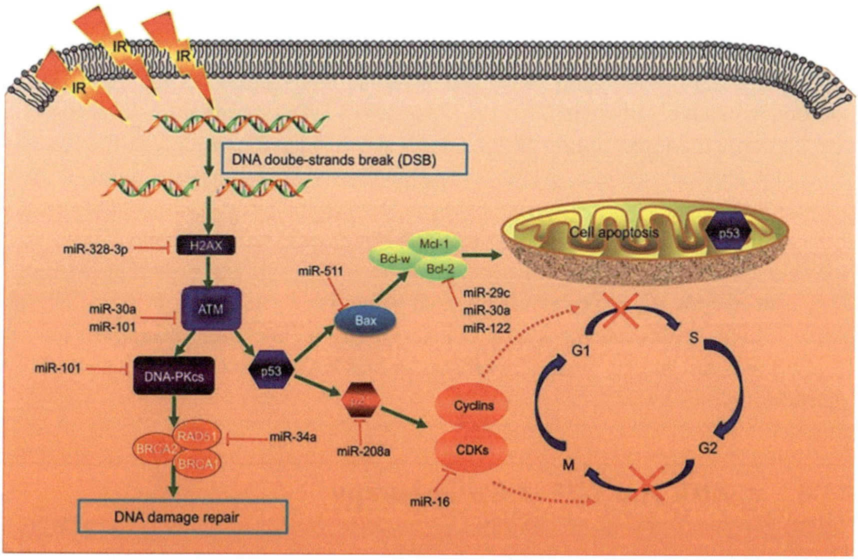


**Fig. 11.2** miRNAs in cell cycle, DNA-damage repair, and apoptosis; the three facets that make up the molecular mechanism of the DNA damage response. (Adapted with permission from Long et al. (2019) under Creative Commons Attribution)

pathways, seems to be a possible regulator of intercellular adhesion molecule-2 (ICAM-2), which has been connected to radiosensitivity in oral squamous cell carcinoma through its interaction with apoptotic signaling pathways (Akt and caspase-3) (Ishigami et al. 2008). Proliferation rates and ICAM-2 levels are reduced when miR-125b is overexpressed. miR-126 overexpression can also result in apoptosis induction, which may be mediated by PI3K-Akt pathway inactivation and increase radiosensitivity (Wang et al. 2011). One potential underlying mechanism in this situation is radiation-induced apoptosis via DNA methyltransferase 3b (DNMT3b) suppression (Wu et al. 2012). Additionally, miR-181a promotes radioresistance by targeting PRKCD, another pro-apoptotic gene, by suppressing radiation-induced apoptosis in cervical cancer cell lines and negatively regulating PRKCD expression (Ke et al. 2013). Additionally, it has been demonstrated that miR-196a overexpression increases radioresistance and proliferation rates in head and neck tumors (Suh et al. 2015). The signaling cascade's key component, the tumor suppressor annexin A1 (ANXA1), is the target gene that miR-196a controls. Interestingly, oncogenic behaviors are identical for ANXA1 knockdown and miR-196a overexpression. It is interesting to note that ANXA1 knockdown and miR-196a overexpression produce comparable carcinogenic effects (25). Lastly, it has recently been shown that miR-100 also influences apoptotic gene expression. The anti-apoptotic expression of Bcl-2 and nuclear Factor kappa-light-chain-enhancer of activated B cells (NF–jB) was reduced in a colorectal cancer cell line by miR-100, while the pro-apoptotic expression of p53 and caspase-3 was increased. Thus, cells were made more sensitive to radiation by upregulating miR-100. In addition to its involvement in other vital functions such as the inflammatory response and immune, the transcription factor NF–jB is considered a crucial early response gene that regulates cellular response and apoptosis in response to damage. It is noteworthy that the active NF–jB pathway is linked to radioresistance; inhibiting this route could offer an additional opportunity to overcome radioresistance. There is a negative correlation between miR-9 expression and NF–jB expression in irradiated lung cancer cells. Overexpression of miR-9 has been shown to decrease NF–jB levels, which in turn increases radiation sensitivity. Comparable results for let-7 g have been demonstrated, suggesting that both miRNAs regulate radiation sensitivity via blocking NF–Jb.

## 11.6 miRNA and TME in Radiotherapy

The TME is made up of a dynamic and intricate web of cancerous cells, stromal components, immune infiltrates, vasculature, and extracellular matrix, collectively orchestrating tumor progression, immune evasion, and metastatic dissemination. Within the TME, miRNAs significantly influence processes such as hypoxia response, angiogenesis, and immune modulation, all of which are critical in determining tumor behavior and response to therapies like radiotherapy (Frediani and Fabbri 2016; Zhao et al. 2012). The hypoxic environment stabilizes hypoxia-inducible factors (HIFs), particularly HIF-1α, which in turn activate the transcription of various genes that promote angiogenesis and alter cellular metabolism to adapt to low-oxygen conditions. miRNAs have been identified as crucial regulators in this hypoxic response (Magar et al. 2024). For instance, miR-210, often termed a "hypoxamir" is upregulated under hypoxic conditions and influences numerous processes, including cell cycle regulation, proliferation, apoptosis, angiogenesis, metastasis, and DNA repair (Dang and Myers 2015). In radiotherapy, tumor hypoxia presents a major barrier, as hypoxic tumor cells exhibit increased resistance to radiation-induced damage. miRNAs modulate radiosensitivity by regulating hypoxia-associated pathways. Notably, miR-210 governs hypoxic adaptation and influences DNA repair mechanisms, thereby impacting radiotherapy efficacy (Khakshour et al. 2025). Radiotherapy induces immunogenic cell death, facilitating tumor antigen release and immune activation. miRNAs also play a crucial role in modulating immune responses within the TME (Zhu et al. 2022; Templin et al. 2011). As such, miRNAs emerge as pivotal regulators of hypoxia responses, angiogenesis, and immune dynamics within TME, collectively shaping tumor progression and radiosensitivity (Cortez et al. 2019). Within TME, the synergistic action of radiation and miRNA therapy leads to enhanced radiosensitization and effective killing of cancer cells through multiple interconnected mechanisms depicted in

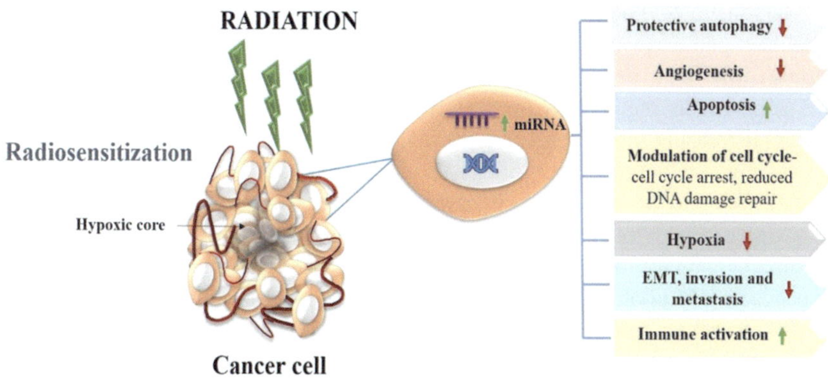


**Fig. 11.3** Mechanism of radiotherapy and miRNA in enhancing radiosensitization of cancer cells and inducing multiple cellular responses

Fig. 11.3. While radiation induces DNA damage, oxidative stress, and apoptosis in tumor cells, its efficacy is often limited by the tumor's intrinsic resistance mechanisms such as hypoxia, enhanced DNA repair, and protective autophagy (Wu et al. 2023). miRNAs counteract these defenses by targeting key regulatory pathways, such as by downregulating genes involved in DNA damage repair and cell cycle progression, promoting radiation-induced cell death. Additionally, miRNAs suppress protective autophagy and promote programmed cell death (apoptosis). Radiation damages tumor-associated endothelial cells thus inhibit angiogenesis. Several miRNAs are also known to suppress angiogenesis by downregulating proangiogenic factors like VEGF (vascular endothelial growth factors) effectively, thereby starving the tumor for nutrients and oxygen supply (Tolue Ghasaban et al. 2023; Natarajan 2016; Tu et al. 2015). Indeed, radiation contributes to a more immunogenic tumor environment, and miRNAs enhance this by activating immune signaling pathways, making the tumor cells more recognizable to immune cells. Furthermore, miRNAs inhibit Epithelial-Mesenchymal Transition (EMT) and thus invasion and metastasis aiding to localizing the tumor and improving radiation targeting (Pei et al. 2022; Du et al. 2019).

## 11.7 Clinical Applications and Therapeutic Strategies in Tumor Therapy

In tumor therapy, miRNAs function as both biomarkers and therapeutic agents, offering novel avenues for treatment, particularly when combined with radiotherapy (Gareev et al. 2020). However, the efficacy of radiotherapy is often compromised by tumor-associated factors such as hypoxia, enhanced DNA repair mechanisms, and intrinsic radioresistance. Integrating miRNA-based therapeutics with radiotherapy has demonstrated potential in improving treatment efficacy by modulating key molecular pathways involved in tumor radiosensitivity (Wu et al. 2023, 2018). Certain miRNAs exhibit radiosensitizing properties, such as miR-210 and miR-15a, which regulate hypoxia-associated pathways and apoptotic signaling, thereby increasing tumor sensitivity to radiation (Wu et al. 2018; Sun et al. 2015). Conversely, radioprotective miRNAs, like miR-21, promote DNA repair mechanisms and inhibit apoptosis, contributing to radioresistance in tumor cells. Targeting or restoring specific miRNAs can modulate tumor radiosensitivity while minimizing damage to surrounding healthy tissues and thus improving the therapeutic efficacy (Hu et al. 2017). Despite their promise in enhancing radiosensitivity, miRNA therapies face challenges in targeted delivery and stability. In this way, advanced delivery platforms such as lipid-based nanoparticles (e.g., liposomes), polymeric nanoparticles (e.g., PLGA), exosomes, and viral vectors help to overcome these barriers by improving miRNA protection, cellular uptake, and sustained release. These systems improve therapeutic efficacy by ensuring efficient delivery of miRNAs to tumor cells with minimal side effects, thereby boosting radiosensitivity with minimal side effects stability, targeted transport, and cellular uptake in tumor tissue (Dasgupta and Chatterjee 2021; Ganju et al. 2017; Wang et al. 2015) (Table 11.2).

**Table 11.2** Roles of various miRNAs in modulating radiosensitivity through distinct molecular pathways across diverse cancer types

| S. No. | Type of miRNA | Type of cancer | Effect on radiosensitivity | Reference |
|---|---|---|---|---|
| 1. | miR-34a | Colorectal cancer cells | Metformin mediated modulation of p53/miR-34a/SIRT1 loop by upregulation of miR-34-a and down regulation of SIRT-1 | Bahadoran et al. (2025) |
| 2. | hsa-miR-7-5p | Prostate cancer | By inhibiting important anti-apoptotic genes, hsa-miR-7-5p increases apoptosis and increases the radiosensitivity of prostate cancer cells | Darvish et al. (2023) |
| 3. | 55 miRNAs (such as miR-16, −27a, −181c, −221, −224 and −1293) | Cervical and head and neck cancer | Reducing DNA damage repair by downregulating key proteins involved in damage response (DDR) | Xu et al. (2025) |
| 4. | miR-92b-3p | Nasopharyngeal carcinoma (NPC) | Direct targeting and binding of miR-92b-3p to the multifunctional scaffold protein FHL2 | Zeng et al. (2025) |

(continued)

**Table 11.2** (continued)

| S. No. | Type of miRNA | Type of cancer | Effect on radiosensitivity | Reference |
|---|---|---|---|---|
| 5. | miR-450b-5p | Breast cancer | By targeting CDK6, inhibiting MCF7 cell proliferation, postponing cell cycle progression, raising the sensitiser enhancement ratio, and decreasing tumor volume after irradiation, miR-450b-5p improves radiosensitivity in HR+ and HER2-breast cancer cells | Wu et al. (2024) |
| 6. | miR-132-3p and miR-576-5p | Esophageal cancer | The four mRNAs and the regulatory network of miR-132-3p and miR-576-5p effectively predict patient outcomes and pathways | Chen et al. (2022) |
| 7. | miR-29a | Cervical cancer | In cervical cancer cells, overexpression of miR-29a increased radiosensitivity, encouraged apoptosis, and suppressed cell migration | Chuang et al. (2022) |
| 8. | miR-640 | Glioma | Increase the radiosensitivity of gliomas by modifying the Wnt-β-catenin pathway and downregulating SLIT1 | Zheng et al. (2022) |
| 9. | miRNA-31 | Colorectal tumors | Through the inhibition of STK40, miRNA-31 improves the radiosensitivity of cancer cells and may be a biomarker and therapeutic target to increase the effectiveness of radiation therapy in colorectal cancer | Zhang et al. (2022) |
| 10. | miR-29b-3p | Lung and breast cancer | Overexpression of miR-29b-3p reduced radioresistance in 3D-cultured cells and tumors in vivo by disrupting DNA damage repair and inhibiting key oncogenes | Pan et al. (2021) |
| 11. | miR-206 | Non-small cell lung cancer (NSCLC) | NSCLC cells' radiosensitivity was increased both in vitro and | Jiang et al. (2021) |

(continued)

**Table 11.2** (continued)

| S. No. | Type of miRNA | Type of cancer | Effect on radiosensitivity | Reference |
|---|---|---|---|---|
| | | | in vivo by CYTOR (long noncoding RNA) knockdown, which increased miR-206 and inhibited a radioresistance component. "PTMA" | |
| 12 | miR-107 | Prostate cancer | miR-107 enhanced radiosensitivity by suppressing granulin, modulating cell cycle, and apoptosis | Lo et al. (2020) |
| 13. | miR-196b | Gastric cancer | miR-196b targets the DNA repair protein RAD23B to increase the radiosensitivity of SNU-638 cells | Shen et al. (2018) |
| 14. | miRNA-200c | Esophageal squamous cancer cells | In vitro and in vivo, overexpression of miRNA-200c increased radiosensitivity by causing G2/M and sub-G1 phase arrest and modifying the expression of cyclin B1, cdc2, and P21 in irradiated Eca-109 cells | Zheng et al. (2017) |
| 15. | Plasma miRNAs | Non-small cell lung cancer (NSCLC) | Radiotherapy objective response rate (ORR) is higher in patients with higher levels of four plasma miRNAs (hsa-miR-98-5p, has-miR-302e, hsa-miR-495-3p, and has-miR-613) | Chen et al. (2016) |
| 16. | miR-18a | Cervical cancer | miR-18a increases radiosensitivity through inhibiting the repair of DNA double-strand breaks, inducing apoptosis upon irradiation, and decreasing ataxia-telangiectasia mutant (ATM) | Liu et al. (2015) |
| 17. | miR-203 | Head and neck squamous cell carcinoma (HNSCC) | Low expression of miR-203 and epithelial-to-mesenchyme transition (EMT) are associated with local recurrence following radiation therapy | de Jong et al. (2015) |

## 11.8 Emerging Trends and Challenges

The integration of miRNA research into radiotherapy marks a transformative advancement in precision oncology, with miRNAs emerging as pivotal regulators of radiosensitivity and radioresistance. miRNAs affect how well radiation therapy works by controlling the cell cycle, apoptosis, and DNA damage response. Advances in technology for miRNA-based treatments, such as mimics to restore tumor-suppressive miRNAs and inhibitors to block oncogenic miRNAs, are creating novel ways to make tumors more sensitive to radiation or protect healthy cells from radiation-induced harm. These therapeutic agents, when combined with chemotherapy, immunotherapy or targeted therapies can modulate multiple oncogenic pathways simultaneously, offering synergistic benefits in managing metastatic or drug-resistant cancers. Despite these advances, several challenges impede its clinical translation, including delivery specificity, off-target toxicity, context-dependent miRNA effects, and the complexity of predicting outcomes due to the "too many targets for miRNA effect" (TMTME). Additional fences include inconsistent tumor vascularization affecting nanoparticle delivery, regulatory hurdles, high production costs, and a lack of standardized detection protocols. To overcome these challenges, novel clinical trial frameworks are being adopted, integrating biomodulatory endpoints, single-cell transcriptomics, epigenetic landscape profiling, and AI-enhanced analytics. These approaches aim to optimize therapeutic timing, improve predictive accuracy of treatment responses, and facilitate longitudinal monitoring of patient-specific molecular dynamics.

## 11.9 Conclusion

In summary, the convergence of miRNA with radiotherapy represents a promising frontier in precision cancer therapeutics. The diverse functions of miRNAs in controlling important physiological reactions to ionizing radiation, such as DNA damage repair, apoptosis, autophagy, cell cycle arrest, and TME modulation, have been discussed throughout this chapter. Throughout this chapter, we have explored the multifaceted roles of miRNAs in governing key cellular responses to ionizing radiation, including DNA damage repair, apoptosis, cell cycle arrest, autophagy, and modulation of the TME. These small, noncoding RNAs act as both mediators and modulators of radiation sensitivity, with certain miRNAs functioning as radiosensitizers while others confer radioresistance. Importantly, miRNAs also influence the interplay between tumor cells and their surrounding microenvironment, shaping response to hypoxia, immune activation, and vascular remodeling post-irradiation. As the landscape of oncogenic research increasingly emphasizes individualized approaches, miRNA profiling offers a valuable tool for predicting therapeutic outcomes and tailoring radiation regimens to patient-specific molecular signatures. While clinical translation remains in its early stages, advancements in delivery technologies, biomarker validation, and integrative analytic frameworks such as AI and single-cell genomics are paving the way for miRNA-guided radiotherapy strategies. Eventually, using miRNA's regulatory potential could greatly improve radiotherapy's efficacy, specificity, reduce normal tissue harm, and achieve long-lasting tumor control. Therefore, it will be essential to conduct continuous multidisciplinary research to fully realize the therapeutic potential of miRNAs in relation to radiation therapy.

## References

Abuetabh Y, Wu HH, Chai C, Al Yousef H, Persad S, Sergi CM, Leng R (2022) DNA damage response revisited: the p53 family and its regulators provide endless cancer therapy opportunities. Exp Mol Med 54(10):1658–1669

Alamilla-Presuel JC, Burgos-Molina AM, González-Vidal A, Sendra-Portero F, Ruiz-Gómez MJ (2022) Factors and molecular mechanisms of radiation resistance in cancer cells. Int J Radiat Biol 98(8):1301–1315

Al-Arafat T, Mao A, Katsube T, Wang B (2024) Exploring the role of p53 in radiosensitivity: a key player in cancer therapy. Radiation 4(4):309–324

Bahadoran E, Zafari Y, Homaei A, Jalilvand M, Moghbelinejad S (2025) Role of the p53/miR-34a/SIRT1 feedback loop in metformin-induced radiosensitivity of colorectal cancer cells. Curr Radiopharm 18(4):340–351

Chandra RA, Keane FK, Voncken FE, Thomas CR (2021) Contemporary radiotherapy: present and future. Lancet 398(10295):171–184

Chaudhry MA (2014) Radiation-induced microRNA: discovery, functional analysis, and cancer radiotherapy. J Cell Biochem 115(3):436–449

Chaudhry MA, Kreger B, Omaruddin RA (2010a) Transcriptional modulation of micro-RNA in human cells differing in radiation sensitivity. Int J Radiat Biol 86(7):569–583

Chaudhry MA, Sachdeva H, Omaruddin RA (2010b) Radiation-induced micro-RNA modulation in glioblastoma cells differing in DNA-repair pathways. DNA Cell Biol 29(9):553–561

Chaudhry MA, Omaruddin RA, Kreger B, De Toledo SM, Azzam EI (2012) Micro RNA responses to chronic or acute exposures to low dose ionizing radiation. Mol Biol Rep 39:7549–7558

Chen X, Xu Y, Liao X, Liao R, Zhang L, Niu K, Li T, Li D, Chen Z, Duan Y (2016) Plasma miRNAs in predicting radiosensitivity in non-small cell lung cancer. Tumour Biol 37(9): 11927–11936

Chen H, Shi X, Ren L, Zhuo H, Zeng L, Qin Q, Wan Y, Sangdan W, Zhou L (2022) Identification of the miRNA-mRNA regulatory network associated with radiosensitivity in esophageal cancer based on integrative analysis of the TCGA and GEO data. BMC Med Genet 15(1):249

Chuang PC, Chen PT, Wang CC, Su WH, Chen YH, Huang EY (2022) MicroRNA-29a manifests multifaceted features to intensify radiosensitivity, escalate apoptosis, and revoke cell migration for palliating radioresistance-enhanced cervical cancer progression. Int J Mol Sci 23(10):5524

Cortez MA, Anfossi S, Ramapriyan R, Menon H, Atalar SC, Aliru M, Welsh J, Calin GA (2019) Role of miRNAs in immune responses and immunotherapy in cancer. Genes Chromosom Cancer 58(4):244–253

Dang K, Myers KA (2015) The role of hypoxia-induced miR-210 in cancer progression. Int J Mol Sci 16(3):6353–6372

Darvish L, Toossi MTB, Azimian H, Shakeri M, Dolat E, Firouzjaei AA, Rezaie S, Amraee A, Aghaee-Bakhtiari SH (2023) The role of microRNA-induced apoptosis in diverse radioresistant cancers. Cell Signal 104:110580

Dasgupta I, Chatterjee A (2021) Recent advances in miRNA delivery systems. Method Protocol 4(1):10

de Jong MC, Ten Hoeve JJ, Grénman R, Wessels LF, Kerkhoven R, Te Riele H, van den Brekel MW, Verheij M, Begg AC (2015) Pretreatment microRNA expression impacting on epithelial-to-mesenchymal transition predicts intrinsic radiosensitivity in head and neck cancer cell lines and patients. Clin Cancer Res 21(24):5630–5638

Derer A, Frey B, Fietkau R, Gaipl US (2016) Immune-modulating properties of ionizing radiation: rationale for the treatment of cancer by combination radiotherapy and immune checkpoint inhibitors. Cancer Immunol Immunother 65:779–786

Dittrick GW, Weber JM, Shridhar R, Hoffe S, Melis M, Almhanna K, Barthel J, McLoughlin J, Karl RC, Meredith KL (2012) Pathologic nonresponders after neoadjuvant chemoradiation for esophageal cancer demonstrate no survival benefit compared with patients treated with primary esophagectomy. Ann Surg Oncol 19:1678–1684

Du M, Wang J, Chen H, Wang S, Chen L, Xu Y, Su F, Lu X (2019) MicroRNA-200a suppresses migration and invasion and enhances the radiosensitivity of NSCLC cells by inhibiting the HGF/c-met signaling pathway. Oncol Rep 41(3):1497–1508

Frediani JN, Fabbri M (2016) Essential role of miRNAs in orchestrating the biology of the tumor microenvironment. Mol Cancer 15:1–11

Ganju A, Khan S, Hafeez BB, Behrman SW, Yallapu MM, Chauhan SC, Jaggi M (2017) miRNA nanotherapeutics for cancer. Drug Discov Today 22(2):424–432

Gao Z, Zhao Q, Xu Y, Wang L (2023) Improving the efficacy of combined radiotherapy and immunotherapy: focusing on the effects of radiosensitivity. Radiat Oncol 18(1):89

Gareev I, Beylerli O, Yang G, Sun J, Pavlov V, Izmailov A, Shi H, Zhao S (2020) The current state of MiRNAs as biomarkers and therapeutic tools. Clin Exp Med 20:349–359

Hashemi M, Mirdamadi MSA, Talebi Y, Khaniabad N, Banaei G, Daneii P, Gholami S, Ghorbani A, Tavakolpournegari A, Farsani ZM (2023) Pre-clinical and clinical importance of miR-21 in human cancers: tumorigenesis, therapy response, delivery approaches and targeting agents. Pharmacol Res 187:106568

He J, Hua J, Ding N, Xu S, Sun R, Zhou G, Xie X, Wang J (2014) Modulation of microRNAs by ionizing radiation in human gastric cancer. Oncol Rep 32(2):787–793

He B, Zhao Z, Cai Q, Zhang Y, Zhang P, Shi S, Xie H, Peng X, Yin W, Tao Y (2020) miRNA-based biomarkers, therapies, and resistance in cancer. Int J Biol Sci 16(14):2628

Hill RP, Bristow RG, Fyles A, Koritzinsky M, Milosevic M, Wouters BG (2015) Hypoxia and predicting radiation response, seminars in radiation oncology. Elsevier, pp 260–272

Hu H, Du L, Nagabayashi G, Seeger RC, Gatti RA (2010) ATM is down-regulated by N-Myc–regulated microRNA-421. Proc Natl Acad Sci 107(4):1506–1511

Hu B, Wang X, Hu S, Ying X, Wang P, Zhang X, Wang J, Wang H, Wang Y (2017) miR-21-mediated radioresistance occurs via promoting repair of DNA double strand breaks. J Biol Chem 292(8):3531–3540

Huang RX, Zhou P-K (2020) DNA damage response signaling pathways and targets for radiotherapy sensitization in cancer. Signal Transduct Target Ther 5(1):60

Ishigami T, Uzawa K, Fushimi K, Saito K, Kato Y, Nakashima D, Higo M, Kouzu Y, Bukawa H, Kawata T (2008) Inhibition of ICAM2 induces radiosensitisation in oral squamous cell carcinoma cells. Br J Cancer 98(8):1357–1365

Jiang G, Yu H, Li Z, Zhang F (2021) lncRNA cytoskeleton regulator reduces non-small cell lung cancer radiosensitivity by downregulating miRNA-206 and activating prothymosin α. Int J Oncol 59(5):88

Jiao Y, Cao F, Liu H (2022) Radiation-induced cell death and its mechanisms. Health Phys 123(5): 376–386

Johnson SM, Grosshans H, Shingara J, Byrom M, Jarvis R, Cheng A, Labourier E, Reinert KL, Brown D, Slack FJ (2005) RAS is regulated by the let-7 microRNA family. Cell 120(5): 635–647

Josson S, Sung SY, Lao K, Chung LW, Johnstone PA (2008) Radiation modulation of microRNA in prostate cancer cell lines. Prostate 68(15):1599–1606

Ke G, Liang L, Yang J, Huang X, Han D, Huang S, Zhao Y, Zha R, He X, Wu X (2013) MiR-181a confers resistance of cervical cancer to radiation therapy through targeting the pro-apoptotic PRKCD gene. Oncogene 32(25):3019–3027

Khakshour E, Shahram MA, Chaman R, Shoorei H, Samami E, Welsh JS, Dhawan G, Javadinia SA, Azimian H (2025) Hypoxia and intrinsic radiosensitivity: exploring mechanisms in radiation resistance of cancers and biomarkers for enhanced treatment strategies. J Radiat Res Appl Sci 18:101450

Kim BM, Hong Y, Lee S, Liu P, Lim JH, Lee YH, Lee TH, Chang KT, Hong Y (2015) Therapeutic implications for overcoming radiation resistance in cancer therapy. Int J Mol Sci 16(11): 26880–26913

Korpela E, Vesprini D, Liu S (2015) MicroRNA in radiotherapy: miRage or miRador? Br J Cancer 112(5):777–782

Lee KM, Choi EJ, Kim IA (2011) microRNA-7 increases radiosensitivity of human cancer cells with activated EGFR-associated signaling. Radiother Oncol 101(1):171–176

Li L, Story M, Legerski RJ (2001) Cellular responses to ionizing radiation damage. Int J Radiat Oncol Biol Phys 49(4):1157–1162

Li Y, Tong Y, Liu J, Lou J (2022) The role of MicroRNA in DNA damage response. Front Genet 13:850038

Liu S, Pan X, Yang Q, Wen L, Jiang Y, Zhao Y, Li G (2015) MicroRNA-18a enhances the radiosensitivity of cervical cancer cells by promoting radiation-induced apoptosis. Oncol Rep 33(6):2853–2862

Lo HC, Hsu JH, Lai LC, Tsai MH, Chuang EY (2020) MicroRNA-107 enhances radiosensitivity by suppressing granulin in PC-3 prostate cancer cells. Sci Rep 10(1):14584

Long L, Zhang X, Bai J, Li Y, Wang X, Zhou Y (2019) Tissue-specific and exosomal miRNAs in lung cancer radiotherapy: from regulatory mechanisms to clinical implications. Cancer Manag Res Volume 11:4413–4424

Lynam-Lennon N, Reynolds JV, Marignol L, Sheils OM, Pidgeon GP, Maher SG (2012) MicroRNA-31 modulates tumour sensitivity to radiation in oesophageal adenocarcinoma. J Mol Med 90:1449–1458

Magar AG, Morya VK, Kwak MK, Oh JU, Noh KC (2024) A molecular perspective on HIF-1$\alpha$ and angiogenic stimulator networks and their role in solid tumors: an update. Int J Mol Sci 25(6): 3313

Majidinia M, Yousefi B (2016) DNA damage response regulation by microRNAs as a therapeutic target in cancer. DNA Repair 47:1–11

Mathan JJ, Simkin SK, Gokul A, McGhee CNJ (2022) Down syndrome and the eye: ocular characteristics and ocular assessment. Surv Ophthalmol 67(6):1631–1646

Meredith KL, Weber JM, Turaga KK, Siegel EM, McLoughlin J, Hoffe S, Marcovalerio M, Shah N, Kelley S, Karl R (2010) Pathologic response after neoadjuvant therapy is the major determinant of survival in patients with esophageal cancer. Ann Surg Oncol 17:1159–1167

Mert U, Özgür E, Tiryakioglu D, Dalay N, Gezer U (2012) Induction of p53-inducible microRNA miR-34 by gamma radiation and bleomycin are different. Front Genet 3:220

Metheetrairut C, Slack FJ (2013) MicroRNAs in the ionizing radiation response and in radiotherapy. Curr Opin Genet Dev 23(1):12–19

Mueller AK, Lindner K, Hummel R, Haier J, Watson DI, Hussey DJ (2016) MicroRNAs and their impact on radiotherapy for cancer. Radiat Res 185(6):668–677

Natarajan V (2016) Regulation of DNA repair by non-coding miRNAs. Non-coding RNA research 1(1):64–68

Niemoeller OM, Niyazi M, Corradini S, Zehentmayr F, Li M, Lauber K, Belka C (2011) MicroRNA expression profiles in human cancer cells after ionizing radiation. Radiat Oncol 6: 1–5

Otmani K, Lewalle P (2021) Tumor suppressor miRNA in cancer cells and the tumor microenvironment: mechanism of deregulation and clinical implications. Front Oncol 11:708765

Otmani K, Rouas R, Berehab M, Lewalle P (2024) The regulatory mechanisms of oncomiRs in cancer. Biomed Pharmacother 171:116165

Pan D, Du Y, Li R, Shen A, Liu X, Li C, Hu B (2021) miR-29b-3p increases radiosensitivity in stemness cancer cells via modulating oncogenes axis. Front Cell Develop Biol 9:741074

Pawlik TM, Keyomarsi K (2004) Role of cell cycle in mediating sensitivity to radiotherapy. Int J Radiat Oncol Biol Phys 59(4):928–942

Pei Q, Shao ZA, Bing W, He JP, Zhang YN, Wei WJ, Hua JR, Heng Z, Dong L, Nan D (2022) MiR-663a inhibits radiation-induced epithelium-to-mesenchymal transition by targeting TGF-β1. Biomed Environ Sci 35(5):437–447

Podralska M, Ciesielska S, Kluiver J, van den Berg A, Dzikiewicz-Krawczyk A, Slezak-Prochazka I (2020) Non-coding RNAs in cancer radiosensitivity: microRNAs and lncRNAs as regulators of radiation-induced signaling pathways. Cancer 12(6):1662

Qu C, Liang Z, Huang J, Zhao R, Su C, Wang S, Wang X, Zhang R, Lee M-H, Yang H (2012) MiR-205 determines the radioresistance of human nasopharyngeal carcinoma by directly targeting PTEN. Cell Cycle 11(4):785–796

Saleh AD, Savage JE, Cao L, Soule BP, Ly D, DeGraff W, Harris CC, Mitchell JB, Simone NL (2011) Cellular stress induced alterations in microRNA let-7a and let-7b expression are dependent on p53. PLoS One 6(10):e24429

Seyhan AA (2024) Trials and tribulations of microRNA therapeutics. Int J Mol Sci 25(3):1469

Shen Z, Wu X, Wang Z, Li B, Zhu X (2015) Effect of miR-18a overexpression on the radiosensitivity of non-small cell lung cancer. Int J Clin Exp Pathol 8(1):643

Shen YN, Bae IS, Park GH, Choi HS, Lee KH, Kim SH (2018) MicroRNA-196b enhances the radiosensitivity of SNU-638 gastric cancer cells by targeting RAD23B. Biomed Pharmacother 105:362–369

Shi Y, Zhang X, Tang X, Wang P, Wang H, Wang Y (2012) MiR-21 is continually elevated long-term in the brain after exposure to ionizing radiation. Radiat Res 177(1):124–128

Simone NL, Soule BP, Ly D, Saleh AD, Savage JE, DeGraff W, Cook J, Harris CC, Gius D, Mitchell JB (2009) Ionizing radiation-induced oxidative stress alters miRNA expression. PLoS One 4(7):e6377

Skafi N, Fayyad-Kazan M, Badran B (2020) Immunomodulatory role for microRNAs: regulation of PD-1/PD-L1 and CTLA-4 immune checkpoints expression. Gene 754:144888

Song CW, Kim H, Kim M-S, Park HJ, Paek S-H, Terezakis S, Cho LC (2025) Role of HIF-1α in the responses of tumors to radiotherapy and chemotherapy. Cancer Res Treat 57(1):1–10

Suh YE, Raulf N, Gäken J, Lawler K, Urbano TG, Bullenkamp J, Gobeil S, Huot J, Odell E, Tavassoli M (2015) MicroRNA-196a promotes an oncogenic effect in head and neck cancer cells by suppressing annexin A1 and enhancing radioresistance. Int J Cancer 137(5):1021–1034

Sun M, Gadad SS, Kim D-S, Kraus WL (2015) Discovery, annotation, and functional analysis of long noncoding RNAs controlling cell-cycle gene expression and proliferation in breast cancer cells. Mol Cell 59(4):698–711

Szatkowska M, Krupa R (2020) Regulation of DNA damage response and homologous recombination repair by microRNA in human cells exposed to ionizing radiation. Cancer 12(7):1838

Taeb S, Rostamzadeh D, Amini SM, Rahmati M, Eftekhari M, Safari A, Najafi M (2024) MicroRNAs targeted mTOR as therapeutic agents to improve radiotherapy outcome. Cancer Cell Int 24(1):233

Templin T, Paul S, Amundson SA, Young EF, Barker CA, Wolden SL, Smilenov LB (2011) Radiation-induced micro-RNA expression changes in peripheral blood cells of radiotherapy patients. Int J Radiat Oncol Biol Phys 80(2):549–557

Tolue Ghasaban F, Maharati A, Akhlaghipour I, Moghbeli M (2023) MicroRNAs as the critical regulators of autophagy-mediated cisplatin response in tumor cells. Cancer Cell Int 23(1):80

Tu Y, Liu L, Zhao D, Liu Y, Ma X, Fan Y, Wan L, Huang T, Cheng Z, Shen B (2015) Overexpression of miRNA-497 inhibits tumor angiogenesis by targeting VEGFR2. Sci Rep 5(1):13827

Visser H, Thomas AD (2021) MicroRNAs and the DNA damage response: how is cell fate determined? DNA Repair 108:103245

Vlatkovic T, Veldwijk MR, Giordano FA, Herskind C (2022) Targeting cell cycle checkpoint kinases to overcome intrinsic radioresistance in brain tumor cells. Cancer 14(3):701

Wagner-Ecker M, Schwager C, Wirkner U, Abdollahi A, Huber PE (2010) MicroRNA expression after ionizing radiation in human endothelial cells. Radiat Oncol 5:1–10

Wang XC, Du LQ, Tian LL, Wu HL, Jiang XY, Zhang H, Li DG, Wang YY, Wu H-Y, She Y (2011) Expression and function of miRNA in postoperative radiotherapy sensitive and resistant patients of non-small cell lung cancer. Lung Cancer 72(1):92–99

Wang H, Jiang Y, Peng H, Chen Y, Zhu P, Huang Y (2015) Recent progress in microRNA delivery for cancer therapy by non-viral synthetic vectors. Adv Drug Deliv Rev 81:142–160

Wang H, Mu X, He H, Zhang X-D (2018) Cancer radiosensitizers. Trends Pharmacol Sci 39(1): 24–48

Weidhaas JB, Babar I, Nallur SM, Trang P, Roush S, Boehm M, Gillespie E, Slack FJ (2007) MicroRNAs as potential agents to alter resistance to cytotoxic anticancer therapy. Cancer Res 67(23):11111–11116

Wu Y, Liu GL, Liu SH, Wang CX, Xu YL, Ying Y, Mao P (2012) MicroRNA-148b enhances the radiosensitivity of non-Hodgkin's lymphoma cells by promoting radiation-induced apoptosis. J Radiat Res 53(4):516–525

Wu Y, Huang J, Xu H, Gong Z (2018) Over-expression of miR-15a-3p enhances the radiosensitivity of cervical cancer by targeting tumor protein D52. Biomed Pharmacother 105: 1325–1334

Wu Y, Song Y, Wang R, Wang T (2023) Molecular mechanisms of tumor resistance to radiotherapy. Mol Cancer 22(1):96

Wu R, Zhang H, Huang X, Zhuang L (2024) MiR-450b-5p enhances the radiosensitivity of HR+ and HER2− breast cancer by targeting CDK6. Oncol Transl Med 10(4):198–203

Xiao-chun W, Wei W, Zhu-Bo Z, Jing Z, Xiao-Gang T, Jian-Chao L (2013) Overexpression of miRNA-21 promotes radiation-resistance of non-small cell lung cancer. Radiat Oncol 8(1):146

Xu M, van de Wiel MA, Martinovičová D, Huseinovic A, van Beusechem VW, Stalpers LJ, Oei AL, Steenbergen RD, Snoek BC (2025) High-throughput 3D spheroid screens identify microRNA sensitizers for improved thermoradiotherapy in locally advanced cancers. Mol Ther Nucleic Acids 36(2)

Yamazaki T, Young KH (2022) Effects of radiation on tumor vasculature. Mol Carcinog 61(2): 165–172

Yan D, Ng WL, Zhang X, Wang P, Zhang Z, Mo Y-Y, Mao H, Hao C, Olson JJ, Curran WJ (2010) Targeting DNA-PKcs and ATM with miR-101 sensitizes tumors to radiation. PLoS One 5(7): e11397

Yang XD, Xu XH, Zhang S, Wu Y, Xing CG, Ru G, Xu HT, Cao JP (2015) Role of miR-100 in the radioresistance of colorectal cancer cells. Am J Cancer Res 5(2):545

Zeng C, Duan S, Zhao L, Jiang J (2025) Hsa-miR-92b-3p targeting FHL2 to enhance radiosensitivity of nasopharyngeal carcinoma. Biochem Genet 63(1):713–729

Zhang W, Zhu Y, Zhou Y, Wang J, Jiang P, Xue L (2022) miRNA-31 increases radiosensitivity through targeting STK40 in colorectal cancer cells. Asia Pac J Clin Oncol 18(3):267–278

Zhang Z, Peng Y, Peng X, Xiao D, Shi Y, Tao Y (2023) Effects of radiation therapy on tumor microenvironment: an updated review. Chin Med J 136(23):2802–2811

Zhao L, Bode AM, Cao Y, Dong Z (2012) Regulatory mechanisms and clinical perspectives of miRNA in tumor radiosensitivity. Carcinogenesis 33(11):2220–2227

Zheng R, Liu Y, Zhang X, Zhao P, Deng Q (2017) miRNA-200c enhances radiosensitivity of esophageal cancer by cell cycle arrest and targeting P21. Biomed Pharmacother 90:517–523

Zheng Y, Xiao M, Zhang J, Chang F (2022) Micro RNA-640 targeting SLIT1 enhances glioma radiosensitivity by restraining the activation of Wnt/β-catenin signaling pathway. Br J Biomed Sci 79:10067

Zhu S, Wang Y, Tang J, Cao M (2022) Radiotherapy induced immunogenic cell death by remodeling tumor immune microenvironment. Front Immunol 13:1074477

# miRNA as Micromanager of Immunotherapy Resistance and Side Effects

# 12

Rekha Jalandra, Anusha Singhal, and Archana Joon

**Abstract**

MicroRNAs (miRNAs) have emerged as crucial regulators of immune system dynamics and cancer progression, positioning them at the forefront of modern immuno-oncology. This chapter explores the multifaceted roles of miRNAs in shaping cancer immunotherapy outcomes, from modulating immune checkpoints and the tumor microenvironment to influencing the efficacy and toxicity of treatment. We examine how specific miRNAs contribute to immunotherapy resistance mechanisms and immune-related adverse events (irAEs), with a focus on their potential as predictive biomarkers. In addition, we highlight therapeutic strategies aimed at restoring or inhibiting miRNA function using novel delivery platforms. Challenges such as off-target effects, biological complexity, and translational barriers are also discussed, alongside emerging technologies that hold promise for overcoming these obstacles. Overall, this chapter underscores the translational value of miRNA biology in advancing precision immunotherapy and improving patient outcomes.

**Learning Objectives**

- *Explain* the role of miRNAs in regulating immune checkpoint pathways, tumor immune evasion, and response to immunotherapy.
- *Analyze* how dysregulated miRNAs contribute to immunotherapy resistance in different cancer types.

R. Jalandra
Medical University of South Carolina, Charleston, SC, USA

A. Singhal · A. Joon (✉)
Albert Einstein College of Medicine, Bronx, NY, USA
e-mail: archana.joon@einsteinmed.edu



S. Tiwari et al. (eds.), *microRNA Modulation: A Promising Combination Therapy for Cancer*, https://doi.org/10.1007/978-981-95-5989-3_12

- *Evaluate* the involvement of specific miRNAs in modulating immune-related adverse effects (irAEs) associated with checkpoint inhibitors and other immunotherapies.
- *Assess* the therapeutic potential of targeting or modulating miRNAs to overcome resistance and reduce side effects in cancer immunotherapy.

## 12.1 Introduction

MicroRNAs (miRNAs) are small, noncoding RNA molecules that play an essential role in regulating gene expression at the posttranscriptional level. These molecules, typically 18–25 nucleotides in length, act as master regulators of cellular processes by binding to complementary sequences in the 3′ untranslated regions (UTRs) of target mRNAs. This binding leads to either the degradation of the mRNA or the inhibition of its translation (O'Brien et al. 2018). miRNAs are involved in critical cellular pathways such as differentiation, proliferation, apoptosis, immune regulation, and stress responses, making them indispensable for maintaining physiological homeostasis (Fleshner and Crane 2017). Dysregulation of miRNAs has been linked to various diseases, including cancer, autoimmune disorders, and infections (Dai and Ahmed 2011).

The advent of immunotherapy has revolutionized the treatment landscape for many diseases, particularly cancer. By leveraging the body's immune system to target and eliminate abnormal cells, immunotherapy offers a promising alternative to conventional therapies such as chemotherapy and radiation (Kaur et al. 2023). Breakthroughs in immune checkpoint inhibitors (ICIs), such as anti-CTLA-4 therapies and anti-programmed cell death protein 1 (PD-1) and programmed death-ligand 1 (PD-L1), have led to remarkable clinical outcomes, achieving durable responses in several cancers, including melanoma, renal cell carcinoma, and non-small cell lung cancer (NSCLC). These therapies have also inspired innovations in adoptive T-cell therapies, cancer vaccines, and cytokine-based treatments (Marei et al. 2023).

However, the success of immunotherapy is not universal. A considerable number of patients fail to respond to treatment, a challenge known as immunotherapy resistance. Resistance can be classified as primary (intrinsic) or acquired and arises due to a variety of factors, including the suppression of T-cell activation, alterations in the tumor microenvironment, and dysregulated immune checkpoints (Varadé et al. 2021). Additionally, even among those who benefit from immunotherapy, many experience immune-related adverse events (irAEs), which occur when the immune system attacks healthy organs and tissues. irAEs can affect multiple systems, including the gastrointestinal tract, endocrine organs, skin, and cardiovascular system, and can range from mild to life-threatening severity (Yin et al. 2023). The dual challenge of resistance and irAEs underscores the need for deeper insights into the mechanisms governing immune modulation during therapy.

Emerging evidence positions miRNAs as pivotal "micro-managers" in these processes, influencing both therapeutic efficacy and adverse effects. miRNAs regulate the expression of key proteins involved in immune checkpoints, such as PD-L1, as well as pathways that govern immune cell infiltration and cytokine production. They also impact the balance between anti-inflammatory and pro-inflammatory signals, modulating both anti-tumor immunity and systemic immune responses (Z. Wang et al. 2024). By acting as fine-tuners of immune pathways, miRNAs can either promote or hinder the success of immunotherapy, depending on their specific roles and targets. Furthermore, circulating miRNAs have demonstrated potential as noninvasive biomarkers for predicting therapy response and the risk of irAEs, paving the way for personalized treatment strategies (Cortez et al. 2019).

This chapter explores the hypothesis that miRNAs serve as critical regulators of immunotherapy resistance and side effects, acting as molecular "micro-managers" that orchestrate these outcomes. Understanding the interplay between miRNAs and immune pathways could unlock new avenues for overcoming resistance, mitigating side effects, and improving the precision and efficacy of immunotherapy.

## 12.2 Biology of miRNAs

### 12.2.1 Biogenesis and Maturation of miRNAs

The canonical biogenesis pathway is the most common way that miRNAs are generated. Most miRNAs are processed from intragenic regions, like introns, and less commonly from exons, while the rest are regulated by their own specific promoters from intergenic regions (O'Brien et al. 2018).

To initiate biogenesis, RNA Polymerase II (RNA Pol II) transcribes miRNA genes to form primary miRNAs (pri-miRNAs), which are long hairpin structures capped at the 5′ end and polyadenylated at the 3′ end tail (Peng and Croce 2016). In the nucleus, the pri-miRNAs are then cleaved at the basal side of the hairpin stem by a microprocessor complex, which has an RNA-binding protein, DiGeorge Syndrome Critical Region 8 (DGCR8), and a ribonuclease III enzyme, Drosha. The pre-miRNA is exported to the cytoplasm by Exportin 5, and the second pri-miRNA cleavage is carried out by Dicer, another RNAse III endonuclease. Dicer interacts with the pre-miRNA at the 3′ or the 5′ end, and cleaves the terminal loop, forming the 20–22 nucleotide mature duplex miRNA (Loomis and Brock 2011). The nomenclature of mature miRNAs is determined by the originating strand of the precursor miRNA hairpin structure: the 5p strand arises from the 5′ end, whereas the 3p strand is derived from the 3′ end (O'Brien et al. 2018) as shown in Fig. 12.1.

The final step in miRNA biogenesis is the loading of the duplex, made up of the guide strand and its complementary passenger strand, into the Argonaute (Ago) complex. Chaperone proteins, HSP70/HSP90, regulate this loading by mediating conformational changes in Argonaute using ATP hydrolysis. The passenger strand is then unwound and removed, completing the formation of the RNA-induced silencing complex (RISC). Once RISC forms, the miRNA guides the complex to target

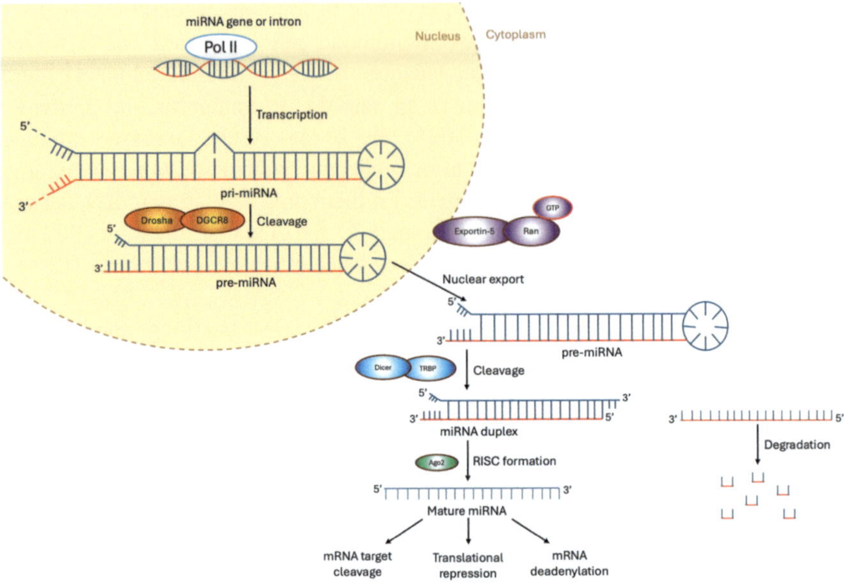


**Fig. 12.1** Canonical biogenesis of miRNA. Transcription of primary miRNAs (pri-miRNAs) is done by RNA polymerase II, followed by sequential cleavage by Drosha in the nucleus and Dicer in the cytoplasm. This produces a mature miRNA duplex, one strand of which is loaded into the RNA-induced silencing complex (RISC) for gene silencing, while the other strand is degraded

mRNAs for mRNA degradation or translational repression (Loomis and Brock 2011).

There are also several noncanonical pathways of miRNA biogenesis, which use different combinations of proteins used in the canonical pathway, like Dicer, Drosha, Exportin 5, and Ago (O'Brien et al. 2018). One such noncanonical miRNA is known as miRtrons, which is transcribed from introns of coding genes. The hairpin structure of miRtrons is shorter than that of canonical miRNAs, so it cannot be processed by the DGCR8/Drosha complex. Instead, it is debranched by debranching enzyme 1 (DRB1), and then follows the regular pathway, involving nuclear export by exportin 5 and cleavage by Dicer. Another example includes pre-miRNA cleavage in a Dicer-independent fashion. These pre-miRNAs are cleaved by Ago2 and subsequently loaded onto RISC (Stavast and Erkeland 2019).

### 12.2.2 Mechanism of Action and Role in Immune System Regulation

Regardless of the biogenesis pathway, once RISC is formed, the miRNA functions as a guide molecule to target and silence specific mRNAs. miRNAs target mRNAs through their 5′ seed region, which is a 7–8 nucleotide long sequence forming Watson–Crick base pairs with the miRNA response element (MRE) at the 3′ UTR of the specific mRNA target (Liu et al. 2018). This binding can cause mRNA degradation if there is perfect complementarity, but this is rare in mammals. More common is imperfect complementarity, which leads to translational repression (O'Brien et al. 2018). There are three possible pathways by which translation can be blocked: (i) Ago2 recruits trinucleotide repeat containing 6 (TNRC6), which subsequently recruits CCR4-NOT deadenylase complex to deadenylate and degrade the mRNA. (ii) TNRC6 recruits Dcp 1/2 decapping complex to cleave the 5′ cap of the mRNA, leading to reduced stability. (iii) When Ago2 binds, it does not allow binding of the ribosome to the mRNA, so translation is effectively blocked (Liu et al. 2018).

miRNAs play a key role in regulating many cellular processes, such as differentiation, cell growth, and apoptosis. Their abnormal expression or dysregulation has been associated with several human diseases, including metabolic disorders, autoimmune conditions, and cancer (Xiao and Rajewsky 2009). miRNAs have been implicated in the adaptive immune response through their role in the regulation and differentiation of immune cells like B cells, T cells, Natural Killer (NK) cells, etc. One example of this is miR-15/16, which modulates the formation of functionally active regulatory T-cells (Tregs) and controls NK cell maturation by targeting the c-Myb transcription factor expressed in immature NK cells (Gaál 2024). miRNAs are also involved in the innate immune response as they recruit factors for the host's defense against viral or bacterial pathogens. Once miR-155 is induced, it activates Interleukin-6 (IL-6), tumor necrosis factor (TNF), mediates the interferon response, influences macrophage polarization and NK cell activity, and supports the production of antiviral antibodies (Nejad et al. 2018). Altogether, this shows the central role of miRNAs as key regulators of both innate and adaptive immune responses, highlighting their potential as diagnostic biomarkers as well as therapeutic targets.

## 12.3 Immunotherapy Resistance: The Role of miRNAs

### 12.3.1 Mechanisms of Resistance

#### 12.3.1.1 Regulation of Immune Checkpoint Expression

Immunotherapy is a form of cancer treatment that upregulates or modulates the body's immune response to effectively target cancer cells. But a common and severe side effect of this therapy is the hyperactivation of the immune system. Thus, there has been extensive research in developing treatments based on immune checkpoints (ICs), and the FDA has approved many such drugs for various types of cancer (Kipkeeva et al. 2022). Immune checkpoints are stimulatory or inhibitory molecules, crucial for maintaining immune homeostasis and preventing the development of autoimmune diseases by severe upregulation of the immune system. miRNAs can modulate the expression of ICs directly by binding to the 3′ UTR of their mRNAs, or indirectly through influencing various signaling pathways like PTEN, JAK-STAT, NF–κB, etc. (Smolle et al. 2017).

The most widely studied ICs are PD-1 and PD-L1, which are expressed on activated immune cells. The normal function of PD-1 is to suppress T-cell activity to prevent any autoimmune response, so it is usually upregulated in exhausted T cells to facilitate programmed cell death (Kipkeeva et al. 2022). Tumors overexpress PD-L1 to evade the immune system by binding to PD-1, leading to the death of cytotoxic lymphocytes and amplifying the activity of regulatory T cells (Tregs). Thus, miRNAs are necessary to keep tight control over the PD-1/PD-L1 expression. miR-28 and miR-138 have been implicated in directly inhibiting the expression of PD-1 by binding to the 3′UTR of its mRNA (Vaxevanis et al. 2024). The miR-34 family targets PD-L1 and downregulates its expression. Other miRNAs binding to the PD-L1 gene include miR-570, miR-200, miR-21, miR-34a, and miR-197 (Q. Wang et al. 2017).

There are several miRNAs that indirectly regulate the PD-1/PD-L1 immune checkpoint by exerting control over the signaling pathway. miR-19, miR-20b, and miR-26 all inhibit the expression of PTEN, leading to an upregulation in PD-L1, and enabling cancer cell proliferation and progression. miR-100, miR-199a-3p, miR-101, and miR-497 suppress mTOR signaling while miR-145 directly regulates STAT1 (Q. Wang et al. 2017).

Like PD-L1, CTLA-4 is also an inhibitory receptor that attenuates T-cell hyperactivity and modulates the function of Tregs, which are important for immune suppression. Blocking the expression of CTLA-4 could serve as a potential strategy to enhance the body's anti-tumor response. miR-155 suppresses CTLA-4 mRNA, which has been shown to lead to a worse prognosis (Vaddi et al. 2023). miR-424 and miR-138 interact with both PD-1/PD-L1 and CTLA-4 in various cancers to modulate their activity (Vaxevanis et al. 2024).

Other immunosuppressive receptors include T-cell immunoglobulin and mucin domain-3 (Tim-3)/Galectin-9 (Gal-9) V-domain Ig-containing suppressor of T-cell activation and VISTA. MiR-128 is a direct inhibitor of Gal-3, while miR-498 inhibits Tim3. Suppression of the Tim-3/Gal9 pathway has led to a better prognosis in most cancers. miR-125a and miR-125b bind to VISTA to degrade its mRNA, increasing antitumor immunity associated with T-cells (Kipkeeva et al. 2022).

### Modulation of TME

Many miRNAs are differentially expressed in tumors, contributing to the development and regulation of malignant transformation, immune escape, and composition of the tumor microenvironment (TME). These immune modulatory miRNAs are usually termed as tumor suppressive or oncogenic based on which class of pathways they regulate. Tumor suppressive miRNAs promote immune surveillance, thus acting as anti-cancer regulators, while oncogenic miRNAs suppress the immune system, enhancing the ability of cancerous cells to escape the immune system (Suzuki et al. 2015). Studies in various cancers have shown that miRNAs of tumor cells can downregulate the expression or loss of MHC-1 surface antigens, impairing recognition of CD8+ T cells, allowing the tumor cells to escape immune surveillance (Eichmüller et al. 2017). miRNA-148-3p downregulates MHC-1 expression in esophageal adenocarcinoma by binding to the 3′ UTR. It also mediates indirect downregulation by binding to calnexin, a chaperone protein ensuring proper folding of MHC-1. miRNAs like miR-30b/-30d have also been implicated in promoting metastasis by upregulating the secretion of immunosuppressive cytokine IL-10, which enhances the activity of Tregs and suppresses the immune system. In breast cancer cells, miR-9 increased tumor invasiveness by targeting E-cadherin, an important regulator of cell adhesion. Loss of E-cadherin allows increased tumor cell invasion into surrounding blood vessels and subsequent dissemination into distant organs. Thus, the oncogenic activity of miRNAs can be directly modulating antigen presentation machinery on immune cells or through distal regulation of key tumoral components like angiogenesis and extracellular matrix remodeling (Annese et al. 2020).

On the other hand, miR-17-5p and miR-20 have been shown to inhibit cell migration and invasion by modulating the secretion of CK8, IL-8, and CXCL1 in breast cancer cells. miR-126 also suppresses metastasis in breast cancer by modulating the endothelial recruitment pathways promoting tumor angiogenesis. miR-29 interacts with the transcriptional factor GATA3 to promote differentiation of cancer cells, modifying the tumor microenvironment to inhibit metastasis. miR-29 also mechanistically downregulates angiogenesis by targeting the VEGF and MMP9 remodeling pathways. Usually, these tumor suppressive miRNAs are downregulated in many cancers, but restoring their expression or upregulating them restores their ability to target the tumor environment (Suzuki et al. 2015).

Several types of cells facilitate the growth and progression of cancer by contributing to the formation of the tumor microenvironment. These include fibroblasts, endothelial cells, immune inflammatory cells of these cancer-associated fibroblasts (CAFs), which make up a major part of the tumor stroma. CAFs promote tumorigenesis by secreting cytokines and chemokines, along with recruiting Treg, myeloid-derived suppressor cells (MDSCs), and tumor-associated macrophages (TAMs) (Eichmüller et al. 2017). Though it is unclear how CAFs arise from normal fibroblasts, one important difference between them is their miRNA expression profile. One example of this is miR-148, which targets WNT10B. WNT10B activates the WNT/β-catenin signaling cascade to mediate proliferation and stemness, and its aberrant expression has been implicated in many cancers (Perkins et al. 2023). miR-148, which downregulates WNT10B, is highly expressed in normal fibroblasts but suppressed in CAFs. miR-31 is another tumor suppressor miRNA that is downregulated in CAFs. It targets SATB2, which plays a role in cell migration and cancer metastasis (Rupaimoole et al. 2016). miR-214 overexpression in CAFs also promotes tumor development by enhancing the secretion of many pro-tumorigenic chemokines, like CCL20, CCL5, and CXCL8/IL-8 (Eichmüller et al. 2017).

MDSCs suppress the immune system through inhibition of T cell activation. These are speculated to arise from aberrant differentiation of myeloid progenitor cells. These cells promote tumor infiltration by suppressing CD4+ and CD8+ T cells through the production of ROS and increasing the number of Tregs (Rupaimoole et al. 2016). In many tumor models, miR-494 is upregulated by TGF-β1, and it leads to suppression of the PTEN pathway and activation of the Akt pathway, sustaining the survival of MDSCs. The JAK–STAT3 axis is important for the tumorigenic properties of MSDCs, and STAT3 also regulates the production of ROS. miR-20a and miR-17-5p have been shown to downregulate STAT3 in MDSCs, and these two miRNAs are usually lowly expressed in tumors (Suzuki et al. 2015). Targeting these miRNAs for upregulation could be a potential way to abrogate MDSC function.

### 12.3.2 Key miRNAs Implicated in Resistance

miRNAs are often regarded as potential biomarkers in cancers since their expression profile can often point to the upregulation of tumorigenic pathways. Oncogenic miRNAs are highly expressed in the TME, while tumor suppressive miRNAs are inhibited. miR-155 is currently one of the most studied targets for immunoediting in cancer due to its involvement in immune cell regulation, inflammatory signaling, and modulation of the TME (Kalkusova et al. 2022).

miR-155 is located on chromosome 21 within the noncoding B cell integration cluster (BIC) gene. It is highly expressed in CD34+ hematopoietic stem cells, as well as the thymus and spleen. Molecules taking part in hematopoietic differentiation are targets for miR-155, which has been validated by the presence of binding regions for its miRNA on their 3′UTR (Kalkusova et al. 2022). miR-155 also regulates myelopoiesis and lymphopoiesis, so dysregulation in its levels would severely impact the development of many cell types. miR-155 enhances tumoral cell proliferation by inhibiting apoptosis through the PTEN/PI3K/AKT pathway. The regulation of this miRNA is controlled by several genes like the tumor suppressive BRCA1 and by SMAD4 via TGF-β1 signaling. NF–κB and TP53 also directly regulate miR-155 expression (Moutabian et al. 2023). miR-155 is usually regarded as an oncogenic miRNA, especially in the context of hematologic malignancies and solid tumors, like breast cancer (Kalkusova et al. 2022). But in certain cases, it can also have antitumor activity by enhancing the immune system activation.

miR-155 overexpression sustains the activated state of immune cells and facilitates the secretion of pro-inflammatory cytokines like IL-1β and TNF-α. miR-155 upregulation also enhances the ability of dendritic cells (DCs) to promote T cell proliferation, and the miR-155-regulated DC maturation has an anti-tumor phenotype through the production of IL-12 and IFN-γ (18). miR-155 also shifts the role of TAMs towards a more tumorigenic M1 phenotype, which promotes tumor targeting by cytotoxic lymphocytes. An abundance of miR-155 has been correlated with increased cytotoxicity of NK cells toward tumors (Moutabian et al. 2023). Thus, miR-155 has a multi-faceted role in tumor progression by upregulating inflammation and skewing the TME from a "cold" to a "hot" environment.

Another important family of miRNAs is the lethal-7 (let-7) family of genes, comprising ten mature miRNAs, arising from thirteen different genes. These include let-7a, let-7b, let-7c, let-7d, let-7e, let-7f, let-7g, let-7i, miR-202, and miR-98 (Ma et al. 2021). These miRNAs are tumor suppressive, inhibiting the proliferation and metastasis of cancer cells. Cancer stem cells (CSCs) are a small subpopulation of cells in the tumor that can self-renew and differentiate into many cell types to maintain and grow the tumor bulk (Chirshev et al. 2019). They heighten the resistance of cancers to immunotherapy, thereby facilitating disease progression and relapse. Let-7 influences the metabolism and tumorigenesis of CSCs by targeting genes involved in self-renewal and stemness, like H-Ras and HMGA2. It is downregulated in many stem-cell-like cells in various cancers like head and neck cancer, pancreatic cancer (Ma et al. 2021).

Let-7 also influences many signaling pathways involved in cancer progression. Overexpression of the let-7 family member miR-98 reduces the activation of Akt and Erk, and in Ewing's sarcoma, let-7 directly blocks STAT3, leading to a less aggressive subtype of the cancer (Chirshev et al. 2019). In breast cancer, let-7 downregulates the Wnt signaling pathway by binding to estrogen receptors, mitigating cellular proliferation and differentiation. Overall, let-7 promotes a more chemo-sensitive, less aggressive phenotype (Ma et al. 2021).

Let-7 is involved in the inhibition of epithelial-to-mesenchymal transition (EMT) in cancer cells, preventing their ability to invade surrounding tissues and blood vessels to metastasize (Ma et al. 2021). Lower let-7 expression is associated with elevated levels of EMT markers such as Twist, Snail, vimentin, and N-cadherin, leading to a more aggressive cancer phenotype characterized by enhanced spheroid formation, migration, invasion, and chemoresistance. In advanced high-grade serous ovarian cancer (HGSOC), overexpressing miR-98 inhibits EMT by downregulating HMGA2 and transcriptionally repressing POSTN, a gene linked to EMT (Chirshev et al. 2019). Thus, let-7 could be used both as a diagnostic biomarker to assess the aggressiveness of the cancer and be targeted for upregulation to inhibit the progression and metastatic capacity of the tumor.

### 12.3.3 Therapeutic Strategies

miRNAs offer a distinct advantage for targeted therapies, as each miRNA can regulate multiple genes simultaneously. While previous studies have emphasized the potential of delivering miRNAs to tumors as a promising strategy to enhance cancer treatment, achieving effective delivery remains a significant challenge (Ishida and Selaru 2013). There are two basic strategies to use miRNAs as therapeutic targets in cancer immunotherapy: (1) To inhibit oncogenic miRNAs, upregulating the tumor-suppressing genes they target, (2) Restoring the inhibition of tumor suppressive miRNAs to enhance their immune-activating function (Menon et al. 2022). miRNAs designed to inhibit endogenous oncogenic miRNAs are known as "anti-miRs." These are single-stranded oligonucleotides that are complementary to the sequence of the target miRNA. Some therapeutic strategies include (i) anti-miRNA oligonucleotides (AMOs), (ii) miRNA sponges, and (iii) small molecule inhibitors (Menon et al. 2022).

AMOs are often chemically modified, for example, by adding Two′-O-methyl (2′-OMe), to make them more resistant to nucleases (Ishida and Selaru 2013). Another modification of oligos are termed peptide nucleic acids (PNAs). These are highly stable and hybridize tightly with DNA/RNA. PNA in conjugation with a photosensitizer chlorin e6 targeting miR-21 was developed by Lee et al. since miR-21 levels are upregulated in cancer cells. Upon binding, the photosensitizer is released, which promotes the death of cancer cells by near-infrared radiation (Menon et al. 2022). miRNA sponges are competitive inhibitors that compete with miRNAs for their target. It is possible to inhibit multiple miRNAs in a family, and the interaction of these sponges is as strong as AMOs (Ishida and Selaru 2013). miRNA sponges have been used to downregulate miR-9 expression in breast cancer to mitigate metastasis (Menon et al. 2022).

Small-molecule inhibitors can be used to inhibit both the biogenesis and the function of miRNAs. One example of this is a small molecule called enoxacin, which binds to TAR RNA-binding protein 2 (TRBP), a protein involved in miRNA synthesis. Another molecule modulating the biogenesis of miRNAs is polylysine, which abrogates Dicer activity (Menon et al. 2022). An important consideration for these types of inhibitors is the evaluation of their availability and efficacy.

Tumor-suppressing miRNAs, which are downregulated in cancers, lead to overactivation of oncogenes. Two methods to increase the levels of these miRNAs are through ectopic expression and by using miRNA mimics. miRNA mimics are small double-stranded RNA molecules that mimic or recapitulate the endogenous miRNA (Menon et al. 2022). One of the earliest studies showing a significant impact on tumorigenesis was through the delivery of miR-34 and let-7 in lung cancer. Another important tumor suppressive miRNA is miR-200, and effective delivery in ovarian, breast, lung, and renal cancer led to the reduction of tumor metastasis and vascularization (Rupaimoole et al. 2016). Thus, using miRNA mimics can enhance anti-cancer therapies.

A major challenge of miRNA therapeutics is designing an effective delivery system since RNA is relatively unstable and vulnerable to nucleases. Lipid-based carriers have been designed since they are able to cross the cellular barrier and are thought to be minimally toxic. MRX34 is a lipid-based nanoparticle that is currently in clinical trials and has shown a significant impact on the miR-34 system. Polyethylenimine particles have also been used in a colon cancer model to deliver miR-145 and miR-33a, leading to a decrease in tumor size (Rupaimoole et al. 2016).

Wu et al. recently developed a lipid nanoparticle loaded with miR-155 for reprogramming of TAMs to a more tumor-suppressive phenotype. Reprogramming TAMs to an M1-like macrophages made the TME more conducive to immune surveillance by increasing the secretion of pro-inflammatory cytokines like TNF-α and IL-6. Intratumoral administration of this miR-155 nanoparticle significantly reduced tumor growth (Wu et al. 2025). This system can also be utilized for the delivery of other tumor-suppressive miRNAs. Thus, the broad signaling regulation exerted by miRNAs could emerge as a key anti-cancer immunotherapy if their off-target effects can be decreased by efficiently targeting them using specific delivery systems.

## 12.4 Immune-Related Adverse Events (irAEs): miRNA's Role in Side Effects

Immune checkpoint inhibitors (ICIs) have revolutionized cancer therapy by enhancing antitumor immunity. However, their mechanism of action—broad activation of immune responses—often disrupts immune tolerance, leading to irAEs (Shiravand et al. 2022). These adverse effects can affect multiple organ systems and significantly impact patient outcomes. In recent years, miRNAs have emerged as critical regulators of immune homeostasis and inflammation. This part explores the role of miRNAs in the pathophysiology of irAEs, their potential as biomarkers for early detection, and implications for personalized risk assessment.

### 12.4.1 Mechanisms behind irAEs

The immune system is tightly regulated by complex networks of signaling molecules and gene expression regulators. Among them, miRNAs—a class of small, noncoding RNAs—play a pivotal role in modulating immune cell differentiation, activation, and cytokine production (O'Brien et al. 2018). Dysregulation of miRNAs can perturb immune equilibrium, leading to excessive or misdirected immune responses.

#### 12.4.1.1 Dysregulation of Immune Homeostasis by miRNAs

Several miRNAs maintain immune tolerance by controlling the expression of immune checkpoint molecules, inflammatory mediators, and transcription factors. For example, miR-146a and miR-155 are known to modulate the NF–κB pathway, a key regulator of inflammation and immunity. The aberrant expression of these miRNAs can predispose individuals to autoimmune-like phenomena, which underlie many irAEs (Boldin and Baltimore 2012; Testa et al. 2017).

#### 12.4.1.2 Influence on T-cell Differentiation and Activation

miRNAs influence the differentiation of T-helper cell subsets, including Th17, Th1, and regulatory T cells (Tregs). miR-21, for instance, promotes Th17 differentiation, contributing to pro-inflammatory environments, while miR-10a supports Treg stability (Murugaiyan et al. 2015; Takahashi et al. 2012). The imbalance between effector and regulatory T cells is a hallmark of irAE development.

#### 12.4.1.3 miRNAs in Cytokine Production and Signaling

Cytokine storms and elevated pro-inflammatory signaling are often associated with severe irAEs. miRNAs such as miR-125b and miR-223 regulate IL-6, TNF-α, and other cytokines by targeting their mRNA transcripts or upstream signaling mediators (Aziz 2016). Dysregulated cytokine expression driven by altered miRNA profiles may exacerbate tissue inflammation and damage during ICI therapy.

### 12.4.2 miRNA Profiles in Specific irAEs

Emerging studies have begun to map miRNA expression changes in tissues and the circulation of patients experiencing specific irAEs. These signatures not only provide insights into pathogenesis but also reveal tissue-specific regulatory patterns.

#### 12.4.2.1 Cardiovascular Toxicities

Cardiotoxic irAEs such as myocarditis have been linked to miRNAs regulating cardiac inflammation and apoptosis. For instance, downregulation of miR-126 and miR-1—known to modulate endothelial integrity and cardiac myocyte survival—has been observed in myocarditis models (Peterlin et al. 2020). miR-155, a pro-inflammatory miRNA, is frequently upregulated in immune-mediated cardiac injury.

#### 12.4.2.2 Gastrointestinal and Hepatic Toxicities

In immune-mediated colitis and hepatitis, miR-29 family members, which suppress IFN-γ signaling, are often reduced, resulting in excessive inflammatory cytokine release (Yao et al. 2025). In the liver, miR-122—a hepatocyte-specific miRNA—is a potential marker of immune hepatotoxicity due to its sensitivity to inflammatory insults (Cione et al. 2022).

#### 12.4.2.3 Endocrinopathies and Other Systemic Effects

Endocrine irAEs, including thyroiditis and hypophysitis, have been associated with altered expressions of miRNAs such as miR-21 and miR-146b, which regulate immune cell infiltration and tissue-specific autoimmunity (Martínez-Hernández et al. 2019). These systemic manifestations underscore the complexity of miRNA regulation in organ-specific immune tolerance.

### 12.4.3 Biomarkers for irAE Prediction

The unpredictable onset and severity of irAEs necessitate the development of reliable biomarkers. Circulating miRNAs offer a minimally invasive approach to monitor immune dysregulation in real time.

#### 12.4.3.1 Circulating miRNAs as Early Indicators of irAEs

Several miRNAs are stably expressed in plasma and serum, allowing for early detection of immune dysregulation. Studies have identified elevated plasma levels of miR-155 and miR-146a in patients prior to the onset of irAEs, suggesting their utility in predicting immune toxicities (Bagheri-Hosseinabadi et al. 2021). These miRNAs may reflect systemic immune activation even before clinical symptoms emerge.

#### 12.4.3.2 Integration of miRNA Profiling into Patient Risk Assessment

Integrating miRNA profiling with existing clinical parameters and genomic data may enhance the precision of risk stratification for irAEs. Multivariate models incorporating miRNA signatures could help tailor immunotherapy regimens to minimize toxicity without compromising efficacy (Qin et al. 2024). Moreover, miRNA-based risk scores could guide pre-treatment counseling and post-treatment surveillance.

## 12.5 Diagnostic and Prognostic Potential of miRNAs in Immunotherapy

Beyond their mechanistic roles in immune modulation and adverse events, microRNAs (miRNAs) are increasingly recognized for their diagnostic and prognostic potential in immuno-oncology. Their small size, remarkable stability in circulation, and specificity to disease states make them attractive candidates as biomarkers—not just for monitoring disease progression, but also for predicting how a patient might respond to therapy.

### 12.5.1 Circulating and Tissue-Specific miRNAs as Biomarkers

miRNAs can be detected in serum, plasma, and even urine, often packaged in protective vesicles such as exosomes or associated with proteins that shield them from degradation (Mori et al. 2019). This stability makes them ideal for liquid biopsy applications, where tissue access is limited. For instance, miR-21 and miR-155 are commonly elevated in various cancers and have been associated with immune cell activation and tumor progression (Ekiz et al. 2019; Rhim et al. 2022). In melanoma, miR-34a levels have been correlated with immune infiltration and response to anti-PD-1 therapy, highlighting the potential of miRNA signatures to stratify patients for immunotherapy (Beg et al. 2017).

In tissue biopsies, differential miRNA expression reflects the immune landscape of the tumor. Tumor-infiltrating lymphocytes, macrophages, and even tumor cells themselves contribute to the overall miRNA signature, providing a molecular snapshot of the tumor microenvironment (TME) (Yadav et al. 2024).

### 12.5.2 Predicting Therapy Response and irAE Risk

miRNA-based prediction models are now being developed to anticipate both efficacy and toxicity of immunotherapy. For example, specific combinations such as miR-146a, miR-125b, and miR-31 have shown promise in predicting patient responses to CTLA-4 and PD-1 blockade (Yadav et al. 2024). Similarly, miRNAs like miR-21 have been implicated in the pathogenesis of immune-related adverse events, including colitis and pneumonitis, making them potential dual-purpose biomarkers for response and risk (Popa et al. 2025).

### 12.5.3 Insights from Case Studies and Clinical Trials

In a prior study involving NSCLC patients treated with nivolumab, serum levels of miR-200c and miR-34a were found to be predictive of progression-free survival (Goh et al. 2023; Meira et al. 2023; Monastirioti et al. 2022). Similarly, Marschner et al. (2020) reported that elevated baseline levels of miR-146a were associated with a lower incidence of immune-related adverse events (Marschner et al. 2020). These findings not only support preclinical data but also highlight the potential of circulating miRNA panels as tools for patient risk stratification and personalized treatment planning.

## 12.6 Therapeutic Potential of Targeting miRNAs

While miRNAs serve as useful markers, they are also active participants in shaping the immune landscape-prompting researchers to explore them as therapeutic targets. Because miRNAs regulate entire gene networks, modulating a single miRNA can influence key pathways involved in immune activation, tumor growth, and immune escape.

### 12.6.1 Modulating miRNA Activity: AntagomiRs and LNA Inhibitors

Therapeutic strategies fall into two main categories: inhibiting oncogenic miRNAs (oncomiRs) and restoring tumor-suppressor miRNAs. AntagomiRs—chemically modified antisense oligonucleotides—bind to target miRNAs and prevent them from interacting with their mRNA targets. Locked nucleic acid (LNA) inhibitors, a more stable class of antisense tools, have been used to block miR-155 in preclinical models, leading to reduced inflammation and improved anti-tumor immunity (Bajan and Hutvagner 2020). Restorative strategies focus on delivering miRNA mimics. For example, synthetic miR-34a mimics were tested in clinical trials for solid tumors, based on the molecule's tumor-suppressive and immunomodulatory roles. Although the first-generation candidate MRX34 was halted due to immune-related toxicities, it provided critical proof-of-concept for miRNA replacement therapy (Mockly and Seitz 2023).

### 12.6.2 Overcoming Delivery Challenges: Liposomes to Exosomes

A key hurdle in miRNA therapeutics has been efficient and safe delivery. Early approaches using liposomal carriers lacked specificity and caused systemic immune activation. More recent work is leveraging exosomes—naturally occurring vesicles secreted by cells—to package and deliver miRNA therapeutics to target tissues. These vesicles exhibit innate biocompatibility and can even be engineered for tissue-specific delivery, significantly advancing the safety profile of miRNA-based therapies (Kim et al. 2024).

### 12.6.3 Synergistic Use with Immunotherapy

Perhaps the most exciting avenue is the synergistic use of miRNA therapies with immune checkpoint inhibitors. By fine-tuning immune cell activity, miRNAs can enhance the efficacy of checkpoint blockade. For example, restoring miR-124 or miR-34a in resistant tumors has been shown to increase CD8+ T cell infiltration and PD-L1 expression, effectively priming tumors for immunotherapy (Yang et al. 2018). Such combination strategies may prove especially valuable in cases of partial or acquired resistance.

### 12.6.4 Risks and Off-Target Effects

Despite these advances, caution is warranted. Because miRNAs often target multiple genes, unintended effects are a real concern. Off-target gene silencing, immunogenicity of synthetic oligonucleotides, and tissue non-specificity can lead to unwanted inflammation or toxicity. Rigorous preclinical screening and controlled delivery will be essential for safely translating these strategies into the clinic.

## 12.7 Challenges and Future Directions

As the field of miRNA research in immunotherapy matures, several technical and biological challenges must be addressed to unlock its full potential.

### 12.7.1 Biological Complexity

miRNAs operate within complex regulatory networks and exhibit cell-type specificity, making it difficult to draw simple causal links between miRNA expressions and functional outcomes. For example, the same miRNA might promote inflammation in macrophages but suppress T cell activation. Disentangling these context-dependent effects is crucial for therapeutic development.

### 12.7.2 Clinical Translation Hurdles

Few miRNA-targeted therapies or diagnostics have received regulatory approval, despite promising preclinical data. The "bench-to-bedside" gap stems from several factors: small sample sizes in clinical trials, lack of long-term toxicity data, and uncertainty about dosing and delivery. Progress will depend on collaborative, multicenter trials and the establishment of regulatory frameworks for nucleic acid-based therapeutics.

### 12.7.3 Emerging Technologies and Personalized Medicine

Excitingly, new tools like single-cell miRNA sequencing and spatial transcriptomics are revolutionizing our ability to study miRNAs in complex tissues. These technologies enable researchers to map miRNA expressions to specific immune cell types within the tumor microenvironment, offering unprecedented granularity. Combined with machine learning and systems biology approaches, these innovations are paving the way for personalized miRNA-guided immunotherapy.

### 12.7.4 Technical Barriers

One persistent issue is standardization. Differences in sample preparation, detection platforms (qPCR, microarrays, RNA-seq), and normalization strategies can lead to inconsistent results. Additionally, the existence of isomiRs—variant forms of miRNAs—complicates the interpretation of miRNA datasets. Without harmonized protocols, translating findings across labs and studies remains difficult.

## 12.8 Conclusion

From regulators of immune checkpoints to predictors of adverse events and promising therapeutic targets, miRNAs are proving to be multifaceted players in the immunotherapy landscape. They bridge the gap between genetic regulation and immune response, offering opportunities for more precise and personalized cancer treatment. As our tools for detecting, analyzing, and manipulating miRNAs become more refined, the integration of miRNA-based diagnostics and therapeutics into routine clinical practice seems increasingly within reach. The path forward will require interdisciplinary collaboration, technological innovation, and thoughtful clinical trial design. But the promise is clear: miRNA research holds the key to unlocking safer, more effective, and personalized immunotherapy strategies, moving us closer to the ideal of precision oncology.

## 12.9 Summary

MicroRNAs serve as critical regulators of immunotherapy outcomes, acting at the intersection of immune modulation, tumor biology, and treatment response. This chapter has reviewed their biogenesis, mechanisms of action, and roles in both therapeutic resistance and immune-related adverse events. It has also explored the diagnostic and therapeutic potential of miRNAs, along with emerging technologies and challenges in clinical translation. Together, these insights underscore the promise of miRNAs in advancing personalized immunotherapy strategies.

## References


Annese T, Tamma R, De Giorgis M, Ribatti D (2020) microRNAs biogenesis, functions and role in tumor angiogenesis. Front Oncol 10:581007. https://doi.org/10.3389/fonc.2020.581007

Aziz F (2016) The emerging role of miR-223 as novel potential diagnostic and therapeutic target for inflammatory disorders. Cell Immunol 303:1–6. https://doi.org/10.1016/j.cellimm.2016.04.003

Bagheri-Hosseinabadi Z, Mirzaei MR, Hajizadeh MR, Asadi F, Rezaeian M, Abbasifard M (2021) Plasma MicroRNAs (miR-146a, miR-103a, and miR-155) as potential biomarkers for rheumatoid arthritis (RA) and disease activity in Iranian patients. Mediterr J Rheumatol 32(4):324–330. https://doi.org/10.31138/mjr.32.4.324

Bajan S, Hutvagner G (2020) RNA-based therapeutics: from antisense oligonucleotides to miRNAs. Cells 9(1):137

Beg MS, Brenner AJ, Sachdev J, Borad M, Kang Y-K, Stoudemire J et al (2017) Phase I study of MRX34, a liposomal miR-34a mimic, administered twice weekly in patients with advanced solid tumors. Investig New Drugs 35(2):180–188. https://doi.org/10.1007/s10637-016-0407-y

Boldin MP, Baltimore D (2012) MicroRNAs, new effectors and regulators of NF-κB. Immunol Rev 246(1):205–220. https://doi.org/10.1111/j.1600-065X.2011.01089.x

Chirshev E, Oberg KC, Ioffe YJ, Unternaehrer JJ (2019) Let-7 as biomarker, prognostic indicator, and therapy for precision medicine in cancer. Clin Transl Med 8(1):24. https://doi.org/10.1186/s40169-019-0240-y

Cione E, Abrego Guandique DM, Caroleo MC, Luciani F, Colosimo M, Cannataro R (2022) Liver damage and microRNAs: an update. Curr Issues Mol Biol 45(1):78–91. https://doi.org/10.3390/cimb45010006

Cortez MA, Anfossi S, Ramapriyan R, Menon H, Atalar SC, Aliru M et al (2019) Role of miRNAs in immune responses and immunotherapy in cancer. Genes Chromosomes Cancer 58(4):244–253. https://doi.org/10.1002/gcc.22725

Dai R, Ahmed SA (2011) MicroRNA, a new paradigm for understanding immunoregulation, inflammation, and autoimmune diseases. Transl Res 157(4):163–179. https://doi.org/10.1016/j.trsl.2011.01.007

Eichmüller SB, Osen W, Mandelboim O, Seliger B (2017) Immune modulatory microRNAs involved in tumor attack and tumor immune escape. JNCI J Natl Cancer Inst 109(10). https://doi.org/10.1093/jnci/djx034

Ekiz HA, Huffaker TB, Grossmann AH, Stephens WZ, Williams MA, Round JL, O'Connell RM (2019) MicroRNA-155 coordinates the immunological landscape within murine melanoma and correlates with immunity in human cancers. JCI Insight 4(6). https://doi.org/10.1172/jci.insight.126543

Fleshner M, Crane CR (2017) Exosomes, DAMPs and miRNA: features of stress physiology and immune homeostasis. Trends Immunol 38(10):768–776. https://doi.org/10.1016/j.it.2017.08.002

Gaál Z (2024) Role of microRNAs in immune regulation with translational and clinical applications. Int J Mol Sci 25(3):1942

Goh KY, Cheng TYD, Tham SC, Lim DW-T (2023) Circulating biomarkers for prediction of immunotherapy response in NSCLC. Biomedicines 11(2):508

Ishida M, Selaru FM (2013) miRNA-based therapeutic strategies. Curr Anesthesiol Rep 1(1):63–70. https://doi.org/10.1007/s40139-012-0004-5

Kalkusova K, Taborska P, Stakheev D, Smrz D (2022) The role of miR-155 in antitumor. Immunity 14(21):5414

Kaur R, Bhardwaj A, Gupta S (2023) Cancer treatment therapies: traditional to modern approaches to combat cancers. Mol Biol Rep 50(11):9663–9676. https://doi.org/10.1007/s11033-023-08809-3

Kim HI, Park J, Zhu Y, Wang X, Han Y, Zhang D (2024) Recent advances in extracellular vesicles for therapeutic cargo delivery. Exp Mol Med 56(4):836–849. https://doi.org/10.1038/s12276-024-01201-6

Kipkeeva F, Muzaffarova T, Korotaeva A, Mansorunov D, Apanovich P, Nikulin M et al (2022) The features of immune checkpoint gene regulation by microRNA in cancer. Int J Mol Sci 23(16):9324

Liu H, Lei C, He Q, Pan Z, Xiao D, Tao Y (2018) Nuclear functions of mammalian MicroRNAs in gene regulation, immunity and cancer. Mol Cancer 17(1):64. https://doi.org/10.1186/s12943-018-0765-5

Loomis AK, Brock GJ (2011) Chapter 21 - MicroRNAs—basic biology and therapeutic potential. In: Macor JE (ed) Annual reports in medicinal chemistry, vol 46. Academic Press, pp 351–365

Ma Y, Shen N, Wicha MS, Luo M (2021) The roles of the Let-7 family of microRNAs in the regulation of cancer stemness. Cells 10(9):2415

Marei HE, Hasan A, Pozzoli G, Cenciarelli C (2023) Cancer immunotherapy with immune checkpoint inhibitors (ICIs): potential, mechanisms of resistance, and strategies for reinvigorating T cell responsiveness when resistance is acquired. Cancer Cell Int 23(1):64. https://doi.org/10.1186/s12935-023-02902-0

Marschner D, Falk M, Javorniczky NR, Hanke-Müller K, Rawluk J, Schmitt-Graeff A et al (2020) MicroRNA-146a regulates immune-related adverse events caused by immune checkpoint inhibitors. JCI Insight 5(6). https://doi.org/10.1172/jci.insight.132334

Martínez-Hernández R, Serrano-Somavilla A, Ramos-Leví A, Sampedro-Nuñez M, Lens-Pardo A, Muñoz De Nova JL et al (2019) Integrated miRNA and mRNA expression profiling identifies novel targets and pathological mechanisms in autoimmune thyroid diseases. EBioMedicine 50: 329–342. https://doi.org/10.1016/j.ebiom.2019.10.061

Meira DD, de Castro e Caetano MC, Casotti MC, Zetum ASS, Gonçalves AFM, Moreira AR et al (2023) Prognostic factors and markers in non-small cell lung cancer: recent progress and future challenges. Genes 14(10):1906

Menon A, Abd-Aziz N, Khalid K, Poh CL, Naidu R (2022) miRNA: a promising therapeutic target in cancer. Int J Mol Sci 23(19):11502

Mockly S, Seitz H (2023) Synthetic miR-34a against solid tumours: a predictable failure. Br J Cancer 128(3):478–480. https://doi.org/10.1038/s41416-022-02123-8

Monastirioti A, Papadaki C, Kalapanida D, Rounis K, Michaelidou K, Papadaki MA et al (2022) Plasma-based microRNA expression analysis in advanced stage NSCLC patients treated with nivolumab. Cancers (Basel) 14(19):4739

Mori MA, Ludwig RG, Garcia-Martin R, Brandão BB, Kahn CR (2019) Extracellular miRNAs: from biomarkers to mediators of physiology and disease. Cell Metab 30(4):656–673. https://doi.org/10.1016/j.cmet.2019.07.011

Moutabian H, Radi UK, Saleman AY, Adil M, Zabibah RS, Chaitanya MNL et al (2023) MicroRNA-155 and cancer metastasis: regulation of invasion, migration, and epithelial-to-mesenchymal transition. Pathology 250:154789. https://doi.org/10.1016/j.prp.2023.154789

Murugaiyan G, da Cunha AP, Ajay AK, Joller N, Garo LP, Kumaradevan S et al (2015) MicroRNA-21 promotes Th17 differentiation and mediates experimental autoimmune encephalomyelitis. J Clin Invest 125(3):1069–1080. https://doi.org/10.1172/JCI74347

Nejad C, Stunden HJ, Gantier MP (2018) A guide to miRNAs in inflammation and innate immune responses. FEBS J 285(20):3695–3716. https://doi.org/10.1111/febs.14482

O'Brien J, Hayder H, Zayed Y, Peng C (2018) Overview of microRNA biogenesis mechanisms of actions, and circulation. 9-2018. https://doi.org/10.3389/fendo.2018.00402

Peng Y, Croce CM (2016) The role of MicroRNAs in human cancer. Signal Transduct Target Ther 1(1):15004. https://doi.org/10.1038/sigtrans.2015.4

Perkins RS, Singh R, Abell AN, Krum SA, Miranda-Carboni GA (2023) The role of WNT10B in physiology and disease: a 10-year update. 11-2023. https://doi.org/10.3389/fcell.2023.1120365

Peterlin A, Počivavšek K, Petrovič D, Peterlin B (2020) The role of microRNAs in heart failure: a systematic review. 7-2020. https://doi.org/10.3389/fcvm.2020.00161

Popa ML, Ichim C, Anderco P, Todor SB, Pop-Lodromanean D (2025) MicroRNAs in the diagnosis of digestive diseases: a comprehensive review. J Clin Med 14(6):2054

Qin Y, Huo M, Liu X, Li SC (2024) Biomarkers and computational models for predicting efficacy to tumor ICI immunotherapy. 15-2024. https://doi.org/10.3389/fimmu.2024.1368749
Rhim J, Baek W, Seo Y, Kim JH (2022) From molecular mechanisms to therapeutics: understanding MicroRNA-21 in cancer. Cells 11(18):2791
Rupaimoole R, Calin GA, Lopez-Berestein G, Sood AK (2016) miRNA deregulation in cancer cells and the tumor microenvironment. Cancer Discov 6(3):235–246. https://doi.org/10.1158/2159-8290.CD-15-0893
Shiravand Y, Khodadadi F, Kashani SMA, Hosseini-Fard SR, Hosseini S, Sadeghirad H et al (2022) Immune checkpoint inhibitors in cancer. Therapy 29(5):3044–3060
Smolle MA, Calin HN, Pichler M, Calin GA (2017) Noncoding RNAs and immune checkpoints—clinical implications as cancer therapeutics. FEBS J 284(13):1952–1966. https://doi.org/10.1111/febs.14030
Stavast CJ, Erkeland SJ (2019) The non-canonical aspects of microRNAs: many roads to gene regulation. Cells 8(11):1465
Suzuki HI, Katsura A, Matsuyama H, Miyazono K (2015) MicroRNA regulons in tumor microenvironment. Oncogene 34(24):3085–3094. https://doi.org/10.1038/onc.2014.254
Takahashi H, Kanno T, Nakayamada S, Hirahara K, Sciumè G, Muljo SA et al (2012) TGF-β and retinoic acid induce the microRNA miR-10a, which targets Bcl-6 and constrains the plasticity of helper T cells. Nat Immunol 13(6):587–595. https://doi.org/10.1038/ni.2286
Testa U, Pelosi E, Castelli G, Labbaye C (2017) miR-146 and miR-155: two key modulators of immune response and tumor development. Non-coding RNA 3(3). https://doi.org/10.3390/ncrna3030022
Vaddi PK, Osborne DG, Nicklawsky A, Williams NK, Menon DR, Smith D et al (2023) CTLA4 mRNA is downregulated by miR-155 in regulatory T cells, and reduced blood CTLA4 levels are associated with poor prognosis in metastatic melanoma patients. 14-2023. https://doi.org/10.3389/fimmu.2023.1173035
Varadé J, Magadán S, González-Fernández Á (2021) Human immunology and immunotherapy: main achievements and challenges. Cell Mol Immunol 18(4):805–828. https://doi.org/10.1038/s41423-020-00530-6
Vaxevanis C, Bachmann M, Seliger B (2024) Immune modulatory microRNAs in tumors, their clinical relevance in diagnosis and therapy. J Immunother Cancer 12(8). https://doi.org/10.1136/jitc-2024-009774
Wang Q, Lin W, Tang X, Li S, Guo L, Lin Y, Kwok HF (2017) The roles of microRNAs in regulating the expression of PD-1/PD-L1 immune checkpoint. 18(12):2540
Wang Z, Wang H, Zhou S, Mao J, Zhan Z, Duan S (2024) miRNA interplay: mechanisms and therapeutic interventions in cancer. MedComm – Oncol 3(4):e93. https://doi.org/10.1002/mog2.93
Wu Y, Park J, Xu E, Kim D, Lee J, Oh Y-K (2025) MicroRNA-induced reprogramming of tumor-associated macrophages for modulation of tumor immune microenvironment. J Control Release 381:113593. https://doi.org/10.1016/j.jconrel.2025.113593
Xiao C, Rajewsky K (2009) MicroRNA control in the immune system: basic principles. Cell 136(1):26–36. https://doi.org/10.1016/j.cell.2008.12.027
Yadav R, Khatkar R, Yap KCH, Kang CY-H, Lyu J, Singh RK et al (2024) The miRNA and PD-1/PD-L1 signaling axis: an arsenal of immunotherapeutic targets against lung cancer. Cell Death Discov 10(1):414. https://doi.org/10.1038/s41420-024-02182-1
Yang Y, Alderman C, Sehlaoui A, Xiao Y, Wang W (2018) MicroRNAs as immunotherapy targets for treating gastroenterological cancers. 2018(1):9740357. https://doi.org/10.1155/2018/9740357
Yao XC, Wu JJ, Yuan ST, Yuan FL (2025) Recent insights and perspectives into the role of the miRNA-29 family in innate immunity (Review). Int J Mol Med 55(3). https://doi.org/10.3892/ijmm.2025.5494
Yin Q, Wu L, Han L, Zheng X, Tong R, Li L et al (2023) Immune-related adverse events of immune checkpoint inhibitors: a review. 14-2023. https://doi.org/10.3389/fimmu.2023.1167975

# Antiangiogenic miRNA-Based Combination Therapy

# 13

Pranjal Agarwal, Jugal Patil, Dhiraj Bhatia, and Amit K. Yadav

**Abstract**

Angiogenesis, the process of forming new blood vessels from existing ones, is essential for normal physiology but becomes dysregulated in diseases like cancer, neovascular eye conditions, and chronic inflammation. This chapter explores the significant potential of microRNAs (miRNAs) as antiangiogenic agents and their promise in combination therapies to enhance treatment efficacy beyond current monotherapies. It begins by outlining critical pro-angiogenic signaling pathways such as vascular endothelial growth factor (VEGF), fibroblast growth factor (FGF), and platelet-derived growth factor (PDGF), which drive endothelial cell proliferation and vessel remodeling. Key criteria for selecting effective, particular antiangiogenic miRNAs are discussed, alongside comprehensive evidence from in vitro and in vivo studies demonstrating their ability to inhibit endothelial proliferation, migration, and tube formation in tumor and retinal models. Detailed coverage of diverse miRNA delivery systems, including lipid and polymer nanoparticles, viral vectors, exosome-based carriers, and innovative stimuli-responsive platforms, is provided, assessing their efficacy, safety, and translational potential along with crucial preclinical and clinical findings.

**Keywords**

microRNA (miRNA) · Antiangiogenesis · Combination therapy · Cancer therapeutics · Tumor microenvironment · Gene regulation

P. Agarwal · J. Patil · D. Bhatia (✉) · A. K. Yadav (✉)
Department of Biological Sciences and Engineering, Indian Institute of Technology Gandhinagar, Gandhinagar, Gujarat, India
e-mail: 24210073@iitgn.ac.in; agarwalpranjal2001@gmail.com; 24210049@iitgn.ac.in; dhiraj.bhatia@iitgn.ac.in; yadav.amit@iitgn.ac.in; amitkmr1194@gmail.com





S. Tiwari et al. (eds.), *microRNA Modulation: A Promising Combination Therapy for Cancer*, https://doi.org/10.1007/978-981-95-5989-3_13

**Learning Objectives**

1. *Explain* the role of pro- and antiangiogenic miRNAs in regulating tumor vascularization and progression.
2. *Analyze* how targeting angiogenesis-related miRNAs can enhance the efficacy of conventional therapies such as chemotherapy, radiotherapy, or targeted agents.
3. *Evaluate* preclinical and clinical evidence supporting antiangiogenic miRNA-based combination therapies in different cancer models.
4. *Assess* the opportunities and challenges of translating antiangiogenic miRNA approaches into clinical practice, including delivery, specificity, and long-term safety.

## 13.1 Introduction

Angiogenesis is the physiological process of forming new blood vessels from preexisting ones, as opposed to vasculogenesis, which includes the formation of blood vessels from endothelial precursor cells (Lugano et al. 2019). This highly regulated process is characterized by endothelial cell activation, proliferation, migration, and tube formation, followed by vessel maturation and stabilization (Zhang et al. 2024). Under normal physiological conditions, angiogenesis is tightly controlled and primarily occurs during embryonic development, wound healing, the female reproductive cycle, muscle growth, and tissue repair following injury (Liu et al. 2023). The angiogenic process involves several phases, including the breakdown of the basement membrane, endothelial cell migration, and proliferation through a remodeled extracellular matrix, ultimately leading to tube formation and vessel maturation (Zhang et al. 2024). Key signaling pathways, including vascular endothelial growth factor (VEGF), platelet-derived growth factor (PDGF), fibroblast growth factor-2 (FGF-2), apelin (APLN), angiopoietins, ephrins, and chemokines, orchestrate these complex cellular events (Lugano et al. 2019). Figure 13.1 illustrates the key differences between normal angiogenesis and pathological angiogenesis, emphasizing the roles of three major growth factors: VEGF, FGF, and PDGF. The balance between pro-angiogenic and anti-angiogenic factors maintains vascular homeostasis in healthy tissues (Chauhan et al. 2025; Aliashrafi and Ebrahimi-Mameghani 2017); however, pathological deviation from this normal vascular development leads to tumor angiogenesis, a phenomenon recognized as a hallmark of cancer since Judah Folkman's pioneering work in 1971. When solid tumors exceed 1–2 $mm^3$ in volume, they require active angiogenesis to sustain growth beyond the diffusion limits for oxygen and nutrients (Ribatti 2008). At this so-called "angiogenic switch," an imbalance in favor of pro-angiogenic signals such as VEGF, FGF, and transforming growth factor-β (TGF-β) over endogenous inhibitors like endostatin and thrombospondin drives endothelial cell proliferation, migration, and invasion through mechanisms including sprouting angiogenesis, intussusceptive angiogenesis, vasculogenesis, vessel co-option, and even vasculogenic mimicry (Zhang et al. 2024; Mabeta et al. 2022). These newly formed

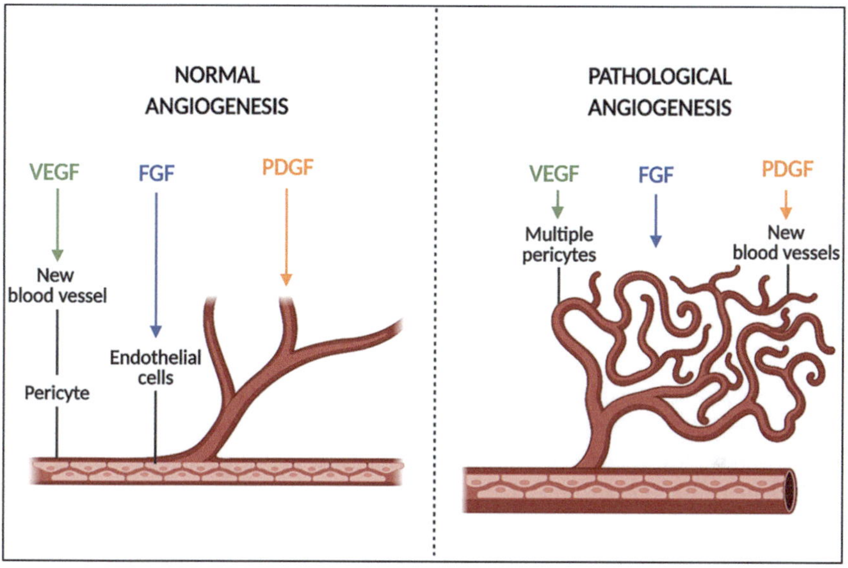


**Fig. 13.1** Schematic of normal vs. pathological angiogenesis with key signaling pathway

tumor vessels are typically abnormal, exhibiting irregular morphology, increased permeability, and poor perfusion, which contribute to the hostile tumor microenvironment (Liu et al. 2023). A similar imbalance underlies ocular neovascular disorders such as wet age-related macular degeneration (AMD), retinal vein occlusion, diabetic retinopathy, retinopathy of prematurity, neovascular glaucoma, and corneal neovascularization (Crawford et al. 2009). In these conditions, upregulation of hypoxia-inducible factor-1α (HIF-1α) in the retina, choroid, or cornea leads to overexpression of VEGF, along with FGF and PDGF, and overwhelms antiangiogenic factors like pigment epithelium-derived factor (PEDF), resulting in vision loss (Zimna and Kurpisz 2015; Yadav et al. 2023; YadavDamini and Solanki 2013). Current antiangiogenic therapies for ocular diseases primarily target the VEGF pathway through intravitreal injections of anti-VEGF agents (Crawford et al. 2009). These insights highlight the therapeutic potential of targeting angiogenic pathways across oncology, ophthalmology, and chronic inflammatory conditions (Walsh and Pearson 2001). Current antiangiogenic drugs primarily target the VEGF pathway through monoclonal antibodies, such as bevacizumab or small-molecule tyrosine kinase inhibitors, including sunitinib and sorafenib (Stacker and Achen 2013; Pezzella 2019). While these agents have improved patient outcomes, acquired resistance frequently develops through activation of alternative angiogenic pathways, highlighting the need for combination approaches that can target multiple pathways simultaneously (Haibe et al. 2020).

### 13.1.1 Rationale for Combination Therapy

The rationale for miRNA-based combination therapy stems from the complex, multi-pathway nature of angiogenesis and the limitations of single-agent approaches. Combination therapies can achieve synergistic effects through several mechanisms, including targeting complementary pathways, overcoming resistance mechanisms, and enhancing overall therapeutic efficacy, as demonstrated by mathematical models and experimental evidence (Chakraborty et al. 2018; Miroshnichenko and Patutina 2019). In cancer therapy, combination approaches targeting multiple angiogenic pathways have shown promise in preclinical studies, with dual inhibitors of VEGF and FGF pathways demonstrating enhanced anti-cancer efficacy. miRNA combination therapy offers unique advantages because individual miRNAs can target multiple genes within the same pathway or across different pathways. Studies have demonstrated that combinations of miRNA mimics can achieve two- to four-fold greater tumor growth inhibition than individual miRNAs. For example, the simultaneous administration of miR-34a and miR-497 mimics showed up to four-fold greater anti-tumor effects than either mimic alone (Miroshnichenko and Patutina 2019; Juan et al. 2010).

This chapter will systematically explore the development and application of antiangiogenic miRNA-based combination therapies. Following this introduction, subsequent sections will examine specific miRNAs involved in angiogenesis regulation (AngiomiRs), delivery systems, formulation strategies, and preclinical evidence for combination approaches.

## 13.2 Antiangiogenic miRNA Therapeutic Candidates

Antiangiogenic microRNAs (miRNAs) are a promising new research area for treating diseases characterized by abnormal blood vessel growth, including cancer, diabetic retinopathy, and various inflammatory conditions. These small noncoding RNA molecules have unique advantages as therapeutic agents because they can simultaneously target multiple genes in angiogenic signaling pathways. This could lead to more comprehensive and effective treatment options than traditional single-target therapies (Lahooti et al. 2021; Tiwari et al. 2018). Table 13.1 provides an overview of key candidate miRNAs implicated in angiogenesis, highlighting their validated targets and the resulting functional outcomes, including pro- or anti-angiogenic effects.

### 13.2.1 Selection Criteria for Therapeutic miRNAs

The development of effective miRNA-based therapeutics requires precise selection criteria that encompass multiple biological and pharmacological parameters. The most critical selection criteria include potency, specificity, expression profiles, and stability characteristics (Thibonnier 2024; Garreau et al. 2024; Ishida and Selaru 2012).

**Table 13.1** Summary of candidate miRNAs with targets and reported effects

| miRNA | Primary targets | Angiogenic effects | Disease models | References |
|---|---|---|---|---|
| miR-15b | VEGF | Inhibits tube formation, reduces proliferation | Diabetic retinopathy | Yang et al. (2020) |
| miR-29a | VEGF, MMPs, collagens | Suppresses neovascularization, reduces vessel density | Ocular angiogenesis, cancer | Peng et al. (2022) |
| miR-126 | Spred-1, PIK3R2, VCAM-1 | Promotes physiological angiogenesis, inhibits inflammation | Myocardial infarction, development | Wu et al. (2009) |
| miR-92a | α5 integrin, EDG1 | Inhibits endothelial sprouting, migration | Ischemic limbs, myocardial infarction | Landskroner-Eiger et al. (2013) |
| miR-140-5p | VEGF-A | Reduces invasion, angiogenesis | Breast cancer | Lu et al. (2017) |
| miR-200 family | IL-8, CXCL1 | Inhibits tumor angiogenesis | Various cancers | Pecot et al. (2013) |
| miR-204 | ANGPT1, TGFβR2, VEGF | Multi-pathway inhibition | Breast cancer | Hussen et al. (2021) |
| miR-199b-5p | ALK1 | Suppresses endothelial migration, tube formation | Breast cancer | Hussen et al. (2021) |
| miR-190 | STC2, multiple angiogenic genes | Inhibits migration, invasion, angiogenesis | Breast cancer | Hussen et al. (2021) |
| miR-17-92 cluster | TSP-1, CTGF, multiple targets | Complex angiogenic regulation | Tumor angiogenesis | Landskroner-Eiger et al. (2013) |

(i) *Potency* represents the fundamental requirement for therapeutic miRNAs, determined by their ability to modulate target gene expression and produce measurable biological effects effectively. Potency assessment involves evaluating the concentration-dependent effects of miRNA mimics or inhibitors on target mRNA levels and downstream protein expression (Ketterl et al. 2015; Deskins et al. 2013).

(ii) *Specificity* is crucial for minimizing off-target effects and ensuring therapeutic safety. This parameter involves both target specificity and tissue/cell-type specificity (Varley and Desaulniers 2021). Advanced bioinformatics approaches and experimental validation techniques are employed to predict and confirm miRNA-target interactions, utilizing databases such as TargetScan, miRanda, and RNAhybrid (Oesinghaus et al. 2024).

(iii) *Expression profiles* provide essential information about candidate miRNAs' endogenous distribution and regulation across different tissues, developmental stages, and disease conditions. Comprehensive expression profiling helps identify miRNAs that are dysregulated in pathological angiogenesis and provides insights into their therapeutic potential (Bargaje et al. 2010; Mannarino et al. 2025).

(iv) *Stability* encompasses both the intrinsic stability of the miRNA molecule and its stability following therapeutic delivery. Chemical modifications, such as 2'-O-methyl nucleotides, phosphorothioate linkages, and locked nucleic acid (LNA) modifications, enhance the stability of miRNA in biological systems (Varley and Desaulniers 2021; Seyhan 2024).

### 13.2.2 Preclinical Efficacy Assessment

Evaluating antiangiogenic miRNAs requires comprehensive preclinical testing using both in vitro and in vivo experimental systems that recapitulate key aspects of pathological angiogenesis.

(i) **In Vitro Assays:**

*Tube Formation Assays* constitute the gold standard for evaluating angiogenic potential in vitro. These assays typically employ human umbilical vein endothelial cells (HUVECs) or other cell types cultured on Matrigel or similar extracellular matrix substrates (Gentile et al. 2019; Carpentier et al. 2020). When exposed to pro-angiogenic stimuli, endothelial cells organize into branched, tubular networks that mimic the formation of capillaries. Antiangiogenic miRNAs are evaluated for their ability to inhibit this process, with quantification parameters including total tube length, the number of branching points, the number of closed loops, and overall network complexity (Mani et al. 2024).

*Migration Assays* assess the ability of endothelial cells to move in response to chemotactic gradients, a critical component of the angiogenic process. Various methodologies include scratch wound assays, transwell chamber assays, and under-agarose assays (Goodwin 2007). In scratch wound assays, a standardized wound is created in a confluent endothelial cell monolayer, and the rate of wound closure is monitored in the presence or absence of test miRNAs (Geng et al. 2020). Transwell assays provide more precise measurements of directed migration by allowing cells to migrate through porous membranes toward chemotactic stimuli (Goodwin 2007).

*Proliferation Assays* measure the impact of miRNAs on endothelial cell growth and division. Multiple methodologies are available, including direct cell counting, bromodeoxyuridine (BrdU) incorporation assays, MTT/CCK-8 colorimetric assays, and flow cytometry-based approaches (Goodwin 2007; Peng et al. 2022).

(ii) **In Vivo Models.**

*Tumor Xenograft Models* involve implanting human tumor cells into immunocompromised mice and monitoring the effects of miRNA treatment on tumor growth, vascularization, and metastasis. The models allow for the assessment of both direct anti-tumor effects and indirect effects mediated through angiogenesis inhibition. Advanced imaging techniques, including the fluorescent labeling of tumor vasculature, enable the precise quantification of angiogenic responses (Hanyu et al. 2009; Thijssen et al. 2004).

*Corneal Micropocket Assay* exploits the naturally avascular cornea as a "screen" for visualizing new blood vessel growth. Slow-release pellets containing pro-angiogenic factors such as VEGF or basic FGF (bFGF) are surgically implanted into corneal micropockets, and the resulting neovascularization is monitored over several days. The assay offers several advantages, including the absence of background vasculature, ease of visualization, and standardized quantification protocols (Grogan et al. 2018; Birsner et al. 2014).

*Oxygen-Induced Retinopathy (OIR) Model* involves exposing neonatal mice to hyperoxia (75% oxygen) for 5 days, which causes retinal vessel loss. Following this the mice are returned to room air, triggering pathological neovascularization (Vähätupa et al. 2020; Connor et al. 2009). This model closely recapitulates the pathophysiology observed in human diseases such as retinopathy of prematurity and proliferative diabetic retinopathy (Zhong et al. 2025).

## 13.3 Design of Combination Therapies

The development of combination therapies represents a pivotal advancement in cancer treatment, particularly through the strategic integration of miRNAs with conventional therapeutics to simultaneously target multiple pathways involved in cancer progression, thereby overcoming the limitations of single-agent therapies and addressing the complex, heterogeneous nature of malignant diseases.

### 13.3.1 miRNA + Chemotherapy

The mechanistic foundation for miRNA-chemotherapy combinations is rooted in the capacity of specific miRNAs to target genes involved in chemoresistance pathways. For example, miR-21, an extensively studied oncomiR, has been identified as a key regulator of chemotherapeutic sensitivity across various cancer types (Vyas et al. 2025). Inhibition of miR-21 sensitizes cancer cells to multiple chemotherapeutic agents, including temozolomide in glioblastoma, topotecan in breast cancer, and enhances the effectiveness of trastuzumab in HER2-positive breast cancer. The mechanism involves downregulating anti-apoptotic proteins, such as B-cell lymphoma 2 (BCL2), modulating the NF–κB signaling pathway, leading to enhanced cellular susceptibility to chemotherapy-induced cell death (van Beijnum et al. 2017). Preclinical studies have provided compelling evidence for the clinical potential of miRNA-chemotherapy combinations. A notable example involves the co-delivery of dacarbazine and miRNA-34a using polypeptide micelles for the treatment of malignant melanoma. This combination demonstrated a 1.84-fold greater cytotoxicity than dacarbazine alone in A375 cells, significantly improving both in vitro and in vivo efficacy parameters. The enhanced therapeutic effect was attributed to miRNA-34a's ability to target multiple oncogenic pathways while simultaneously potentiating the cytotoxic effects of dacarbazine (Ding et al. 2025). Similarly, innovative delivery systems have been developed to optimize miRNA-chemotherapy combinations. The development of dual miRNA-triggered drug release systems exemplifies this approach, where miRNA-21 and miRNA-10b serve as molecular triggers for the catalytic release of doxorubicin and short interfering RNA (siRNA) in cancer cells. This system operates through a thermodynamically driven entropy gain process, ensuring targeted drug release specifically in cancerous tissues while minimizing off-target effects (Yue et al. 2020). Such approaches demonstrate the sophisticated engineering solutions being developed to maximize the therapeutic potential of miRNA-chemotherapy combinations (Fig. 13.2a).

### 13.3.2 miRNA + Targeted Agents

The scientific rationale for miRNA-targeted agent combinations is exemplified by studies involving Epidermal Growth Factor Receptor (EGFR)-targeted therapies and miRNA modulators. Research has demonstrated that T790M mutation-associated resistance to EGFR tyrosine kinase inhibitors can be overcome through combination therapy involving gefitinib and miR-21 inhibitors delivered via chitosan-derived nanoparticles. This approach achieves synergistic anti-tumor effects by simultaneously blocking EGFR signaling and modulating downstream anti-apoptotic pathways through the regulation of miRNA. The combination results in enhanced sensitivity of EGFR-T790M mutated non-small cell lung cancer cells, significantly improving both in vitro cytotoxicity and in vivo tumor regression (Huang et al. 2021). Clinical evidence supporting miRNA-targeted agent combinations is derived from studies examining the combination of antiangiogenic therapies with miRNA modulators. The combination of chlorotoxin-coupled anti-miR-21 oligonucleotides with sunitinib has demonstrated enhanced anti-tumor efficacy in glioblastoma models, resulting in superior tumor growth inhibition compared to either agent alone. This approach exploits the dual mechanisms of antiangiogenic pathway inhibition and oncomiR suppression to achieve more comprehensive tumor control (van Beijnum et al. 2017). The development of resistance to targeted agents often involves the upregulation of compensatory pathways, which can be effectively addressed through miRNA-based interventions. Studies in clear cell renal cell carcinoma (RCC) have identified specific miRNA signatures associated with

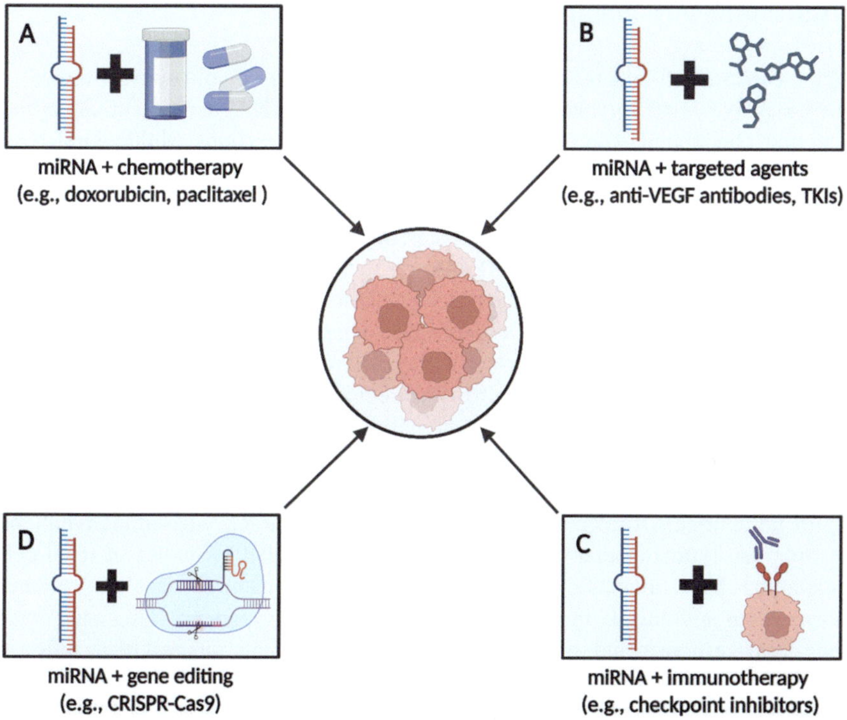


**Fig. 13.2** A schematic of miRNA-based combination therapies. (**a**) miRNA sensitizes cells to DNA damage, (**b**) miRNA downregulates pro-angiogenic pathways, (**c**) miRNA boosts antigen presentation, (**d**) miRNA + CRISPR for dual-pathway knockdown

sunitinib resistance, revealing that Programmed Death-Ligand 1 (PD-L1) upregulation contributes to therapeutic resistance and can be overcome through combination approaches targeting both angiogenic pathways and immune checkpoint mechanisms. This demonstrates how miRNA profiling can inform rational combination design by identifying resistance mechanisms that can be targeted through complementary therapeutic approaches (Armesto et al. 2021). Advanced delivery systems have been developed to optimize the co-administration of miRNAs and targeted agents. Polymeric micelle formulations have been designed to simultaneously deliver hedgehog pathway inhibitors and tumor suppressor miRNAs, such as let-7b, for pancreatic cancer treatment. These systems achieve enhanced pharmacokinetic properties for both components, significantly improving tumor targeting and therapeutic efficacy compared to conventional formulations (Kumar et al. 2018) (Fig. 13.2b).

### 13.3.3 miRNA + Immunotherapy

The mechanistic foundation for miRNA-immunotherapy combinations is rooted in the extensive regulatory networks through which miRNAs control immune cell function, checkpoint expression, and tumor-immune interactions. Multiple miRNAs have been identified as direct regulators of PD-L1 expression, including miR-34a, miR-138-5p, miR-424-5p, miR-200, miR-570, miR-513, and miR-197, all of which function as tumor suppressors by suppressing immune checkpoint proteins. The therapeutic delivery of these miRNAs or their mimics can enhance immune responses against tumors by reducing checkpoint inhibition and promoting T-cell activation (Xing et al. 2021). Research has demonstrated that anti-PD1 therapy induces the release of lymphocyte-derived exosomal miRNA-4315, which promotes apoptosis resistance in cancer cells by downregulating Bim. This resistance mechanism can be circumvented through combination therapy with BH3 mimetic agents, providing a rational basis for triple combination approaches involving miRNAs, checkpoint inhibitors, and pro-apoptotic drugs. Such studies highlight the complexity of immune-tumor interactions and the potential for miRNA-based interventions to optimize immunotherapy outcomes (Guyon et al. 2020). Studies in renal cell carcinoma have identified "lymphocyte miRNA signatures" that can predict response to nivolumab therapy, with specific miRNA patterns associated with long-term response and overall survival. These findings suggest that miRNA profiling could aid in selecting patients who are most likely to benefit from immunotherapy and inform the design of personalized combination approaches (Incorvaia et al. 2020). Innovative approaches to enhance immunotherapy efficacy include utilizing miRNAs to modulate the polarization of tumor-associated macrophage and T cell infiltration. Research has demonstrated that modulation of miR-146b can enhance anti-PD-1 therapy in colorectal cancer by remodeling the tumor microenvironment and promoting anti-tumor immunity. The combination approach resulted in enhanced T cell infiltration, reduced immunosuppression, and improved therapeutic outcomes compared to either monotherapy. This exemplifies how miRNA-based interventions can prime the tumor microenvironment for enhanced responsiveness to immunotherapy (He et al. 2023). Integrating miRNAs with adoptive cell therapies represents another promising direction in combination immunotherapy. Studies have shown that miR-340-5p can enhance CD8+ T cell infiltration and function in diffuse large B-cell lymphoma, with potential applications in optimizing CAR-T cell therapies and other adoptive immunotherapies. The approach demonstrates how miRNA modulators can enhance the effectiveness of engineered immune cells, potentially reducing the risk of immune-related adverse events (Xu et al. 2020) (Fig. 13.2c).

### 13.3.4 miRNA + Gene Editing or Other RNA Therapeutics

The integration of miRNAs with CRISPR–Cas9 gene editing has emerged as a particularly sophisticated approach that enables tissue-specific and condition-responsive genetic modifications. Recent developments in CRISPR MiRAGE (miRNA-activated genome editing) technology demonstrate how endogenous miRNA expression can control CRISPR activity in specific cellular contexts. These systems utilize miRNA-responsive guide RNAs that adopt a trigger hairpin conformation to conceal their targeting sequence until they encounter a particular miRNA. Upon binding by an Argonaute–miRNA complex, the hairpin is unwound, exposing the guide RNA and thereby enabling precise, spatially and temporally controlled gene editing. The approach addresses one of the significant challenges in gene-editing therapeutics, namely achieving specificity for diseased tissues while minimizing off-target effects in healthy cells (Ijee et al. 2024; Garcia-Guerra et al. 2025). The combination of miRNAs with other RNA therapeutics, particularly siRNAs, has demonstrated synergistic therapeutic potential in multiple preclinical models. Studies in ovarian cancer have shown that dual targeting of EphA2 using both miR-520d-3p and siRNA achieves synergistic inhibition with significantly enhanced tumor regression compared to either monotherapy alone. The combination demonstrated synergistic effects in cell invasion assays and provided additive benefits in reducing tumor angiogenesis, proliferation, and promoting apoptosis (Nishimura et al. 2013; Kumar et al. 2025). Advanced delivery systems have been developed to enable the simultaneous administration of multiple RNA therapeutic modalities. Mesoporous silica nanoparticles have been engineered to co-deliver siRNA and miRNA combinations, incorporating photosensitizers to facilitate endosomal escape and tumor-penetrating peptides to enhance tissue distribution and penetration. These systems demonstrate combined cell-killing activity and significant suppression of primary tumor growth and metastasis in preclinical models (Wang et al. 2020) (Fig. 13.2d).

## 13.4 Delivery Strategies for miRNA Combinations

The development of effective delivery strategies for miRNA combinations has emerged as a critical area of research, encompassing various approaches from nonviral systems to sophisticated biomimetic carriers (Forterre et al. 2020), as shown in Fig. 13.3.

### 13.4.1 Nonviral Systems

Nonviral delivery systems have gained considerable attention due to their superior safety profile compared to viral vectors. These systems offer several advantages, including reduced immunogenicity, lower toxicity, and ease of manufacturing, making them attractive for clinical applications. The primary nonviral approaches include lipid-based nanoparticles, polymeric carriers, and inorganic nanoparticles (Jiang et al. 2022; Bai et al. 2019; Toma et al. 2022).

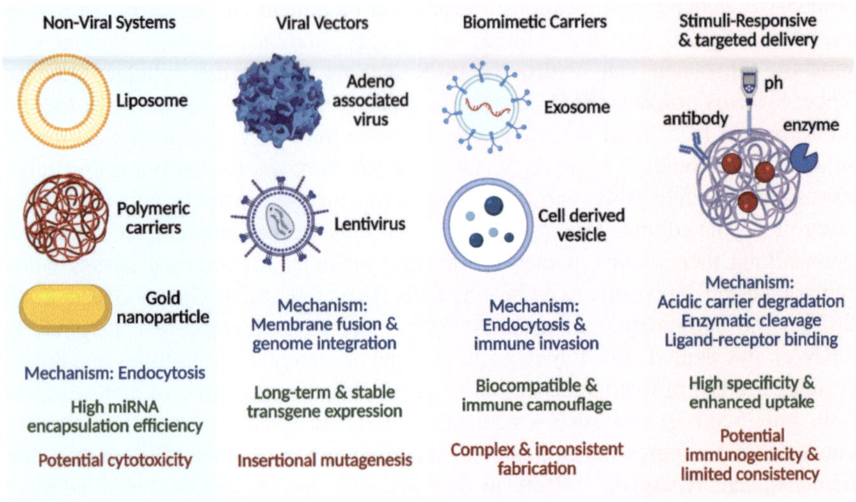


**Fig. 13.3** Comparative schematic of delivery platforms for miRNA and drug co-delivery

*Lipid-based nanoparticles* represent the most widely used nonviral delivery platform for miRNA therapeutics (Dogheim et al. 2025). Liposomes and lipoplexes utilize cationic lipids to electrostatically complex with negatively charged miRNAs, protecting from nuclease degradation (Yang 2015). Solid lipid nanoparticles (SLNs) have effectively delivered miRNAs, such as miR-34a, for lung cancer treatment (Moraes et al. 2021). The regulatory approval of Onpattro® has validated lipid nanoparticles as a viable platform for RNA therapeutics, paving the way for miRNA-based therapies (He et al. 2020).

*Polymeric delivery systems* offer structural flexibility and customizable properties for miRNA delivery (Holjencin and Jakymiw 2022). Polyethylenimine (PEI) is among the most efficient polymeric carriers, utilizing the proton sponge effect to facilitate endosomal escape (Conte et al. 2020). However, high-molecular-weight PEI exhibits significant cytotoxicity, leading to the development of modified versions with improved biocompatibility (Gareev et al. 2023). Poly(lactic-co-glycolic acid) (PLGA) nanoparticles offer sustained release capabilities and have been successfully utilized for the co-delivery of miRNAs with chemotherapeutic agents (Devulapally et al. 2016). Biodegradable polymeric systems, such as polyhydroxybutyrate-polyethylenimine (PHB–PEI) nanoparticles, have demonstrated superior delivery efficiency compared to commercial transfection agents (Conte et al. 2020).

*Inorganic nanoparticles* offer unique advantages, including high stability and the potential for multimodal therapy. Gold nanoparticles (AuNPs) have been extensively studied for miRNA delivery, with the ability to be functionalized with targeting ligands and combined with photothermal therapy (Wang et al. 2019).

Silica-based nanoparticles provide excellent biocompatibility and can be engineered for controlled release. Metal-organic frameworks (MOFs) represent an emerging platform that combines high loading capacity and stimuli-responsive properties for controlled miRNA release (Zhao et al. 2021).

### 13.4.2 Viral Vectors

Despite safety concerns, viral vectors remain highly efficient delivery systems for miRNA therapeutics due to their natural ability to transduce cells and achieve stable expression (Fu et al. 2019). The primary viral platforms include lentiviruses, adenoviruses, and adeno-associated viruses (AAVs).

*Adenoviral vectors* offer high transduction efficiency and can accommodate relatively large genetic constructs (Scholz et al. 2022). These vectors enable accurate dosing of delivered miRNAs and provide strong short-term transgene expression. Recent developments have focused on engineering adenoviral vectors with miRNA-responsive elements to improve specificity and safety (Geisler and Fechner 2016).

*Lentiviral vectors* offer the advantage of integrating into the host genome, resulting in stable and long-term expression of therapeutic miRNAs. These vectors can transduce both dividing and non-dividing cells, making them suitable for targeting differentiated tissues. However, concerns regarding genomic integration and potential insertional mutagenesis limit their clinical application (Yang 2015).

*Adeno-associated virus (AAV) vectors* represent the most promising viral platform for miRNA delivery. AAVs are non-pathogenic, exhibit low immunogenicity, and can achieve long-term expression without genomic integration. AAV-mediated delivery of miR-7 has demonstrated significant therapeutic efficacy in glioblastoma models (Gareev et al. 2023).

### 13.4.3 Biomimetic Carriers

Biomimetic delivery systems utilize natural biological processes to enhance targeting and reduce immunogenicity. The most prominent biomimetic approach utilizes extracellular vesicles, particularly exosomes, as natural carriers for miRNA delivery.

*Exosomes* represent an ideal biomimetic platform due to their inherent biocompatibility, natural targeting properties, and ability to cross biological barriers, including the blood-brain barrier. These naturally occurring nanovesicles can be loaded with therapeutic miRNAs through various methods, including electroporation, sonication, and transfection. Exosome-mediated delivery has demonstrated superior stability compared to synthetic carriers and a reduced toxicity profile (Kursunluoglu et al. 2025; Beuzelin and Kaeffer 2018). However, recent studies have challenged the efficiency of exosome-mediated miRNA delivery, suggesting that only a small fraction of extracellular miRNAs are contained within exosomes. The majority of circulating miRNAs appear to be complexed with RNA-binding proteins rather than encapsulated in vesicles (Albanese et al. 2021).

*Cell-derived membrane vesicles* and *artificial exosome-like nanoparticles* represent alternative biomimetic approaches that combine the advantages of natural vesicles with improved loading efficiency. These systems can be engineered to express specific targeting ligands while maintaining the biocompatibility of natural membranes (Holjencin and Jakymiw 2022; Beuzelin and Kaeffer 2018).

### 13.4.4 Stimuli-Responsive and Targeted Delivery

Advanced delivery systems incorporate stimulus-responsive elements to achieve controlled release and enhanced specificity. These systems respond to various triggers, including pH changes, enzymatic activity, redox conditions, and specific biomarkers (Zhou et al. 2013).

*pH-responsive systems* exploit the acidic tumor microenvironment for the selective release of drugs. Tumor tissues typically exhibit lower pH values than normal tissues, allowing for pH-triggered release of therapeutic miRNAs (Karimi et al. 2016). Metal-organic frameworks with pH-sensitive linkers have demonstrated effective miRNA delivery with minimal off-target effects (Zhao et al. 2021).

*Redox-responsive carriers* utilize the elevated glutathione levels in tumor cells for selective payload release. Gold nanoparticle aggregates cross-linked with disulfide bonds can release miRNAs in response to the reducing environment of cancer cells, while maintaining their structural integrity for continued therapeutic activity (Wang et al. 2019).

*miRNA-responsive systems* represent a sophisticated approach in which endogenous miRNAs serve as molecular triggers for drug release. These systems can achieve dual functionality by responding to disease-specific miRNA signatures while delivering therapeutic payloads. DNA nanoswitches responsive to cancer-associated miRNAs have demonstrated high sensitivity and specificity for targeted therapy (Ye et al. 2022).

*Targeting ligands* enhance the specificity of miRNA delivery systems by facilitating receptor-mediated uptake. Aptamers, antibodies, and peptides can be conjugated to delivery vehicles to achieve cell-type-specific targeting. Elastin-like polypeptides modified with targeting peptides have demonstrated enhanced tumor penetration and improved therapeutic efficacy (Hong et al. 2024).

### 13.4.5 Safety and Immunogenicity Considerations

The safety profile of miRNA delivery systems is paramount for clinical translation. Immunogenicity concerns vary significantly among delivery platforms. Viral vectors can trigger both innate and adaptive immune responses, potentially limiting the effectiveness of repeated administration. However, they also offer the advantage of being well-characterized systems with established safety profiles (Gilbert et al. 2020). Nonviral systems, while generally safer, may still induce inflammatory responses depending on their composition and surface properties (Ullah et al. 2025). Biocompatibility assessments must consider both the carrier material and the delivered miRNA payload. Cationic lipids and polymers can cause membrane disruption and cytotoxicity at high concentrations (Conte et al. 2020). Careful optimization of formulation parameters, including surface charge, particle size, and dose, is essential for maintaining therapeutic efficacy while minimizing toxicity (Teo et al. 2021). Long-term safety considerations include the potential for off-target effects and unintended gene regulation. miRNAs can target multiple mRNAs, potentially affecting unintended cellular pathways (Teo et al. 2021). Regulatory considerations for miRNA therapeutics require demonstrating both safety and efficacy through rigorous clinical testing. The complexity of combination delivery systems may increase regulatory requirements, emphasizing the importance of early consideration of clinical translatability in system design (Godsey et al. 2013).

## 13.5 Preclinical and Clinical Studies

Substantial preclinical research has demonstrated the therapeutic potential of miRNA-based antiangiogenic combination strategies across multiple cancer models. In breast cancer studies, modulation of miR-223 expression enhanced the antiangiogenic effects of propolis extract-loaded nanoparticles when combined with 5-fluorouracil. This combination demonstrated superior anti-tumor activity by suppressing angiogenesis, inflammatory pathways, and mTOR signaling while inducing apoptotic mechanisms (Shaker et al. 2023). Cardiovascular applications have shown particularly promising results. MiR-519e demonstrated remarkable pro-angiogenic and anti-apoptotic effects in myocardial infarction models, promoting cardiomyocyte proliferation and vascular recovery (Gong et al. 2024). Similarly, studies utilizing miR-29a in ocular neovascularization models revealed potent antiangiogenic properties, effectively inhibiting pathological blood vessel formation across choroidal neovascularization, oxygen-induced retinopathy, and corneal neovascularization models (Peng et al. 2022). Combination strategies targeting multiple pathways have shown synergistic effects. Research combining antiangiogenic tyrosine kinase inhibitors with PERK inhibitor HC-5404 in renal cell carcinoma models demonstrated enhanced tumor growth inhibition, with the combination resulting in tumor stasis or regression across multiple RCC models. The enhanced antiangiogenic effects were mediated by inhibiting new vasculature formation and mature tumor blood vessel function (https://aacrjournals.org/cancerres/article/83/7_Supplement/4010/724706/Abstract-4010-Inhibition-of-PERK-by-HC-5404 2023). Mechanistic studies have revealed the multi-target advantage of miRNA therapeutics. Unlike single-target approaches, miRNAs can simultaneously regulate multiple pro-angiogenic factors, including VEGF, PDGF, and MMP-2. This "one-hit-multiple-target" approach provides superior efficacy compared to conventional single-pathway inhibitors and reduces the likelihood of resistance development (van Beijnum et al. 2017). The clinical translation of miRNA-based antiangiogenic therapies has progressed significantly, with multiple compounds currently in various phases of clinical development (Table 13.2).

## 13.6 Conclusion

In summary, the application of miRNA modulators to inhibit pathological angiogenesis represents a highly promising strategy for managing oncologic, ophthalmologic, and chronic inflammatory disorders. These small RNAs can concurrently regulate multiple pro-angiogenic targets, thereby enhancing therapeutic precision and adaptability. When combined with conventional modalities such as cytotoxic chemotherapeutics, monoclonal anti-VEGF antibodies, immune checkpoint inhibitors, or tyrosine kinase inhibitors, miRNA regimens have demonstrated synergistic efficacy, allowing for dose reduction and mitigating the development of resistance. Advancements in delivery technologies, including lipid and polymer-based nanoparticles, virally derived vectors, and biomimetic exosomes, have substantially improved miRNA stability, biodistribution, and controlled release, overcoming key pharmacokinetic barriers. Preclinical studies have provided compelling evidence of antiangiogenic activity, and early-phase clinical trials are currently assessing safety and biological efficacy.

**Acknowledgments** D. B. thanks SERB, GoI for the Ramanujan Fellowship, DST-Nidhi Prayas for the start-up grant, and Gujcost-DST, GSBTM, BRNS-BARC, and HEFA-GoI, MoES for STARS for research grants. D. B. is a member of the Indian National Young Academy of Sciences (INYAS). P.A. thanks to the Ministry of Education, Govt. of India for the GATE fellowship. A.K.Y expresses thankful to the Indian Institute of Technology Gandhinagar (IITGN) to provide financial support through the Early Career Fellowship (ECF).

**Conflicts of Interest** There are no conflicts to declare.

**Table 13.2** Ongoing clinical trials of miRNA-based antiangiogenic combinations

| Therapeutic | Trial identifier (s) | Indication | Phase | Primary endpoints | Status | Key findings |
|---|---|---|---|---|---|---|
| MRG-110 (Anti-miR-92a) (Delving into the Latest Updates on MRG-110 with Synapse) | NCT03603431 | Angiogenesis Enhancement | Phase 1 | Safety, tolerability, and biomarker assessment of angiogenesis | Completed | Increased perfusion and histological markers of neoangiogenesis with good safety profile |
| MRX34 (miR-34a mimic) (Hong et al. 2020) | NCT01829971 | Cancer Treatment (Advanced solid tumors) | Phase 1 | Safety, tolerability, and pharmacodynamics | Terminated due to serious immune-mediated adverse events | Proof-of-concept for miRNA delivery and target modulation despite safety concerns |
| Lademirsen (Anti-miR-21) (Gale et al. 2024) | NCT02855268, NCT03373786 | Fibrotic Diseases (Alport syndrome) | Phase 1/2 | Rate of eGFR decline and safety assessment | Completed with mixed efficacy results | Mixed efficacy results |
| CDR132L (Anti-miR-132) (Bayrak et al. 2025) | NCT05350969 | Cardiovascular Applications (Myocardial infarction) | Phase 1 | Cardiac function improvement and vascular recovery | Active and recruiting | Focus on cardiac function improvement and vascular recovery |

## References

Albanese M et al (2021) MicroRNAs are minor constituents of extracellular vesicles that are rarely delivered to target cells. PLoS Genet 17:e1009951

Aliashrafi S, Ebrahimi-Mameghani M (2017) 7: a systematic review on vitamin d and angiogenesis. BMJ Open 7(Suppl 1). https://doi.org/10.1136/bmjopen-2016-015415.7

Armesto M et al (2021) Integrated mRNA and miRNA transcriptomic analyses reveals divergent mechanisms of sunitinib resistance in clear cell renal cell carcinoma (ccRCC). Cancer 13:4401

Bai Z et al (2019) Non-viral nanocarriers for intracellular delivery of microRNA therapeutics. J Mater Chem B 7:1209–1225

Bargaje R, Hariharan M, Scaria V, Pillai B (2010) Consensus miRNA expression profiles derived from interplatform normalization of microarray data. RNA 16:16–25

Bayrak H, Sharafi P, Özketen AÇ, Kılıç M (2025) Effects of miRNAs in inborn error of metabolism and treatment strategies. Postgrad Med J 101(1199):791–803. https://doi.org/10.1093/postmj/qgae201

Beuzelin D, Kaeffer B (2018) Exosomes and miRNA-loaded biomimetic nanovehicles, a focus on their potentials preventing type-2 diabetes linked to metabolic syndrome. Front Immunol 9:2711

Birsner AE, Benny O, D'Amato RJ (2014) The corneal micropocket assay: a model of angiogenesis in the mouse eye. J Vis Exp:51375. https://doi.org/10.3791/51375

Carpentier G et al (2020) Angiogenesis analyzer for ImageJ — a comparative morphometric analysis of "endothelial tube formation assay" and "fibrin bead assay". Sci Rep 10:11568

Chakraborty C, Sharma AR, Sharma G, Sarkar BK, Lee S-S (2018) The novel strategies for next-generation cancer treatment: miRNA combined with chemotherapeutic agents for the treatment of cancer. Oncotarget 9:10164–10174

Chauhan D, Yadav AK, Bhatia D, Solanki PR (2025) Global impact of vitamin D deficiency and innovative biosensing technologies. Chem Eng J 506:159790

Connor KM et al (2009) Quantification of oxygen-induced retinopathy in the mouse: a model of vessel loss, vessel regrowth and pathological angiogenesis. Nat Protoc 4:1565–1573

Conte R et al (2020) Cationic polymer nanoparticles-mediated delivery of miR-124 impairs tumorigenicity of prostate cancer cells. Int J Mol Sci 21:869

Crawford TN, Alfaro DV, Kerrison JB, Jablon EP (2009) Diabetic retinopathy and angiogenesis. Curr Diabetes Rev 5:8–13

Delving into the Latest Updates on MRG-110 with Synapse. https://synapse-patsnap-com.libproxy1.nus.edu.sg/drug/6758519866734670a6b321a36c9b7bdd

Deskins DL et al (2013) Human mesenchymal stromal cells: identifying assays to predict potency for therapeutic selection. Stem Cells Transl Med 2:151–158

Devulapally R, Foygel K, Sekar TV, Willmann JK, Paulmurugan R (2016) Gemcitabine and antisense-microRNA co-encapsulated PLGA–PEG polymer nanoparticles for hepatocellular carcinoma therapy. ACS Appl Mater Interfaces 8:33412–33422

Ding B, Li M, Zhang J, Zhang X, Gao H, Gao J, Shen C, Zhou Y, Li F, Liu A (2025) Co-delivery of Dacarbazine and miRNA-34a combinations to synergistical. Drug Des Dev Ther 19:553–568. https://www.dovepress.com/co-delivery-of-dacarbazine-and-mirna-34a-combinations-to-synergistical-peer-reviewed-fulltext-article-DDDT

Dogheim GM, Chinnam S, Amralla MT (2025) Lipid nanoparticles as a platform for miRNA and siRNA delivery in hepatocellular carcinoma. *Curr Drug Deliv* 22:837–861

Forterre A, Komuro H, Aminova S, Harada M (2020) A comprehensive review of cancer microRNA therapeutic delivery strategies. Cancers. 12(7):1852. https://www.mdpi.com/2072-6694/12/7/1852

Fu Y, Chen J, Huang Z (2019) Recent progress in microRNA-based delivery systems for the treatment of human disease. ExRNA 1:24. https://exrna.biomedcentral.com/articles/10.1186/s41544-019-0024-y

Gale DP et al (2024) A randomized controlled clinical trial testing effects of Lademirsen on kidney function decline in adults with Alport syndrome. Clin J Am Soc Nephrol 19:995

Garcia-Guerra A et al (2025) Tissue-specific modulation of CRISPR activity by miRNA-sensing guide RNAs. Nucleic Acids Res 53:gkaf016

Gareev I et al (2023) Methods of miRNA delivery and possibilities of their application in neuro-oncology. Noncoding RNA Res 8:661–674

Garreau M, Weidner J, Hamilton R, Kolosionek E, Toki N, Stavenhagen K, Paris C, Bonetti A, Czechtizky W, Gnerlich F, Rydzik A (2024) Chemical modification patterns for microRNA therapeutic mimics: a structure-activity relationship (SAR) case-study on miR-200c. Nucleic Acids Res 52(6):2792–2807. https://academic.oup.com/nar/article/52/6/2792/7616333?login=true

Geisler A, Fechner H (2016) MicroRNA-regulated viral vectors for gene therapy. World J Exp Med 6:37–54

Geng T et al (2020) Exosome derived from coronary serum of patients with myocardial infarction promotes angiogenesis through the miRNA-143/IGF-IR pathway. Int J Nanomedicine 15:2647–2658

Gentile MT, Pastorino O, Bifulco M, Colucci-D'Amato L (2019) HUVEC tube-formation assay to evaluate the impact of natural products on angiogenesis. J Vis Exp:e58591. https://doi.org/10.3791/58591

Gilbert AT et al (2020) Safety, immunogenicity, and efficacy of intramuscular and oral delivery of ERA-G333 recombinant rabies virus vaccine to big brown bats (Eptesicus fuscus). J Wildl Dis 56:620–630

Godsey ME, Suryaprakash S, Leong KW (2013) Materials innovation for co-delivery of diverse therapeutic cargos. RSC Adv. 3(47):24794–24811. https://pmc.ncbi.nlm.nih.gov/articles/PMC4012692/

Gong Y, Generali M, Renikunta HV, Balbi C, Wang Y, Mohammed SA, Gorica E, Mongelli A, Ruschitzka F, Landmesser U, Costantino S, Paneni F, Jakob P (2024) MiRNA-519e promotes cardiomyocyte cell division accompanied by pro-angiogenic and anti-apoptotic effects. Eur Heart J. 45(Suppl_1):ehae666.3719. Oxford Academic. https://academic.oup.com/eurheartj/article/45/Supplement_1/ehae666.3719/7838365?login=true

Goodwin AM (2007) In vitro assays of angiogenesis for assessment of angiogenic and anti-angiogenic agents. Microvasc Res 74:172–183

Grogan JA, Connor AJ, Pitt-Francis JM, Maini PK, Byrne HM (2018) The importance of geometry in the corneal micropocket angiogenesis assay. PLoS Comput Biol 14:e1006049

Guyon N, Garnier D, Briand J, Nadaradjane A, Bougras-Cartron G, Raimbourg J, Campone M, Heymann D, Vallette FM, Frenel J-S, Cartron P-F (2020) Anti-PD1 therapy induces lymphocyte-derived exosomal miRNA-4315 release inhibiting Bim-mediated apoptosis of tumor cells. Cell Death Dis 11:1048. https://www.nature.com/articles/s41419-020-03224-z

Haibe Y et al (2020) Resistance mechanisms to anti-angiogenic therapies in cancer. Front Oncol 10:221

Hanyu A et al (2009) Functional in vivo optical imaging of tumor angiogenesis, growth, and metastasis prevented by administration of anti-human VEGF antibody in xenograft model of human fibrosarcoma HT1080 cells. Cancer Sci 100:2085–2092

He W, Turkeshi A, Li X, Zhang H (2020) Progress in systemic co-delivery of microRNAs and chemotherapeutics for cancer treatment by using lipid-based nanoparticles. Ther Deliv 11:591–603

He S, Song W, Cui S, Li J, Jiang Y, Chen X, Peng L (2023) Modulation of miR-146b by N6-methyladenosine modification remodels tumor-associated macrophages and enhances anti-PD-1 therapy in colorectal cancer. Cell Oncol 46:1731–1746. https://link.springer.com/article/10.1007/s13402-023-00839-0

Holjencin C, Jakymiw A (2022) MicroRNAs and their big therapeutic impacts: delivery strategies for cancer intervention. Cells 11:2332

Hong DS et al (2020) Phase 1 study of MRX34, a liposomal miR-34a mimic, in patients with advanced solid tumours. Br J Cancer 122:1630–1637

Hong J, Sim D, Lee B-H, Sarangthem V, Park R-W (2024) Multifunctional elastin-like polypeptide nanocarriers for efficient miRNA delivery in cancer therapy. J Nanobiotechnol 22:293. https://jnanobiotechnology.biomedcentral.com/articles/10.1186/s12951-024-02559-5

Stokes ME, Calvo V, Dudgeon C, Fujisawa S, Huang S, Shen L, Ballal N, McGinley J, Liu D, Mulvihill MJ, Rigby AC, Bose N, Lightcap ES, Surguladze D (2023) Abstract 4010: Inhibition of PERK by HC-5404 sensitizes clear cell renal cell carcinoma tumor models to anti-angiogenic tyrosine kinase inhibitors. Cancer Res. 83(7_Suppl):4010. American Association for Cancer Research. https://aacrjournals.org/cancerres/article/83/7_Supplement/4010/724706/Abstract-4010-Inhibition-of-PERK-by-HC-5404

Huang G et al (2021) Chitosan-derived nanoparticles impede signal transduction in T790M lung cancer therapy. Biomater Sci 9:7412–7419

Hussen BM et al (2021) MicroRNAs: important players in breast cancer angiogenesis and therapeutic targets. Front Mol Biosci 8

Ijee S et al (2024) Efficient deletion of microRNAs using CRISPR/Cas9 with dual guide RNAs. Front Mol Biosci 10:1295507

Incorvaia L et al (2020) A "lymphocyte MicroRNA signature" as predictive biomarker of immunotherapy response and plasma PD-1/PD-L1 expression levels in patients with metastatic renal cell carcinoma: pointing towards epigenetic reprogramming. Cancer 12:3396

Ishida M, Selaru FM (2012) miRNA-based therapeutic strategies. Curr Anesthesiol Rep (1, 1): 63–70. https://pmc.ncbi.nlm.nih.gov/articles/PMC3601800/

Jiang Y et al (2022) Recent advances in nanotechnology approaches for non-viral gene therapy. Biomater Sci 10:6862–6892

Juan X, Li C-X, Li Y-S, Lv J-Y, Ma Y, Shao T-T, Liang-De X, Wang Y-Y, Lei D, Zhang Y-P, Jiang W, Li C-Q, Xiao Y, Li X (2010) MiRNA–miRNA synergistic network: construction via co-regulating functional modules and disease miRNA topological features. Nucleic Acids Res 39(3):825–836. https://pmc.ncbi.nlm.nih.gov/articles/PMC3035454/

Karimi M et al (2016) pH-Sensitive stimulus-responsive nanocarriers for targeted delivery of therapeutic agents. Wiley Interdiscip Rev Nanomed Nanobiotechnol 8:696–716

Ketterl N et al (2015) A robust potency assay highlights significant donor variation of human mesenchymal stem/progenitor cell immune modulatory capacity and extended radio-resistance. Stem Cell Res Ther 6:236

Kumar V et al (2018) Pharmacokinetics and biodistribution of polymeric micelles containing miRNA and small-molecule drug in orthotopic pancreatic tumor-bearing mice. Theranostics 8:4033–4049

Kumar A, Patel DD, Banerjee J, Sharma M, Bhatia D, Banerjee S, Yadav AK (2025) Next-generation cancer theragnostic: applications of carbon quantum dots. ChemNanoMat 11: e202500061. Wiley Online Library. https://aces.onlinelibrary.wiley.com/doi/full/10.1002/cnma.202500061

Kursunluoglu G, Aydin O, Yerer MB (2025) Abstract 6587: plasma-derived exosomes in the frontier of gene therapy development. Cancer Res 85:6587

Lahooti B, Poudel S, Mikelis CM, Mattheolabakis G (2021) MiRNAs as anti-angiogenic adjuvant therapy in cancer: synopsis and potential. Front Oncol 11:705634

Landskroner-Eiger S, Moneke I, Sessa WC (2013) miRNAs as modulators of angiogenesis. Cold Spring Harb Perspect Med 3:a006643

Liu Z-L, Chen H-H, Zheng L-L, Sun L-P, Shi L (2023) Angiogenic signaling pathways and anti-angiogenic therapy for cancer. Signal Transduct Target Ther 8:198

Lu Y, Qin T, Li J, Wang L, Zhang Q, Jiang Z, Mao J (2017) MicroRNA-140-5p inhibits invasion and angiogenesis through targeting VEGF-A in breast cancer. Cancer Gene Ther 24:386–392. https://www.nature.com/articles/cgt201730

Lugano R, Ramachandran M, Dimberg A (2019) Tumor angiogenesis: causes, consequences, challenges and opportunities. Cell Mol Life Sci CMLS 77:1745–1770

Mabeta P, Hull R, Dlamini Z (2022) LncRNAs and the angiogenic switch in cancer: clinical significance and therapeutic opportunities. Genes 13:152

Mani AM et al (2024) miRNA-6236 regulation of postischemic skeletal muscle angiogenesis. J Am Heart Assoc 13:e035923

Mannarino S et al (2025) The role of miRNA expression in congenital heart disease: insights into the mechanisms and biomarker potential. Children 12:611

Miroshnichenko S, Patutina O (2019) Enhanced inhibition of tumorigenesis using combinations of miRNA-targeted therapeutics. Frontiers:10. https://www.frontiersin.org/journals/pharmacology/articles/10.3389/fphar.2019.00488/full

Moraes FC, Pichon C, Letourneur D, Chaubet F (2021) miRNA delivery by nanosystems: state of the art and perspectives. Pharmaceutics 13:1901

Nishimura M et al (2013) Therapeutic synergy between microRNA and siRNA in ovarian cancer treatment. Cancer Discov 3. https://doi.org/10.1158/2159-8290.CD-13-0159

Oesinghaus L, Castillo-Hair S, Ludwig N, Keller A, Seelig G (2024) Quantitative design of cell type-specific mRNA stability from microRNA expression data. bioRxiv [Preprint]. 2024.10.28.620728. https://doi.org/10.1101/2024.10.28.620728

Pecot CV et al (2013) Tumour angiogenesis regulation by the miR-200 family. Nat Commun 4: 2427

Peng D-W, Lan C-L, Dong L-Q, Jiang M-X, Xiao H, D'Amato RJ, Chi Z-L (2022) Anti-angiogenic properties of microRNA-29a in preclinical ocular models. Proc Natl Acad Sci U S A. 119(45): e2204795119. https://www.pnas.org/doi/10.1073/pnas.2204795119

Pezzella F (2019) Mechanisms of resistance to anti-angiogenic treatments. Cancer Drug Resist 2: 595–607

Ribatti D (2008) Judah Folkman, a pioneer in the study of angiogenesis. Angiogenesis 11:3–10

Scholz J et al (2022) An adenoviral vector as a versatile tool for delivery and expression of miRNAs. Viruses 14:1952

Seyhan AA (2024) Trials and tribulations of microRNA therapeutics. Int J Mol Sci 25:1469

Shaker SA, Alshufta SM, Gowayed MA, El-Salamouni NS, Bassam SM, Megahed MA, El-Tahan RA (2023) Propolis-loaded nanostructured lipid carriers halt breast cancer progression through miRNA-223 related pathways: an in-vitro/in-vivo experiment. Sci Rep 13(1):15752. https://www.nature.com/articles/s41598-023-42709-7

Stacker SA, Achen MG (2013) The VEGF signaling pathway in cancer: the road ahead. Chin J Cancer 32:297–302

Teo AYT et al (2021) Tiny miRNAs play a big role in the treatment of breast cancer metastasis. Cancer 13:337

Thibonnier M (2024) Criteria for developing active cellular targeting miRNA oligonucleotide therapeutics with a peptide nucleic acid backbone: combating cardiometabolic pandemics. INNOSC Theranostics Pharmacol Sci 7:3025

Thijssen VLJL, Brandwijk RJMGE, Dings RPM, Griffioen AW (2004) Angiogenesis gene expression profiling in xenograft models to study cellular interactions. Exp Cell Res 299:286–293

Tiwari A, Mukherjee B, Dixit M (2018) MicroRNA key to angiogenesis regulation: MiRNA biology and therapy. Curr Cancer Drug Targets 18:266–277

Toma I, Porfire AS, Tefas LR, Berindan-Neagoe I, Tomuță I (2022) A quality by design approach in pharmaceutical development of non-viral vectors with a focus on miRNA. Pharmaceutics 14: 1482

Ullah A et al (2025) Advancing therapeutic strategies with polymeric drug conjugates for nucleic acid delivery and treatment. Int J Nanomedicine 20:25–52

Vähätupa M, Järvinen TAH, Uusitalo-Järvinen H (2020) Exploration of oxygen-induced retinopathy model to discover new therapeutic drug targets in retinopathies. Front Pharmacol 11:873

van Beijnum JR, Giovannetti E, Poel D, Nowak-Sliwinska P, Griffioen AW (2017) miRNAs: micro-managers of anticancer combination therapies. Angiogenesis 20:269–285

Varley AJ, Desaulniers J-P (2021) Chemical strategies for strand selection in short-interfering RNAs. RSC Adv 11:2415–2426

Vyas A et al (2025) Advancing the frontier of artificial intelligence on emerging technologies to redefine cancer diagnosis and care. Comput Biol Med 191:110178

Walsh DA, Pearson CI (2001) Angiogenesis in the pathogenesis of inflammatory joint and lung diseases. Arthritis Res 3:147–153

Wang X et al (2019) MiRNA delivery system based on stimuli-responsive gold nanoparticle aggregates for multimodal tumor therapy. ACS Appl Bio Mater 2:2833–2839

Wang Y, Xie Y, Kilchrist KV, Li J, Duvall CL, Oupický D (2020) Endosomolytic and tumor-penetrating mesoporous silica nanoparticles for siRNA/miRNA combination cancer therapy. ACS Appl Mater Interfaces 12(4):4308–4322. https://pubs.acs.org/doi/10.1021/acsami.9b21214

Wu F, Yang Z, Li G (2009) Role of specific microRNAs for endothelial function and angiogenesis. Biochem Biophys Res Commun 386:549

Xing Y et al (2021) MicroRNAs: immune modulators in cancer immunotherapy. Immunother Adv 1:ltab006

Xu Y, Liu Z, Lv L, Li P, Xiu B, Qian W, Liang A (2020) MiRNA-340-5p mediates the functional and infiltrative promotion of tumor-infiltrating CD8+ T lymphocytes in human diffuse large B cell lymphoma. J Exp Clin Cancer Res 39:238. https://jeccr.biomedcentral.com/articles/10.1186/s13046-020-01752-2

Yadav AK, Verma D, Kumar A, Bhatt AN, Solanki PR (2023) Biocompatible epoxysilane substituted polymer-based nano biosensing platform for label-free detection of cancer biomarker SP17 in patient serum samples. Int J Biol Macromol 239:124325

YadavDamini AK, Solanki VPR (2013) Enhanced electrochemical biosensing of the Sp17 cancer biomarker in serum samples via engineered two-dimensional MoS2 nanosheets on the reduced graphene oxide interface. ACS Appl Bio Mater 7(10) https://pubs.acs.org/doi/10.1021/acsabm.3c00464

Yang N (2015) An overview of viral and nonviral delivery systems for microRNA. Int J Pharm Investig 5:179–181

Yang Y, Liu Y, Li Y, Chen Z, Xiong Y, Zhou T, Tao W, Fan X, Yang H, Ylä-Herttuala S, Chaurasia SS, Adam W-C, Yang K (2020) MicroRNA-15b targets VEGF and inhibits angiogenesis in proliferative diabetic retinopathy. J Clin Endocrinol Metab. 105(11):3404–3415. https://pmc.ncbi.nlm.nih.gov/articles/PMC7947967/

Ye S, Ma L, Zhang J, Li R, Cheng H (2022) Amplified drug delivery system with a pair of master keys triggering precise drug release for chemo-photothermal therapy. Anal Chem 94:11538–11548

Yue R, Chen M, Ma N (2020) Dual microRNA-triggered drug release system for combined chemotherapy and gene therapy with logic operation. ACS Appl Mater Interfaces 12:32493–32502

Zhang R, Yao Y, Gao H, Hu X (2024) Mechanisms of angiogenesis in tumour. Front Oncol 14

Zhao H, Li T, Yao C, Zi G, Liu C, Li J, Yang D (2021) Dual roles of metal–organic frameworks as nanocarriers for miRNA delivery and adjuvants for chemodynamic therapy. ACS Appl Mater Interfaces 13(5):6034–6042. https://pubs.acs.org/doi/10.1021/acsami.0c21006

Zhong Z et al (2025) Phytochemicals from Salvia substolonifera with anti-angiogenic properties and substolide H decreased oxygen-induced retinal neovascularization. Chem Biodivers 22:e202401427

Zhou H, Chen DS, Hu CJ, Hong X, Shi J, Xiao Y (2013) Stimuli-responsive nanotechnology for RNA delivery. Adv Sci 10:2303597. Wiley Online Library. https://advanced.onlinelibrary.wiley.com/doi/full/10.1002/advs.202303597

Zimna A, Kurpisz M (2015) Hypoxia-inducible factor-1 in physiological and pathophysiological angiogenesis: applications and therapies. BioMed Res Int 2015:549412

# Combination of miRNA and Photodynamic and Photothermal Anti-cancer Therapy

# 14

Payal Sharma and Shikha Tiwari

**Abstract**

The integration of microRNA (miRNA) based strategies with photodynamic therapy (PDT) and photothermal therapy (PTT) offers a novel combinational approach to cancer treatment. miRNA regulates multiple oncogenic pathways, including cell proliferation, apoptosis, angiogenesis, and therapy resistance, while PDT and PTT provide targeted, minimally invasive methods to kill tumor cells via reactive oxygen species or localized hyperthermia. Combining these modalities enhances the therapeutic efficacy by sensitizing cancer cells, overcoming resistance, and reducing recurrence. Nanoparticle-based delivery systems further facilitate co-administration of miRNA with PDT or PTT, improving the tumor-specific uptake and minimizing off-target effects. Despite the challenges related to delivery efficiency, tissue penetration, and safety, chapter also discusses the molecular basis, experimental advances, and therapeutic opportunities integrating miRNAs with light-based therapies, emphasizing their promise for precision oncology and future clinical applications.

**Keywords**

miRNA therapy · Photodynamic · Photothermal · Nanoparticle · Drug delivery

**Learning Objectives**

1. The principles of photodynamic, and photothermal therapy in cancer treatment.
2. miRNA impact on the sensitivity and efficacy of PDT and PTT.

P. Sharma
Centre of Biotechnology, Nana JI Deshmukh Veterinary Science University, Jabalpur, India

S. Tiwari (✉)
School of Sciences, Sanjeev Agrawal Global Educational University, Bhopal, India
e-mail: shikha.t@sageuniversity.edu.in


S. Tiwari et al. (eds.), *microRNA Modulation: A Promising Combination Therapy for Cancer*, https://doi.org/10.1007/978-981-95-5989-3_14

3. Nanotechnlogy based delivery systems enabling miRNA, and light-thermal therapy combinations.
4. The therapeutic potential, and clinical challenges of integrating miRNA with the PDT/PTT approaches.

## 14.1 Introduction

In every nation on earth, cancer is a major cause of death and a significant obstacle to raising life expectancy. In 2019, the World Health Organization (WHO) estimated that cancer ranks first or second in 112 out of 183 nations for causes of mortality before the age of 70 and third or fourth in another 23 (Bray et al. 2021). A major challenge to modern medicine is cancer's complicated biology, heterogeneity, and resistance to traditional therapy, which make it one of the world's top causes of illness and mortality (Sung et al. 2021). Depending on the type and severity of the cancer, there are several therapy approaches available. For example, surgery aids in the removal of cancerous bulk or tumors. Drugs are used in chemotherapy to destroy certain cancer cells. Other cancer treatment possibilities include immunotherapy, hormone therapy, targeted medication therapy, radiation therapy, bone marrow transplantation, and cryoablation. These treatments cause serious adverse effects even though they have been shown to be beneficial in a number of cases (Bhole 2021).

In recent years, the landscape of cancer treatment has rapidly evolved, with a focus on precision medicine and combinatorial therapeutic approaches (Tsimberidou et al. 2020). Among these, microRNAs (miRNAs), which are tiny, noncoding RNAs that control gene expression after transcription, have become important regulators of oncogenesis, tumor growth, and resistance to treatment. Their potential as indicators and therapeutic targets is highlighted by their abnormal expression profiles in a variety of malignancies (El-Daly et al. 2017).

Simultaneously, photothermal therapy (PTT) and photodynamic therapy (PDT) have become more popular as focused, noninvasive anti-cancer treatments. PDT uses photosensitizers that are triggered by particular light wavelengths to create reactive oxygen species (ROS), which kill tumor cells locally. In contrast, PTT uses photothermal agents to convert near-infrared light into heat, which causes cancer cells to thermally ablate. By providing temporal and geographical control, both modalities improve treatment precision and reduce off-target effects (Duan et al. 2023).

Combining miRNA therapy with photodynamic and photothermal techniques offers a new, complementary cancer treatment platform. By tackling tumor heterogeneity and resistance mechanisms, this combination not only improves therapeutic efficacy but also makes it possible to fine-tune molecular pathways involved in immune evasion and cancer cell survival. Additionally, co-delivery of miRNAs and photosensitizers/photothermal agents has been made possible by sophisticated nanotechnology-based delivery systems, guaranteeing tumor-specific accumulation,

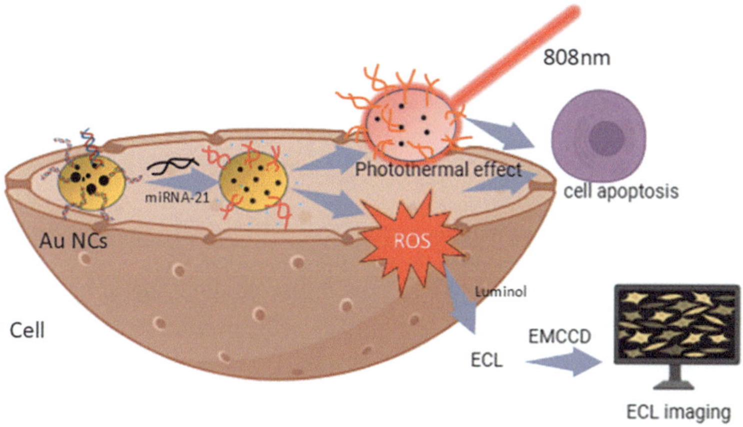


**Fig. 14.1** ECL imaging and combined therapy

enhanced cellular uptake, and less systemic toxicity (El-Daly et al. 2017). This chapter explores the rationale, mechanisms, and therapeutic potential of combining miRNA modulation with PDT and PTT. It highlights recent advances in nanocarrier design, preclinical outcomes, and translational challenges, offering a comprehensive overview of this promising multimodal strategy in oncological precision medicine (Fig. 14.1).

## 14.2 Photothermal Therapy (PTT)

Thermal therapies for cancer cells have existed since the eighteenth century. When the temperature rises over the physiological range, usually to 40–45 °C, it is referred to as hyperthermia. Creating an environment that aids in the removal of tumors while protecting the healthy tissues used in cancer treatment is the primary objective of hyperthermia. This is accomplished via hyperthermia through the induction of direct cytotoxic and physiological consequences. In clinical settings, heat can work in concert with chemotherapy and radiation therapy. As a result of protein denaturation and cell membrane deterioration, cancer cells are permanently harmed by hyperthermia. Nevertheless, the normal tissues are frequently impacted by this treatment. Applying photothermal therapy for more targeted cancer treatment can be made easier by incorporating laser radiation treatment into cancer therapy.

Therefore, laser-induced hyperthermia appears to be beneficial for treating choroidal or retinal tumors. The requirement for a high-power laser to kill the tumor cells is a major drawback of this treatment. A PTT with a photothermal agent has been suggested as a selective heating method in the meantime. The main prerequisites for PTT are an NIR light source, an NIR area, and a biocompatible photothermal agent with a high absorption coefficient. The absorption of the NIR wavelength and the light-excitation coefficient determine how much the PTT's temperature rises. Cancer cells in the main tumor or in the tumor at the early metastatic stage may be killed by PTT alone or in combination. However, with the following methods, PTT offers an essential advantage in reducing metastasis in several cancer types:

### 14.2.1 Direct Removal of Cancer Cells with PTT

An NIR laser can penetrate soft tissues up to 2 cm. PTT on NIR-laser irradiation may harm or kill primary tumor cancer cells in addition to causing lymph node metastases or ablation of primary tumors (Okuno et al. 2013). Additionally, it eliminates the cancer-causing stem cells and tumor-triggering cells, preventing the spread of the disease to other organs (Liang et al. 2015).

### 14.2.2 Imaging Guide

Details for a better PTT treatment plan are provided by the picture guidance, ensuring the safety and effectiveness of photothermal ablation. Computed tomography [CT], magnetic resonance imaging, photoacoustic imaging, and X-rays are examples of multimodal imaging that can be used to determine the location of lymph node metastases and to improve treatment plans during PTT. The current method is restricted to mapping metastases from lymph nodes, which is uncommon in deep tissue metastatic imaging of the brain, liver, and lung. Photoacoustic imaging and X-ray CT incorporate image-guided PTT cancer metastases (Bhole 2021). The use of PTT in radiotherapy prevents tumor development and increases the survival rate of the treated patients.

### 14.2.3 Combination of PTT in Chemotherapy

Chemotherapy is frequently used to treat metastatic cancer. Scientists have recently noticed that using PTT and chemotherapy together has synergistic effects on cancer spread. Doxorubicin (DOX) and gold nanostructures have been utilized in chemotherapy, mostly to treat cancer. A synergistic combination therapy is offered by DOX, an anti-cancer medication used in chemotherapy, and gold nanoclusters or nanorods, a photothermal agent used in PTT (Wang et al. 2014). A synergistic combination therapy is offered by DOX, an anti-cancer medication used in chemotherapy, and gold nanoclusters or nanorods, a photothermal agent used in PTT (Wang et al. 2014). Following NIR radiation, DOX-loaded DNA was wrapped around gold nanorods primed for combination therapy in metastatic breast cancer. This combination therapy with gold nanoparticles and doxorubicin nanoparticles markedly reduced the growth of breast tumors and lung metastases (Peng et al. 2014). The use of combination therapy effectively prevents pulmonary and mediastinal metastases, which contributes to animal survival.

### 14.2.4 Gold Nanostructures as Photothermal Agents in Cancer Treatment

Photothermal and ultrasensitive detection benefit from the unique optical properties of gold nanoparticles. The conduction band on the surface of gold nanoparticles oscillates with one another when light of a specific wavelength strikes them, creating a phenomenon known as surface plasmon resonance. The light emitted by the gold nanoparticles is heated by this event. The size, shape, and composition of the nanoparticles determine the wavelength of light at which they scatter and absorb light energy. The peak, adjustable in the NIR region, penetrates tissues more efficiently than other light regions and can also be altered by varying the size and structure of gold nanoparticles (Bhole 2021).

## 14.3 Photodynamic Therapy (PDT)

Photodynamic therapy (PDT) has gained popularity as a tumor treatment method since the late 1970s. Many nations, including the US, UK, France, Germany, and Japan, have approved it for cancer treatment. To differentiate photooxidation from the sensitization process in photography, Hermann von Tappeiner first used the word "photodynamic" in 1904. The destruction of cancer cells by light irradiation with eosin-coated tumors was first documented in 1903. After that, a variety of hematoporphyrin derivatives were used in photodynamic therapy (Cengel et al. 2016). For the diagnosis of cancer, PDT consists of two parts: a PS and a light source (often in the red spectral zone because red light penetrates deeper into the tissue). PDT has the advantage of being able to be repeated numerous times without causing immunosuppressive or myelosuppressive side effects, and it can be given even following chemotherapy, radiation therapy, or surgery (Bhole 2021).

### 14.3.1 Photosensitizer (PS)

A single pure chemical that permits quality assurance research with affordable production costs and respectable stability should be the ideal PS agent. Since photon absorption with a wavelength >800 nm lacks the energy to excite oxygen to its singlet state and generate a significant yield, the optimal PS agent should have a high absorption peak in the 600–800 nm (red to dark red) range. Phthalocyanine, bacteriochlorins, and chlorines, for instance, can enhance tumor regulation. Furthermore, allowing dark-red light to enter tissues at the right wavelength aids in lowering toxicity and facilitating quick clearance from healthy tissues, which lessens the negative consequences of phototoxicity (Allison and Sibata 2010). The first PS used for cancer therapy was a water-soluble porphyrin mixture called the hematoporphyrin derivative (HPD), a refined porfimer sodium form, which later came to be known as photofrin (Clichici and Filip 2011). Some PSs have been used to treat cancer (Table 14.1) (Bhole 2021).

**Table 14.1** Diverse photosensitizers used in cancer treatment strategies

| S. No. | Sensitizer | Trade name | Cancer type | Wavelength |
|---|---|---|---|---|
| 1. | Hematoporphyrin derivatives (HqD) | Photofrin | Endobronchial, esophageal, bladder, and gastric cancers and cervical and brain tumors. | 630 nm |
| 2. | Benzoporphyrin derivative- monoacid ringA (BPD-MA) | Verteporfin | Basal-cell carcinoma | 689 nm |
| 3. | Meta-tetrahydroxyphenylcholin (m-THPC) | Foscan | Neck and head tumors, prostate and pancreatic tumors | 654 nm |
| 4. | 2-(1-hexyloxyethyl)-2-devinyl pyropheophorbide-alpha (HPPH) | Photochlor | Basal-cell carcinoma | 665 nm |
| 5. | Taporfin sodium | Talaporfin | Solid tumor from diverse origins | 664 nm |

#### 14.3.1.1 Mechanism of Action

Photosensitizers (PS) used in photodynamic therapy (PDT) can be delivered topically or through intravenous injection, but their distribution in the body changes over time, making the timing of light exposure critical. When activated by light, the PS transitions from its ground state to a short-lived singlet state and then to a longer-lasting triplet state. In this state, it can cause damage through two main mechanisms: *Type I*, where it generates free radicals that react with oxygen to form toxic products, and *Type II*, where it directly transfers energy to oxygen, forming reactive singlet oxygen species (ROS). Both reactions depend on oxygen availability, the type of PS, substrate, and its binding affinity. Since ROS are highly reactive and short-lived (e.g., singlet oxygen has a half-life of less than 0.04 μs and acts within 0.02 μm), PDT mainly affects cells very close to the PS location. The overall cytotoxic effect depends on multiple factors such as PS type, location, dosage, light dose, oxygen levels, and timing between drug and light exposure (Dougherty et al. 1998; Jiang et al. 2023). The PDT mechanism is summarized in Figs. 14.2 and 14.3.

### 14.3.2 Effect of PDT on Tumor

There are three primary ways that PDT mediates tumor elimination. Initially, tumor infarction results from PDT-generated ROS's direct destruction of tumor cells and harm to the tumor-associated vasculature. The immune system's reaction to tumor cells is eventually triggered by PDT. Additionally, the three forms may influence one another. All of these elements work together to manage tumors over the long run. In 1913, Friedrich Meyer-Betz, a German scientist, administered porphyrins to humans for the first time. He applied 200 mg of hematoporphyrin to his skin to check for any

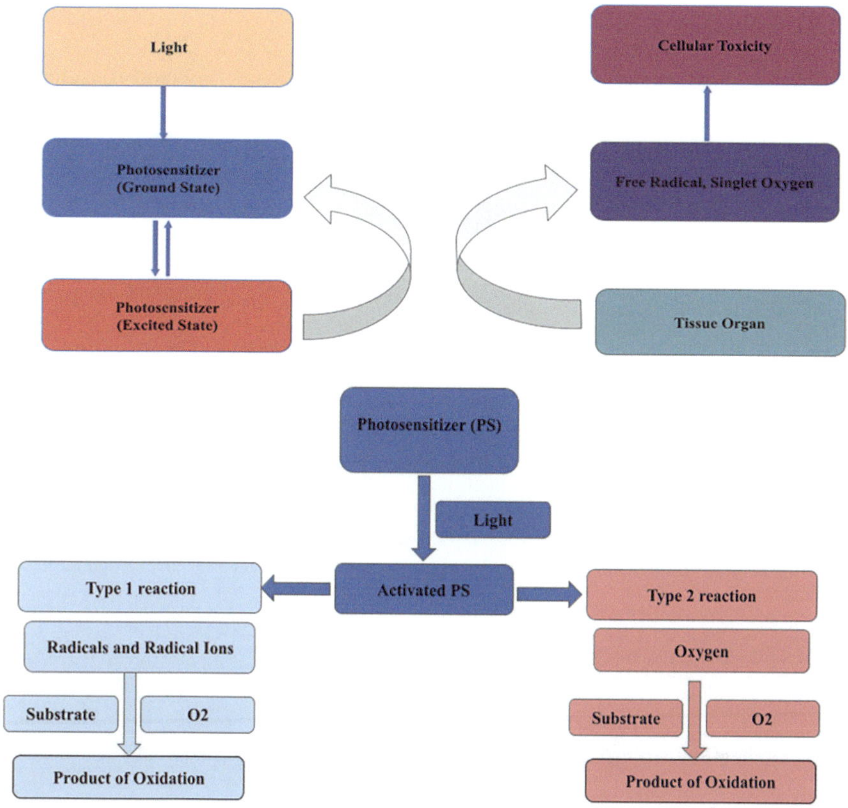


**Fig. 14.2** Mechanism of PDT with type I and type II reactions

side effects. Pain and edema were noted in the skin regions that were exposed to light. Thomas Dougherty and associates reported in 1975 that mice's breast tumors were completely destroyed when the derivative hematoporphyrin and red light were administered. J. F. Kelly and associates discovered that hematoporphyrin light activation may completely eradicate bladder cancer in rats that same year (Bhole 2021). Table 14.2 (Bhole 2021; Al-Alousi 2016).

## 14.4 Anti-cancer Therapy

### 14.4.1 Combination of miRNA with Photodynamic Therapy

Most studies exploring the role of miRNAs (small molecules that regulate gene activity) in photodynamic therapy (PDT) focus on how PDT affects miRNA levels in cancer cells. One of the earliest studies was by Kushibiki and his team, who looked at six miRNAs related to cell death in HeLa cervical cancer cells after PDT

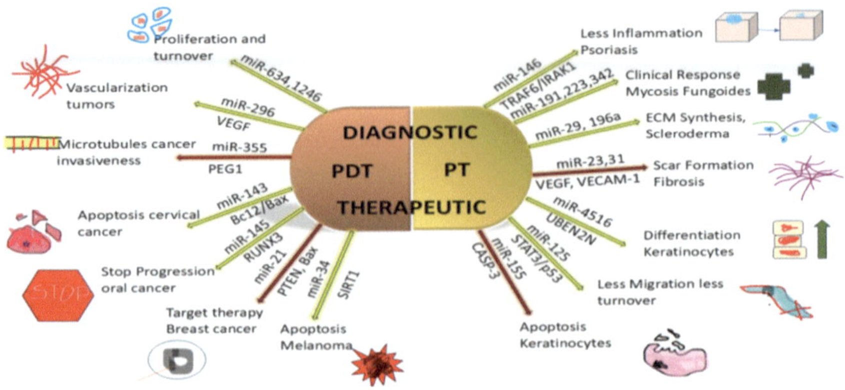


**Fig. 14.3** Photodynamic therapy and phototherapy affect the expression of miRNA in various cancers. Red arrows indicate downregulated levels of miRNA, while yellow arrows indicate upregulated levels of miRNA in the bloodstream or locally. miRNAs are used as a diagnostic tool in the upper portion of the figure, while when down- or upregulated, they offer therapeutic targets or have restorative effects in the lower portion. In one section of the image, miRNAs are used as a diagnostic tool; in the other, when down- or upregulated, miRNAs offer therapeutic targets or have restorative effects

**Table 14.2** Key difference between the PDT and PTT

| S. No. | PDT | PTT |
|---|---|---|
| 1. | It show low penetration into biological tissue | It can deeply penetrate into biological tissue |
| 2. | PDT is useful for lining organs or problems on or just under the skin | It can be used to treat different types of cancer |
| 3. | The drugs used in PDT leave people very sensitive to light for some time | No sensitivity problem |
| 4. | It is denaturation of proteins and disruption of cell membranes | Its destruction of localized organelles |
| 5. | Photosensitizers use as photodynamic agents, such as porfimer sodium and Taporfin sodium Boronated protoporphyrin | Photothermal agents such as gold nanoshells, nanorods, nanocages, and nanospheres |

treatment. They used a photosensitizer called talaporfin sodium and exposed the cells to light. After 1 h, they found that two miRNAs—miR-210 and miR-296—had increased significantly. This increase was linked to hypoxia (low oxygen levels), which happens because PDT quickly uses up oxygen in the tumor. Hypoxia activates a protein called HIF-1α, which helps cells survive in low-oxygen conditions. Interestingly, HIF-1α boosts miR-210, and in return, miR-210 helps keep HIF-1α stable—creating a positive feedback loop. Similarly, miR-296 levels went up due to the action of VEGF, a molecule that helps form new blood vessels and also increases during hypoxia. This suggests that PDT may unintentionally activate miRNAs that help cancer cells survive. Although the study proposed that blocking miR-210 or miR-296 might make PDT more effective, they did not test this idea in the lab.

In a study by Lee and colleagues, researchers looked at how photodynamic therapy (PDT) using protoporphyrin IX (PpIX) affects microRNA (miRNA) levels in liver cancer cells. They were especially interested in whether miRNAs could be responsible for reducing the levels of HIF-1α, a protein that helps cancer cells survive under low-oxygen (hypoxic) conditions. Since HIF-1α is known to be active after PDT, they focused on 10 specific miRNAs predicted to target it. Their experiments showed that miR-199a-5p increased significantly after PpIX treatment, especially in SK-Hep1 liver cancer cells, and this increase was also seen in two other liver cancer cell lines (SNU878 and SNU449)(Borgia et al, 2021). Higher levels of miR-199a-5p reduced HIF-1α levels and also decreased E2F3, a protein involved in cell growth. Additionally, PpIX lowered the levels of several other proteins linked to cancer spread, including ZEB1/2, SNAIL, SLUG, and TWIST (which are involved in EMT, the process that helps cancer cells become more aggressive). As a result, the treated cancer cells became more sensitive to chemotherapy drugs like cisplatin and doxorubicin. Overall, the study suggests that miR-199a-5p plays an important role in weakening cancer cell defenses and could be key in making PDT and chemotherapy more effective in liver cancer (El-Daly et al. 2017). Additionally, microRNAs have been employed to improve PDT's specificity. Recently demonstrated this strategy by choosing to use the nearly universal overexpression of miR-21 in tumor tissues as a "tool" to improve PDT target specificity. As a result, effectiveness. The authors connected a PS, chlorin e6 (Ce6), to a sequence-specific peptide nucleic acid (PNA) that is complementary to miR-21. The drug delivery vehicle (Dex-RGON), which also served as a fluorescence quencher, carried the Ce6-PNA21 complex to the cancer cells. The PS (Ce6) was released as a result of the Ce6-PNA complex's hybridization with the miR-21 miRNA after it had entered the cancer cells. Light irradiation is then used to activate the PS. The unique overexpression of miR-21 in cancer cells was successfully used in this study to improve PDT's specificity and cause a selective phototoxic effect on cancer cells while preserving healthy cells.

### 14.4.2 Combination of miRNA with Photothermal Therapy

miRNA has been utilized in several studies as a biomarker for tumor diagnosis in conjunction with medications, which may prove to be a very effective way to treat cancer. Owing to these drawbacks, our team created a novel miRNA-activated chemotherapy-PPT system that activated the combination therapy and effectively killed cancer cells by using in situ miRNA-21 rather than an external light source (Zhang et al. 2019). Even SNP associated with cancer can also provide valuable insight in combination therapy (Tiwari et al. 2023; Tiwari et al. 2024).

## 14.5 Future Direction and Conclusion

As a promising and minimally intrusive method of treating cancer, phototherapy—which includes both photodynamic therapy (PDT) and photothermal therapy (PTT)—continues to gain popularity since it provides localized targeting with few adverse effects. In contrast to PDT, which uses light-activated photosensitizers to produce reactive oxygen species (ROS), PTT uses photothermal agents, like gold nanoparticles, to produce heat and kill cancer cells only. PTT is becoming more and more popular despite PDT's significant promise because of its ability to target precisely, support imaging-guided therapy, and work well in conjunction with chemotherapy. As a result of these developments, microRNAs (miRNAs) have become important modulators of pathways linked to cancer and are becoming more well acknowledged for their ability to affect the results of light-based chemotherapy.

Current research indicates that miRNA expression is routinely changed after therapy, despite our limited understanding of the precise molecular pathways behind PDT and PTT. However, finding a miRNA signature that is universally sensitive has been challenging due to variations in photosensitizers, light conditions, and cell models. However, miRNAs have demonstrated promise as therapeutic agents and prognostic biomarkers, with the ability to improve tumor targeting and apoptosis. It has been suggested that miRNA mimics and inhibitors can predict efficacy and enhance the therapeutic response, particularly in chronic illnesses like skin problems.

All things considered, future studies should focus on expanding the study of important miRNAs in light-based therapies and standardizing experimental conditions. In addition to advancing our knowledge of the healing pathways, investigating miRNA-targeted tactics in carefully planned clinical studies may make PDT and PTT more accurate, effective, and individualized cancer treatment techniques (Fig. 14.3).

## References

Al-Alousi MAH (2016) Synthesis and characterization of the golden nanoparticles and their medical applications by using (LALP) technique. [cited 2025 Jun 12]; Available from: http://rgdoi.net/10.13140/RG.2.2.25730.89282

Allison RR, Sibata CH (2010) Oncologic photodynamic therapy photosensitizers: a clinical review. Photodiagn Photodyn Ther 7(2):61–75

Bhole R (2021) A comprehensive review on photodynamic therapy (PDT) and photothermal therapy (PTT) for cancer treatment. Turk J Oncol [Internet]. [cited 2025 Jun 10]; Available from: http://onkder.org/pdf/pdf_TOD_1195.pdf

Borgia F, Custurone P, Peterle L, Pioggia G, Guarneri F, Gangemi S (2021) Involvement of microRNAs as a response to phototherapy and photodynamic therapy: a literature review. Antioxidants 10(8):1310

Bray F, Laversanne M, Weiderpass E, Soerjomataram I (2021) The ever-increasing importance of cancer as a leading cause of premature death worldwide. Cancer 127(16):3029–3030

Cengel KA, Simone CB, Glatstein E (2016) PDT: what's past is prologue. Cancer Res 76(9): 2497–2499

Clichici S, Filip GA. 2011 Photodynamic therapy in skin cancer. IntechOpen

Dougherty TJ, Gomer CJ, Henderson BW, Jori G, Kessel D, Korbelik M et al (1998) Photodynamic therapy. JNCI J Natl Cancer Inst 90(12):889–905

Duan S, Hu Y, Zhao Y, Tang K, Zhang Z, Liu Z et al (2023) Nanomaterials for photothermal cancer therapy. RSC Adv 13(21):14443–14460

El-Daly SM, Abba ML, Gamal-Eldeen AM (2017) The role of microRNAs in photodynamic therapy of cancer. Eur J Med Chem 142:550–555

Jiang W, Liang M, Lei Q, Li G, Wu S (2023) The current status of photodynamic therapy in cancer treatment. Cancer 15(3):585

Liang C, Song X, Chen Q, Liu T, Song G, Peng R et al (2015) Magnetic field-enhanced photothermal ablation of tumor sentinel lymph nodes to inhibit cancer metastasis. Small 11(37):4856–4863

Okuno T, Kato S, Hatakeyama Y, Okajima J, Maruyama S, Sakamoto M et al (2013) Photothermal therapy of tumors in lymph nodes using gold nanorods and near-infrared laser light. J Control Release 172(3):879–884

Peng J, Qi T, Liao J, Chu B, Yang Q, Qu Y et al (2014) Mesoporous magnetic gold "nanoclusters" as theranostic carrier for chemo-photothermal co-therapy of breast cancer. Theranostics 4(7): 678–692

Sung H, Ferlay J, Siegel RL, Laversanne M, Soerjomataram I, Jemal A et al (2021) Global cancer statistics 2020: GLOBOCAN estimates of incidence and mortality worldwide for 36 cancers in 185 countries. CA Cancer J Clin 71(3):209–249

Tiwari S, Pandey R, Kumar V et al (2023) Association of single nucleotide polymorphism miRNA-146a (rs2910164) with increased predisposition to oral squamous cell carcinoma in Central India population. Cancer Biomark 38(2):203–214. https://doi.org/10.3233/CBM-230064

Tiwari S, Pandey R, Kumar V, Das S, Gupta V, Nema R, Kumar A (2024) miRNA genetic variations associated with the predisposition of oral squamous cell carcinoma in central Indian population. Noncoding RNA Res 4:1333–1341. https://doi.org/10.1016/j.ncrna.2024.07.002

Tsimberidou AM, Fountzilas E, Nikanjam M, Kurzrock R (2020) Review of precision cancer medicine: evolution of the treatment paradigm. Cancer Treat Rev 86:102019

Wang D, Xu Z, Yu H, Chen X, Feng B, Cui Z et al (2014) Treatment of metastatic breast cancer by combination of chemotherapy and photothermal ablation using doxorubicin-loaded DNA-wrapped gold nanorods. Biomaterials 35(29):8374–8384

Zhang H, Gao W, Liu Y, Sun Y, Jiang Y, Zhang S (2019) Electrochemiluminescence microscopy for microRNA imaging in single cancer combined with chemotherapy-photothermal therapy. Anal Chem 91(19):12581–12586

# Exosomal miRNAs and Their Therapeutic Implications in Cancer 15

Garima Rathee, Sahil Kohli, and Antonio Puertas-Segura

**Abstract**

Exosomes, nanoscale extracellular vesicles released by several cell types, have become essential facilitators of intercellular communication in both physiological and pathological contexts. MicroRNAs (miRNAs) are prominent components of their varied payload, significantly affecting gene expression and cellular processes. In cancer, exosomal miRNAs serve dual functions as either oncogenic or tumor-suppressive factors, influencing cancer cell proliferation, invasion, metastasis, and therapeutic resistance. This chapter comprehensively examines the biogenesis and molecular processes that support exosomal miRNA-mediated signaling pathways. The potential of exosomal miRNAs as biomarkers for early cancer diagnosis and prognosis is examined, alongside the formulation of treatment techniques that utilize their regulatory roles. The chapter finishes by addressing the obstacles and future prospects of using exosomal miRNAs in cancer therapy, emphasizing their scalability, delivery mechanisms, and translational promise in precision medicine.

**Keywords**

Exosomes · miRNAs · Cancer therapy · Oncogenic miRNAs · Tumor-suppressive miRNAs · Tumor microenvironment

G. Rathee (✉) · A. Puertas-Segura
Department of Chemical Engineering, Universitat Politècnica de Catalunya, Barcelona, Spain

S. Kohli
Department of Chemistry, School of Engineering and Technology, Manav Rachna International Institute of Research and Studies, Faridabad, Haryana, India



S. Tiwari et al. (eds.), *microRNA Modulation: A Promising Combination Therapy for Cancer*, https://doi.org/10.1007/978-981-95-5989-3_15

**Learning Objectives**

- Explain the biogenesis and role of exosomes as carriers of miRNAs in intercellular communication within the tumor microenvironment.
- Analyze how exosomal miRNAs contribute to cancer progression, metastasis, immune evasion, and therapy resistance.
- Evaluate the potential of exosomal miRNAs as non-invasive diagnostic and prognostic biomarkers through liquid biopsy approaches.
- Assess therapeutic strategies that exploit or target exosomal miRNAs for cancer treatment, along with associated challenges in delivery, specificity, and safety.

## 15.1 Introduction

Cellular communication is essential in tumor development, allowing cancer cells to synchronize survival, proliferation, and metastasis. Tumor cells engage with one another and the adjacent stromal cells to alter the tumor microenvironment, promoting immune evasion, invasion, and therapeutic resistance (Hanahan and Weinberg 2011; Brücher and Jamall 2014). These connections transcend local tissue, as tumors initiate contact with other organs to facilitate the formation of metastatic habitats. Numerous signaling systems participate in this intricate interplay, with exosomes, small extracellular vesicles (EVs) derived from endosomes, emerging as pivotal mediators of intercellular communication in cancer (Zhang et al. 2007).

MicroRNAs (miRNAs), a category of bioactive molecules found in exosomes, have garnered considerable interest for their capacity to influence gene expression posttranscriptionally. miRNAs are brief, noncoding RNAs that modulate essential cellular processes, including proliferation, differentiation, and death. In cancer, their expression is often dysregulated, resulting in the inhibition of tumor-suppressor pathways or the activation of oncogenic signaling networks (Zhang et al. 2007; Tickner et al. 2014; Valadi et al. 2007; Bartel 2004). Recent findings demonstrate that exosomal miRNAs (exo-miRNAs) have a role in several facets of cancer development, encompassing angiogenesis, immunological regulation, and resistance to treatment. In addition to their function in tumor biology, exosomal-miRNAs exhibit significant potential for therapeutic applications (Thind and Wilson 2016; Ingenito et al. 2019). Their durability in bodily fluids and distinct expression patterns across many cancer types provide them with appealing indicators for noninvasive cancer detection and prognosis. The ability to modify exosomal-miRNAs composition using synthetic exosomes has created new opportunities for targeted cancer therapy, either by administering tumor-suppressive miRNAs or inhibiting oncogenic miRNAs. Nonetheless, obstacles such as efficient miRNA loading, targeted distribution, and immune clearance persist as significant barriers to translating these results into viable therapeutic therapies (Zhang et al. 2007; Taylor and Gercel-Taylor 2008; Kowal et al. 2016; Minciacchi et al. 2015).

This chapter examines the therapeutic significance of exo-miRNAs in cancer, outlining their functions in tumor growth, their potential as diagnostic and prognostic biomarkers, and new developments in miRNA-based treatment approaches.

Researchers seek to comprehend and utilize exosomal-miRNAs to create novel medicines that augment or improve cancer therapies, facilitating more targeted and efficacious interventions.

## 15.2 miRNAs and Exosomes

### 15.2.1 Extracellular miRNAs

The human genome harbors a complexity far beyond the traditional understanding of genetic information. Beyond protein-coding regions lies a vast landscape of non-coding DNA, home to molecular regulators that challenge our most fundamental concepts of genetic expression. Among these, long noncoding RNAs (lncRNAs) and miRNAs have emerged as critical players in cellular communication, operating through mechanisms both intricate and profound.

MiRNAs represent a revolutionary class of genetic regulators, typically spanning just 20–25 nucleotides in length yet capable of influencing approximately 50% of protein-coding genes. These diminutive molecules function as precise genetic editors, predominantly binding to the 3′ untranslated regions of messenger RNAs. Their impact is dramatic and decisive: they can either suppress protein translation or trigger complete messenger RNA degradation, effectively rewriting cellular instructions with remarkable specificity (Buechner et al. 2011; Li et al. 2013; Chan et al. 2005).

In cancer biology, miRNAs reveal themselves as molecular switches with extraordinary power. Some function as oncogenic agents, oncomiRs that actively promote tumor progression. Molecules like miR-221 and miR-21 act as cellular saboteurs, suppressing apoptosis and driving tumor cell proliferation. Conversely, other miRNAs, such as miR-34a and let-7 family members, assume the role of cellular guardians, promoting programmed cell death and inhibiting tumor growth (Buechner et al. 2011; Li et al. 2013; Chan et al. 2005). The clinical significance of miRNAs extends far beyond simple gene regulation, particularly in the context of treatment resistance. In breast cancer stem cells, miR-24 remarkably enhances cisplatin resistance by targeting pro-apoptotic proteins (Roscigno et al. 2017). Glioblastoma provides an even more complex narrative: miR-221/222 simultaneously facilitates cancer cell migration by reducing tyrosine phosphatase expression and increasing resistance to temozolomide through sophisticated manipulation of DNA repair mechanisms (Quintavalle et al. 2012).

Perhaps most revolutionary is the discovery of circulating miRNAs, molecular messengers that defy traditional understanding of genetic material. Unlike fragile synthetic RNA molecules, these miRNAs exist with extraordinary stability in various body fluids, including serum, plasma, saliva, urine, and cerebrospinal fluid (Mause and Weber 2010; Chim et al. 2008; Arroyo et al. 2011; Noferesti et al. 2015; Chen et al. 2008). Their survival mechanism has intrigued researchers, revealing a sophisticated protective strategy ensuring continued functionality. The protective packaging of these extracellular miRNAs is a marvel of biological engineering. They travel within complex carriers, including membranous vesicles like exosomes, protein complexes such as Argonaute proteins, and lipid carriers like high-density lipoproteins (Zernecke et al. 2009; Callis et al. 2009). A landmark study by Valadi et al. first demonstrated that exosomes carry miRNAs, sometimes at concentrations exceeding those in donor cells (Valadi et al. 2007), with Hunter et al. subsequently confirming miRNA presence in blood vesicles (Hunter et al. 2008). These discoveries suggested a non-random, potentially selective packaging mechanism that challenges previous understanding of genetic transmission. This revelation has fundamentally reshaped our comprehension of cellular communication. MiRNAs are no longer viewed as mere intracellular regulators but as mobile molecular messengers capable of complex signaling. They operate through multiple communication channels, autocrine (within the same cell), paracrine (between nearby cells), and even in an endocrine-like manner across broader distances. The implications are profound, suggesting a dynamic and fluid genetic information transfer model extending beyond traditional cellular boundaries.

The unique characteristics of exo-miRNAs, their tissue specificity, remarkable stability, and ease of collection position them as promising noninvasive biomarkers. Researchers are actively exploring their potential in early disease detection across a spectrum of conditions, from cancer to neurodegenerative diseases (Callis et al. 2009). What was once considered cellular noise has emerged as a critical language of biological communication, offering unprecedented insights into disease mechanisms and potential therapeutic interventions. As our understanding continues to evolve, miRNAs stand as a testament to the complexity of life at its most fundamental level. These tiny molecules challenge our basic assumptions about genetic regulation, cellular communication, and disease progression. They remind us that the most significant biological mechanisms often operate in the smallest packages, waiting to be deciphered by the persistent curiosity of scientific investigation.

### 15.2.2 Exo-miRNAs: Nature's Tiny Messengers in Cellular Communication and Disease

Cells function like bustling cities, constantly communicating with each other. Among their most fascinating messaging systems are tiny bubbles called exosomes, which carry important genetic instructions in the form of miRNAs. These microscopic messengers, only about 30 to 150 nanometers in size (smaller than a human hair is thick!), are released by virtually every cell in the body. Exosomes begin their journey inside specialized compartments called multivesicular bodies (MVBs). These compartments act as sorting facilities where cells package important molecular mail; as these packages mature, the membrane folds inward, creating small bubbles inside bubbles. When the outer bubble fuses with the cell's outer wall, these inner packages are released into the bloodstream and other body fluids as exosomes. Their miRNAs are short instruction manuals, about 22 letters (nucleotides) long. Each starts life in the cell's command center, the nucleus, where it is transcribed as a rough draft. This draft gets edited by a protein complex

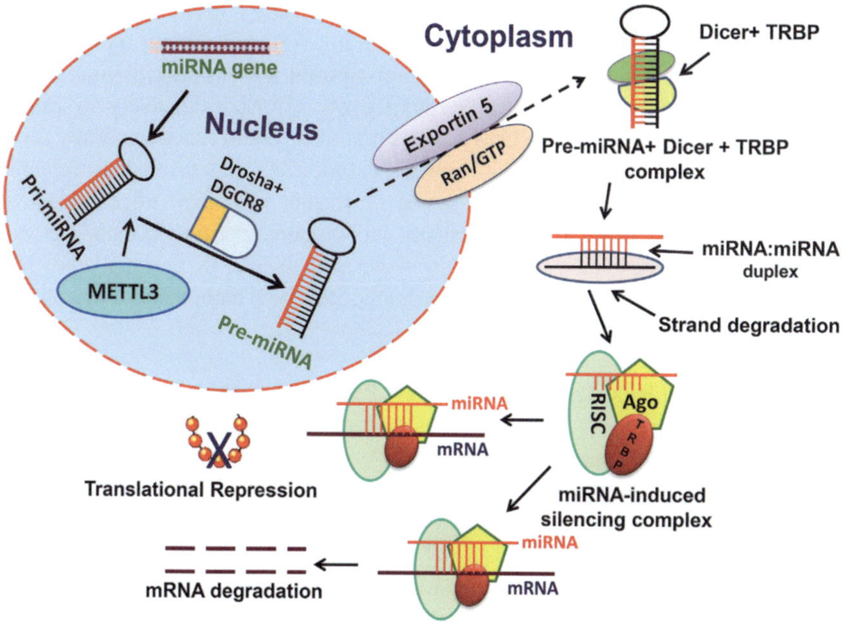


**Fig. 15.1** Overview of miRNA biogenesis and gene silencing mechanism. miRNA genes are transcribed by RNA polymerase II into primary transcripts (pri-miRNAs), which METTL3 methylates to facilitate recognition by the Drosha–DGCR8 complex. This microprocessor cleaves pri-miRNAs into precursor stem-loop structures (pre-miRNAs), which are exported to the cytoplasm via Exportin-5. Dicer further processes pre-miRNAs into ~22 nucleotide miRNA duplexes in the cytoplasm. One strand is degraded, while the other forms the mature miRNA. This mature strand is loaded into the RNA-induced silencing complex (RISC), which guides it to complementary mRNA sequences to repress translation or induce mRNA degradation (Jafri et al. 2017)

called Drosha–DGCR8, creating a precursor miRNA exported from the nucleus via a transporter called Exportin-5. Once in the cell's main body, another editor called Dicer trims it to its final form. The finished miRNA then joins a team of proteins called RISC, which uses the miRNA as a guide to regulate gene expression (Fig. 15.1) (Tétreault and De Guire 2013; Jafri et al. 2017).

What is fascinating is that cells do not randomly stuff these miRNAs into exosomes. It is like carefully choosing what messages to send to different recipients. Several systems control this selection process. One involves an enzyme called neutral sphingomyelinase 2 (nSMase2), which helps produce a fatty substance called ceramide that influences exosome formation and miRNA loading. Another involves special proteins that recognize specific patterns (called EXO-motifs) in the miRNA sequence. These proteins, with names like hnRNPA2B1, YBX1, and AGO2, function as postal workers who know exactly which packages should go where. Once released, these exosomes travel through blood, urine, saliva, and other body fluids until they reach their target cells. The recipient cells take them in through various methods, sometimes by engulfing them, sometimes by fusing with them, or sometimes through specialized receptors. Once inside, the miRNAs regulate genes, influencing crucial cellular processes like inflammation, cell growth, programmed cell death, and cell specialization. This cellular postal system plays a vital role in health and disease. Scientists are studying exo-miRNAs extensively in cancer research, brain disorders, heart diseases, and immune system functions. They serve dual purposes, actively influencing biological processes and serving as valuable markers to help doctors diagnose diseases or predict how they might progress; understanding how these tiny messengers work opens exciting possibilities for medicine. Researchers might develop new ways to diagnose, monitor, and treat various conditions by learning to interpret or engineer these natural communications.

### 15.2.3 Passive Mechanism of miRNA Packaging into Exosomes

Several studies have demonstrated that miRNAs can be packaged into exosomes via a passive mechanism dependent on the cellular miRNA expression profile. In this process, miRNAs are packaged passively into exosomes in proportion to their abundance within the originating cell. Pigati et al. were the first to explore this phenomenon in depth by performing a microarray analysis comparing intracellular and exo-miRNAs in MCF7 cells (Pigati et al. 2010). They found that approximately 66% of exosome miRNA levels were proportional to the intracellular miRNA content. For example, the miR-638 detected in exosomes closely matched its intracellular level. Moreover, transcriptomic changes in cancer or activated cells may further influence the quantity of miRNAs loaded into exosomes. A high number of miRNA target transcripts, for instance, may affect how much of a particular miRNA is engaged in gene silencing activities or multivesicular body (MVB) biogenesis, which impacts its release into exosomes (Squadrito et al. 2014). Pigati and colleagues also noted that specific miRNAs, such as miR-451 and miR-1246, were released in higher quantities by cancerous mammary cells, while most miRNAs in non-malignant cells remained intracellular (Pigati et al. 2010). This observation suggests that the dynamics of miRNA release may vary depending on cell state or pathology, although a direct correlation with intracellular expression levels remains a key aspect of the passive loading mechanism.

### 15.2.4 Functional Roles of Exo-miRNAs in Cancer

Cancer cells demonstrate a remarkable capacity for intercellular communication through exosomes, nanoscale extracellular vesicles that play a pivotal role in tumor biology. Quantitative analyses reveal a significant disparity in exosome production between normal and malignant cells (Huang and Deng 2019). Riches et al. documented that normal breast epithelial cells generate approximately $4.5 \pm 2.3 \times 10^8$ exosomes per million cells over 24 hours, in stark contrast to breast cancer cell lines, which release about $53.2 \pm 1.6 \times 10^8$ exosomes under identical conditions (Riches et al. 2014). Similar observations have been made in other cancer types, with Zhang and colleagues noting distinct exosome secretion patterns in ovarian cells, where cancerous cells sustain exosome release for extended periods compared to normal cells (Zhang et al. 2016).

Cancer cells' enhanced exosome production stems from intrinsic cellular alterations and extrinsic microenvironmental factors. Hypoxic conditions significantly modulate exosome release, as demonstrated by King et al., who observed that exposure to 1% and 0.1% oxygen levels increased exosome output in breast cancer cells by $32.3 \pm 4.8\%$ and $90.9 \pm 7.1\%$, respectively (King et al. 2012). Moreover, alterations in microenvironmental pH contribute to exosome dynamics, with acidic conditions enhancing exosome release and uptake in melanoma cells through increased membrane rigidity and altered lipid composition (Parolini et al. 2009).

In glioblastoma (GBM), exosomes are critical mediators of tumor progression and intercellular communication. Skog et al. identified various pro-angiogenic proteins within GBM-derived exosomes, including VEGF, IL-6, IL-8, TIMP-1/2, FGF, and angiogenin. These proteins facilitate tumor expansion by interacting with endothelial receptors and promoting angiogenesis (Taraboletti et al. 2006). Notably, hypoxic stress induces the secretion of VEGF-A-enriched exosomes that compromise blood–brain barrier integrity by reducing claudin-5 expression and increasing endothelial permeability (Zhao et al. 2018).

Emerging studies indicate that exosomes are crucial in transporting miRNAs from malignant donor cells to adjacent stromal or immune cells, functioning as essential mediators within the tumor microenvironment (Donnarumma et al. 2017). This two-way communication can profoundly influence gene expression in recipient cells, thereby facilitating key oncogenic processes such as tumor progression, metastasis, and epithelial–mesenchymal transition (EMT) (He et al. 2019), neovascularization, and modulation of immune responses (Fig. 15.2) (Schwarzenbach 2015). The oncogenic potential of exosomes extends beyond protein transfer. The horizontal transfer of mutated receptors, such as EGFRvIII, represents a critical mechanism of cancer progression. These modified receptors, detected in circulating microvesicles, can integrate into recipient cell membranes, activating downstream oncogenic pathways, including ERK1/2 and AKT (Al-Nedawi et al. 2008; Skog et al. 2008).

Immunomodulation represents another crucial function of tumor-derived exosomes. Research by Domenis et al. revealed that glioblastoma stem cell-derived exosomes suppress T cell activation while stimulating the release of inflammatory cytokines, effectively creating an immunosuppressive microenvironment (Domenis et al. 2017). Similar mechanisms have been observed in other cancer types, with exosomes playing a significant role in metastasis and immune evasion.

The role of exosomes in metastasis and drug resistance is particularly noteworthy. In gastric cancer, tumor-derived exosomes induce mesothelial apoptosis and facilitate the mesothelial-to-mesenchymal transition, promoting peritoneal dissemination (Deng et al. 2017). Exo-miRNAs have been shown to activate epithelial–mesenchymal transition pathways, enhancing cellular invasiveness (Hu et al. 2019). Furthermore, exosomes contribute to chemotherapeutic resistance through multiple mechanisms, including the transfer of drug-resistance-associated molecules such

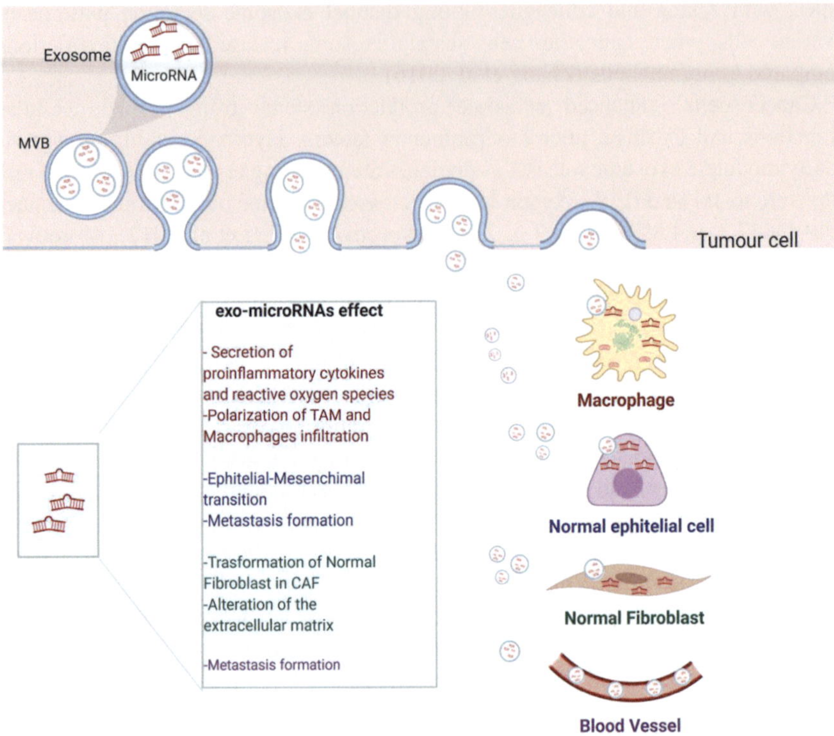


**Fig. 15.2** Schematic illustration showing how tumor-derived exo-miRNAs influence nearby and distant cells. These vesicles transfer oncogenic cargo to local macrophages, epithelial cells, and fibroblasts, promoting cellular reprogramming and tumor progression. They can also circulate through the bloodstream to distant sites (Ingenito et al. 2019)

as miR-501 in gastric cancer (Liu et al. 2019) and EphA2 in pancreatic cancer (Fan et al. 2018).

Beyond proteins and miRNAs, exosomal nucleic acids emerge as critical mediators of tumor progression. Long noncoding RNAs, such as MALAT1, can be transferred via exosomes to modify the behavior of recipient cells, demonstrating the complex molecular landscape of exosomal communication in cancer (Qiu et al. 2018).

A summary of additional tumor-suppressive and oncogenic miRNAs identified in exosomes is provided in Table 15.1.

**Table 15.1** Oncogenic and tumor-suppressive exo-miRNAs and their functional roles in cancer progression

| Exo-miRNAs | Tumor | Effect | Reference |
|---|---|---|---|
| Oncogenic miR | | | |
| miR-10b | Breast cancer | Invasion increase | Singh et al. (2014) |
| miR-223 | Pancreatic cancer | Invasion increase | Ma et al. (2019) |
| miR-32-5p | Hepatocellular carcinoma | Drug resistance increase | Fu et al. (2018) |
| miR-99a-5p, miR-125b-5p | Large B-cell lymphoma | Drug resistance increase | Feng et al. (2019) |
| miR-9 | Glioma | Angiogenesis increase | Chen et al. (2019) |
| miR-93-5p | Oesophageal cancer | Growth increase | Liu et al. (2018) |
| miR-21,miR-378e, miR-143 | Breast cancer | EMT and stemness increase | Donnarumma et al. (2017) |
| Oncosuppressor miR | | | |
| miR-192 | Lung adenocarcinoma | Metastasis decrease | Valencia et al. (2014) |
| miR-451a | Hepatocellular carcinoma | Apoptosis increase, angiogenesis decrease | Zhao et al. (2019) |
| miR-9 | Nasopharyngeal carcinoma | Angiogenesis decrease | Lu et al. (2018) |
| miR-8073 | Colorectal cancer | Growth decrease | Mizoguchi et al. (2018) |
| miR-100 | Breast cancer | Angiogenesis decrease | Pakravan et al. (2017) |
| miR-16-5p | Mesothelioma | Growth, migration, invasion decrease | Munson et al. (2019) |
| miR-199a | Glioma | Growth, migration, invasion decrease | Yu et al. (2019) |

## 15.3 Clinical Potential of Exo-miRNAs in Cancer Detection and Monitoring

The quest for accurate, noninvasive cancer detection has led researchers to an elegant approach: liquid biopsy. This method, focused on analyzing disease markers in body fluids rather than tissue samples, revolutionizes how we might detect and monitor cancer. The concept isn't entirely new, dating back to 1965 when Gold and colleagues demonstrated we could detect tumor cells in blood (Gold and Freedman 1965). Traditional serum biomarkers like CA-125 for ovarian cancer and PSA for prostate cancer emerged from this thinking, though their clinical usefulness has been limited by accuracy challenges (Kulasingam et al. 2010).

Recent scientific advances have dramatically expanded the liquid biopsy's potential. Rather than looking solely at circulating cells, researchers now examine multiple tumor-derived components floating in bodily fluids, including tiny cellular messengers called exosomes. Among the various molecular cargo these exosomes carry, miRNAs have emerged as particularly valuable. These exo-miRNAs offer several compelling advantages as cancer biomarkers; they're selectively concentrated in exosomes released by tumor cells (Srivastava et al. 2015), they remain remarkably stable under various storage and handling conditions (Ge et al. 2014), and their protective exosome coating shields them from enzymes that would normally degrade RNA (Koga et al. 2011).

Scientists have developed several methods to analyze these exo-miRNAs. Quantitative real-time PCR remains the gold standard for detecting specific miRNAs with high sensitivity. For broader screening, microarray platforms allow simultaneous detection of numerous miRNAs. More recently, next-generation sequencing has transformed the field by enabling comprehensive analysis of both known and previously undiscovered miRNAs (Hu et al. 2017; Nelson et al. 2004; Chugh and Dittmer 2012). The evidence supporting exo-miRNAs' clinical value continues to grow. In non-small cell lung cancer, researchers identified four miRNAs (let-7b-5p, let-7e-5p, miR-23a-3p, and miR-486-5p) that could detect early-stage disease with over 80% accuracy (Jin et al. 2017). In glioblastoma, a panel of three RNA markers (miR-320, miR-574-3p, and RNU6-1) effectively distinguished patients from healthy individuals (Liu et al. 2016). Prostate cancer studies found high levels of exosomal miR-182 and miR-183 in tumor-derived exosomes (Mihelich et al. 2016), while miR-1290 and miR-375 correlated with poorer outcomes (Huang et al. 2015; Shi et al. 2015).

However, challenges remain. Some exo-miRNAs appear in multiple cancer types, limiting their diagnostic specificity. For example, exo-miR-21 is elevated not only in breast cancer (Hannafon et al. 2016) but also in glioma (Shi et al. 2015), colorectal cancer (Mousavi et al. 2019), esophageal cancer (Tanaka et al. 2013), and ovarian cancer (Au Yeung et al. 2016). This suggests it may be a general marker of malignancy rather than a cancer-specific indicator. Beyond simple detection, exo-miRNA profiles offer insights into cancer staging and progression. In lung cancer, exo-miR-451a levels correlate with disease stages I to III (Kanaoka et al. 2018). In breast cancer, exo-miR-223-3p levels reflect tumor stage, lymphatic invasion, and nuclear grade, particularly in early-stage disease (Yoshikawa et al. 2018). While promising, these findings require validation in larger studies. Interestingly, some markers like exo-miR-301a in glioblastoma can track treatment response, decreasing after surgical tumor removal and rising again if the cancer returns (Lan et al. 2018). In ovarian cancer, miR-200b and miR-200c appear significantly elevated in advanced stages compared to early disease (Meng et al. 2016).

Some exo-miRNAs can even distinguish between cancer subtypes. For instance, exo-miR-373 appears more abundant in triple-negative breast cancer compared to other subtypes. In lung cancer, researchers identified distinct miRNA panels for adenocarcinoma versus squamous cell carcinoma (Santos et al. 2018). The potential clinical significance of these findings has prompted multiple clinical trials. In 2018, a prospective study began examining exo-miRNAs and other non-coding RNAs as potential markers in high-grade serous ovarian cancer (NCT03738319). Two ongoing trials are investigating prostate-derived exosomes ("prostasomes") by analyzing their RNA content through next-generation sequencing (NCT03694483 and NCT03911999).

Exo-miRNAs represent a promising frontier in cancer diagnostics. Their stability, accessibility through simple blood draws, and ability to reflect various aspects of cancer biology position them as potentially transformative tools for early detection, monitoring, and personalized treatment strategies. While further refinement is needed to enhance specificity and standardize testing protocols, these tiny molecular messengers may soon play a crucial role in cancer care.

## 15.4 Therapeutic Applications of Exo-miRNAs

Exosomes have become a hot topic in medical research lately, and for good reason. These tiny bubble-like structures that cells release naturally are showing promise as a new way to deliver treatments throughout the body. What makes them special? They're compatible with our biology, they stick around in the bloodstream for a good while, and they're naturally programmed to find specific cells. This means they can carry important molecules, like RNA, proteins, and smaller compounds, directly to target cells that need treatment.

Cancer researchers are particularly excited about using exosomes to transport miRNAs. These small RNA fragments play important roles in how tumors grow, evade the immune system, form new blood vessels, and spread to other parts of the body. By packing therapeutic miRNAs into exosomes, scientists are developing a new cell-free approach to influence gene activity in cancer cells. Recent research is backing up this potential. For instance, Schmittgen's team discovered that exosomes from natural killer cells carrying a specific miRNA (miR-186) could restore the cancer-fighting abilities of these immune cells against neuroblastoma, a type of childhood cancer (Schmittgen 2019). In another study, Le and colleagues showed how engineered exosomes could deliver molecules that interfere with RNA to leukemia and breast cancer cells, slowing down their reproduction (Usman et al. 2018).

Interestingly, exosomes aren't just delivery trucks, they're potential targets themselves. Since they help cancer spread and carry cancer-promoting miRNAs, scientists are looking at ways to block their release or uptake. One approach involves silencing a protein called Rab27a that helps exosomes leave cells. When researchers reduced Rab27a in melanoma models, they saw less miR-494 (a problematic miRNA) and, consequently, reduced tumor growth and spread (Li et al. 2019). Another compound called GW4869 has been used to decrease exosome production in various cancers (Liu et al. 2019). Researchers have also tried blocking the receiving end by using drugs that prevent exosomes from entering new cells, which helped combat multiple myeloma cell growth and drug resistance (Zheng et al. 2019).

While many clinical trials are examining exo-miRNAs for diagnosis and prognosis, their therapeutic use is still in early stages. Only a handful of treatment-focused studies have made it to clinical trial registration. One example involves patients with malignant brain tumors. In this phase I trial (NCT02507583), researchers use exosomes from glioblastoma cells that were treated with molecules targeting a specific growth factor receptor. These exosomes are delivered through implanted chambers, with the idea that local delivery might trigger a stronger whole-body immune response with fewer side effects than traditional treatments.

Another promising trial (NCT03608631) focuses on patients with metastatic pancreatic cancer. This study tests exosomes from special supportive cells to deliver RNA that silences a mutated gene (KRAS) driving pancreatic tumor growth. The goal is to prove that exosome-based gene silencing can work in aggressive cancers. Meanwhile, lab researchers continue improving how we load medications into exosomes, make them more targeted, and ensure they work well with the immune system. New techniques like microfluidic engineering and click chemistry are making the process more precise and scalable for clinical use.

In the big picture, while exo-miRNA therapies are still emerging, their dual potential as both targets and carriers makes them exciting for personalized medicine. Since they're involved in nearly every aspect of tumor development and immune responses, they could transform cancer treatment approaches. As we get better at isolating and engineering exosomes, and as regulatory frameworks develop, these tiny biological packages might soon move from experimental concept to clinical reality.

## 15.5 Conclusion

Exo-miRNAs represent a transformative frontier in cancer research, offering powerful tools for both diagnosis and therapy. Their inherent stability, accessibility through liquid biopsies, and ability to reflect the molecular landscape of tumors make them highly promising biomarkers for early detection, disease monitoring, and treatment response prediction. Furthermore, their role in mediating intercellular communication and tumor progression positions them as valuable candidates for therapeutic delivery. While substantial progress has been made, further clinical validation, technological advancements in detection, and refined methods to distinguish tumor-derived signals from the surrounding microenvironment are critical for their translation into routine oncology practice. Harnessing the full potential of exo-miRNAs may ultimately drive the shift toward more precise, personalized, and less invasive cancer care.

## References

Al-Nedawi K, Meehan B, Micallef J, Lhotak V, May L, Guha A et al (2008) Intercellular transfer of the oncogenic receptor EGFRvIII by microvesicles derived from tumour cells. Nat Cell Biol 10. https://doi.org/10.1038/ncb1725

Arroyo JD, Chevillet JR, Kroh EM, Ruf IK, Pritchard CC, Gibson DF et al (2011) Argonaute2 complexes carry a population of circulating microRNAs independent of vesicles in human plasma. Proc Natl Acad Sci USA 108. https://doi.org/10.1073/pnas.1019055108

Au Yeung CL, Co NN, Tsuruga T, Yeung TL, Kwan SY, Leung CS et al (2016) Exosomal transfer of stroma-derived miR21 confers paclitaxel resistance in ovarian cancer cells through targeting APAF1. Nat Commun 7. https://doi.org/10.1038/ncomms11150

Bartel DP (2004) MicroRNAs: genomics, biogenesis, mechanism, and function. Cell 116. https://doi.org/10.1016/S0092-8674(04)00045-5

Brücher BLDM, Jamall IS (2014) Cell-cell communication in the tumor microenvironment, carcinogenesis, and anticancer treatment. Cell Physiol Biochem 34:213–243. https://doi.org/10.1159/000362978

Buechner J, Tømte E, Haug BH, Henriksen JR, Løkke C, Flægstad T et al (2011) Tumour-suppressor microRNAs let-7 and mir-101 target the proto-oncogene MYCN and inhibit cell proliferation in MYCN-amplified neuroblastoma. Br J Cancer 105. https://doi.org/10.1038/bjc.2011.220

Callis TE, Pandya K, Hee YS, Tang RH, Tatsuguchi M, Huang ZP et al (2009) MicroRNA-208a is a regulator of cardiac hypertrophy and conduction in mice. J Clin Invest 119. https://doi.org/10.1172/JCI36154

Chan JA, Krichevsky AM, Kosik KS (2005) MicroRNA-21 is an antiapoptotic factor in human glioblastoma cells. Cancer Res 65. https://doi.org/10.1158/0008-5472.CAN-05-0137

Chen X, Ba Y, Ma L, Cai X, Yin Y, Wang K et al (2008) Characterization of microRNAs in serum: a novel class of biomarkers for diagnosis of cancer and other diseases. Cell Res 18. https://doi.org/10.1038/cr.2008.282

Chen X, Yang F, Zhang T, Wang W, Xi W, Li Y et al (2019) MiR-9 promotes tumorigenesis and angiogenesis and is activated by MYC and OCT4 in human glioma. J Exp Clin Cancer Res 38. https://doi.org/10.1186/s13046-019-1078-2

Chim SSC, Shing TKF, Hung ECW, Leung TY, Lau TK, Chiu RWK et al (2008) Detection and characterization of placental microRNAs in maternal plasma. Clin Chem 54. https://doi.org/10.1373/clinchem.2007.097972

Chugh P, Dittmer DP (2012) Potential pitfalls in microRNA profiling. Wiley Interdiscip Rev RNA 3. https://doi.org/10.1002/wrna.1120

Deng G, Qu J, Zhang Y, Che X, Cheng Y, Fan Y et al (2017) Gastric cancer-derived exosomes promote peritoneal metastasis by destroying the mesothelial barrier. FEBS Lett 591. https://doi.org/10.1002/1873-3468.12722

Domenis R, Cesselli D, Toffoletto B, Bourkoula E, Caponnetto F, Manini I et al (2017) Systemic T cells immunosuppression of glioma stem cell-derived exosomes is mediated by monocytic myeloid-derived suppressor cells. PLoS One 12. https://doi.org/10.1371/journal.pone.0169932

Donnarumma E, Fiore D, Nappa M, Roscigno G, Adamo A, Iaboni M et al (2017) Cancer-associated fibroblasts release exosomal microRNAs that dictate an aggressive phenotype in breast cancer. Oncotarget 8. https://doi.org/10.18632/oncotarget.14752

Fan J, Wei Q, Koay EJ, Liu Y, Ning B, Bernard PW et al (2018) Chemoresistance transmission via exosome-mediated EphA2 transfer in pancreatic cancer. Theranostics 8. https://doi.org/10.7150/thno.26650

Feng Y, Zhong M, Zeng S, Wang L, Liu P, Xiao X et al (2019) Exosome-derived miRNAs as predictive biomarkers for diffuse large B-cell lymphoma chemotherapy resistance. Epigenomics 11. https://doi.org/10.2217/epi-2018-0123

Fu X, Liu M, Qu S, Ma J, Zhang Y, Shi T et al (2018) Exosomal microRNA-32-5p induces multidrug resistance in hepatocellular carcinoma via the PI3K/Akt pathway. J Exp Clin Cancer Res 37. https://doi.org/10.1186/s13046-018-0677-7

Ge Q, Zhou Y, Lu J, Bai Y, Xie X, Lu Z (2014) MiRNA in plasma exosome is stable under different storage conditions. Molecules 19. https://doi.org/10.3390/molecules19021568

Gold P, Freedman SO (1965) Demonstration of tumor-specific antigens in human colonic carcinomata by immunological tolerance and absorption techniques. J Exp Med 121. https://doi.org/10.1084/jem.121.3.439

Hanahan D, Weinberg RA (2011) Hallmarks of cancer: the next generation. Cell 144. https://doi.org/10.1016/j.cell.2011.02.013

Hannafon BN, Trigoso YD, Calloway CL, Zhao YD, Lum DH, Welm AL et al (2016) Plasma exosome microRNAs are indicative of breast cancer. Breast Cancer Res 18. https://doi.org/10.1186/s13058-016-0753-x

He S, Li Z, Yu Y, Zeng Q, Cheng Y, Ji W et al (2019) Exosomal miR-499a-5p promotes cell proliferation, migration and EMT via mTOR signaling pathway in lung adenocarcinoma. Exp Cell Res 379. https://doi.org/10.1016/j.yexcr.2019.03.035

Hu Y, Lan W, Miller D (2017) Next-generation sequencing for microRNA expression profile. Methods Mol Biol 1617. https://doi.org/10.1007/978-1-4939-7046-9_12

Hu Y, Qi C, Liu X, Zhang C, Gao J, Wu Y et al (2019) Malignant ascites-derived exosomes promote peritoneal tumor cell dissemination and reveal a distinct miRNA signature in advanced gastric cancer. Cancer Lett 457. https://doi.org/10.1016/j.canlet.2019.04.034

Huang T, Deng CX (2019) Current progresses of exosomes as cancer diagnostic and prognostic biomarkers. Int J Biol Sci 15. https://doi.org/10.7150/ijbs.27796

Huang X, Yuan T, Liang M, Du M, Xia S, Dittmar R et al (2015) Exosomal miR-1290 and miR-375 as prognostic markers in castration-resistant prostate cancer. Eur Urol 67. https://doi.org/10.1016/j.eururo.2014.07.035

Hunter MP, Ismail N, Zhang X, Aguda BD, Lee EJ, Yu L et al (2008) Detection of microRNA expression in human peripheral blood microvesicles. PLoS One 3. https://doi.org/10.1371/journal.pone.0003694

Ingenito F, Roscigno G, Affnito A, Nuzzo S, Scognamiglio I, Quintavalle C et al (2019) The role of Exo-miRNAs in cancer: a focus on therapeutic and diagnostic applications. Int J Mol Sci 20. https://doi.org/10.3390/ijms20194687

Jafri MA, Al-Qahtani MH, Shay JW (2017) Role of miRNAs in human cancer metastasis: implications for therapeutic intervention. Semin Cancer Biol 44. https://doi.org/10.1016/j.semcancer.2017.02.004

Jin X, Chen Y, Chen H, Fei S, Chen D, Cai X et al (2017) Evaluation of tumor-derived exosomal miRNA as potential diagnostic biomarkers for early-stage non-small cell lung cancer using next-generation sequencing. Clin Cancer Res 23. https://doi.org/10.1158/1078-0432.CCR-17-0577

Kanaoka R, Iinuma H, Dejima H, Sakai T, Uehara H, Matsutani N et al (2018) Usefulness of plasma exosomal microRNA-451a as a noninvasive biomarker for early prediction of recurrence and prognosis of non-small cell lung cancer. Oncologia 94. https://doi.org/10.1159/000487006

King HW, Michael MZ, Gleadle JM (2012) Hypoxic enhancement of exosome release by breast cancer cells. BMC Cancer 12. https://doi.org/10.1186/1471-2407-12-421

Koga Y, Yasunaga M, Moriya Y, Akasu T, Fujita S, Yamamoto S et al (2011) Exosome can prevent RNase from degrading microRNA in feces. J Gastrointest Oncol 2. https://doi.org/10.3978/j.issn.2078-6891.2011.015

Kowal J, Arras G, Colombo M, Jouve M, Morath JP, Primdal-Bengtson B et al (2016) Proteomic comparison defines novel markers to characterize heterogeneous populations of extracellular vesicle subtypes. Proc Natl Acad Sci USA 113. https://doi.org/10.1073/pnas.1521230113

Kulasingam V, Pavlou MP, Diamandis EP (2010) Integrating high-throughput technologies in the quest for effective biomarkers for ovarian cancer. Nat Rev Cancer 10. https://doi.org/10.1038/nrc2831

Lan F, Qing Q, Pan Q, Hu M, Yu H, Yue X (2018) Serum exosomal miR-301a as a potential diagnostic and prognostic biomarker for human glioma. Cell Oncol 41. https://doi.org/10.1007/s13402-017-0355-3

Li L, Yuan L, Luo J, Gao J, Guo J, Xie X (2013) MiR-34a inhibits proliferation and migration of breast cancer through down-regulation of Bcl-2 and SIRT1. Clin Exp Med 13. https://doi.org/10.1007/s10238-012-0186-5

Li J, Chen J, Wang S, Li P, Zheng C, Zhou X et al (2019) Blockage of transferred exosome-shuttled miR-494 inhibits melanoma growth and metastasis. J Cell Physiol 234. https://doi.org/10.1002/jcp.28234

Liu Y, Luo F, Wang B, Li H, Xu Y, Liu X et al (2016) STAT3-regulated exosomal miR-21 promotes angiogenesis and is involved in neoplastic processes of transformed human bronchial epithelial cells. Cancer Lett 370. https://doi.org/10.1016/j.canlet.2015.10.011

Liu MX, Liao J, Xie M, Gao ZK, Wang XH, Zhang Y et al (2018) miR-93-5p transferred by exosomes promotes the proliferation of esophageal cancer cells via intercellular communication by targeting PTEN. Biomed Environ Sci 31. https://doi.org/10.3967/bes2018.023

Liu X, Lu Y, Xu Y, Hou S, Huang J, Wang B et al (2019) Exosomal transfer of miR-501 confers doxorubicin resistance and tumorigenesis via targeting of BLID in gastric cancer. Cancer Lett 459. https://doi.org/10.1016/j.canlet.2019.05.035

Lu J, Liu QH, Wang F, Tan JJ, Deng YQ, Peng XH et al (2018) Exosomal miR-9 inhibits angiogenesis by targeting MDK and regulating PDK/AKT pathway in nasopharyngeal carcinoma. J Exp Clin Cancer Res 37. https://doi.org/10.1186/s13046-018-0814-3

Ma J, Cao T, Cui Y, Zhang F, Shi Y, Xia J et al (2019) miR-223 regulates cell proliferation and invasion via targeting PDS5B in pancreatic cancer cells. Mol Ther Nucleic Acids 14. https://doi.org/10.1016/j.omtn.2019.01.009

Mause SF, Weber C (2010) Microparticles: protagonists of a novel communication network for intercellular information exchange. Circ Res 107. https://doi.org/10.1161/CIRCRESAHA.110.226456

Meng X, Müller V, Milde-Langosch K, Trillsch F, Pantel K, Schwarzenbach H (2016) Diagnostic and prognostic relevance of circulating exosomal miR-373, miR-200a, miR-200b and miR-200c in patients with epithelial ovarian cancer. Oncotarget 7. https://doi.org/10.18632/oncotarget.7850

Mihelich BL, Dambal S, Lin S, Nonn L (2016) miR-182, of the miR-183 cluster family, is packaged in exosomes and is detected in human exosomes from serum, breast cells and prostate cells. Oncol Lett 12. https://doi.org/10.3892/ol.2016.4710

Minciacchi VR, Freeman MR, Di Vizio D (2015) Extracellular vesicles in cancer: exosomes, microvesicles and the emerging role of large oncosomes. Semin Cell Dev Biol 40. https://doi.org/10.1016/j.semcdb.2015.02.010

Mizoguchi A, Takayama A, Arai T, Kawauchi J, Sudo H (2018) MicroRNA-8073: tumor suppressor and potential therapeutic treatment. PLoS One 13. https://doi.org/10.1371/journal.pone.0209750

Mousavi S, Moallem R, Hassanian SM, Sadeghzade M, Mardani R, Ferns GA et al (2019) Tumor-derived exosomes: potential biomarkers and therapeutic target in the treatment of colorectal cancer. J Cell Physiol 234. https://doi.org/10.1002/jcp.28080

Munson PB, Hall EM, Farina NH, Pass HI, Shukla A (2019) Exosomal miR-16-5p as a target for malignant mesothelioma. Sci Rep 9. https://doi.org/10.1038/s41598-019-48133-0

Nelson PT, Baldwin DA, Scearce LM, Oberholtzer JC, Tobias JW, Mourelatos Z (2004) Microarray-based, high-throughput gene expression profiling of microRNAs. Nat Methods 1. https://doi.org/10.1038/nmeth717

Noferesti SS, Sohel MMH, Hoelker M, Salilew-Wondim D, Tholen E, Looft C et al (2015) Controlled ovarian hyperstimulation induced changes in the expression of circulatory miRNA in bovine follicular fluid and blood plasma. J Ovarian Res 8. https://doi.org/10.1186/s13048-015-0208-5

Pakravan K, Babashah S, Sadeghizadeh M, Mowla SJ, Mossahebi-Mohammadi M, Ataei F et al (2017) MicroRNA-100 shuttled by mesenchymal stem cell-derived exosomes suppresses in vitro angiogenesis through modulating the mTOR/HIF-1α/VEGF signaling axis in breast cancer cells. Cell Oncol 40. https://doi.org/10.1007/s13402-017-0335-7

Parolini I, Federici C, Raggi C, Lugini L, Palleschi S, De Milito A et al (2009) Microenvironmental pH is a key factor for exosome traffic in tumor cells. J Biol Chem 284. https://doi.org/10.1074/jbc.M109.041152

Pigati L, Yaddanapudi SCS, Iyengar R, Kim DJ, Hearn SA, Danforth D et al (2010) Selective release of MicroRNA species from normal and malignant mammary epithelial cells. PLoS One 5. https://doi.org/10.1371/journal.pone.0013515

Qiu JJ, Lin XJ, Tang XY, Zheng TT, Lin YY, Hua KQ (2018) Exosomal metastasis-associated lung adenocarcinoma transcript 1 promotes angiogenesis and predicts poor prognosis in epithelial ovarian cancer. Int J Biol Sci 14. https://doi.org/10.7150/ijbs.28048

Quintavalle C, Garofalo M, Zanca C, Romano G, Iaboni M, Del Basso De Caro M et al (2012) MiR-221/222 overexpession in human glioblastoma increases invasiveness by targeting the protein phosphate PTP. Oncogene 31. https://doi.org/10.1038/onc.2011.280

Riches A, Campbell E, Borger E, Powis S (2014) Regulation of exosome release from mammary epithelial and breast cancer cells-a new regulatory pathway. Eur J Cancer 50. https://doi.org/10.1016/j.ejca.2013.12.019

Roscigno G, Puoti I, Giordano I, Donnarumma E, Russo V, Affinito A et al (2017) MiR-24 induces chemotherapy resistance and hypoxic advantage in breast cancer. Oncotarget 8. https://doi.org/10.18632/oncotarget.14470

Santos JC, Lima NDS, Sarian LO, Matheu A, Ribeiro ML, Derchain SFM (2018) Exosome-mediated breast cancer chemoresistance via miR-155 transfer. Sci Rep 8. https://doi.org/10.1038/s41598-018-19339-5

Schmittgen TD (2019) Exosomal miRNA cargo as mediator of immune escape mechanisms in neuroblastoma. Cancer Res 79. https://doi.org/10.1158/0008-5472.CAN-19-0021

Schwarzenbach H (2015) The clinical relevance of circulating, exosomal miRNAs as biomarkers for cancer. Expert Rev Mol Diagn 15. https://doi.org/10.1586/14737159.2015.1069183

Shi R, Wang PY, Li XY, Chen JX, Li Y, Zhang XZ et al (2015) Exosomal levels of miRNA-21 from cerebrospinal fluids associated with poor prognosis and tumor recurrence of glioma patients. Oncotarget 6. https://doi.org/10.18632/oncotarget.4699

Singh R, Pochampally R, Watabe K, Lu Z, Mo YY (2014) Exosome-mediated transfer of miR-10b promotes cell invasion in breast cancer. Mol Cancer 13. https://doi.org/10.1186/1476-4598-13-256

Skog J, Würdinger T, van Rijn S, Meijer DH, Gainche L, Curry WT et al (2008) Glioblastoma microvesicles transport RNA and proteins that promote tumour growth and provide diagnostic biomarkers. Nat Cell Biol 10. https://doi.org/10.1038/ncb1800

Squadrito ML, Baer C, Burdet F, Maderna C, Gilfillan GD, Lyle R et al (2014) Endogenous RNAs modulate MicroRNA sorting to exosomes and transfer to acceptor cells. Cell Rep 8. https://doi.org/10.1016/j.celrep.2014.07.035

Srivastava A, Filant J, Moxley K, Sood A, McMeekin S, Ramesh R (2015) Exosomes: a role for naturally occurring nanovesicles in cancer growth, diagnosis and treatment. Curr Gene Ther 15. https://doi.org/10.2174/1566523214666141224100612

Tanaka Y, Kamohara H, Kinoshita K, Kurashige J, Ishimoto T, Iwatsuki M et al (2013) Clinical impact of serum exosomal microRNA-21 as a clinical biomarker in human esophageal squamous cell carcinoma. Cancer 119. https://doi.org/10.1002/cncr.27895

Taraboletti G, D'Ascenzo S, Giusti I, Marchetti D, Borsotti P, Millimaggi D et al (2006) Bioavailability of VEGF in tumor-shed vesicles depends on vesicle burst induced by acidic pH 1. Neoplasia 8. https://doi.org/10.1593/neo.05583

Taylor DD, Gercel-Taylor C (2008) MicroRNA signatures of tumor-derived exosomes as diagnostic biomarkers of ovarian cancer. Gynecol Oncol 110. https://doi.org/10.1016/j.ygyno.2008.04.033

Tétreault N, De Guire V (2013) MiRNAs: their discovery, biogenesis and mechanism of action. Clin Biochem 46. https://doi.org/10.1016/j.clinbiochem.2013.02.009

Thind A, Wilson C (2016) Exosomal miRNAs as cancer biomarkers and therapeutic targets. J Extracell Vesicles 5. https://doi.org/10.3402/jev.v5.31292

Tickner JA, Urquhart AJ, Stephenson SA, Richard DJ, O'Byrne KJ (2014) Functions and therapeutic roles of exosomes in cancer. Front Oncol. https://doi.org/10.3389/fonc.2014.00127

Usman WM, Pham TC, Kwok YY, Vu LT, Ma V, Peng B et al (2018) Efficient RNA drug delivery using red blood cell extracellular vesicles. Nat Commun 9. https://doi.org/10.1038/s41467-018-04791-8

Valadi H, Ekström K, Bossios A, Sjöstrand M, Lee JJ, Lötvall JO (2007) Exosome-mediated transfer of mRNAs and microRNAs is a novel mechanism of genetic exchange between cells. Nat Cell Biol 9. https://doi.org/10.1038/ncb1596

Valencia K, Luis-Ravelo D, Bovy N, Antón I, Martínez-Canarias S, Zandueta C et al (2014) MiRNA cargo within exosome-like vesicle transfer influences metastatic bone colonization. Mol Oncol 8. https://doi.org/10.1016/j.molonc.2014.01.012

Yoshikawa M, Iinuma H, Umemoto Y, Yanagisawa T, Matsumoto A, Jinno H (2018) Exosome-encapsulated microRNA-223-3p as a minimally invasive biomarker for the early detection of invasive breast cancer. Oncol Lett 15. https://doi.org/10.3892/ol.2018.8457

Yu L, Gui S, Liu Y, Qiu X, Zhang G, Zhang X et al (2019) Exosomes derived from microRNA-199a-overexpressing mesenchymal stem cells inhibit glioma progression by down-regulating AGAP2. Aging (Albany NY) 11. https://doi.org/10.18632/aging.102092

Zernecke A, Bidzhekov K, Noels H, Shagdarsuren E, Gan L, Denecke B et al (2009) Delivery of microRNA-126 by apoptotic bodies induces CXCL12-dependent vascular protection. Sci Signal 2. https://doi.org/10.1126/scisignal.2000610

Zhang B, Pan X, Cobb GP, Anderson TA (2007) microRNAs as oncogenes and tumor suppressors. Dev Biol 302. https://doi.org/10.1016/j.ydbio.2006.08.028

Zhang W, Yang J, Cao D, You Y, Shen K, Peng P (2016) Regulation of exosomes released from normal ovarian epithelial cells and ovarian cancer cells. Tumour Biol 37. https://doi.org/10.1007/s13277-016-5394-2

Zhao C, Wang H, Xiong C, Liu Y (2018) Hypoxic glioblastoma release exosomal VEGF-A induce the permeability of blood-brain barrier. Biochem Biophys Res Commun 502. https://doi.org/10.1016/j.bbrc.2018.05.140

Zhao S, Li J, Zhang G, Wang Q, Wu C, Zhang Q et al (2019) Exosomal MIR-451a functions as a tumor suppressor in hepatocellular carcinoma by targeting LPIN1. Cell Physiol Biochem 53. https://doi.org/10.33594/000000118

Zheng Y, Tu C, Zhang J, Wang J (2019) Inhibition of multiple myeloma-derived exosomes uptake suppresses the functional response in bone marrow stromal cell. Int J Oncol 54. https://doi.org/10.3892/ijo.2019.4685

# Clinical Trials and Patent Updates of miRNA Therapy in Cancer 16

Rudraksh Mukherjee, Abhi Tyagi, Shikha Tiwari,
and Aayushi Arora

**Abstract**

MicroRNA (miRNA)-based therapeutics have advanced from preclinical studies to early-phase clinical trials, highlighting their promise in oncology. Several miRNA mimics and inhibitors are currently under evaluation for cancers such as liver, lung, and hematological malignancies, aiming to restore tumor suppressor miRNA function or silence oncogenic miRNAs. Clinical outcomes so far demonstrate both therapeutic potential and challenges, including delivery efficiency, off-target effects, and immune responses. Parallel to clinical developments, a growing number of patents reflect the rapid innovation in miRNA drug design, delivery platforms, and combinatorial approaches with chemotherapy, radiotherapy, and immunotherapy. Patent filings also indicate significant global interest from academia, biotechnology firms, and pharmaceutical industries. Together, clinical trial progress and patent landscapes provide critical insights into the translational potential of miRNA therapy. This chapter reviews key clinical updates, patient trends, and future perspectives in miRNA-based cancer therapeutics, emphasizing opportunities and barriers for clinical application.

R. Mukherjee · A. Tyagi
Department of Mathematics, Bioinformatics and Computer Applications, Maulana Azad National Institute of Technology (MANIT), Bhopal, Madhya Pradesh, India

S. Tiwari (✉)
School of Sciences, Sanjeev Agrawal Global Educational University, Bhopal, Madhya Pradesh, India
e-mail: shikha.t@sageuniversity.edu.in

A. Arora (✉)
Department of Microbiology, Anatomy, Physiology, and Pharmacology, La Trobe Institute for Molecular Science, School of Agriculture, Biomedicine and Environment, La Trobe University, Melbourne, VIC, Australia



S. Tiwari et al. (eds.), *microRNA Modulation: A Promising Combination Therapy for Cancer*, https://doi.org/10.1007/978-981-95-5989-3_16

**Keywords**

miRNA · Clinical trial · Patents · miRNA therapy

**Learning Objectives:**
- Describe the current status of clinical trials investigating miRNA-based therapeutics in cancer.
- Analyze the global patent landscape related to miRNA therapy, including innovations in design and delivery.
- Evaluate the opportunities and challenges revealed by clinical trial outcomes and intellectual property trends.
- Assess the translational potential of miRNA-based therapies for integration into future cancer treatment strategies.

## 16.1 Introduction

MicroRNAs (miRNAs) represent one of the most significant discoveries in modern molecular biology, fundamentally transforming our understanding of gene regulation and cellular control mechanisms (O'Brien et al. 2018; Lee et al. 1993). These small, endogenous, noncoding RNA molecules, typically 19–25 nucleotides in length, function as critical posttranscriptional regulators that orchestrate diverse biological processes throughout life (O'Brien et al. 2018; Wightman et al. 1993; Catalanotto et al. 2016). The 2024 Nobel Prize in Physiology or Medicine, awarded to Victor Ambros and Gary Ruvkun for discovering microRNAs and their role in posttranscriptional gene regulation, underscores the profound impact of this field on biomedical science (Press release: the Nobel prize in physiology or medicine 2024 2025; Blood science 2025; MIT News 2025).

## 16.2 The Fundamental Role of miRNAs in Gene Regulation

The discovery of miRNAs revealed an entirely new dimension of gene regulation that was previously unknown to science (Lee et al. 1993; Wightman et al. 1993). Unlike the traditional central dogma of molecular biology—where DNA is transcribed to messenger RNA (mRNA) and then translated into proteins—miRNAs introduced a sophisticated layer of posttranscriptional control that allows cells to fine-tune gene expression with remarkable precision (Bartel 2004). This regulatory mechanism is essential for multicellular organisms, enabling cells containing identical genetic information to develop into vastly different cell types with specialized functions (Ambros 2004). The miRNA biogenesis pathway involves three precisely coordinated enzymatic steps that transform primary transcripts into functional regulatory molecules (Winter et al. 2009). The process begins when RNA polymerase II transcribes miRNA genes into long primary transcripts (pri-miRNAs) containing characteristic hairpin structures. In the nucleus, the Drosha-DGCR8 microprocessor complex recognizes and cleaves these pri-miRNAs to generate precursor miRNAs (pre-miRNAs) of approximately 55–70 nucleotides (Lee et al. 1993; Winter et al. 2009). These pre-miRNAs are subsequently transported to the cytoplasm via Exportin-5, where the RNase III enzyme Dicer processes them into mature 22-nucleotide miRNA duplexes (Lee et al. 1993; Bernstein et al. 2001).

Following processing, mature miRNA duplexes are loaded into the RNA-induced silencing complex (RISC), where the guide strand directs target recognition through sequence-specific base-pairing interactions, primarily involving the "seed region" comprising nucleotides 2–8 of the miRNA's 5′ end (Bartel 2004; Lewis et al. 2005). The degree of complementarity between miRNA and target mRNA determines the silencing mechanism, with perfect complementarity leading to mRNA cleavage and partial complementarity resulting in translational repression and mRNA destabilization (Bartel 2004; Filipowicz et al. 2008). A distinguishing feature of miRNA regulation is its capacity for broad targeting, where a single miRNA can regulate hundreds of target mRNAs, while individual mRNAs can be targeted by multiple miRNAs, creating complex regulatory networks (Lim et al. 2005).

## 16.3 miRNAs as Oncogenes and Tumor Suppressors

The role of miRNAs in cancer biology has revealed their dual functionality as either oncogenes (oncomiRs) or tumor suppressors, depending on their target genes and cellular context (Esquela-Kerscher and Slack 2006; He et al. 2007). This dichotomy reflects the complexity of miRNA regulatory networks and their intimate integration into cancer-relevant pathways.

### 16.3.1 Oncogenic miRNAs

Promote tumorigenesis when overexpressed by silencing tumor suppressor genes, cell cycle inhibitors, or pro-apoptotic factors (He et al. 2007; Volinia et al. 2006). The miR-17-92 cluster, also known as "oncomiR-1," comprises six miRNAs frequently overexpressed in various cancers, including lymphomas, lung, breast, colon, pancreatic, prostate, and gastric cancers (He et al. 2007; Takamizawa et al. 2004). This cluster promotes oncogenesis by targeting tumor suppressors PTEN and p21, inhibiting pro-apoptotic proteins such as Bim, and regulating E2F transcription factors (O'Donnell et al. 2005; Xiao et al. 2008). Similarly, miR-155 is overexpressed in various hematopoietic and solid malignancies, targeting multiple tumor suppressor genes including Bach1, Sla, Cutl1, Csf1r, and PU.1 (Seyhan 2024), while miR-21 functions across multiple cancer types by targeting tumor suppressors such as PDCD4, PTEN, and RECK (Meng et al. 2007; Asangani et al. 2008).

### 16.3.2 Tumor Suppressor miRNAs

Inhibit oncogenesis by targeting oncogenes, growth-promoting factors, or anti-apoptotic proteins, and are characteristically downregulated or deleted in cancer (Calin and Croce 2006). The miR-15a/16–1 cluster holds historical significance as the first identified tumor suppressor miRNAs, discovered through studies of chronic lymphocytic leukemia where these miRNAs function by targeting the anti-apoptotic protein Bcl-2 (Calin et al. 2002; Cimmino et al. 2005). The let-7 family represents another paradigmatic example, comprising ten mature isoforms that target multiple oncogenes, including RAS, HMGA2, and MYC (Takamizawa et al. 2004; Johnson et al. 2005). The miR-34 family functions as tumor suppressors directly regulated by the p53 pathway, targeting numerous oncogenes and PD-L1 in non-small cell lung cancer and acute myeloid leukemia (He et al. 2007; Chang et al. 2007).

## 16.4 Therapeutic Rationale for miRNA-Based Cancer Treatment

The therapeutic potential of miRNAs in cancer arises from their unique capacity to simultaneously regulate multiple genes within oncogenic or tumor suppressive networks, offering a paradigm shift from traditional single-target approaches to network-level intervention (Van Rooij and Kauppinen 2014; Bader et al. 2010). This multi-target capability addresses the fundamental challenge of cancer's multi-factorial nature, where dysregulation of numerous genes and pathways contributes to the malignant phenotype (Lam et al. 2015).

miRNA-based cancer therapeutics employ two complementary strategies (Bader et al. 2010; Rupaimoole and Slack 2017). miRNA replacement therapy utilizes synthetic miRNA mimics to restore the function of downregulated tumor suppressor miRNAs, while **antagomiRs** or anti-miRs employ antisense oligonucleotides to sequester and inhibit overexpressed oncogenic miRNAs (Krützfeldt et al. 2005; Stenvang et al. 2012). Clinical validation has been achieved through pioneering trials such as TargomiRs, which deliver miR-16 mimics using targeted delivery systems in patients with recurrent malignant pleural mesothelioma and non-small cell lung cancer (Van Zandwijk et al. 2017; Reid et al. 2016).

## 16.5 Challenges and Opportunities Compared to Other Nucleic Acid Therapies

MiRNA-based therapeutics present distinct advantages and challenges compared to other nucleic acid-based therapeutic modalities (Lam et al. 2015; Rupaimoole and Slack 2017). The fundamental distinction between miRNAs and small interfering RNAs (siRNAs) lies in their targeting philosophy: siRNAs provide highly specific single-gene targeting through perfect complementarity, while miRNAs exhibit partial complementarity and regulate multiple targets simultaneously (Lam et al. 2015).

This broader targeting increases off-target risk but provides therapeutic advantages for complex diseases requiring modulation of multiple pathways (Lam et al. 2015).

Compared to antisense oligonucleotides (ASOs), miRNAs leverage endogenous cellular machinery for target recognition, potentially providing advantages in cellular compatibility (Stenvang et al. 2012). CRISPR-based gene editing offers permanent genetic modifications, while miRNA therapeutics provide reversible effects that can be modulated through dosing (Rupaimoole and Slack 2017). This fundamental difference carries important safety implications, as adverse effects from miRNA therapeutics are potentially reversible (Rupaimoole and Slack 2017).

### 16.5.1 Current Challenges and Future Prospects

Despite their significant therapeutic potential, miRNA-based therapeutics face substantial translational challenges (Rupaimoole and Slack 2017; Beg et al. 2017). Delivery remains the most formidable challenge, as naked miRNAs suffer from rapid degradation, poor cellular uptake, and limited tissue specificity (Rupaimoole and Slack 2017). Off-target effects represent a fundamental challenge inherent to the broad regulatory networks of miRNAs, potentially leading to unintended biological consequences (Beg et al. 2017). The early termination of several clinical trials, including the prominent MRX34 study, due to severe immune-related adverse events, highlights these critical challenges (Beg et al. 2017; Hong et al. 2020).

The field continues advancing with enhanced understanding of miRNA biology, improved delivery technologies, and refined regulatory guidelines (Rupaimoole and Slack 2017). The 2024 Nobel Prize recognition provides renewed momentum and highlights the scientific significance of this field (Takamizawa et al. 2004; Rupaimoole and Slack 2017). With over 200 clinical studies registered for miRNA-based cancer treatments, the field demonstrates sustained commitment to clinical translation (Rupaimoole and Slack 2017). Success will require addressing current limitations while leveraging the unique advantages these molecules offer for network-level therapeutic intervention in cancer treatment (Van Rooij and Kauppinen 2014).

## 16.6 Clinical Trials of miRNA Therapy in Cancer

Despite strong preclinical promise, miRNA therapeutics have only just entered the clinic, with no agents approved to date (What will it take to get miRNA therapies to market 2025; Seyhan 2024). Seyhan (2024) notes that *"only a few"* miRNA-based therapies have progressed to clinical trials, none beyond early phases (Seyhan 2024). The global landscape comprises mostly Phase I/II trials targeting various cancers, led by biotech companies and academic consortia in the US, Europe, and Australia. The first-in-human miRNA cancer trial was MRX34 (a liposomal miR-34a mimic) in 2013 (What will it take to get miRNA therapies to market 2025). To date, many trials remain early-stage, and several have been terminated due to safety or funding issues

**Table 16.1** Summary of representative clinical trials of miRNA-based therapeutics in cancer

| Trial ID(s) | miRNA target (Type) | Cancer indication | Sponsor | Phase |
|---|---|---|---|---|
| NCT01829971 | miR-34a mimic (MRX34) | Advanced solid tumors | Mirna Therapeutics, Inc. | I (with expansion) |
| NCT02369198 | miR-16 mimic (TargomiRs) | Mesothelioma, NSCLC | EnGeneIC Ltd. | I |
| NCT04675996 | miR-193b mimic (INT-1B3) | Advanced solid tumors | (INT/Unknown) | I |
| NCT04811898 | miR-221 inhibitor (LNA) | Advanced solid tumors | Phio Pharmaceuticals | I |
| NCT02580552<br/ >NCT03713320 | miR-155 inhibitor (Cobomarsen) | CTCL, DLBCL (hematologic) | Miragen Therapeutics | I/II |
| NCT05908773<br/ >NCT06260774 | miR-10b inhibitor (TTX-MC138) | Metastatic solid tumors | TransCode Therapeutics | 0/IB |

(Seyhan 2024; Full Text 2025). Common challenges—low specificity, immune toxicity, and delivery hurdles—have emerged as recurring themes (Seyhan 2024; Full Text 2025). Nevertheless, these trials provide critical lessons and validate technical approaches (e.g., delivery systems) for advancing miRNA drugs (Full Text 2025). Table 16.1 summarizes key trials (mimics and inhibitors) of miRNA therapies in cancer, including their targets, indications, phase, sponsors, and outcomes.

- *MRX34 (miR-34a mimic)*: A liposomal miR-34a mimic developed by Mirna Therapeutics, MRX34 was tested in a Phase I dose-escalation study (NCT01829971) in patients with refractory advanced solid tumors (including hepatocellular carcinoma) (Full Text 2025). Initial results showed a manageable safety profile at low doses and some antitumor activity (e.g., partial responses) (Full Text 2025). However, severe immune-mediated toxicities emerged: the trial was halted in 2016 after four patients died from immune-related adverse events (Full Text 2025; Van Zandwijk et al. 2017). These outcomes underscored the risk of cytokine release and inflammation with systemic miRNA mimics.
- *TargomiRs (miR-16 mimic)*: TargomiRs are bacterially derived minicells (EnGeneIC Dream Vectors) loaded with a synthetic miR-16 mimic, designed to replace lost tumor-suppressor miR-15/16 family activity (Full Text 2025; Van Zandwijk et al. 2017). In a Phase I trial (NCT02369198) in malignant pleural mesothelioma and non-small cell lung cancer, escalating IV doses were well tolerated (Full Text 2025; Van Zandwijk et al. 2017). Of 27 heavily pretreated patients, one (5%) achieved a partial response, and ~ 68% had stable disease (Full Text 2025). The safety profile and signs of anti-tumor effect (inhibition of growth signaling) were encouraging, supporting further development (Full Text 2025).
- *INT-1B3 (miR-193b mimic)*: INT-1B3 is a lipid-nanoparticle–formulated mimic of tumor-suppressive miR-193b. A Phase I/Ib study (NCT04675996) in advanced

solid tumors was initiated based on potent preclinical anti-leukemic effects (Full Text 2025). Patients received twice-weekly IV infusions of INT-1B3. However, this study was terminated early for nonclinical reasons (insufficient funding), with no clinical outcomes reported (Full Text 2025). Thus, no human efficacy data are available, though preclinical findings remain promising.

- *LNA-i-miR-221 (miR-221 inhibitor)*: Locked nucleic acid (LNA) antisense oligonucleotides against miR-221 have been explored for advanced cancers. In a Phase I dose-escalation trial (NCT04811898) in patients with refractory solid tumors, LNA-i-miR-221 was administered IV (Full Text 2025). The drug was very well tolerated (no dose-limiting toxicities or grade 3–4 adverse events reported) (Full Text 2025). Clinically, 50% of patients achieved disease control (stable disease), and one colorectal cancer patient had a durable partial response lasting >3 years (Full Text 2025; Martino et al. 2025). Correlative studies confirmed on-target effects: miR-221 levels fell, and its tumor-suppressor targets (e.g., CDKN1B/p27, PTEN) were upregulated (Full Text 2025). These positive safety and activity signals support advancing LNA-221 inhibitors into Phase II trials.
- *Cobomarsen (MRG-106, miR-155 inhibitor)*: Cobomarsen is an LNA-based antisense against oncogenic miR-155 (involved in lymphomas). In a first-in-human Phase I study (NCT02580552) in cutaneous T-cell lymphoma (CTCL) patients, intratumoral or IV dosing (75–900 mg) was safe (Full Text 2025). An expansion Phase II (SOLAR, NCT03713320) in CTCL, DLBCL, CLL, and other T-cell malignancies treated patients at 300–600 mg IV (Full Text 2025; Martino et al. 2025). Responses were seen (improvements in skin lesion scores and clinical parameters), and toxicity remained low. Despite these encouraging outcomes, the program was discontinued due to strategic business decisions (not for safety or efficacy reasons) (Full Text 2025; Martino et al. 2025).
- *TTX-MC138 (miR-10b inhibitor)*: TTX-MC138 is an LNA antagomir targeting miR-10b (a driver of metastasis), conjugated to aminated iron-oxide nanoparticles for tumor targeting. A Phase 0 microdose study (NCT05908773) has been completed to assess biodistribution in metastatic solid-tumor patients (Martino et al. 2025). A follow-on Phase I/II dose-escalation study (NCT06260774) is ongoing, with IV dosing of 0.4–3.2 mg/kg every 28 days (Martino et al. 2025). This trial aims to suppress metastatic progression by silencing miR-10b. No outcomes are yet reported, but the novel delivery strategy illustrates current innovation in miRNA targeting.

These case studies illustrate the current miRNA clinical pipeline: mostly early-phase trials with various delivery platforms (liposomes, nanoparticles, exosomes) and both miRNA replacement and inhibition strategies. Many trials have reported manageable safety profiles, but *immune-related toxicities* and *delivery issues* have limited efficacy (e.g., MRX34) (What will it take to get miRNA therapies to market 2025; Martino et al. 2025). On the other hand, successes include demonstration of on-target engagement (LNA-221) and hints of clinical benefit in refractory cancers (Cobomarsen, LNA-221) (Full Text 2025; Martino et al. 2025). Overall, the global trial landscape is dynamic, with dozens of trials registered worldwide (in the US, EU, China, Australia, etc.) exploring miRNA mimics for tumor-suppressors and antimiRs for oncomiRs. Table 16.1 summarizes representative trials, including phase, target miRNA, cancer type, sponsor, and key outcomes.

The clinical trials above highlight several lessons: miRNA mimics (like MRX34) can trigger immune reactions, necessitating modified dosing or delivery; conversely, chemically stabilized antisense (LNA) miRNA inhibitors have shown good tolerability (Full Text 2025; Martino et al. 2025). Trials also underscore the importance of delivery vectors: TargomiRs used bacterial minicells to enhance tumor targeting (Martino et al. 2025), and TTX-MC138 uses nanoparticle carriers. The overall "current state" is one of cautious optimism: no miRNA cancer drug is yet approved, but multiple Phase I/II programs are investigating diverse miRNAs, and some (e.g., LNA-221) have progressed on safety (Martino et al. 2025; Menon et al. 2022). Researchers continue to refine chemistry (2′-O modifications, LNAs) and delivery (nanoparticles, EVs) to overcome barriers (Seyhan 2024; Martino et al. 2025).

In summary, the global trial landscape for miRNA cancer therapy is expanding but remains early-stage (Seyhan 2024). Key examples like MRX34 and TargomiRs illustrate both promise and pitfalls. The field has learned that careful attention to specificity, dosing, and immune effects is crucial; forthcoming trials will determine whether miRNA-based drugs can achieve clinical success on par with other RNA therapies.

## 16.7 Sample Case Studies on Clinical Trial of microRNA

**Case Study 1**

This case study looks at the first cancer treatment that used MRX34 (miR-34a mimic). Researchers are testing the first microRNA-based therapy, miRNA MRX34, in a phase I clinical trial for advanced solid cancers like melanoma, renal cell carcinoma, and hepatocellular carcinoma (HCC). The aim of this liposomal formulation of a synthetic miR-34a mimic was to restore tumor suppressor capabilities impaired in cancer by inhibiting different oncogenes, including BCL2, FOXP1, and CTNNB1. The study found that the recommended doses for phase 2 for HCC and other malignancies are 70 $mg/m^2$ and 93 $mg/m^2$, respectively.

Three patients, along with others whose illnesses remained stable, had partial remissions that lasted up to 65 weeks. Pharmacodynamic data showed that the target genes in samples from patients were blocked in a way that depended on the dose. The medicine proved safe at low doses, but four patients died from major side effects that affected their immune systems, so the trial had to stop. This discovery showed how difficult it is to transfer miRNA mimics around the body. These mimics can help the immune system and work on various things. MRX34 defined some basic rules on how safe and useful miRNA may be for treating cancer (Hong et al. 2020).

**Case Study 2**

TargomiRs are small cells made from bacteria that are filled with synthetic miR-16 mimics. They are meant to replace the tumor-suppressing activity that is lost in malignant pleural mesothelioma and non-small cell lung cancer. The goal of this tailored delivery system is to make tumors more selective while reducing negative effects throughout the body.

A phase I dose-escalation experiment with 26 patients showed that the safety profile was acceptable and that the infusion-related side effects were controllable. A clinical benefit was noted, featuring one verified partial response lasting 32 weeks, with 68% of patients exhibiting stable illness. The research validated the effective transport of miR-16 mimics to EGFR-expressing tumor cells, hence endorsing the viability of minicell-mediated targeted miRNA therapy. TargomiRs is a new way to navigate around problems with delivering miRNA therapies (Van Zandwijk et al. 2017).

## 16.8 Patent Landscape of miRNA Cancer Therapeutics

### 16.8.1 Global Trends

Patent activity around cancer technologies has surged in the past decade. The European Patent Office (EPO) reports ~70% growth in cancer-fighting inventions between 2015 and 2021 (measured by international patent families), and over 140,000 cancer-related inventions disclosed over the last 50 years. While this statistic covers oncology broadly (not miRNAs alone), it captures the strong backdrop into which miRNA IP has expanded (Patents and innovation against cancer).

Globally, the US leads oncology IP output, with China rising rapidly and Europe remaining strong but growing more slowly in several high-growth therapeutics domains.

What gets patented in miRNA oncology?

Three major categories dominate:

- Therapeutics—mimics (restore tumor-suppressor miRNAs).
- Therapeutics—inhibitors/antagomiRs (silence oncogenic miRNAs).
- Diagnostics/theranostics—signatures/panels (circulating or tissue miRNA profiles). (Overview from patent practice commentary.) (Kilburn and Strode 2025)

Delivery IP (e.g., LNPs, nanoparticles, engineered carriers) increasingly underpins therapeutic claims and often defines the protectable moat.

Representative miRNA cancer patients (examples)

These illustrate typical claim scopes: replacement (miR-34), inhibition (miR-22/antagomiRs), and diagnostics (solid-tumor signatures, circulating panels) (Table 16.2).

**Table 16.2** Patents filed based on miRNA therapy

| Patent/App. ID | Title/Focus | Applicant | Cancer/Application |
|---|---|---|---|
| US9371526B2 | Synthetic mimics of miR-34 (methods & compositions) | Mirna Therapeutics | Tumor-suppressor replacement-liver, lung, colon |
| EP2487253A2 | miRNA-based methods & compositions for solid cancers (diagnosis/treatment) | Applications incl. Ohio State Univ. et al. | Prostate, colon, pancreas, breast, stomach |
| US7709616B2 | miRNAs & precursors for diagnosis/prognosis/treatment | (Assignee in record) | Prostate & lung; detection & therapeutic modulation |
| WO2018187673A1 | Circulating miRNA signatures for early cancer biomarkers | (Applicant in record) | Broad diagnostic/ prognostic across cancers |
| US20210017521A1 | Inhibitors of microRNA-22 (chemically modified ASOs incl. LNA) | (Applicant in record) | Platform for miRNA inhibition; oncology-relevant biology |

## 16.9 IP Challenges Specific to miRNAs

- Patent-eligibility boundaries (US vs EU): In the US, natural sequences are not patent-eligible (Myriad decision); cDNA and genuinely man-made constructs can be. Europe allows patents on biological material when technical criteria are met (per Biotech Directive/EPO Guidelines). Therapeutic claims typically focus on engineered sequences, formulations, delivery, and uses.
- Thickets & overlap: With many players staking claims on similar sequences, uses, and delivery motifs, freedom-to-operate (FTO) hinges on design-around of sequence claims and securing differentiated delivery IP.
- Delivery platforms as IP moats: Given sequence-eligibility constraints and crowded biomarkers, LNP/nanoparticle/exosome delivery patents often become the decisive, enforceable edge (Bakshi 2014).

### 16.9.1 Practical Implications for Researchers and Sponsors

1. Anchor composition/use claims to engineered chemistry (e.g., LNA, 2′-O-Me) and specific delivery architectures.
2. Pair therapeutic claims with companion-diagnostic logic where evidence supports it (strengthens clinical narrative and IP story).
3. Plan early FTO in the US & China; align prosecution strategy across USPTO/ EPO/CNIPA families.

## 16.10 Translational and Regulatory Challenges in miRNA Cancer Therapeutics

### 16.10.1 Delivery, Off-Target Effects, and Immune Responses

Unmodified miRNAs degrade rapidly in circulation, are cleared quickly, and show poor tissue penetration. Lipid nanoparticles (LNPs) and exosomes improve stability and tumor targeting, yet precise delivery and organ selectivity are still difficult to achieve.

The promiscuity of miRNA targeting complicates therapeutic safety: false-positive rates in RNAi screens can reach ~30% [PMC], and unintended repression of essential genes has been linked to toxicity. The siRNA drug revusiran was terminated in Phase III after fatalities attributed to off-target effects.

Synthetic oligonucleotides can also trigger immune responses. Sequence motifs and delivery vehicles may activate innate immune receptors (TLR3, TLR7, TLR8), leading to cytokine release. Formulations such as LNPs and antibody–oligonucleotide conjugates have been reported to provoke cytokine storms or CARPA (complement activation-related pseudo-allergy) (Moraes et al. 2021; Zhang et al. 2021; Nigar and Shimosato 2019).

### 16.10.2 Manufacturing and Large-Scale Production

Scaling from gram-scale laboratory synthesis to tonne-scale GMP production remains a key bottleneck. The UK Medicines Manufacturing Innovation Centre (MMIC) aims to achieve GMP-grade oligonucleotide capacity by 2025.

The oligonucleotide manufacturing market is projected to grow at a 17.5% CAGR, reaching $6.53 billion by 2030 [PwC]. This reflects widening adoption: in 2022, 15 oligonucleotide drugs were approved (covering ~7% of the UK population), while Phase III candidates target diseases affecting 68% (Pharma manufacturing of the future: harnessing the oligonucleotide opportunity).

Technical bottlenecks include:

- Purification—high purity standards drive cost and complexity.
- Solid-phase synthesis—Effective at small scale, inefficient and costly at large scale.
- Process control—Mechanistic modelling of LNP formation is being developed to ensure reproducibility (Chemistry, manufacturing and controls strategies for using novel excipients in lipid nanoparticles 2025).

### 16.10.3 Regulatory Considerations

The FDA's *Clinical Pharmacology Considerations for Oligonucleotide Therapeutics* (2024) highlights the need to evaluate prolongation, immunogenicity, hepatic/

renal impairment, and drug–drug interactions. Complementary non-clinical safety guidance emphasizes assessing off-target hybridization, exaggerated pharmacology, and model suitability (C for D E and Research 2025).

The EMA's draft guideline stresses manufacturing consistency, analytical methods, conjugation chemistry, and comparability standards. Efforts are ongoing to harmonize FDA and EMA requirements. Both agencies also provide Orphan Drug Designation and paediatric programs applicable to rare oncological indications.

### 16.10.4 Bioethical and Legal Aspects

The 2013 Myriad ruling in the US declared that naturally occurring DNA and RNA sequences are not patentable unless significantly modified or applied in a novel industrial context. For miRNA-based patents, this requires focusing claims on engineered mimics, chemically modified constructs, delivery systems, or therapeutic applications.

This restriction has sparked ethical debate. Proponents argue that engineered constructs deserve protection to encourage investment, while critics caution against over-patenting fundamental biological molecules that could limit research accessibility (Fig. 16.1).

## 16.11 Future Directions in miRNA-Based Cancer Therapeutics

The delivery of miRNA-based therapeutics remains the most formidable obstacle to clinical translation. Conventional carriers such as liposomes and viral vectors often suffer from nuclease degradation, off-target effects, and limited biodistribution. Future strategies are moving toward bioengineered systems with improved stability, targeting precision, and biocompatibility (Fig. 16.2).

Among these, engineered exosomes are gaining significant attention. Derived from the body's own cellular communication machinery, exosomes can be loaded with therapeutic miRNAs and surface-modified to home into tumor microenvironments. Preclinical studies using exosomes loaded with miR-146a demonstrated antitumor efficacy while reducing systemic toxicity, highlighting their translational promise.

Parallel innovations involve integrating CRISPR technologies with miRNA platforms. Cas13 and Cas9 systems are being explored both for the detection of oncogenic miRNAs and for editing miRNA regulators. These approaches could enable "smart vesicles" that combine diagnostic and therapeutic functions, pioneering the concept of theranostic nanomedicine in oncology (Liu et al. 2023).

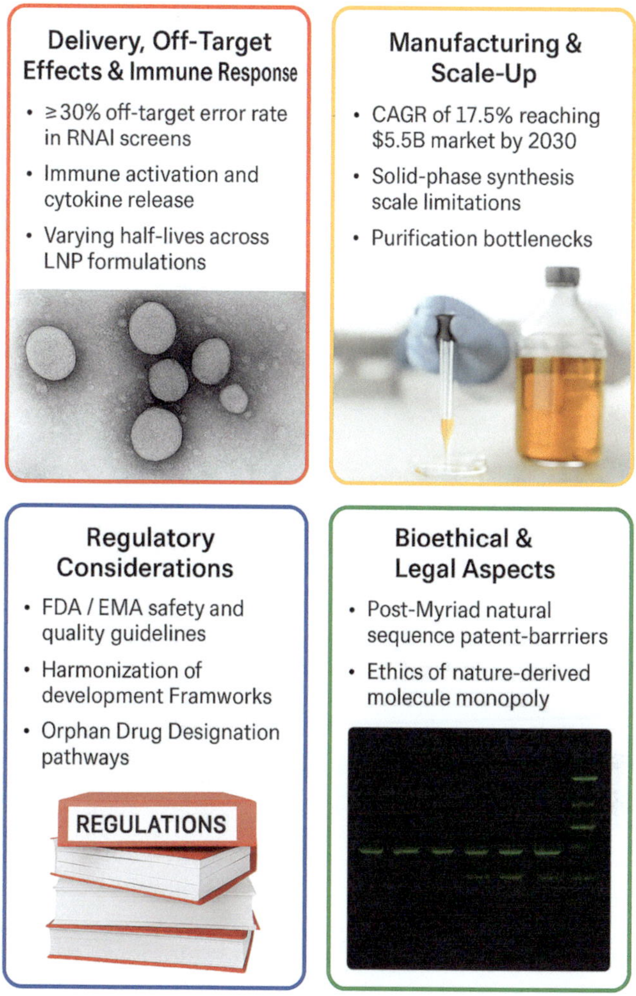


**Fig. 16.1** Translational and regulatory challenges in miRNA cancer therapeutics

### 16.11.1 Personalized Cancer Therapy via miRNA Profiling

miRNAs exhibit context-specific expression, varying across cancer types and between patients. This heterogeneity offers a unique opportunity for precision oncology. High-throughput sequencing combined with bioinformatics enables patient-specific miRNA signatures to guide prognosis, predict drug response, and personalize treatment selection.

Analyses of The Cancer Genome Atlas (TCGA) have shown that machine learning models trained on miRNA expression can predict response to immune

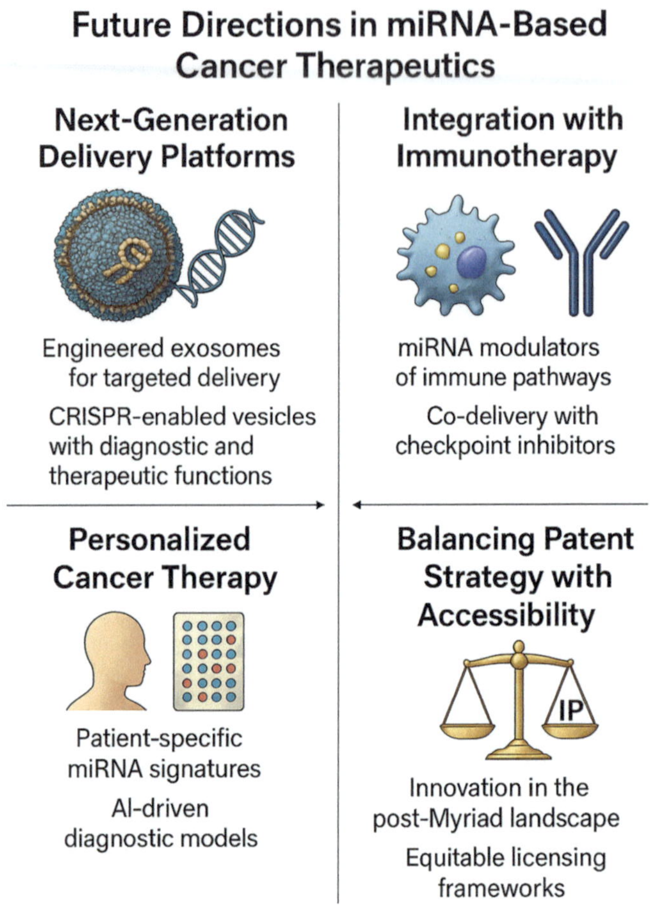


**Fig. 16.2** Future directions shaping miRNA-based cancer therapeutics

checkpoint blockade with high accuracy. This suggests that integrating miRNA profiling into clinical workflows could refine therapeutic decision-making.

In future oncology practice, miRNA panels may be incorporated alongside genomic and proteomic markers in diagnostic assays. Such personalization could enable adaptive treatment plans tailored to each patient's regulatory landscape, increasing therapeutic efficacy while minimizing unnecessary toxicity (Saumet 2014).

### 16.11.2 Integration with Immunotherapy

The interplay between miRNAs and the immune system presents a promising avenue for innovation. miRNAs regulate multiple immune pathways, including antigen presentation, T-cell activation, and cytokine signaling. Dysregulated miRNAs such as miR-21 and miR-155 have been implicated in shaping the tumor immune microenvironment.

Targeting these immune-modulating miRNAs could enhance responses to immunotherapy. Early evidence suggests that co-delivery of miRNA modulators with checkpoint inhibitors may reduce resistance and expand treatment benefits to non-responders. Other strategies involve embedding miRNA modulators into immune cell–targeted nanoparticles, thereby amplifying antitumor immunity.

Although these approaches are still at the preclinical stage, they illustrate the potential of synergistic regimens combining miRNA therapeutics with established immunotherapies. This integration could reshape the immunotherapeutic landscape by broadening the patient population that benefits from checkpoint blockade (Zabeti Touchaei and Vahidi 2024).

### 16.11.3 Balancing Patent Strategy with Accessibility

The future of miRNA therapeutics is shaped not only by scientific advances but also by intellectual property (IP) and ethical considerations. In the post-*Myriad* legal environment, naturally occurring RNA sequences are not patentable unless modified or applied in novel ways. This necessitates inventive patent strategies focusing on chemically modified constructs, delivery technologies, and therapeutic methods.

Pharmaceutical developers are increasingly adopting balanced IP approaches:

- *Offensive strategies*: securing patents to build competitive advantage and licensing revenue.
- *Defensive strategies*: using open-access patents or strategic publications to prevent monopolization of foundational biology.

Transparent licensing frameworks and equitable access models will be crucial to ensure that miRNA therapies reach patients globally. Overly restrictive IP practices could hinder innovation and widen disparities in cancer care, while balanced policies could stimulate both innovation and accessibility (Montenegro et al. 2016; Kumar et al. 2024).

## 16.12 Outlook

The convergence of miRNA biology with synthetic biology, immunotherapy, and precision medicine is catalyzing a new era in cancer therapeutics. Although obstacles remain in delivery, off-target effects, immune activation, and regulatory complexity, the field is increasingly equipped with tools to overcome them.

Future success will require scientific ingenuity, patient-centered design, and cross-sector collaboration. By integrating bioengineering, artificial intelligence, and clinical insight, miRNA therapeutics are well-positioned to become an integral component of personalized, multi-modal cancer care.

## 16.13 Conclusion

MicroRNAs (miRNAs) have evolved from a basic biological discovery into one of the most compelling frontiers in cancer therapeutics. Their ability to function as both oncogenes and tumor suppressors underscores their central role in tumor biology and validates them as therapeutic targets. Over the past decade, early-phase clinical trials ranging from miRNA mimics such as MRX34 to inhibitors like Cobomarsen and LNA-i-miR-221 have offered proof-of-concept for clinical translation while simultaneously revealing the formidable challenges of off-target effects, immune toxicity, and delivery barriers.

Beyond the clinic, the patent landscape reflects both the intensity of global innovation and the complexity of intellectual property rights in the post-*Myriad* era. The sharp rise in oncology-related filings, particularly in the US and China, highlights growing investment, yet also emphasizes the need for thoughtful IP strategies that balance innovation with equitable access.

Translational progress now hinges on resolving key obstacles: scalable and reproducible manufacturing, harmonized FDA/EMA regulatory pathways, and ethical frameworks that preserve access while incentivizing innovation. Encouragingly, advances in bioengineered exosomes, CRISPR-enabled vesicles, and AI-driven patient profiling are beginning to redefine the boundaries of what is clinically achievable.

Looking ahead, the convergence of miRNA therapeutics with precision oncology and immunotherapy offers a powerful opportunity to transform cancer care. Success will depend on coordinated efforts—scientific, regulatory, and ethical that ensure these therapies progress from experimental promise to accessible, life-changing interventions. In this trajectory, miRNA-based therapeutics are poised not merely as adjuncts, but as integral components of next-generation, multi-modal cancer treatment strategies.

## References

Ambros V (2004) The functions of animal microRNAs. Nature 431(7006):350–355

Asangani IA et al (2008) MicroRNA-21 (miR-21) post-transcriptionally downregulates tumor suppressor Pdcd4 and stimulates invasion, intravasation and metastasis in colorectal cancer. Oncogene 27(15):2128–2136

Bader AG, Brown D, Winkler M (2010) The promise of microRNA replacement therapy. Cancer Res 70(18):7027

Bakshi AM (2014) Gene patents at the Supreme Court: Association for Molecular Pathology v. myriad genetics. J Law Biosci 1(2):183–189. https://doi.org/10.1093/jlb/lsu007

Bartel DP (2004) MicroRNAs: genomics, biogenesis, mechanism, and function. Cell 116(2): 281–297

Beg MS et al (2017) Phase I study of MRX34, a liposomal miR-34a mimic, administered twice weekly in patients with advanced solid tumors. Investig New Drugs 35(2):180–188. https://doi.org/10.1007/s10637-016-0407-y

Bernstein E, Caudy AA, Hammond SM, Hannon GJ (2001) Role for a bidentate ribonuclease in the initiation step of RNA interference. Nature 409(6818):363–366

Blood science [Online]. Available: https://journals.lww.com/bls/fulltext/2024/10000/micrornas__discovery,_breakthrough,_and_innovation.12.aspx. Accessed 22 Aug 2025

C for D E and Research. Clinical pharmacology considerations for the development of oligonucleotide therapeutics [Online]. Available: https://www.fda.gov/regulatory-information/search-fda-guidance-documents/clinical-pharmacology-considerations-development-oligonucleotide-therapeutics. Accessed 25 Aug 2025.

Calin GA, Croce CM (2006) MicroRNA signatures in human cancers. Nat Rev Cancer 6(11):857–866

Calin GA et al (2002) Frequent deletions and down-regulation of micro- RNA genes *miR15* and *miR16* at 13q14 in chronic lymphocytic leukemia. Proc Natl Acad Sci USA 99(24):15524–15529. https://doi.org/10.1073/pnas.242606799

Catalanotto C, Cogoni C, Zardo G (2016) MicroRNA in control of gene expression: an overview of nuclear functions. Int J Mol Sci 17(10):1712

Chang T-C et al (2007) Transactivation of miR-34a by p53 broadly influences gene expression and promotes apoptosis. Mol Cell 26(5):745–752

Chemistry, manufacturing and controls strategies for using novel excipients in lipid nanoparticles. Nat Nanotechnol [Online]. Available: https://www.nature.com/articles/s41565-024-01833-9. Accessed 25 Aug 2025

Cimmino A et al (2005) *miR-15* and *miR-16* induce apoptosis by targeting BCL2. Proc Natl Acad Sci USA 102(39):13944–13949. https://doi.org/10.1073/pnas.0506654102

Esquela-Kerscher A, Slack FJ (2006) Oncomirs—microRNAs with a role in cancer. Nat Rev Cancer 6(4):259–269

Filipowicz W, Bhattacharyya SN, Sonenberg N (2008) Mechanisms of post-transcriptional regulation by microRNAs: are the answers in sight? Nat Rev Genet 9(2):102–114

Full Text. MicroRNA in cancer therapy: breakthroughs and challenges in early clinical applications. J Exper Clin Cancer Res [Online]. Available: https://jeccr.biomedcentral.com/articles/10.1186/s13046-025-03391-x. Accessed 22 Aug 2025

He L et al (2007) A microRNA component of the p53 tumour suppressor network. Nature 447(7148):1130–1134

Hong DS et al (2020) Phase 1 study of MRX34, a liposomal miR-34a mimic, in patients with advanced solid tumours. Br J Cancer 122(11):1630–1637

Johnson SM et al (2005) RAS is regulated by the let-7 microRNA family. Cell 120(5):635–647

Kilburn & Strode. Nobel Prize awarded for the discovery of microRNA and its role in post-transcriptional gene regulation. [Online]. Available: https://www.kilburnstrode.com/knowledge/european-ip/mirna-nobel-prize?utm_source=chatgpt.com. Accessed 25 Aug 2025

Krützfeldt J et al (2005) Silencing of microRNAs in vivo with 'antagomirs'. Nature 438(7068):685–689

Kumar R, Tiwari S, Harilal S (2024) MicroRNA-based clinical trials for head and neck cancer. In: Diagnostic, prognostic, and therapeutic role of microRNAs in head and neck cancer. Academic Press, pp 325–335

Lam JK, Chow MY, Zhang Y, Leung SW (2015) siRNA versus miRNA as therapeutics for gene silencing. Mol Ther Nucleic Acids 4 [Online]. Available: https://www.cell.com/molecular-therapy-family/nucleic-acids/fulltext/S2162-2531(16)30037-3. Accessed 22 Aug 2025

Lee RC, Feinbaum RL, Ambros V (1993) The C. elegans heterochronic gene lin-4 encodes small RNAs with antisense complementarity to lin-14. Cell 75(5):843–854

Lewis BP, Burge CB, Bartel DP (2005) Conserved seed pairing, often flanked by adenosines, indicates that thousands of human genes are microRNA targets. Cell 120(1):15–20

Lim LP et al (2005) Microarray analysis shows that some microRNAs downregulate large numbers of target mRNAs. Nature 433(7027):769–773

Liu X et al (2023) Hierarchically tumor-activated nanoCRISPR-Cas13a facilitates efficient microRNA disruption for multi-pathway-mediated tumor suppression. Theranostics 13(9): 2774–2786. https://doi.org/10.7150/thno.81776

Martino MTD, Tagliaferri P, Tassone P (2025) MicroRNA in cancer therapy: breakthroughs and challenges in early clinical applications. J Exp Clin Cancer Res 44(1):126. https://doi.org/10.1186/s13046-025-03391-x

Meng F, Henson R, Wehbe–Janek H, Ghoshal K, Jacob ST, Patel T (2007) MicroRNA-21 regulates expression of the PTEN tumor suppressor gene in human hepatocellular cancer. Gastroenterology 133(2):647–658

Menon A, Abd-Aziz N, Khalid K, Poh CL, Naidu R (2022) miRNA: a promising therapeutic target in cancer. Int J Mol Sci 23(19):11502. https://doi.org/10.3390/ijms231911502

Victor Ambros '75, PhD '79 and Gary Ruvkun share Nobel prize in Physiology or Medicine. MIT News. Massachusetts Institute of Technology [Online]. Available: https://news.mit.edu/2024/victor-ambros-gary-ruvkun-share-nobel-prize-physiology-medicine-1007. Accessed 22 Aug 2025

Montenegro JH, Romero AF, Tejada PA, Olaya SX, Rubiano AM (2016) Perceived versus observed patient safety measures in a critical care unit from a teaching hospital in Southern Colombia. Crit Care Res Pract 2016:2175436. https://doi.org/10.1155/2016/2175436

Moraes FC, Pichon C, Letourneur D, Chaubet F (2021) miRNA delivery by nanosystems: state of the art and perspectives. Pharmaceutics 13(11):1901. https://doi.org/10.3390/pharmaceutics13111901

Nigar S, Shimosato T (2019) Cooperation of Oligodeoxynucleotides and synthetic molecules as enhanced immune modulators. Front Nutr 6:140. https://doi.org/10.3389/fnut.2019.00140

O'Brien J, Hayder H, Zayed Y, Peng C (2018) Overview of microRNA biogenesis, mechanisms of actions, and circulation. Front Endocrinol 9:402

O'Donnell KA, Wentzel EA, Zeller KI, Dang CV, Mendell JT (2005) C-Myc-regulated microRNAs modulate E2F1 expression. Nature 435(7043):839–843

Patents and innovation against cancer

Pharma manufacturing of the future: harnessing the oligonucleotide opportunity

Press release: the Nobel prize in physiology or medicine 2024. NobelPrizeorg [Online]. Available: https://www.nobelprize.org/prizes/medicine/2024/press-release/. Accessed 22 Aug 2025

Reid G et al (2016) Clinical development of TargomiRs, a miRNA mimic-based treatment for patients with recurrent thoracic cancer. Epigenomics 8(8):1079–1085. https://doi.org/10.2217/epi-2016-0035

Rupaimoole R, Slack FJ (2017) MicroRNA therapeutics: towards a new era for the management of cancer and other diseases. Nat Rev Drug Discov 16(3):203–222

Saumet A (2014) The potential of microRNAs in personalized medicine against cancers. BioMed Res Int. Wiley Online Library. [Online]. Available: https://onlinelibrary.wiley.com/doi/10.1155/2014/642916. Accessed 25 Aug 2025

Seyhan AA (2024) Trials and tribulations of MicroRNA therapeutics. Int J Mol Sci 25(3):1469. https://doi.org/10.3390/ijms25031469

Stenvang J, Petri A, Lindow M, Obad S, Kauppinen S (2012) Inhibition of microRNA function by antimiR oligonucleotides. Silence 3(1):1. https://doi.org/10.1186/1758-907X-3-1

Takamizawa J et al (2004) Reduced expression of the let-7 microRNAs in human lung cancers in association with shortened postoperative survival. Cancer Res 64(11):3753–3756

Van Rooij E, Kauppinen S (2014) Development of micro RNA therapeutics is coming of age. EMBO Mol Med 6(7):851–864. https://doi.org/10.15252/emmm.201100899

Van Zandwijk N et al (2017) Safety and activity of microRNA-loaded minicells in patients with recurrent malignant pleural mesothelioma: a first-in-man, phase 1, open-label, dose-escalation study. Lancet Oncol 18(10):1386–1396. https://doi.org/10.1016/S1470-2045(17)30621-6

Volinia S et al (2006) A microRNA expression signature of human solid tumors defines cancer gene targets. Proc Natl Acad Sci USA 103(7):2257–2261. https://doi.org/10.1073/pnas.0510565103

What will it take to get miRNA therapies to market. Nat Biotechnol [Online]. Available: https://www.nature.com/articles/s41587-024-02480-0. Accessed 22 Aug 2025

Wightman B, Ha I, Ruvkun G (1993) Posttranscriptional regulation of the heterochronic gene lin-14 by lin-4 mediates temporal pattern formation in C. elegans. Cell 75(5):855–862

Winter J, Jung S, Keller S, Gregory RI, Diederichs S (2009) Many roads to maturity: microRNA biogenesis pathways and their regulation. Nat Cell Biol 11(3):228–234

Xiao C et al (2008) Lymphoproliferative disease and autoimmunity in mice with increased miR-17-92 expression in lymphocytes. Nat Immunol 9(4):405–414

Zabeti Touchaei A, Vahidi S (2024) MicroRNAs as regulators of immune checkpoints in cancer immunotherapy: targeting PD-1/PD-L1 and CTLA-4 pathways. Cancer Cell Int 24(1):102. https://doi.org/10.1186/s12935-024-03293-6

Zhang S, Cheng Z, Wang Y, Han T (2021) The risks of miRNA therapeutics: in a drug target perspective. DDDT 15:721–733. https://doi.org/10.2147/DDDT.S288859

# The Trials and Tribulations of the microRNA Therapy Market 17

Swetha Vaishnavi and Satheesh Ellipilli

**Abstract**

The development of miRNA therapeutics has emerged as a promising yet challenging frontier in precision medicine. By enabling targeted regulation of gene expression, miRNA-based therapies hold significant potential for treating cancer, genetic disorders, and various other diseases. However, despite the entry of several miRNA-based drugs into clinical trials, only a few have progressed to significant human trials. This is due to the key challenges in high-throughput screening of targets, delivery insufficiencies, off-target effects, immune responses, and high costs. Nevertheless, advancements in delivery technologies, synthetic modifications, and biomarker-driven patient selection may enable miRNA therapies to carve out a niche in modern medicine. This chapter explores the trials and tribulations of miRNA therapy market, examining experimental and developmental setbacks, and regulatory hurdles, also highlights the evolving landscape of miRNA therapeutics and its future prospects.

**Keywords**

MiRNA therapy · Noncoding RNAs · Preclinical and clinical trials · MiRNA mimics · Anti-miRNAs · OncomiRs · AntagomiRs

**Learning Objectives**

- *Explain* the major scientific and technological barriers to successful miRNA therapy development.
- *Analyze* the economic and regulatory factors shaping the miRNA therapy market.

S. Vaishnavi · S. Ellipilli (✉)
Chemical Biology Laboratory, Department of Chemistry, School of Engineering and Sciences, SRM University-AP, Amaravati, Andhra Pradesh, India
e-mail: satheesh.e@srmap.edu.in





S. Tiwari et al. (eds.), *microRNA Modulation: A Promising Combination Therapy for Cancer*, https://doi.org/10.1007/978-981-95-5989-3_17

- *Evaluate* case studies of clinical and commercial setbacks in miRNA therapeutics.
- *Assess* future opportunities and strategies to overcome challenges in bringing miRNA therapies to market.

## 17.1 Introduction

MicroRNA (miRNA), a small noncoding RNA discovered in 1993, plays a pivotal role in regulating gene expression. It is involved in numerous cellular processes, including cell growth, proliferation, differentiation, metabolism, stress response, and apoptosis (Bhaskaran and Mohan 2014; Peng and Croce 2016). The discovery of miRNA marked a breakthrough in molecular biology, earning Victor Ambros and Gary Ruvkun the 2024 Nobel Prize in Physiology or Medicine. This milestone fundamentally transformed our understanding of gene regulation and its profound influence on cellular functions and development (Calin et al. 2024). miRNAs regulate gene expression by binding complementarily to the 3′-UTR of target mRNAs, leading to translational repression or degradation. They are transcribed by RNA polymerase II as primary miRNAs (pri-miRNAs), processed by Drosha-DGCR8 into precursor miRNAs (pre-miRNAs), exported to the cytoplasm via Exportin-5, and further cleaved by Dicer into mature duplexes with 2-nt 3′ overhangs. One strand (guide strand also called as antisense strand) is loaded into AGO2, forming the RNA-induced silencing complex (RISC) with cofactors like GW182, while the passenger strand (also known as sense strand) is degraded. In the canonical pathway, the guide strand binds via its 6-nt seed region at the 5′ end, whereas in target-directed miRNA degradation (TDMD), extended 3′ complementarity destabilizes AGO2 binding, promoting miRNA decay (Seyhan 2024).

MiRNA either represses the gene expression temporarily by binding or permanently by cleaving. Several complex genetic networks and cellular signaling cascades are regulated by miRNA expression (Peng and Croce 2016). Any physiological changes in miRNAs expression play a central role in disease diagnosis and can be served as a biomarker for diseases like cancer, cardiovascular disease, autoimmune diseases, neurological disorders, and hepatitis (Kurakula et al. 2023). Besides, miRNAs serving as promising biomarkers for disease diagnosis, these molecules can also be targeted as interesting therapeutic interventions by regulating protein expression linked with disease severity. MiRNA can serve as a therapeutic intervention either in the form of a miRNA mimic to supplement the downregulated miRNAs or as a targeting agent in the form of an anti-miRNA (antagomir) to target the disease-linked upregulated miRNA (Diener et al. 2022).

In cancer cells, the expression of key oncogenes is negatively regulated by tumor suppressor miRNAs (TS miRNA) (Cannataro and Cione 2023). The downregulation or loss of these TS miRNAs contributes to tumor initiation and progression. For example, a potent TS miRNA-34 has been reported to be frequently downregulated in many disease types, which directly target genes involved in critical pathways implicated in cancer progression, including the MAPK, Notch, Wnt, PI3K/AKT, p53, and Ras pathways (Pan et al. 2023). Moreover, miR-34 s targets are critical for cancer migration, invasion, and EMT (Zhang et al. 2019). MRX34, a miRNA mimic, restores the tumor suppressor pathway to normal, inducing apoptosis in tumor cells in vitro and in mouse models of cancer (Zhang et al. 2019). In certain cases, persistent upregulation of oncogenic miRNAs termed oncomiRs leads to the direct suppression of critical tumor suppressor genes such as *TP53*, *RB1*, and *BRCA1*. This aberrant overexpression contributes to tumorigenesis. Anti-miRNA therapies have progressed to early-stage clinical evaluation, with several candidates entering Phase 1 trials (NCT00688012, NCT00979927, and NCT01200420), as summarized in Table 17.1. These trials primarily target miR-122, a liver-specific miRNA implicated in hepatocellular carcinoma (HCC) and hepatitis C virus (HCV) replication (Burnett and Rossi 2012). Beyond miR-122, other anti-miRNA candidates are also under clinical investigation. For example, MRG-106 (cobomarsen), an inhibitor of miR-155, has indicated encouraging results in cutaneous T-cell lymphoma (CTCL) and is being assessed under clinical trial NCT02580552. Similarly, MRG-110, targeting miR-92, has been evaluated for its role in tissue repair and angiogenesis (NCT03603431) (Seto et al. 2018). These trials highlight the translational potential of miRNA-based therapeutic strategies and emphasize the growing clinical interest in targeting dysregulated miRNAs as a therapeutic modality in cancer and other diseases.

Though clinical progress has been achieved, several key challenges continue to hinder the widespread application of miRNA-based therapies. One of the foremost obstacles is immune activation. The introduction of synthetic RNA molecules can stimulate innate immune responses, particularly through Toll-like receptors (TLRs), leading to unintended inflammation or toxicity (Uehata and Takeuchi 2020). This immune stimulation not only limits the therapeutic window but also complicates dose optimization and delivery strategies. Additionally, off-target effects remain the major concern. Other challenges include target identification and screening, delivery insufficiencies, regulatory and developmental constraints, and economic and competitive challenges. This book chapter will discuss the trials of miRNA-based drugs and its translational and clinical challenges.

## 17.2 The Promise of miRNA Therapeutics and Its Clinical Progress

Although there are numerous drugs available for cancer but not even one can completely eradicate the cancer from the patient. The advances in molecular biology have paved the way to understand the root problems and to develop a method to treat. In the recent arena of RNA therapeutics, miRNA-based therapeutics have taken the high-value approach in disease management due to remarkable miRNA biology linked to disease control.

The pivotal discovery in 2002 of the association between aberrant miRNA expression and human disease marked the commencement of a new era in therapeutic research, introducing novel strategies aimed at correcting dysregulated gene

**Table 17.1** Clinical research progress of miRNA-related drugs in treatment of cancer

| Drug name for miRNA | Target miRNA | Disease/condition | Treatment modalities | Clinical Trail nos. | Phase | status | References |
|---|---|---|---|---|---|---|---|
| *miRNA mimics* | | | | | | | |
| MRX34 | miRNA-34a | Solid tumors (e.g., hepatocellular carcinoma, melanoma, SCLC, NSCLC, lymphoma, multiple myeloma, renal cell carcinoma) | Intravenous/liposomal delivery | NCT02862145 | Phase-I<br>Phase II | Withdrawn (5 immune related serious adverse events in Phase I) | Hong et al. (2020) |
| MRX34 | miRNA-34a | Primary liver cancer<br>SCLC<br>Lymphoma<br>Melanoma<br>Multiple myeloma<br>Renal cell carcinoma<br>NSCLC | Intravenously/vehicle transfer (liposomal) | NCT01829971 | Phase I | Terminated (5 immune related serious adverse events) | |
| MesomiR-1 | miRNA-16 | Malignant pleural mesothelioma and non-small cell lung cancer | Intravenously/vehicle transfer of nonliving bacterial nanocells (EDVsor TargomiRs) | NCT02369198 | Phase I | Completed | van Zandwijk et al. (2017) |
| INT-1B3 | miR-193a-3p | Advanced solid tumors | | NCT04675996 | PhaseI | | Duurland et al. (2024) |
| *Anti-miRNA* | | | | | | | |
| miR-10b | miR-10b | Astrocytoma | | NCT01849952 | Observational | | |
| Cobomarsen/ MRG 106 | miR-155 | Mycosis fungoides (MF)<br>Cutaneous T-cell Lymphoma (CTCL)<br>Chronic Lymphocytic Leukemia (CLL) | Intravenously/ chemical modification (LNA) | NCT02580552,<br>NCT03713320,<br>NCT03837457 | Phase I<br>Phase II<br>Phase II | Completed<br>Terminated<br>Terminated | Anastasiadou et al. (2021) |

| | | | | | | | |
|---|---|---|---|---|---|---|---|
| | | Diffuse large B-cell Lymphoma (DLBCL), ABC subtypeAdult T-cell leukemia/lymphoma (ATLL) | | | | | |
| LNA-i-miR-221 | miR-221 | Advanced cancer | Intravenously/ chemical modification (LNA) | NCT04811898 | Phase I | Completed | Tassone et al. (2023) |

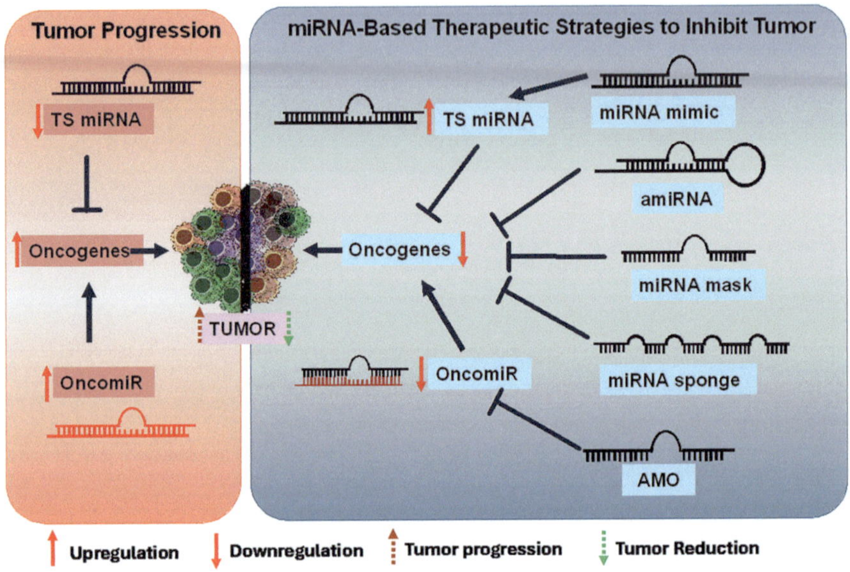


**Fig. 17.1** Pictorial representation of tumor progression and miRNA-based therapeutic strategies to inhibit tumor

expression through miRNA modulation. Since then, significant advancements in chemistry, molecular biology, and medical science have spurred growing interest and substantial investment in the development of miRNA-based therapeutics. These undertakings have culminated in a variety of miRNA-targeted approaches entering preclinical and clinical development. Table 17.1 highlights recent miRNA-based therapies that have progressed into clinical trials in the treatment of cancer, reflecting the translation of this research into practical implementations. miRNA-based therapeutics can be largely classified into two main strategies (Fig. 17.1). The first involves *miRNA mimics*, which are synthetic oligonucleotides designed to recover the activity of downregulated TS miRNA. These mimics compensate for the loss of endogenous miRNA function and help re-establish proper regulation of gene networks involved in cell growth and apoptosis. The second approach utilizes *anti-miRNAs*, which are engineered to inhibit the activity of overexpressed oncogenic miRNAs (oncomiRs). By silencing these miRNAs, anti-miRNAs can relieve the repression of tumor suppressor genes, thereby inhibiting cancer progression (Diener et al. 2022). In this section, we will explore both therapeutic strategies in greater detail, highlighting their mechanisms of action, design considerations, and clinical progress in the context of cancer treatment.

### 17.2.1 miRNA Mimics

Overexpression of oncogenes exerts a major influence on cellular behavior promoting tumorigenesis, angiogenesis, and metastasis. However, these oncogenes expressions are controlled by TS miRNA. In cancer, this critical role-playing TS miRNAs are frequently downregulated which results in dysregulation or overexpression of oncogenes and thereby results in cancer progression.

The miR-34 family, a well-characterized TS miRNA group, is frequently downregulated across various cancer types such as HCC, melanoma, prostate, colorectal, breast and lung carcinoma (Zhang et al. 2019). These miRNAs play crucial roles in controlling cell cycle progression, apoptosis, and cellular metabolism by targeting multiple oncogenic mRNAs, thereby exerting broad anti-tumor effects. Several studies have demonstrated that programmed cell death ligand-1 (PDL-1) allows cancer cell to evade immune system and is associated with cancer aggressiveness (Wang et al. 2015). In many cancers such as non-small cell lung cancer (NSCLC), AML, lung and breast cancer, *PDL1* is directly regulated by miRNA-34. MiRNA-34 targets multiple genes in breast cancer such as *BCL-2, CCND1, NOTCH1, PRKD1, E2F1, E2F3,* and *Axl*, impacting its proliferation, migration and drug resistance, also helps in controlling EMT (Imani et al. 2017). It is also observed to influence pathways like Wnt, Notch, and NF–κB, impacting processes like cell proliferation, migration, and invasion (Fu et al. 2023). Discussing the wide range of oncogenic targets, miRNA-34 serves as a promising therapeutic target for treatment of cancer. Due to its therapeutic capability, miRNA-34 has been developed into the first miRNA-based drug, MRX34. MRX34 is a synthetic double stranded miRNA mimic, which is 23-nt long and encapsulated in liposomal nanoparticle developed by miRNA Therapeutics Inc. to mimic the activity of TS-miRNA-34a for the treatment of advanced solid tumors comprising melanoma, renal carcinoma, HCC and NSCLC. Researchers conducted a first-in-human (Phase 1) trial in patients with advanced solid tumor, to determine a safe and effective dose for future studies (recommended Phase 2 dose) (NCT01829971 & NCT02862145) as mentioned in Table 17.1 (Hong et al. 2020; Martino et al. 2025).

Another miRNA mimic that has advanced to Phase 1 clinical trials is MesomiR-1. A TS miRNA15/16 family has been found to be downregulated in malignant pleural mesothelioma (MPM), lung cancer, prostate, chronic lymphocytic leukemia (CLL); hence, it serves as a potential therapeutic target. Researchers have developed TargomiRs which are engineered, non-living bacterial minicells designed to encapsulate miRNA mimics and deliver them specifically to cancer cells through targeted delivery mechanisms. These nanocells, known as EnGeneIC Dream Vectors (EDV™), are coated with EGFR-specific antibodies to facilitate targeted uptake by tumor cells. In the clinical trial (NCT02369198), TargomiRs were loaded with miR-15/16 mimics aimed to recover the activity of these downregulated TS miRNAs and suppress the expression of oncogene. This therapeutic strategy was evaluated for its potential to treat malignant MPM and NSCLC (van Zandwijk et al. 2017).

Similarly, miR-193a-3p is another TS miRNA, downregulated in various cancers. Researchers formulated a miRNA-based drug named INT-1B3, an innovative lipid

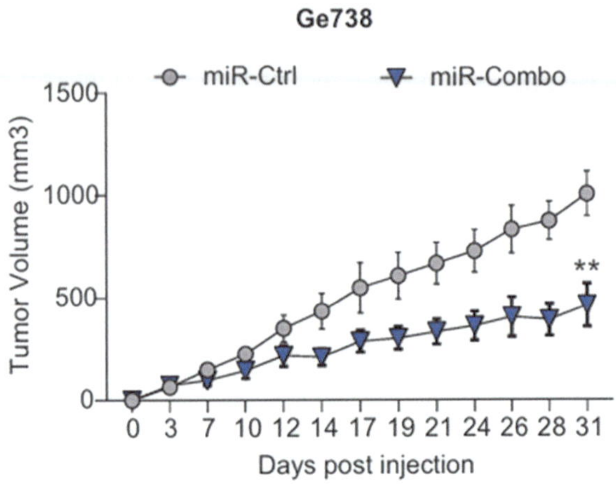


**Fig. 17.2** Repeated administration of miR-Combo suppresses tumor growth in vivo. Nude mice bearing Ge738 xenografts ($n = 5$ per group) were treated with repeated injections of a combinatorial miRNA formulation (miR-17-3p, miR-222, and miR-340; termed miR-Combo). (Adapted from Bassot et al. (2023). Copyright 2023 Springer Nature)

nanoparticle carrying miRNA mimic (1B3), a 22-nt long chemically modified double-stranded miR-193a-3p mimic. INT-1B3 showed notable inhibition in tumor progression modulating the immune tumor microenvironment (TME) from "cold" tumors into "hot" tumors. This was accomplished via upregulation of cytokines, reduction of immunosuppressive cells/Treg, and triggering cytotoxic CD8+ T cell-mediated long-term memory immune protection in resistance to re-challenge with tumor cells (NCT04675996) (Martino et al. 2025). Furthermore, a group of researchers identified miRNA multi-targeting strategy in the treatment of glioblastoma (GBM). In an integrative and comprehensive analysis of the TCGA database and the transcriptome of GBM biopsies, three clinically relevant miRNAs for GBM have been found (miR-17-3p, miR-222, and miR-340). These three miRNAs in combination efficiently inhibited several biological processes in patient-derived GBM cells (PDC). Figure 17.2 describes delayed tumor upon treatment with miR-combo (miR-17-3p, miR-222, and miR-34) in PDC model, Ge738 (Bassot et al. 2023). The resulting preclinical data suggests strong therapeutic potential across multiple cancer types, with a reduced likelihood of developing drug resistance and delayed tumor growth.

In addition to currently studying therapeutic candidates, several other TS microRNAs have been identified across various cancer types. For instance, miR-221/222, miR-99a, and miR-100 are implicated in triple-negative breast cancer (TNBC) (Li et al. 2019; Long et al. 2019; Xie et al. 2021), while miR-122, miR-26a, miR-195, miR-214, miR-211, and miR-142 have been associated with HCC (Ha et al. 2019; Zhao et al. 2019; Yan et al. 2017; Yang et al. 2015; Jiang et al. 2015; Chai et al. 2014). In liver cancer, frequent downregulation of miR-122, miR-146a, miR-26a, miR-191, miR-150, miR-194, miR-152, miR-205, and miR-223 has also been observed. Notably, certain miRNAs like the *let-7* family are consistently downregulated across multiple cancer types (Chirshev et al. 2019). This loss of *let-7* expression contributes to oncogenesis by permitting the overexpression of essential oncogenes such as *KRAS*, *MYC*, *LIN28B*, and *HMGA2*, which promote tumor growth and metastasis. Although many miRNA-based therapeutics remain in the preclinical stage, they show promising results. For example, the combined restoration of miR-34a and let-7b in neuroblastoma preclinical models significantly reduced tumor size compared to untreated controls (Di Paolo et al. 2020). These findings strongly support the therapeutic promise of miRNA mimics in restoring tumor-suppressor function and repressing oncogenic pathways. miRNA mimic-based therapies are steadily gaining recognition as a precise and effective strategy for cancer treatment.

### 17.2.2 Anti-miRNAs

Cancer severity is influenced not only by the downregulation of TS miRNA but also by the aberrant upregulation of certain oncomiR, which can promote tumor progression and aggressiveness. OncomiR is usually upregulated in cancer and mostly targets tumor suppressor genes that result in tumorigenesis and its progression. Hence, there is a need to silence upregulated oncomiR, to bring down the oncomiR level to normal. A synthesized miRNA usually single-stranded which shares complementarity with the endogenous oncomiR is referred to as "anti-miRNA." There are different therapeutic strategies to target oncomiRs, including anti-miRNA oligonucleotides (AMO), miRNA sponges, miRNA masks, and amiRNA.

#### 17.2.2.1 Anti-miRNA Oligonucleotide (AMO)

A chemically modified single-stranded oligonucleotide of 17–22 nt long, anti-miRNA is designed to target miRNA based on complementarity, specifically on its seed region hence bind to the mature miRNA and thereby preventing its target gene expression. Further leads in the expression of target gene, especially tumor suppressor genes resulting in inhibition of tumor growth (Diener et al. 2022). In therapeutics, AMOs are selected and designed with high efficacy to silence the target miRNA with enhanced binding strength. The therapeutic robustness of the AMOs is intrinsically linked with their potency to efficiently suppress the expression of the target miRNA. while the binding affinity is associated to the stability of the double-stranded molecule upon AMO binding to its target and thus related to the melting temperature ($T_m$) of the oligonucleotide. A higher $T_m$ value represents an improved binding affinity and a stronger duplex, because more energy is required to destabilize the connection between both molecules (Lima et al. 2018). Overall, it is desirable for AMOs to possess enhanced extracellular and intracellular stability, rendering them resistant to both exonuclease and endonuclease degradation. They should also exhibit improved cellular uptake rates to be efficiently delivered to cells and tissues, and have no toxicity or other side effects, such as the knockdown of non-target molecules important to maintain biological function or the stimulation of undesired auto-immune responses (Lima et al. 2018).

In the evolution of AMOs, the unmodified AMOs or any other nucleic acid-based therapy present several disadvantages that preclude their applicability as antisense technology. Due to advances in nucleic acid chemistry, several modifications were

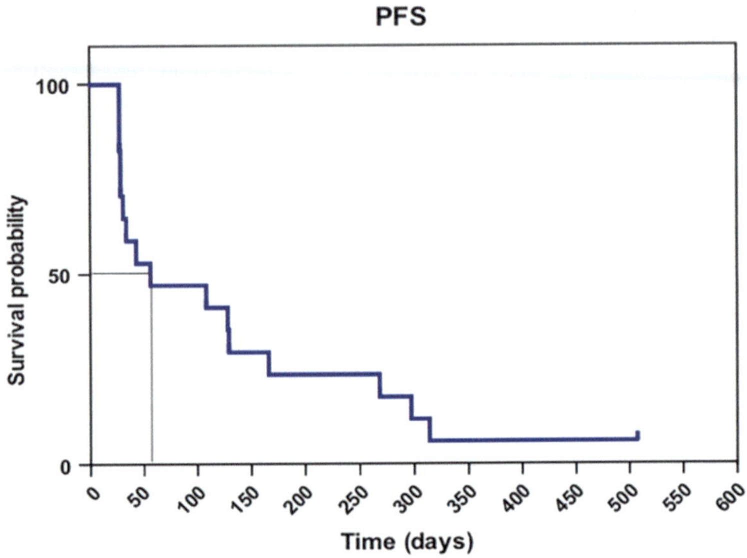


**Fig. 17.3** Kaplan–Meier curve of progression-free survival. (Adapted from Tassone et al. (2023). Copyright 2023 Springer Nature)

introduced to improve the structural properties of the AMOs. Chemical modifications are primarily introduced into the sugar ring—most notably at the C2′ position—and/or into the oligonucleotide backbone. Backbone modifications may involve altering the internucleotide linkages or replacing the entire backbone with structures analogous to DNA or RNA. These internucleotide changes are often combined with sugar modifications to enhance potency and improve resistance to nucleases. Among these, the most widely used modifications in miRNA research target the 2′ carbon of the ribose. It is the case of 2′-OMe, 2′-MOE, and 2′-F alterations, which are all reported to confer AMOs increased binding affinity (ordered by increased potency: 2′-OMe ≅ 2′-MOE < 2′-F) (Swayze et al. 2007). The use of locked nucleic acids (LNA) has become popular modified AMO designs, due to their enhanced structural stability, high specificity and reduced toxicity. For example, preclinical studies, mice were treated with single intravenous dose of LNA-anti-miRNA targeting miR-221, a liver specific miRNA with suggested role in cholesterol, fatty acid and lipid metabolism. It has been reported that miRNA-221 facilitates the viral genome replication of hepatitis C virus in the host cell. This mice study resulted in low levels of miRNA-221 in liver with no hepatotoxicity suggesting that LNA-anti-miRNA is a valuable tool with enhanced potential in miRNA silencing in vivo and increasing overall survival rates, as shown in Fig. 17.3. Due to its potential therapeutic approach LNA-anti-miRNA targeting miRNA-221 have been explored in clinical trials (NCT04811898) (Tassone et al. 2023).

Furthermore, LNA-anti-miRNA in trials includes for the treatment of mycosis fungoides (MF), a most common subtype of CTCL (NCT02580552, NCT03713320, NCT03837457) (Seyhan 2024; Anastasiadou et al. 2021). miRNA-155 is linked to poor prognosis in CTCL. To counter this, researchers developed *cobomarsen*, an LNA-based anti-miRNA designed to inhibit miR-155. Their findings demonstrated the clinical potential of this approach, showing that cobomarsen effectively suppresses miR-155 activity. Notably, pharmacodynamic gene signatures that were elevated and activated in clinical MF biopsies particularly those related to T-cell activation and survival signaling were significantly downregulated and inactivated in both MF and HTLV-1+ CTCL cells following cobomarsen treatment. Conversely, genes involved in checkpoint regulation, which were suppressed and inactive in MF biopsies, were upregulated and activated in treated cell lines. These results indicate that multiple biological pathways disrupted in MF due to elevated miR-155 expression can be restored through targeted inhibition with cobomarsen. Therefore, a microRNA-based therapy, such as cobomarsen, might restore homeostasis to multiple dysregulated gene pathways and networks in MF (Anastasiadou et al. 2021). Therefore, anti-miRNA therapies have demonstrated promising preclinical and early clinical outcomes by selectively silencing oncogenic miRNAs. As more candidates advance through clinical trials, these agents hold strong potential to emerge as targeted and durable treatments for cancer and other miRNA-driven diseases.

#### 17.2.2.2 miRNA Sponges

Introduced in 2007, miRNA "sponge" technology was developed to achieve effective loss-of-function of specific miRNAs in both cell-based models and transgenic organisms. These sponges are artificially designed transcripts engineered with multiple tandem, high-affinity binding sites for the target miRNA, thereby sequestering it and preventing its interaction with endogenous mRNA targets. These microRNA-binding molecules are competitive regulators that sequester specific microRNAs, thus preventing miRNA/mRNA interaction (Ebert and Sharp 2010). MicroRNA sponges are commonly employed in loss-of-function studies to achieve robust and sustained inhibition of specific miRNAs. For instance, to generate a transgenic mouse model exhibiting the loss-of-function phenotype of miR-483-5-5p, researchers designed a miR-483-5p sponge containing six tandem repeats of an imperfectly complementary sequence, enabling effective in vivo sequestration of the target miRNA (Zhang et al. 2018).

Beyond research applications, miRNA sponges are also being investigated for therapeutic purposes. For example, a multifunctional artificial circular RNA (circRNA) was engineered to act simultaneously as a molecular sponge for miR-25 and as a vehicle for expressing an anti-PD-1 single-chain variable fragment (scFv). This dual-function circRNA inhibited tumor growth by suppressing the oncogenic activity of miR-25 in HCC cells, while concurrently blocking PD-1/PD-L1 signaling within the tumor microenvironment, thereby producing a synergistic antitumor effect (Lai et al. 2025). These findings facilitate compelling experimental evidence that synthetic circRNAs acting as miRNA sponges can serve as versatile platforms for anticancer therapy and hold promise as next-generation molecular therapeutic strategies.

#### 17.2.2.3 miRNA Masking

The miRNA-masking antisense oligonucleotide (miR-mask) approach is another AMO-based strategy designed to inhibit miRNA function. Unlike conventional AMOs that directly bind and inactivate the target miRNA, a miR-mask is a single-stranded, chemically modified oligoribonucleotide—commonly 2′-O-methyl–modified—complementary to the specific miRNA binding site within the 3′ untranslated region (UTR) of a target mRNA. By binding perfectly on this site, the miR-mask physically blocks the miRNA from interacting with the mRNA, thereby preventing repression and restoring target gene expression. The anti-miRNA action of an miR-mask is gene-specific because it is designed to be fully complementary to the target mRNA sequence of a miRNA (Wang 2011). The miR-mask strategy serves as a complementary tool to the AMO approach. While AMOs are essential for investigating the broad, global functions of a given miRNA across its multiple targets, miR-masks are particularly suited for dissecting the precise regulatory effect of that miRNA on a specific target gene. miRNA masks are oligonucleotides designed to bind specific miRNA recognition sites on target mRNAs, thereby preventing miRNA binding without altering overall miRNA levels. For example, in zebrafish, target protector morpholinos (a type of miRNA mask) were used to block miR-430 binding to either *squint* or *lefty* mRNAs, revealing how miR-430 balances Nodal signaling by fine-tuning agonist and antagonist expression (Choi et al. 2007).

#### 17.2.2.4 amiRNAs

Artificial microRNAs (amiRNAs) represent a next-generation approach to therapeutic miRNA design. These constructs combine the target specificity of siRNAs with the structural features of primary miRNA transcripts, allowing them to harness the cell's endogenous miRNA processing machinery for efficient maturation and function. By integrating an siRNA sequence within a pri-miRNA scaffold, amiRNAs achieve precise gene silencing while maintaining physiological processing pathways (Diener et al. 2022). A notable example is AMT-130, an amiRNA incorporating a siRNA sequence against the Huntingtin gene within a pri-miR-451 backbone, which is currently being evaluated in a clinical trial for Huntington's disease (NCT04120493) (Kotowska-Zimmer et al. 2022). amiRNAs have also shown potential in cancer therapy. For instance, in pancreatic ductal adenocarcinoma (PDAC), a triple amiRNA construct targeting *VEGFR1*, *VEGFR2*, and *VEGFR3* not only reduced proliferation, migration, and invasion of pancreatic cancer cells in vitro but also significantly suppressed tumor growth and enhanced chemotherapy efficacy in mouse xenograft models. Mechanistically, this approach inhibited epithelial–mesenchymal transition (EMT), highlighting amiRNAs as versatile tools for both prognostic and therapeutic applications in cancer (Huang et al. 2017).

Together, anti-miRNA strategies—including AMOs, sponges, miRNA masks, and amiRNAs—provide diverse platforms to selectively suppress oncogenic miRNAs and re-establish balanced gene regulatory networks. Each platform provides unique advantages: AMOs enable potent and chemically optimized silencing; sponges allow long-term inhibition of miRNA families; miRNA masks provide gene-specific derepression; and amiRNAs combine siRNA-like precision with endogenous processing for efficient and stable targeting. Preclinical and early clinical studies, such as LNA-anti-miR-221 for liver diseases, cobomarsen targeting miR-155 in cutaneous T-cell lymphoma, and amiRNA-based therapies for PDAC and Huntington's disease, underscore their translational promise. However, challenges such as delivery, off-target effects, and immune activation remain key hurdles to clinical success. With continued advances in chemical modifications, delivery systems, and biomarker-driven patient selection, anti-miRNA therapeutics are poised to evolve into a powerful class of targeted interventions for cancer and other miRNA-driven diseases.

## 17.3 Core Challenges in Therapeutic Development

miRNAs hold significant promise as therapeutic agents in cancer, functioning either as miRNA mimics to restore tumor-suppressive activity or as miRNA inhibitors to silence oncogenic miRNAs. Over the past decade, miRNA-based therapeutics have made substantial progress, particularly at the preclinical level, where numerous studies have demonstrated their ability to modulate key oncogenic pathways and suppress tumor growth in various cancer models. Encouragingly, several candidates have progressed into early-phase clinical trials, reflecting the growing maturity of this field. However, despite these advances, no miRNA-based therapy has yet received regulatory approval, underscoring the complexity of translating these molecules from bench to bedside.

A major hurdle in developing effective miRNA therapeutics lies in identifying the most suitable miRNA candidates or target networks for each cancer type, given the multifaceted roles of miRNAs and their context-dependent effects. Furthermore, designing efficient and safe delivery systems remains a critical challenge. An ideal delivery platform must protect the miRNA cargo from degradation, achieve tissue- or tumor-specific targeting, and minimize systemic toxicities and immune responses. Additionally, off-target effects due to the promiscuous binding nature of miRNAs pose significant safety concerns, necessitating advanced strategies to enhance specificity and reduce unintended gene modulation. This section will explore these challenges in detail, highlighting current strategies to overcome them and discussing how ongoing innovations in delivery technologies, target validation, and molecular engineering are shaping the future of miRNA-based therapeutics for cancer.

### 17.3.1 Target Identification and Screening

Accurate identification of miRNA targets is a critical step in understanding their biological functions and developing miRNA-based therapeutics. However, this process remains a significant challenge due to the complex nature of miRNA–mRNA interactions and the limitations of both computational and experimental approaches. Unlike the one-to-one relationships often observed in classical molecular interactions, miRNA regulation is highly complex and context dependent. A single miRNA can bind and regulate dozens to hundreds of different genes, while a single gene can be modulated by multiple miRNAs, often acting in combinations. This creates a multilayered regulatory network that makes it difficult to pinpoint which interactions are truly functionally relevant in each disease setting, particularly in cancers characterized by molecular heterogeneity.

From an experimental perspective, several barriers complicate the process of miRNA target identification. Many commonly used approaches, such as pull-down assays or high-throughput sequencing-based techniques, require large quantities of cells or tissue, which may not always be feasible when working with rare cell populations or clinical samples (Li et al. 2020; Maji et al. 2025). Furthermore, distinguishing direct targets from indirect downstream effects presents a major challenge. Overexpression or inhibition of miRNA can cause widespread changes in gene expression, but not all of these changes reflect direct binding interactions. Adding to this complexity, miRNA–mRNA interactions are often cell- and tissue-specific, meaning that a target validated in one biological context may not hold functional significance in another (Afonso-Grunz and Muller 2015; Schafer and Ciaudo 2020). Even advanced techniques often fail to map miRNA binding sites with nucleotide-level precision, leaving uncertainties about which predicted sites are truly functional (Schafer and Ciaudo 2020). Moreover, experimental artifacts such as non-physiological binding caused by miRNA overexpression can further confound interpretation.

Computational approaches have provided valuable insights into miRNA targeting by predicting interactions based on sequence complementarity, seed region matching, and evolutionary conservation. However, these methods are far from perfect. High false-positive and false-negative rates are a persistent issue, as algorithms may predict targets that are not biologically relevant or fail to identify genuine ones. Most prediction models also fail to account for tissue-specific gene expression or context-dependent factors such as RNA secondary structure, RNA-binding proteins, and the presence of competing endogenous RNAs (ceRNAs), all of which can profoundly influence miRNA activity. As a result, computational predictions alone are insufficient for establishing functional relevance (Akbari Moqadam et al. 2013).

To overcome these challenges, integrated approaches are increasingly being adopted. Combining computational prediction tools with high-throughput experimental techniques such as crosslinking immunoprecipitation sequencing (CLIP-seq) or RNA sequencing following miRNA modulation can provide a more comprehensive view of potential targets (Li et al. 2020). These findings then require targeted functional validation using assays such as luciferase reporter systems, qRT–PCR, and Western blotting to confirm direct regulation. Despite these advances, miRNA target identification remains a complex and evolving field, and careful experimental design is essential to minimize artifacts and ensure biological relevance.

Ultimately, accurate identification and screening of miRNA targets is not only fundamental to understanding the molecular underpinnings of cancer but also critical for advancing miRNA-based therapeutics (Schafer and Ciaudo 2020). Selecting the most functionally significant targets helps avoid off-target effects, improves therapeutic specificity, and increases the likelihood of clinical success. As new computational models and experimental tools continue to evolve, more precise mapping of miRNA–mRNA interactions will likely accelerate the translation of miRNA research from bench to bedside (Akbari Moqadam et al. 2013).

### 17.3.2 Delivery Barriers

The successful delivery of miRNAs for therapeutic applications remains one of the most significant barriers to their clinical translation, particularly in oncology. Unlike conventional small-molecule drugs, miRNAs are inherently unstable, prone to degradation, and require precise delivery to specific cell populations to achieve therapeutic efficacy while minimizing off-target effects. These challenges arise from the need to balance specificity, stability, safety, and efficacy within the complex biological environment of tumors (Fig. 17.3). Key delivery challenges can be broadly categorized into: (i) specificity of delivery, (ii) therapeutic window and toxicity, (iii) endosomal escape, (iv) delivery strategies, and (v) TME associated barriers.

#### 17.3.2.1 Specificity of Delivery

Effective miRNA therapeutics must selectively target tumor cells without affecting healthy tissues (Menon et al. 2022). However, the heterogeneity of tumor cells arising from genetic, epigenetic, and phenotypic diversity makes precise targeting difficult (Zhu et al. 2021). Achieving specificity often requires functionalization of delivery vehicles with ligands or antibodies that exploit ligand–receptor interactions unique to tumor cells. Nanocarriers decorated with molecules such as folate, transferrin, or peptides can improve uptake by cancer cells, yet optimizing their selectivity remains a challenge (Akhter et al. 2021). Furthermore, off-target distribution can lead to undesired effects, reducing both safety and therapeutic efficiency.

#### 17.3.2.2 Therapeutic Window and Toxicity

An ideal delivery system must exhibit a wide therapeutic window, wherein miRNA doses are sufficient to achieve therapeutic effects without inducing toxicity. Chemical modifications to miRNAs—such as 2'-O-methylation, LNAs, or phosphorothioate linkages—are often employed to enhance nuclease resistance and reduce immune activation (Swayze et al. 2007). However, striking a balance between enhanced stability, reduced immunogenicity, and preserved biological activity remains difficult (Kuroda and Tsumoto 2020). Additionally, the delivery vehicle itself (e.g., cationic lipids or polymers) can introduce cytotoxic effects or trigger immune responses, further complicating clinical translation (Lv et al. 2006).

#### 17.3.2.3 Endosomal Escape

After internalization by target cells, miRNA-loaded carriers are often trapped within endosomes, where degradation can occur, drastically reducing the amount of functional miRNA reaching the cytoplasm (Kurakula et al. 2023). Achieving efficient endosomal escape is therefore a critical hurdle. Strategies to address this include designing pH-sensitive carriers that destabilize in acidic endosomal environments, incorporating fusogenic lipids or polymers, and using peptides or nanoparticles that disrupt endosomal membranes. Despite progress, ensuring safe and reproducible endosomal escape without inducing cytotoxicity is still a major challenge (Qiu et al. 2023).

#### 17.3.2.4 Delivery Strategies and Clinical-Stage Platforms

Multiple delivery platforms are being developed to address the challenges of miRNA therapeutics, each offering unique advantages for stability, targeting, and efficacy. *Lipid-based nanoparticles (LNPs),* which have already demonstrated clinical success with siRNA drugs such as Patisiran (Onpattro®) the first FDA-approved RNAi therapeutic are now being adapted for miRNAs, as seen with MRX34, a liposomal miR-34a mimic that advanced to Phase I clinical trials for liver cancer and other solid tumors (Hong et al. 2020). *Polymer-based carriers*, including biodegradable systems such as poly (lactic-co-glycolic acid) (PLGA) and polyethyleneimine (PEI) derivatives, provide tunable release profiles and are widely explored in preclinical studies (Patil and Panyam 2009). *Exosome-mediated* delivery leverages naturally secreted vesicles to enhance biocompatibility and facilitate efficient cellular uptake, with approaches like Exo-miR-21 inhibitors showing promise in early research (Ellipilli et al. 2023). In addition, *nanomedicine*-based systems such as hyaluronic acid nanoparticles and pH-sensitive micelles enable tumor-selective targeting and have yielded encouraging preclinical results. For liver-specific applications, GalNAc-conjugates, already approved for siRNA therapeutics like Givlaari®, offer a blueprint for developing hepatocyte-targeted miRNA therapies (Kurakula et al. 2023). Although challenges remain, including immune-related toxicities observed with MRX34, these advances highlight the potential of rationally engineered delivery systems to drive miRNA-based therapeutics toward clinical translation.

#### 17.3.2.5 Tumor Microenvironment (TME) Challenges

The TME presents substantial obstacles to effective miRNA delivery due to its complex composition and pathological characteristics. The TME is comprised of cancer cells, stromal cells, immune cells, extracellular matrix components, and abnormal vasculature, the TME exhibits profound biochemical and physical alterations compared to normal tissues, including acidic pH, elevated reactive oxygen species (ROS), dysregulated enzyme activities, hypoxia, and imbalanced ATP and glutathione (GSH) levels (de Visser and Joyce 2023). Conventional delivery systems, which rely on passive drug release, often fail to adapt to these conditions, leading to suboptimal targeting and reduced therapeutic efficacy. To overcome these barriers, advanced stimuli-responsive nanocarriers are being engineered to exploit TME-specific triggers, enabling active targeting and controlled release of miRNAs (Adepu and Ramakrishna 2021). These smart carriers can undergo structural or surface modifications such as changes in size, shape, or charge upon exposure to pathological stimuli, thereby enhancing tumor accumulation and improving pharmacokinetic properties. Strategies include pH-sensitive systems for acidic environments, ROS-responsive platforms for oxidative regions, enzyme-activated carriers targeting tumor-associated proteases, ATP/GSH-sensitive nanoparticles exploiting redox gradients, hypoxia-responsive constructs, and dual or multi-responsive systems integrating multiple triggers for greater precision (Wang et al. 2022). Collectively, these innovations aim to maximize therapeutic efficiency while minimizing systemic toxicity, thereby rationally optimizing the anti-tumor potential of miRNA-based therapeutics.

While advances in delivery technologies have significantly improved the stability, targeting, and functional efficiency of miRNA therapeutics, challenges remain in translating these approaches to the clinic. Lessons learned from siRNA-based drugs demonstrate that careful optimization of formulation, dosing, and safety profiles can lead to regulatory success. For miRNAs, integrating smart nanocarriers with context-specific targeting strategies may provide the breakthrough needed to overcome current barriers. As delivery systems evolve, they will form the foundation for next-generation miRNA therapeutics capable of addressing the complexity of cancer (Fig. 17.4).

### 17.3.3 Immune Response, Off-target Effects, and Toxicity

miRNAs have emerged as a promising class of therapeutic agents capable of restoring disrupted cellular functions, particularly in malignancies such as cancer, by regulating multiple genes and pathways simultaneously. miRNAs offer a unique therapeutic advantage over conventional single-target approaches such as siRNA-based drug which binds only to its target gene. However, this broad activity is a double-edged sword: while beneficial for reprogramming dysregulated networks, it also increases the risk of off-target effects in non-target tissues and immune-related toxicities. Synthetic miRNA molecules are often recognized as foreign entities, activating innate immune responses and triggering inflammatory pathways that complicate treatment outcomes (Chandan et al. 2019). Consequently, mitigating immunogenicity, enhancing tissue-specific delivery, and optimizing dosing strategies are critical for realizing the clinical potential of miRNA therapeutics.

One of the most notable examples of miRNA-based drug-induced immune response is MRX34, a liposomal miR-34a mimic (NOV40) evaluated in a first-in-human Phase I trial for advanced solid tumors, including melanoma, NSCLC, HCC, and renal cell carcinoma (NCT01829971). MRX34 provided compelling proof-of-concept, demonstrating dose-dependent modulation of disease-associated target genes and meaningful clinical activity, with three patients achieving prolonged confirmed partial responses and 14 others maintaining stable disease for a median duration of 136 days. However, the trial was terminated prematurely due to severe

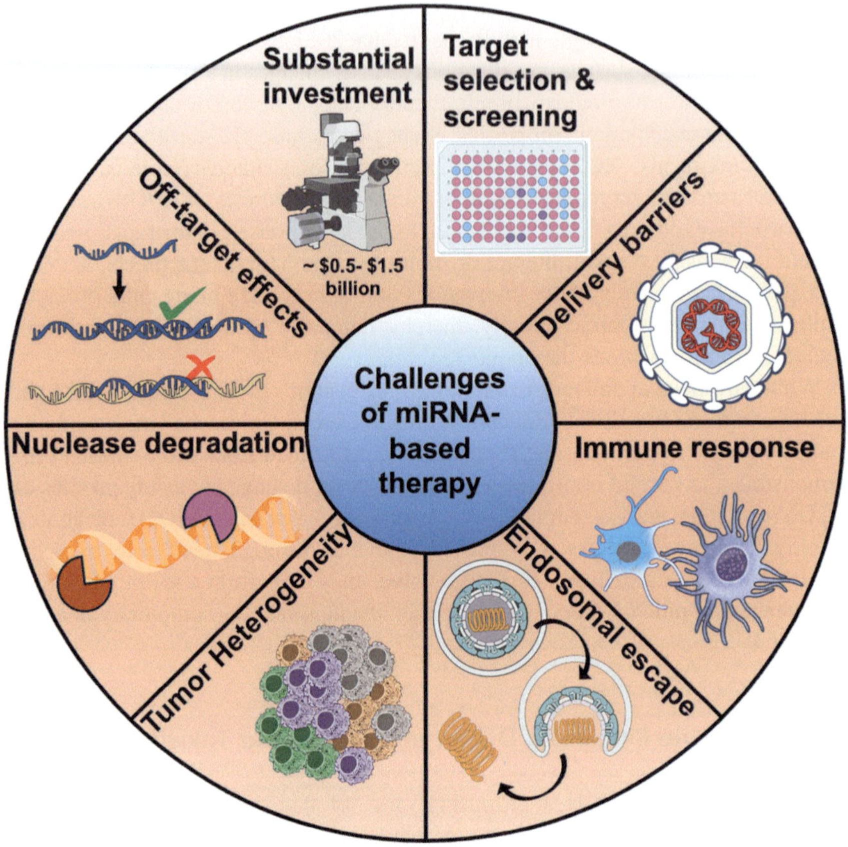


**Fig. 17.4** Challenges of miRNA-based therapeutics

immune-mediated adverse events, including four treatment-related deaths, underscoring the need for safer delivery strategies (Hong et al. 2020).

Other clinical-stage programs have faced similar challenges. TargomiRs, a miR-16 mimic delivered via bacterial-derived minicells, advanced to a Phase I trial for malignant pleural mesothelioma and NSCLC (NCT02369198) (van Zandwijk et al. 2017). While demonstrating a favorable safety profile compared to MRX34 and evidence of target engagement, therapeutic efficacy remained modest, highlighting the need for improved tumor-targeting strategies. Likewise, RG-101, an anti-miR-122 oligonucleotide conjugated to GalNAc for liver-specific delivery, showed potent antiviral effects in HCV infection and reached Phase II trials (NCT02559727). However, its development was halted due to cases of severe hyperbilirubinemia, reflecting the complexity of balancing efficacy and safety even with organ-targeted approaches (van der Ree et al. 2017).

These experiences collectively illustrate both the promise and complexity of miRNA-based therapeutics. While clinical trials confirm that miRNAs can effectively modulate disease-relevant targets and produce measurable therapeutic benefits, immune activation, off-target effects, and delivery limitations remain formidable barriers. Overcoming these challenges will require next-generation delivery systems that combine precise targeting, immune evasion, and context-specific release, thereby enabling safer and more effective translation of miRNA therapeutics into the clinic.

### 17.3.4 Regulatory and Developmental Constraints

The development and regulation of miRNA-based therapeutics are hindered by interconnected biological, technical, and regulatory challenges that complicate their path to clinical translation. Biologically, the pleiotropic nature of miRNAs where a single miRNA can regulate multiple mRNA targets offers therapeutic advantages but also increases the risk of toxicity, immune activation, and off-target effects in non-target tissues, often leading to discontinuation of otherwise promising candidates. These risks are amplified by difficulties in target identification, moderate efficacy of current inhibitors, and the context-dependent roles of miRNAs in different tissues and disease states. From a delivery standpoint, miRNAs' hydrophilic nature, negative charge, and susceptibility to nuclease degradation result in poor systemic stability and limited tumor cell uptake, while physical barriers such as disorganized vasculature, dense extracellular matrices, and fibrotic microenvironments in solid tumors further restrict therapeutic penetration (Brillante et al. 2024). Regulatory agencies, including the U.S. Food and Drug Administration (FDA), European Medicines Agency (EMA), and Indian Council of Medical Research (ICMR), require a comprehensive characterization of both the miRNA molecule and its delivery system as a combined product, alongside rigorous preclinical studies on biodistribution, immunogenicity, and long-term toxicity to capture these risks (Vervaeke et al. 2022). Demonstrating target engagement, mechanism of action, and biomarker-driven patient selection is critical for approval, particularly in oncology, where combination strategies require additional interaction studies. Ultimately, regulatory success for miRNA therapeutics will depend on integrating advanced delivery technologies with rational therapeutic design and aligning them with evolving regulatory frameworks to ensure safety, efficacy, and long-term clinical benefit.

### 17.3.5 Economic and Competitive Challenges

The advancement of miRNA-based therapeutics is constrained by substantial economic and competitive pressures that limit their seamless transition from bench to bedside. While miRNAs hold considerable promise for the treatment of cancer and other diseases, their development requires rigorous target validation, advanced synthetic design, precise chemical modifications, their characterization, and in vitro validation, which are the factors that collectively demand substantial financial investment. Furthermore, extensive funding demands preclinical validation, large-scale clinical trials, and strict regulatory compliance. A notable example is INT-1B3, a chemically modified double-stranded miR-193a-3p mimic, which advanced to a Phase I/Ib clinical trial (NCT04675996) following promising preclinical results. This first-in-human study employed a two-phase dose-escalation design to assess safety, pharmacokinetics, pharmacodynamics, and preliminary efficacy in patients with advanced solid tumors, with participants receiving 60-minute intravenous infusions twice weekly in 21-day cycles (Duurland et al. 2024). Despite its scientific rationale, the trial was prematurely terminated due to insufficient funding, and no clinical outcomes have been disclosed to date, underscoring the financial vulnerabilities that can impede the progress of innovative miRNA-based therapies (Schmidt 2017; Kumar et al. 2024).

Furthermore, the biotechnology landscape is highly competitive, compelling miRNA-based drugs to demonstrate clear therapeutic and commercial advantages over established modalities, such as small molecules, monoclonal antibodies, and immunotherapies. Addressing these challenges will be essential to position miRNA therapeutics as viable and competitive candidates in the evolving pharmaceutical market.

## 17.4 Conclusion and Future Perspectives

The trajectory of miRNA therapeutics reflects both the promise of precision medicine and the complexity of translating molecular insights into viable treatments. Over the past two decades, significant advances have been made in understanding the role of miRNAs in disease, establishing their potential as both biomarkers and therapeutic agents. Yet, the path to clinical success has been fraught with biological, technological, and economic hurdles that have slowed their integration into mainstream medicine. Despite these challenges, the clinical progress achieved thus far demonstrates the feasibility of miRNA-based interventions. For instance, MRX34, a liposomal mimic of the tumor-suppressor miR-34a, was the first miRNA mimic to enter clinical trials for advanced solid tumors (NCT01829971). Although the trial was discontinued due to immune-related toxicities, it provided valuable insights into delivery systems and immune modulation that are informing second-generation designs. Similarly, cobomarsen has shown promising activity in CTCL, where it not only reduced miR-155 levels but also reversed disease-associated gene expression patterns (NCT03713320). These examples illustrate that both miRNA mimics and anti-miRNAs can achieve meaningful biological effects in humans, even if refinement is required for safety and durability. Emerging platforms such as amiRNAs and miRNA sponges expand the therapeutic toolkit, offering more flexible and precise strategies for targeting oncogenic and tumor-suppressor pathways. Preclinical findings, such as the use of triple amiRNAs targeting VEGFR1/2/3 to inhibit tumor growth and enhance chemotherapy in pancreatic ductal adenocarcinoma, suggest that combinational designs may overcome redundancy in cancer signaling networks. Likewise, circular RNA-based miRNA sponges are being explored for their dual roles as inhibitors of oncogenic miRNAs and as vehicles for immunomodulatory payloads, offering novel opportunities to integrate miRNA modulation with immune checkpoint blockade.

Future progress will depend on solving key developmental bottlenecks. Improving delivery technologies remains paramount, with lipid nanoparticles, exosome-based carriers, and ligand-conjugated oligonucleotides being actively optimized for tissue-specific targeting. Advances in chemical modifications, such as LNAs and phosphorothioate backbones, have already enhanced the stability and binding affinity of miRNA drugs, reducing degradation and off-target effects. Furthermore, biomarker-driven patient selection is likely to be transformative, ensuring that miRNA therapeutics are deployed in populations most likely to benefit based on tumor genotype, miRNA expression profiles, and resistance mechanisms. Economic and competitive pressures also shape the future of the miRNA market. The early termination of INT-1B3, a miR-193a-3p mimic for advanced solid tumors (NCT04675996), due to insufficient funding, underscores the need for sustainable investment models, strategic industry partnerships, and regulatory frameworks that balance innovation with feasibility. Demonstrating clear advantages over established therapies, whether through superior efficacy, combinational synergy, or reduced toxicity, will be critical to secure market viability. In conclusion, miRNA therapeutics are poised to transition from experimental tools to clinically meaningful treatments as science and technology converge to address their limitations. By integrating improved delivery systems, synthetic innovations, and precision medicine approaches, these therapies may ultimately carve out a distinct niche in oncology and beyond. The coming decade will determine whether miRNAs fulfill their potential as master regulators of disease, offering not just incremental improvements but transformative changes in how we treat cancer and other complex disorders.

**Acknowledgement** S.E. acknowledges the DST–SERB for the Ramanujan fellowship (RJF/2021/000066) and CRG/2022/002814, India. S.E. also acknowledges SRM University-AP for supporting research scholars and providing infrastructure (SRMAP/URG/E&PP/2022-23/007).

## References

Adepu S, Ramakrishna S (2021) Controlled drug delivery systems: current status and future directions. Molecules 26(19)

Afonso-Grunz F, Muller S (2015) Principles of miRNA-mRNA interactions: beyond sequence complementarity. Cell Mol Life Sci 72(16):3127–3141

Akbari Moqadam F, Pieters R, den Boer ML (2013) The hunting of targets: challenge in miRNA research. Leukemia 27(1):16–23

Akhter MH, Beg S, Tarique M, Malik A, Afaq S, Choudhry H, Hosawi S (2021) Receptor-based targeting of engineered nanocarrier against solid tumors: recent progress and challenges ahead. Biochim Biophys Acta Gen Subj 1865(2):129777

Anastasiadou E, Seto AG, Beatty X, Hermreck M, Gilles ME, Stroopinsky D, Pinter-Brown LC, Pestano L, Marchese C, Avigan D, Trivedi P, Escolar DM, Jackson AL, Slack FJ (2021) Cobomarsen, an oligonucleotide inhibitor of miR-155, slows DLBCL tumor cell growth in vitro and in vivo. Clin Cancer Res 27(4):1139–1149

Bassot A, Dragic H, Haddad SA, Moindrot L, Odouard S, Corlazzoli F, Marinari E, Bomane A, Brassens A, Marteyn A, Hibaoui Y, Petty TJ, Chalabi-Dchar M, Larrouquere L, Zdobnov EM, Legrand N, Tamburini J, Lincet H, Castets M, Yebra M, Migliorini D, Dutoit V, Walker PR, Preynat-Seauve O, Dietrich PY, Cosset E (2023) Identification of a miRNA multi-targeting therapeutic strategy in glioblastoma. Cell Death Dis 14(9):630

Bhaskaran M, Mohan M (2014) MicroRNAs: history, biogenesis, and their evolving role in animal development and disease. Vet Pathol 51(4):759–774

Brillante S, Volpe M, Indrieri A (2024) Advances in MicroRNA therapeutics: from preclinical to clinical studies. Hum Gene Ther 35(17–18):628–648

Burnett JC, Rossi JJ (2012) RNA-based therapeutics: current progress and future prospects. Chem Biol 19(1):60–71

Calin GA, Hube F, Ladomery MR, Delihas N, Ferracin M, Poliseno L, Agnelli L, Alahari SK, Yu AM, Zhong XB (2024) The 2024 Nobel Prize in physiology or medicine: microRNA takes center stage. Noncoding RNA 10(6)

Cannataro R, Cione E (2023) miRNA as drug: antagomir and beyond. Curr Pharm Des 29(6): 462–465

Chai S, Tong M, Ng KY, Kwan PS, Chan YP, Fung TM, Lee TK, Wong N, Xie D, Yuan YF, Guan XY, Ma S (2014) Regulatory role of miR-142-3p on the functional hepatic cancer stem cell marker CD133. Oncotarget 5(14):5725–5735

Chandan K, Gupta M, Sarwat M (2019) Role of host and pathogen-derived MicroRNAs in immune regulation during infectious and inflammatory diseases. Front Immunol 10:3081

Chirshev E, Oberg KC, Ioffe YJ, Unternaehrer JJ (2019) Let-7 as biomarker, prognostic indicator, and therapy for precision medicine in cancer. Clin Transl Med 8(1):24

Choi WY, Giraldez AJ, Schier AF (2007) Target protectors reveal dampening and balancing of nodal agonist and antagonist by miR-430. Science 318(5848):271–274

de Visser KE, Joyce JA (2023) The evolving tumor microenvironment: from cancer initiation to metastatic outgrowth. Cancer Cell 41(3):374–403

Di Paolo D, Pastorino F, Brignole C, Corrias MV, Emionite L, Cilli M, Tamma R, Priddy L, Amaro A, Ferrari D, Marotta R, Ferretti E, Pfeffer U, Ribatti D, Sementa AR, Brown D, Ikegaki N, Shimada H, Ponzoni M, Perri P (2020) Combined replenishment of miR-34a and let-7b by targeted nanoparticles inhibits tumor growth in neuroblastoma preclinical models. Small 16(20):e1906426

Diener C, Keller A, Meese E (2022) Emerging concepts of miRNA therapeutics: from cells to clinic. Trends Genet 38(6):613–626

Duurland CL, Gunst T, Boer HCD, Bosch M, Telford BJ, Vos RM, Xie X, Zang M, Wang F, Shao Y, An X, Wang J, Cai J, Bourre L, Pinxteren L, Schaapveld RQJ, Janicot M, Yahyanejad S (2024) INT-1B3, an LNP formulated miR-193a-3p mimic, promotes anti-tumor immunity by enhancing T cell mediated immune responses via modulation of the tumor microenvironment and induction of immunogenic cell death. Oncotarget 15:470–485

Ebert MS, Sharp PA (2010) MicroRNA sponges: progress and possibilities. RNA 16(11): 2043–2050

Ellipilli S, Wang H, Binzel DW, Shu D, Guo P (2023) Ligand-displaying-exosomes using RNA nanotechnology for targeted delivery of multi-specific drugs for liver cancer regression. Nanomedicine 50:102667

Fu J, Imani S, Wu MY, Wu RC (2023) MicroRNA-34 family in cancers: role, mechanism, and therapeutic potential. Cancers (Basel) 15(19)

Ha SY, Yu JI, Choi C, Kang SY, Joh JW, Paik SW, Kim S, Kim M, Park HC, Park CK (2019) Prognostic significance of miR-122 expression after curative resection in patients with hepatocellular carcinoma. Sci Rep 9(1):14738

Hong DS, Kang YK, Borad M, Sachdev J, Ejadi S, Lim HY, Brenner AJ, Park K, Lee JL, Kim TY, Shin S, Becerra CR, Falchook G, Stoudemire J, Martin D, Kelnar K, Peltier H, Bonato V, Bader AG, Smith S, Kim S, O'Neill V, Beg MS (2020) Phase 1 study of MRX34, a liposomal miR-34a mimic, in patients with advanced solid tumours. Br J Cancer 122(11):1630–1637

Huang J, Mei H, Tang Z, Li J, Zhang X, Lu Y, Huang F, Jin Q, Wang Z (2017) Triple-amiRNA VEGFRs inhibition in pancreatic cancer improves the efficacy of chemotherapy through EMT regulation. J Control Release 245:1–14

Imani S, Wei C, Cheng J, Khan MA, Fu S, Yang L, Tania M, Zhang X, Xiao X, Zhang X, Fu J (2017) MicroRNA-34a targets epithelial to mesenchymal transition-inducing transcription factors (EMT-TFs) and inhibits breast cancer cell migration and invasion. Oncotarget 8(13): 21362–21379

Jiang G, Cui Y, Yu X, Wu Z, Ding G, Cao L (2015) miR-211 suppresses hepatocellular carcinoma by downregulating SATB2. Oncotarget 6(11):9457–9466

Kotowska-Zimmer A, Przybyl L, Pewinska M, Suszynska-Zajczyk J, Wronka D, Figiel M, Olejniczak M (2022) A CAG repeat-targeting artificial miRNA lowers the mutant huntingtin level in the YAC128 model of Huntington's disease. Mol Ther Nucleic Acids 28:702–715

Kumar R, Tiwari S, Harilal S (2024) MicroRNA-based clinical trials for head and neck cancer. In: Diagnostic, prognostic, and therapeutic role of MicroRNAs in head and neck cancer. Academic Press, pp 325–335

Kurakula H, Vaishnavi S, Sharif MY, Ellipilli S (2023) Emergence of small interfering RNA-based gene drugs for various diseases. ACS Omega 8(23):20234–20250

Kuroda D, Tsumoto K (2020) Engineering stability, viscosity, and immunogenicity of antibodies by computational design. J Pharm Sci 109(5):1631–1651

Lai Y, Wang F, Cai G, Li Y, Weng J, Cai F, Cai L, Cheng N, Zhao B, Zeng Y (2025) Utilization of artificial circular RNAs as miRNA sponges and anti-PD-1 scFv expression platforms to suppress hepatocellular carcinoma progression. Front Immunol 16:1609165

Li S, Li Q, Lu J, Zhao Q, Li D, Shen L, Wang Z, Liu J, Xie D, Cho WC, Xu S, Yu Z (2019) Targeted inhibition of miR-221/222 promotes cell sensitivity to cisplatin in triple-negative breast cancer MDA-MB-231 cells. Front Genet 10:1278

Li X, Pritykin Y, Concepcion CP, Lu Y, La Rocca G, Zhang M, King B, Cook PJ, Au YW, Popow O, Paulo JA, Otis HG, Mastroleo C, Ogrodowski P, Schreiner R, Haigis KM, Betel D, Leslie CS, Ventura A (2020) High-resolution in vivo identification of miRNA targets by halo-enhanced Ago2 pull-down. Mol Cell 79(1):167–179 e11

Lima JF, Cerqueira L, Figueiredo C, Oliveira C, Azevedo NF (2018) Anti-miRNA oligonucleotides: a comprehensive guide for design. RNA Biol 15(3):338–352

Long X, Shi Y, Ye P, Guo J, Zhou Q, Tang Y (2019) MicroRNA-99a suppresses breast cancer progression by targeting FGFR3. Front Oncol 9:1473

Lv H, Zhang S, Wang B, Cui S, Yan J (2006) Toxicity of cationic lipids and cationic polymers in gene delivery. J Control Release 114(1):100–109

Maji RK, Leisegang MS, Boon RA, Schulz MH (2025) Revealing microRNA regulation in single cells. Trends Genet 41(6):522–536

Martino MTD, Tagliaferri P, Tassone P (2025) MicroRNA in cancer therapy: breakthroughs and challenges in early clinical applications. J Exp Clin Cancer Res 44(1):126

Menon A, Abd-Aziz N, Khalid K, Poh CL, Naidu R (2022) miRNA: a promising therapeutic target in cancer. Int J Mol Sci 23(19)

Pan W, Chai B, Li L, Lu Z, Ma Z (2023) p53/MicroRNA-34 axis in cancer and beyond. Heliyon 9(4):e15155

Patil Y, Panyam J (2009) Polymeric nanoparticles for siRNA delivery and gene silencing. Int J Pharm 367(1–2):195–203

Peng Y, Croce CM (2016) The role of MicroRNAs in human cancer. Signal Transduct Target Ther 1:15004

Qiu C, Xia F, Zhang J, Shi Q, Meng Y, Wang C, Pang H, Gu L, Xu C, Guo Q, Wang J (2023) Advanced strategies for overcoming endosomal/lysosomal barrier in nanodrug delivery. Research (Wash D C) 6:0148

Schafer M, Ciaudo C (2020) Prediction of the miRNA interactome - established methods and upcoming perspectives. Comput Struct Biotechnol J 18:548–557

Schmidt MF (2017) miRNA targeting drugs: the next blockbusters? Methods Mol Biol 1517:3–22

Seto AG, Beatty X, Lynch JM, Hermreck M, Tetzlaff M, Duvic M, Jackson AL (2018) Cobomarsen, an oligonucleotide inhibitor of miR-155, co-ordinately regulates multiple survival pathways to reduce cellular proliferation and survival in cutaneous T-cell lymphoma. Br J Haematol 183(3):428–444

Seyhan AA (2024) Trials and tribulations of MicroRNA therapeutics. Int J Mol Sci 25(3)

Swayze EE, Siwkowski AM, Wancewicz EV, Migawa MT, Wyrzykiewicz TK, Hung G, Monia BP, Bennett CF (2007) Antisense oligonucleotides containing locked nucleic acid improve potency but cause significant hepatotoxicity in animals. Nucleic Acids Res 35(2):687–700

Tassone P, Di Martino MT, Arbitrio M, Fiorillo L, Staropoli N, Ciliberto D, Cordua A, Scionti F, Bertucci B, Salvino A, Lopreiato M, Thunarf F, Cuomo O, Zito MC, De Fina MR, Brescia A, Gualtieri S, Riillo C, Manti F, Caracciolo D, Barbieri V, Di Paola ED, Di Francesco AE, Tagliaferri P (2023) Safety and activity of the first-in-class locked nucleic acid (LNA) miR-221 selective inhibitor in refractory advanced cancer patients: a first-in-human, phase 1, open-label, dose-escalation study. J Hematol Oncol 16(1):68

Uehata T, Takeuchi O (2020) RNA recognition and immunity-innate immune sensing and its posttranscriptional regulation mechanisms. Cells 9(7)

van der Ree MH, de Vree JM, Stelma F, Willemse S, van der Valk M, Rietdijk S, Molenkamp R, Schinkel J, van Nuenen AC, Beuers U, Hadi S, Harbers M, van der Veer E, Liu K, Grundy J, Patick AK, Pavlicek A, Blem J, Huang M, Grint P, Neben S, Gibson NW, Kootstra NA, Reesink HW (2017) Safety, tolerability, and antiviral effect of RG-101 in patients with chronic hepatitis C: a phase 1B, double-blind, randomised controlled trial. Lancet 389(10070):709–717

van Zandwijk N, Pavlakis N, Kao SC, Linton A, Boyer MJ, Clarke S, Huynh Y, Chrzanowska A, Fulham MJ, Bailey DL, Cooper WA, Kritharides L, Ridley L, Pattison ST, MacDiarmid J, Brahmbhatt H, Reid G (2017) Safety and activity of microRNA-loaded minicells in patients with recurrent malignant pleural mesothelioma: a first-in-man, phase 1, open-label, dose-escalation study. Lancet Oncol 18(10):1386–1396

Vervaeke P, Borgos SE, Sanders NN, Combes F (2022) Regulatory guidelines and preclinical tools to study the biodistribution of RNA therapeutics. Adv Drug Deliv Rev 184:114236

Wang Z (2011) The principles of MiRNA-masking antisense oligonucleotides technology. Methods Mol Biol 676:43–49

Wang X, Li J, Dong K, Lin F, Long M, Ouyang Y, Wei J, Chen X, Weng Y, He T, Zhang H (2015) Tumor suppressor miR-34a targets PD-L1 and functions as a potential immunotherapeutic target in acute myeloid leukemia. Cell Signal 27(3):443–452

Wang X, Li C, Wang Y, Chen H, Zhang X, Luo C, Zhou W, Li L, Teng L, Yu H, Wang J (2022) Smart drug delivery systems for precise cancer therapy. Acta Pharm Sin B 12(11):4098–4121

Xie H, Xiao R, He Y, He L, Xie C, Chen J, Hong Y (2021) MicroRNA-100 inhibits breast cancer cell proliferation, invasion and migration by targeting FOXA1. Oncol Lett 22(6):816

Yan JJ, Chang Y, Zhang YN, Lin JS, He XX, Huang HJ (2017) miR-195 inhibits cell proliferation via targeting AEG-1 in hepatocellular carcinoma. Oncol Lett 13(5):3118–3126

Yang Y, Chang S, Zhao Z, Hou NI, He K, Wang X, Gao L, Wang L, Cai D, Guo BO, Tong D, Song T, Huang C (2015) MicroRNA-214 suppresses the proliferation of human hepatocellular carcinoma cells by targeting E2F3. Oncol Lett 10(6):3779–3784

Zhang Y, Wu D, Zhang X, Ma N, Liu Y (2018) MicroRNA sponge knockdowns miR-483-5p and upregulates serum ALT/AST in transgenic mice. Biochemistry (Mosc) 83(1):54–59

Zhang L, Liao Y, Tang L (2019) MicroRNA-34 family: a potential tumor suppressor and therapeutic candidate in cancer. J Exp Clin Cancer Res 38(1):53

Zhao WT, Lin XL, Liu Y, Han LX, Li J, Lin TY, Shi JW, Wang SC, Lian M, Chen HW, Sun Y, Xu K, Jia JS, Luo RC, Xiao D (2019) miR-26a promotes hepatocellular carcinoma invasion and metastasis by inhibiting PTEN and inhibits cell growth by repressing EZH2. Lab Investig 99(10):1484–1500

Zhu L, Jiang M, Wang H, Sun H, Zhu J, Zhao W, Fang Q, Yu J, Chen P, Wu S, Zheng Z, He Y (2021) A narrative review of tumor heterogeneity and challenges to tumor drug therapy. Ann Transl Med 9(16):1351

MIX
Papier aus verantwortungsvollen Quellen
Paper from responsible sources
FSC® C105338

If you have any concerns about our products,
you can contact us on
**ProductSafety@springernature.com**

In case Publisher is established outside the EU,
the EU authorized representative is:
**Springer Nature Customer Service Center GmbH**
**Europaplatz 3, 69115 Heidelberg, Germany**

Printed by Libri Plureos GmbH
in Hamburg, Germany